Advances in Geographical and Environmental Sciences

Series Editor

R. B. Singh, University of Delhi, Delhi, India

Advances in Geographical and Environmental Sciences synthesizes series diagnostigation and prognostication of earth environment, incorporating challenging interactive areas within ecological envelope of geosphere, biosphere, hydrosphere, atmosphere and cryosphere. It deals with land use land cover change (LUCC), urbanization, energy flux, land-ocean fluxes, climate, food security, ecohydrology, biodiversity, natural hazards and disasters, human health and their mutual interaction and feedback mechanism in order to contribute towards sustainable future. The geosciences methods range from traditional field techniques and conventional data collection, use of remote sensing and geographical information system, computer aided technique to advance geostatistical and dynamic modeling.

The series integrate past, present and future of geospheric attributes incorporating biophysical and human dimensions in spatio-temporal perspectives. The geosciences, encompassing land-ocean-atmosphere interaction is considered as a vital component in the context of environmental issues, especially in observation and prediction of air and water pollution, global warming and urban heat islands. It is important to communicate the advances in geosciences to increase resilience of society through capacity building for mitigating the impact of natural hazards and disasters. Sustainability of human society depends strongly on the earth environment, and thus the development of geosciences is critical for a better understanding of our living environment, and its sustainable development.

Geoscience also has the responsibility to not confine itself to addressing current problems but it is also developing a framework to address future issues. In order to build a 'Future Earth Model' for understanding and predicting the functioning of the whole climatic system, collaboration of experts in the traditional earth disciplines as well as in ecology, information technology, instrumentation and complex system is essential, through initiatives from human geoscientists. Thus human geosceince is emerging as key policy science for contributing towards sustainability/survivality science together with future earth initiative.

Advances in Geographical and Environmental Sciences series publishes books that contain novel approaches in tackling issues of human geoscience in its broadest sense—books in the series should focus on true progress in a particular area or region. The series includes monographs and edited volumes without any limitations in the page numbers.

Rubén C. Lois González ·
Marco Antonio Mitidiero Junior
Editors

Brazilian Geography

In Theory and in the Streets

Springer

Editors
Rubén C. Lois González
Department of Geography
University of Santiago de Compostela
Santiago de Compostela, La Coruña, Spain

Marco Antonio Mitidiero Junior
Federal University of Paraíba
João Pessoa, Paraíba, Brazil

ISSN 2198-3542 ISSN 2198-3550 (electronic)
Advances in Geographical and Environmental Sciences
ISBN 978-981-19-3703-3 ISBN 978-981-19-3704-0 (eBook)
https://doi.org/10.1007/978-981-19-3704-0

This Springer imprint is published by the registered company Springer Nature Singapore Pte Ltd.
The registered company address is: 152 Beach Road, #21-01/04 Gateway East, Singapore 189721, Singapore

Contents

Chapter 1
Introduction

Rubén C. Lois-González and Marco Antonio Mitidiero Junior

Editors

Geography as a social science focused on spatial and territorial analysis has often achieved greater academic relevance in the so-called *new countries*. With this expression, we refer to nation-states located outside Europe, arising from decolonization processes as opposed to old metropolises, and with a vast territory to map, describe, occupy as well as rich natural resources to exploit. Indeed, the discipline of geography has proven highly useful in creating a national feeling thanks to maps, facilitating the colonization of new lands, getting closer to the indigenous peoples and maximizing the capitalist profitability of peripheral regions (Capel 1981; Santos 2002). Undoubtedly, in countries such as the USA, Canada, Australia, the Republic of South Africa or Brazil, among others, Geography is presented as an essential science that has been institutionalized in various fields.

This opening paragraph helps to understand that geographical knowledge in Brazil has historically been valued. A knowledge that has helped systematize the various information arriving from the vast inland and Amazon region, which has been considered fundamental in the process of the political-administrative delimitation of the country, in the setting up of a Brazilian Institute of Geography and Statistics (IBGE) and in the academic institutionalization of the discipline. It could be affirmed that Geography has facilitated the configuration of the modern country, once the control of the Empire was over. A polysemic science valued by the military and explorers, applied to carry out broad demographic and social sampling (very necessary in such an unequal nation) and claimed as superior knowledge in universities and geographical societies. In this way, spatial and territorial analysis has been present in very

R. C. Lois-González · M. A. Mitidiero Junior (✉)
Universities of Santiago de Compostela, Galicia, Spain
e-mail: thomazjrgeo@terra.com.br

Federal of Paraiba, Paraiba, Brazil

R. C. Lois González and M. A. Mitidiero Junior (eds.), *Brazilian Geography*,
Advances in Geographical and Environmental Sciences,
https://doi.org/10.1007/978-981-19-3704-0_1

diverse areas of Brazilian life and, imitating the French example of European culture and civilization in the construction of Brazilian Geography, has always been marked by the French disciplinary tradition (Borzacchielo 2016). Professionals such as J. Tricart, P. Deffontaines or P. Monbeig, among other names, had visited and conditioned the rules of Brazilian Geography since the mid-twentieth century (Monbeig 1975; Boletim Paulista de Geografia 2004). At the same time, the work of prominent authors such as J. de Castro or M. Santos could not be understood without their years of training and professional experience in France (Borzacchielo 2016).

Starting from the aforementioned important historical tradition, since the 1970s the institutionalization of Geography spread throughout Brazil and largely facilitated the construction of a democratic society, yet one which was always fragile and marked by inequality. This is the period covered in this book. If in the 1960s the academic presence of the discipline was evident in Rio de Janeiro, Sâo Paulo, Salvador de Bahía and some cities in the South of the country, the following decades supposed the creation of Geography undergraduate and postgraduate degrees inland, in the Northeast, the North and in the confines of the Amazon (Da Silva 2011; Spósito, 2016a, b). This explains why Geography is currently a university degree in more than 250 towns, is offered as a postgraduate degree in 77 universities and is part of other academic studies such as territorial planning, tourism or cartography (ANPEGE 2010–16; Sposito 2016). Along with this undoubted expansion of the discipline in Brazil, becoming one of the nations in the world where geographic analyses are more explanatory and widespread, the very character of the science has been changing. A milestone in this transformation has been the groundbreaking stage initiated by the Asociaçâo de GeografosBrasileiros (AGB) in the early 1980s, which is discussed in one of the chapters of this book. The AGB stopped being just another traditional and academic society to become the voice of a critical Geography, which claimed direct contact amongst Geography lectures at different educational levels, and understood spatial and territorial analysis as a tool for deep political and social change (AGB 1980–82; Rodrigues 2004). The trajectory of the AGB added to the role played by the *Associaçâo Nacional de Pós-Graduaçâo e PesquisaemGeografía* (ANPEGE) has conditioned the research developed in recent decades, always imbued with a critical zeal against the existing reality and promoting a new type of socio-spatial relationships.

As already mentioned, since the 1980s, Geography has largely become an expression of the demands for profound changes in the main nation of South America and, by extension, of Latin America. This role of Geography justifies the direct link with the widespread social movements at the end of the twentieth century (Geography *on the street*) and with the rise to power of progressive or leftist governments since the early 2000s. This role adopted by the majority of the geographical community, far from reducing the discipline to academic knowledge, contributed to its popularization, to the increase of its theoretical prestige, along with other disciplines such as Anthropology, Sociology, Critical Economics and Environmental Sciences. In addition, its popularity spread throughout Latin America, since the feeling of unity of the continent against the interference of the North has always been a vindication of the left in contrast to an apparently more nationalist right, defender of bilateral relations with the USA or the European Union. Brazilian Geography is presented as the main reference of what should be done in the discipline in neighbouring countries. Its books and research are beginning to be perceived as important at an international level, although their scope and meaning beyond their linguistic community are often not well understood, as translations into English are scarce. To increase its academic influence, Brazilian Geography has reinforced its theoretical content (similarly to what was learned in the UK), its involvement with social and political movements (a Geography for action), while neglecting the applied character, of planning the territory, of its work, often considered a collaborator of power. In Physical Geography, the protection of biodiversity, exemplified in the defence of the Amazon, becomes one of the hallmarks of the scientific work carried out (Santos 1996a, b; Porto-Gonçalves 2002).

The complete renewal of the AGB and the alignment of Brazilian Geography with the ideologies and policies of social, environmental and territorial change at the end of the twentieth century and the beginning of the twenty first do not hide the enormous role that certain personalities in the discipline played in structuring the spatial and territorial knowledge. These individualities, without a doubt, were very relevant in placing the discipline among one of the great social and human sciences developed in the country. In addition to the work of geographers in the exploration, description and mapping of the Amazon and the entire vast inland, as well as the very creation of the IGBE, the first great reference for the collective was the medical doctor and geographer J. de Castro who lived the last years of his life in exile in France. To him we owe the work *Geopolitics* of *Hunger*, where the denunciation of social and territorial inequalities in accessing health services constitutes the basis of a very well-articulated, critical and scientifically innovative discourse (De Castro 1946, 1965). Josué de Castro is the first great moral example of Brazilian intellectuals who had to go into exile after the Military Coup of 1964, a coup that would inaugurate a series of authoritarian governments until the presidential elections of 1985. Another illustrious exile was Milton Santos, a true icon and scientific testimony to the honesty of Brazilian Geography. Santos lived in France, where he was among the geographical elite of the country, and in Africa, which would allow him to make a fundamental contribution to the study of cities and the urbanization process in Third World countries (Santos 1971, 1979). With his research on Urban and Economic Geography, M. Santos initiated a pioneering epistemological reflection on decoloniality and the abandonment of Eurocentric paradigms that had characterized territorial analysis until then. In addition, after his return to Brazil, the renowned scholar deepened his theoretical contributions to the discipline (Santos 1978, 1996), a good part of which are systematized in a chapter in this book. Another author who deserves a special mention in this work is C. A. Figueiredo Monteiro, for his contribution to the construction of a Brazilian School of Physical Geography, from a fundamentally climatological approach (Monteiro 1991). In Geomorphology, the classic referent was Aziz Ab'Saber, mentioned in different parts of this work, such as in the chapter authored by D. Suertegaray.

Alongside these great names, and also including some prematurely taken scholars such as M. Abreu, a good part of the generation that led Geography's great change and leap ahead in democratic Brazil are represented by the authors participating in this book. A book that we wanted to subtitle "Geography in the streets", because one of the main distinctive elements of the discipline in this country is its strong commitment to social movements. Geographers have collaborated in campaigns for the occupation of underexploited large estates in large rural regions of the country with the *Movemento sem Terra* (MST), in public demonstrations in defence of biodiversity and for the preservation of the Amazon, in the occupation of farms and abandoned urban buildings with the *Movemento dos Sem Teto*, in demands of residents in popular neighbourhoods and favelas, in union campaigns or in demands for better public services and social welfare (Mitidiero et al. 2015; Gledhill et al. 2017). Even in some of the main events of Brazilian Geography such as the SINGA (National Symposium of Agrarian Geography) meetings held regularly and attracting more than two thousand participants each time, a demonstration is scheduled on the last day of the symposium in the city where it is held so that members of landless communities, small farmers, environmental movements, organizations against mining projects or large infrastructures, and indigenous movements, among others, make their claims heard and address the public gathered in the street to explain their arguments and reason for being (SINGA 2019). In more modest events, such as regional geographical congresses, this use of the academic forum is repeated as an amplifier platform for social and territorial struggles. In short, Brazilian Geography cultivates a permanent relationship with the most activist part of society, a fact that is reflected in the content of numerous academic works such as theses, articles, books or more and more didactic videos. At the same time, Geography permeates the contents of dissemination materials that organizations as powerful as the MST or the *Pastoral da Terra* publish in yearbooks, atlases or catalogues of the conflicts throughout the territory (MST 2001–2014; Mitidiero 2010).

Until now we have referred to Geography as a unitary knowledge, only slightly differentiating from Physical Geography. However, in Brazil the existence of different branches is important to understand the importance of the entire discipline. Indeed, in the country departments, specialists in Agrarian, Urban and Political Geography, Theory and History, Environmental studies, Geomorphology, Climatology and Biogeography are usually identified. If we begin by referring to Agrarian Geography (not Rural as it is usually called in Northern countries), it should be noted that in this scientific field the relationships between the territorial analysis carried out by academia and the discourses and practices of social movements are more clearly expressed. Brazilian Agrarian Geography has a strong theoretical base taken from Marxist thought. Thus, the countryside is often studied as an important setting for social conflict, a conflict caused by the unequal distribution of land ownership and the spectacular growth of agribusiness throughout the country since the mid-twentieth century. Therefore, it is very important that the concepts of "territorial dispute" and "territorial conflict" appear as basic instruments in the analysis of Agrarian Geography in this country.

Faced with the dominant discourse, which insists on the contribution of agricultural and livestock exports to the country's economic growth and its financial balance, geographers (in collaboration with anthropologists, sociologists or critical economists), defend that another organization of rural space is possible. Firstly, the countryside must provide good living conditions for the majority of its inhabitants. Therefore, as an alternative to commercial agriculture, community models, more ecological and agroecological, and focused on self-sufficiency, are defended. Secondly, the aggressive systems for increasing soil productivity (through the excessive use of fertilizers, transgenics, etc.) are questioned. Lastly, support is given to a genuine democratization of the peasant communities, which still suffer from daily employment contracts or work performed by children and pregnant women, etc. Geography is agrarian, since the land is the foundation of food production, on which a new social contract that can be extended to the entire country must be articulated (Oliveira 1991; Fernandes 1996; Thomaz 2018).

As far as Urban Geography is concerned, it is possible to differentiate between two different approaches, although both claim to originate from critical epistemes. On the one hand, a significant group of scholars (and in this work we find some examples) have maintained the classic Marxist reference for understanding the city, incorporating the contributions of H. Lefebvre as a great theoretician of the right to the city. A right that makes possible the joint work of the *Movemento sem Teto* with the resistance of neighbourhoods and favelas to large urban renewal projects, and with neighbourhood leaders who fight to improve living conditions in their districts (De Queiroz-Ribeiro and Júnior 1994; Fani and Oliveira 2009). On the other hand, an important segment of urban geographers has evolved towards more reformist readings of the city. In these cases, approaching the study of city systems and metropolitanization processes, the dynamics of rehabilitation of historic centres, the logic of creating new urban centres and medium-sized cities or new sustainable mobility policies within the urban space, among other topics, are of great interest. An Urban Geography closer to certain international examples is thus configured, with highly influential authors such as E. Beltrao Spósito (Spósito 2001, 2016). A particular fact of Brazilian Urban Geography is its little interest in urban governance issues in a country where the growth of cities has generated all kinds of management problems and the need for novel urban projects, of profound change, are a matter of urgency.

From the teaching of authors such as B. Becker or I. Castro, an important group of experts in Political Geography has been consolidated in recent times. Its dynamism is remarkable, with successful congresses being held every few years under the acronym CONGEO (de Azevedo et al. 2021). This disciplinary branch deals with very diverse research topics such as border analysis (with frequent events and notable publicity), territorial rebalancing policies (addressing the Northeast or Amazon issues), Electoral Geography or the relations between Geography and Political Science, among other topics (Castro 2005; Dorfman et al. 2021). Additionally, the Theory of Geography is frequently cultivated in university departments, as is shown in several chapters of this book. It is approached from the classic tradition of Marxist Geography to newer readings of critical thought. In this sense, Brazilian Geography follows the epistemological debates coming from the North, especially from the USA, the UK and, in keeping with a long tradition, from France (Harsbaert 2010; Geographia 2012; Moreira 2015).

The recent evolution of Physical Geography shows fewer features than the evolution of certain branches of Human Geography. Geomorphology and Climatology have become autonomous fields of knowledge, with the collaboration of geographers with geologists, soil scientists and other experimental scientists, stepping away from the rest of the discipline. In this sense, the location of many classic departments of Geography in Institutes of Geosciences has facilitated this connection, in a country where land surveys, the study of droughts or water sources, mineral wealth or surveys of potentially hydrocarbon-producing maritime sectors, among other issues, are still very important. Geomorphology, Climatology and Biogeography in strictly academic terms have enormous potential throughout the territory (Mendonça and Danni-Oliveira 2007; Suertegaray 2017). From a more critical perspective, environmental approaches have insisted on the protection of biodiversity, with a special focus on the Amazon, the *Cerrado* and the *Mata Atlántica*, and on land deforestation for agricultural purposes in vast inland regions (the example of Mato Grosso, etc.). Also in the mapping of risks in cities with a large extension of informal settlements, very vulnerable to intense precipitation events or climate changes of various kinds (Machado et al. 2019; Mendonça and Del Vechio 2020). In these cases, Geography's urban social perspective acquires enormous significance in environmental-based territorial studies.

Another important question to deal with in this introduction is that which refers to the different schools of Brazilian Geography that can possibly be found. The answer to this statement is clear: the discipline maintains a marked unity and common identity in the country as a whole, although large universities or department networks display certain state or regional awareness. Thus, the traditional Carioca school focused on Rio de Janeiro maintains its significance and prestige, represented by the Federal University of Rio de Janeiro (UFRJ) (more classical), the Federal Fluminense (with a series of theoretical contributions from some of its most representative professors, who are included in this work), the Rio de Janeiro State University (young and critical) and the Pontifical Catholic University of Rio (PUC). In São Paulo, a certain differentiation can be seen between the highly influential Geography department of the University of São Paulo (USP), the dynamic department of University of Campinas (UNICAMP) and the different branches of the State University of São Paulo (UESP), where the academic production of the Presidente Prudente department stands out. In the three states of the South, the presence of Geography is significant both in federal and state universities in Paraná, Rio Grande do Sul and Santa Catarina, with the development of agrarian research, Geopolitics, Geomorphology and gender studies.

Inland, the number of centres that offer undergraduate and postgraduate courses in Geography is quite high in Minas Gerais, Goiás, Brasilia or the two Mato Grosso, among others. Since the mid-twentieth century, the tradition in the Northeast has highlighted the research work and geographical dissemination of the Federal Universities of Bahia, Pernambuco, Sergipe, Ceará or Paraíba, to name a few, and a significant number of areas, departments, and postgraduate courses dependent on the state universities located in medium-sized cities and inland, throughout the Sertâo region. Something similar occurs in the North and the Amazon, with centres located from

Amapá to Rondônia, with growing territorial research on agrarian, urban, geopolitical or environmental issues. The emerging nature and personality of Geography in the North and Northeast has promoted some supra-regional coordination initiatives of postgraduate courses, even under the unitary authority of the National Geography Postgraduate Association (ANPEGE), one of the main Geography organizations in Brazil.

This book is divided into two sections. The first brings together 13 works that represent, at least in part, the theoretical contributions and challenges of Brazilian Geography. The second section brings together 9 works that exemplify the concerns of Brazilian Geography regarding social and territorial problems of a country marked by contradictions and injustices.

At the opening of the book, the work by Alexandrina Luz and Sócrates Menezes (A Critical Geography: from the Cabinet to the Streets) points out the historical paths followed by Brazilian Geography in creating a critical science concerned with social transformations. A Geography highly connected with the transformations that have taken place outside its territory and with a foreign academic production. An approach that has contributed to the construction of a scientific field in Brazil that conceives the social space as a space of political struggle, in which the commitment and militancy of geographers has been an integral part of the construction of this discipline, which has made it hit "the streets".

Marcelo Lopes invites geographers to produce knowledge in this scientific discipline which refers to the relationship between society and nature. (Far Beyond the 'Natural Environment': geography at the crossroads of the Capitalocene). The critical distinction between the concepts of Anthropocene and Capitalocene is the starting point that leads to the deepening of the subject, placing capitalism, with its specificities on the periphery of the system, as the object of an environmental geography. To do this, the work calls for the need to integrate natural and social knowledge, as the author notes: "In the last fifteen years, the emergence of a certain environmental geography has raised the hope that many radical geographers have finally begun to understand that, in order to preserve a high level of sophistication and coherence from the point of view of social research, it is not necessary to give up a genuine interest in geobiophysical knowledge".

The idea of Brazilian "territorial formation" is a contribution of the work by Manoel Fernandes (Brazilian Geography and the Study of Territorial Formation). The author affirms the necessary relationship between social formation and territorial formation, arguing that these conceptual notions can only be valid when observing the real processes that characterize and determine the formation of a society and a territory. In the Brazilian case, its formation is the result of the European colonization process. Therefore, the task of this Geography is to expose "the contradictions of the process by stating that the violence was not only material, but also symbolic and epistemological, expressed in a range of social conflicts derived from the colonial logic of territorial appropriation".

Ruy Moreira proposes a theoretical deepening of the relationship between society and space as a condition of the existence of the human being in his geography (Man is his Being-Being in the World. Geography and Geographicity). Starting from the concept of man, the author places him as subject and object of himself, as a condition of being in the world in his obligatory relationship between society and nature, where "the man makes space and the space makes man".

In Dirce Suertegaray's work (Physical Geography and the Study of Environmental Problems: the Brazilian Contribution), there is an excellent account of the role of Brazilian Geography in conducting research on nature, with an emphasis on environmental issues. The author shows that Physical Geography, by focusing on environmental issues, has deepened, as well as subverted and redefined, the concept of the environment addressing "the streets" (social processes), that is, approaching issues such as territorial disputes and social justice from nature studies. As the author concludes: "The critical perspective emphasizes, in the environmental issue, the analysis of social processes, the commercial exploitation of nature, above all, through the demands of an economically globalized world".

Geographical studies on Brazilian cities have a long history and are part of the construction of Geography in Brazil. This is what the work by Pedro Vasconcelos (The Study of Cities in Brazilian Geography) discusses. Its objective is to reassemble and highlight the formation of a specific field of geographic studies, passing through themes and events of history. Vasconcelos presents readers with a detailed inventory of Brazilian Urban Geography, highlighting the role of researchers considered pioneers and specialists in urban studies, such as Milton Santos, Maurício de Abreu and Ana Fani; and highlights the importance of having scientific meetings on urban issues, such as the National Symposia on Urban Geography (Simpurbs).

In Ana Fani's work (The Production of Urban Space and "Critical Geography") we find the meaning of a discipline from an intellectual starting point, which is the production of space. Anchored in the dialectical method, Fani defines space "as a condition, medium and product of social reproduction" to base her reflections on a critical-radical Urban Geography. The author maintains that the "construction of geographic thought is the result of various theoretical perspectives, indicating that there is not a single path", but places critical geography as a powerful possibility to understand and explain the world and its contradictions in movement, especially in realities like Brazil.

Adriana Dorfman and Licio Monteiro present a concise and fundamental evaluation of the production of knowledge by Brazilian Political Geography (Dialogues on Brazilian Political Geography and its Perspectives in the Twenty First Century). The authors propose 10 themes to characterize the history and relevance of Political Geography produced in Brazil, among which stand out: the link between Political Geography and Geopolitics; intradisciplinary exchanges; the "territorial turn" with the popularity of the concept of territory; the articulation of Political Geography with the State, the management of the territory and public policies; and the recognition of actors beyond the State and spaces for the production of political knowledge beyond academia that challenge and nurture scientific production in Political Geography.

The work by Tadeu Arrais (The Consensual Divorce of Geography. Adherence to neoliberalism, the cult of freedom and the overthrow of democracy) criticizes Geography for abandoning the issue of the State as a fundamental concern and with it the concern for the challenge of building a national project. A social science that does not produce a scathing critique of neoliberalism, especially in the current Brazilian and world situation, is surely no longer close to "the street", that is, it does not fulfil the role of being close to social transformations. Arrais makes a critical assessment of a Geography that became critical, but that divorced from essential themes in its constitution.

Ângela Katuta and Maria Martins Albuquerque present the development of an increasingly solid field of research in Brazilian academic Geography that focuses on the Teaching of Geography, especially Geography as a school subject. (Scientific Research and the Construction of the Field of Geography Teaching from Academia: Trends and Challenges) The work addresses the history of the research topic on the teaching of Geography to the analysis, and condemns the neoliberal educational reforms that "through prescriptive curricula, focused on active methodologies and on skills and abilities, aimed at training for the labor market in the context of flexible accumulation of capital". A strategy that produces the alienation, control and subordination of students and teachers.

The work by Mônica Arroyo and Fabio Contel aims to present and discuss the fundamental role of the geographer Milton Santos in the construction of Brazilian Geography (Milton Santos' contribution to the theoretical formation of Brazilian Geography). As stated in the first lines of the work, "Milton Santos' academic and political career makes him a central author for the theoretical production of Brazilian Geography". In other words, to know the theoretical contributions and challenges of Brazilian Geography, as the title of this section of the book points out, it is essential to know the academic production of this author, with more than 40 books and more than two hundred articles. The theoretical development provided by Milton Santos and his critical independence and emancipatory vision is one of the most important features of a Geography that "hits the streets".

Francisco Mendonça's text highlights the role and importance of another unique geographer in the history of this discipline in Brazil (Carlos Augusto de Figueiredo Monteiro and the Construction of Brazilian Geographic Climatology). Carlos Augusto Figueiredo Monteiro played a pioneering role in studies on the relationship between society and nature in his research in the field of climatology, especially for his theoretical-methodological rationale for the inclusion of urban climate in city planning. As Mendonça points out, Monteiro was one of the first to apply "the approach of atmospheric circulation and air mass dynamics for the genetic analysis and classification of climates in Brazil. This contribution initiated a new (1960s) approach to climate based on atmospheric dynamics and its interaction with human activities in the Brazilian territory". Thanks to this, it was possible to think about one of the dimensions of the chaotic Brazilian urbanization and its consequences.

Yet another geographer is highlighted in this book for the importance of his academic production in the construction of a Brazilian Geography. In Antonio Carlos Vitte's work (Aziz Nacib Ab'Saber and the Professionalisation of Geomorphology Research in Brazilian Geography Courses), the contributions of Aziz Ab'Saber in studies of the relationship between society and nature through the development of Geomorphology are presented and discussed. In the country, the sharpness and depth of the studies carried out by this geomorphologist "marked not only the reinvention of Ab'Sáber himself in his conception of geomorphology, but also enabled him to be a leader in the renewal of Brazilian geomorphology from the referential perspective and using the epistemological structure of the Brazilian geographic science".

The text by Arlete Moysés Rodrigues opens the second section of the book, highlighting the approach of the academic production of Brazilian Geography to social struggles, pointing out one of the ways in which Geography "hit the streets" (The Right to the City and Housing in Brazilian Cities). Rodrigues discusses the consequences of housing as a commodity in the capitalist system and analyses the general conditions of housing in Brazilian cities and housing policies with an emphasis on the current situation. It discusses the issues of "right to housing" and "right to the city" as a "relentless" struggle. With the aim of criticizing the contradictory evolution of reality, the author concludes: "the historical struggles for the right to housing and to the city that had been institutionalised have 'disappeared', but the struggle for housing as a human right remains and has no respite, because it is about the struggle for life."

In the chapter by Ariovaldo Umbelino de Oliveira, we find a theme that is very present in Brazilian Agrarian Geography, which is the study of social movements and the struggle for agrarian reform (The Long March of the Brazilian Peasantry: Socio-territorial Movements, Conflicts and Agrarian Reform). Starting from the theoretical scope that capitalist evolution occurs in an unequal and contradictory manner, the author's arguments highlight the role of social struggles led by peasants in a "long march" that mark their presence in Brazilian history. For the author, the concentration of private land ownership in Brazil must be understood as a constitutive part of the evolution of capitalism in that country, and Geography can contribute to explain the social struggles that go far beyond the struggle for agrarian reform, that is, a struggle "understood as a proposal for the transformation of this society, in search of justice, dignity and solidarity", in which "the landless peasants are re-teaching the ideals of nation, homeland and patriotism at the beginning of the twenty first century".

Next, Bernardo Mançano Fernandes (Land and Food: the New Struggles of the Rural Landless Workers Movement-MST) presents the meaning of the largest and most important social movement in the history of Brazil, the Rural Landless Workers Movement (MST). This author, of great importance in the construction of a Geography that studies social movements, develops in his work a more current issue in the MST campaigns and experiences, which is the production of healthy food. The struggle for land (Agrarian Reform) is simultaneously the struggle for food production, which can, according to the author, "show the creation of a new peasant food system resulting in a sustainable modernisation of agriculture".

Márcia Yukari Mizusaki and José Gilberto de Souza address a dramatic theme in Brazilian history, the indigenous question (Geography and Indigenous Peoples: Struggles of Resistance). Firstly, it should be highlighted that Geography has become more relevant as a scientific field that contributes to the understanding and monitoring of the struggles of indigenous peoples. This work reflects on the impacts of capitalist society on the lives of indigenous peoples, highlighting questions about their resistance struggles. In these struggles, according to the authors, "indigenous identity becomes an identity of resistance, and socio-spatial practices aim at the preservation of knowledge. Such knowledge and practices become relevant and reaffirm the condition of being and living in the world, opposing the metabolic split between man and nature, although the commodification of life and common goods leads to the self-destruction of the human species".

In the work by Antônio Thomaz Junior, we find the presentation and theorisation of an increasingly strong field in Brazilian Geography, Labor Geography (Labor Geography in Construction: Theoretical Challenges and Research Praxis). From the formation of a network of researchers around the Centro de Estudos de Geografía do Trabalho, a part of Brazilian Geography has begun to investigate the contradictions and territorial dynamics of class conflicts from (de)realization, exploitation, subordination of male and female workers situations revealed through the plasticity of work. Also the systemic degradation of work and the territorial movement of existing work, personified in formal and informal workers, temporary, self-employed, self-employed contractors, peasants, quilombolas, men, women, etc.

Carlos Walter Porto-Gonçalves presents an academic biographical work of an intellectual geographer and activist concerned with the debate on environmental issues seen from the bottom, and in the establishment of a popular environmentalism (A Popular Environmentalism to Fight for Life, Dignity and the Land—An Autobiographical Contribution of an Activist Geographer). For this reason, he positioned himself as a geographer "on the street" who produces a Geography that hits "the streets". After combining the account of his academic career with theoretical debate and political practice, the author concludes "that social groups in a situation of oppression/exploitation/subordination are showing us, in their struggles to overcome this situation, other horizons in the meaning of life. I invite everyone to walk with them".

In Rogério Haesbaert's work, the debate and the challenges of developing a Geography with colonial heritage rooted in European and North American hegemonic thinking are addressed (Decolonising Challenges of the Brazilian/Latin American Geography(s)). The author highlights that the struggle here is undertaken in the academic field to forge a decolonized Brazilian (and Latin American) Geography, original and closer to the dilemmas of its space and time, and "the dissemination of post-colonial approaches, and/or decolonial, that have created new conditions for international and interdisciplinary dialogue, especially by focusing on the recognition of local or regional knowledge and on the political commitment to social transformation, not only in analytical-critical terms, but also in practical-participatory terms".

Joseli Maria Silva and Marcio José Ornat present the contributions of a Feminist Geography, affirming that, even being part of an "epistemological movement with little legitimacy in Brazil", feminist geographies have been gaining ground in the geographical thought in that country (Brazilian Feminist Geographies: Occupying Space, Resisting Denial and Generating Challenges for Geography). Topics such as feminist geographies, sexualities, epistemology and the geopolitics of knowledge are addressed by the authors and a series of researchers who are part of the research group Observatory of Brazilian Geographic Production. The conclusion is that there is a "growing feminisation of Brazilian geography", but it is emphasized "that the mere bodily presence of women is insufficient to produce geographies that make gender and sexual relations visible in this scientific field".

The book ends with the authors Charlles da França Antunes and Paulo Alentejano in a work aimed at presenting the leading role of the Associação dos Geografos-Brasileiros (AGB) in the construction of a Geography that "hits the streets" (Associação dos GeografosBrasileiros (AGB): the Construction of a Geography of Struggle). The proximity, exchange and organicity of AGB with social movements led Brazilian Geography to simultaneously and inseparably merge academic production and political practice, so that the authors intend to "discuss the forms, means and processes that assigned the AGB a prominent role within this scientific community in Brazil, and thus seek to understand to what extent and in what way the AGB and the movement around it have been responsible for the consolidation in Brazilian Geography of a Geography of struggle".

With this work, we intend to present to readers, at least in part, a Geography that has become unique precisely because it gets closer to "the streets". "Uma Geografia naRua" is the figurative expression of concerns and real practices of a body of researchers who organically produce knowledge from and with the subjects of social and territorial processes. In this sense, Brazilian Geography is not unitary, since it is full of contradictions and is defined by a plurality and diversity of methodological theoretical positions. However, there is no doubt that a significant part of the production of knowledge in Brazilian Geography has come from a critical perspective in the defence of social and subaltern groups; also from a concern for the appropriation and destructive use of natural resources. Indigenous people, landless peasants, homeless, marginalized people in the peripheral of urban centres, exploited and deprived workers, unemployed, populations affected and impacted by deforestation and environmental devastation caused, mainly, by agribusiness and mining, etc., are the objects and subjects of this Geography, in which the production of knowledge seeks to promote transformative processes.

References

Anpege (2010–2016) Memórias da Anpege. http://www.anpege.ggf.br/

Boletim Paulista de Geografia (2004) Ediçâo Histórica, N°. 81. AGB Seçâo Local de Sâo Paulo, Sâo Paulo

Borzaccgiello J (2016) French-Brazilian geography. The influence of French geography in Brazil. Springer, Baton Touge

Capel H (1981) Filosofía y ciencia en la Geografía contemporánea. Una introducción a la Geografía. Barcanova, Barcelona

Da Silva MA (2011) 10 Anossem Milton Santos. Asamblea Legislativa do Estado da Bahia & Universidade Federal da Bahia, SanbadordaBahia

de Azevedo DA, Castro IE, Winter Ribeiro R (orgs.) (2021) Os desafios e os novos debates na Geografia Polñitica Contemporânea. Terra escrita, Rio de Janeiro

de Castro IE (2005) Geografia e política: territórios, escalas de ação e instituições. Bertrand Brasil, Rio de Janeiro

De Castro J (1946) Geografia da Fome: A Fome no Brasil. O Cruzeiro, Rio de Janeiro
De Castro J (1965) Ensayos sobre el Sub-Desarrollo. Siglo Veinte, Buenos Aires
De Oliveira AU (1991) A agricultura camponesado Brasil. Contexto, Sâo Pulo
De Queiroz-Ribeiro LC, Santos Júnior OA (orgs.) (1994) Globalizaçâo, Fragmentaçâo e Reforma Urbana. O futuro das cidades brasileiras em crise. Civilizaçâo Brasileira, Rio de Janeuro
Dorfman A, Filizola R, Felix JM (Orgs.) (2021) Ensinando Fronteiras. Diadorim Editora, Porto Alegre
Fani Alessandri Carlos A, De Oliveira AU (Orgs.) (2009) Geografia das metrópolis. Contexto, Sâo Paulo
Fernandes BM (1996) MST: formação e territorialização. Hucitec, São Paulo
Gledhill J, Hita MG, Perelman M (Orgs.) (2017) Disputas em torno do espaço urbano. Processos de (re)produçâo e apropiaçâo da cidade. Edufba, Salvador de Bahia
Geographia (2012) Artigos Internacionais. Edição Especial. Universidade Federal Fluminense, Rio de Janeiro
Haersbaert R (2010) Regional-.Global. Dilemas da Regiâo e da Regionalizaçâo na Geografia Contemporânea. Bertrand Brasil, Rio de Janeiro
Machado RAS, Oliveira AG, Lois-GonzÁlez RC (2019) Urban ecological infrastructure: the importance of vegetation cover in the control of floods and landslides in Salvador/Bahia, Brasil. Land Use Policy 89:8 p. https://doi.org/10.1016/j.landusepol.2019.104180
Mendonça F, Danni-Oliveira IM (2007) Climatologia: Noções básicas e climas do Brasil. Oficina de Textos, São Paulo
Mendonça F, DelVechio M (Orgs.) (2020) A cidade e os problemas socioambientais urbanos - Uma perspectiva interdisciplinar. UFPR, Curitiba
Mitidiero Júnior M (2010) A ação Teritorial de uma Igrreja Radical. Teologia da libertação, luta pela terra e atuação da Comissâo Pastoral da Terra no estado da Paraíba. Editora CRV, Curitiba
Mitidiero Júnior M, Franco Garcia M, GuedesViana PC (Org.) (2015) A qurestâo agrário no século XXI. Outras Expressôes, Sâo Paulo
Monbeig P (1975) O Brasil. DIFEL, Sâo Paulo
Monteiro CAF (1991) Clima e excepcionalismo – Conjecturas sobre o desempenho da atmosfera como fenômeno geográfico. UFSC, Florianópolis
Moreira R (2015) Pensar e Ser em Geografia. Contexto, Sâo Paulo
MST-Movemento Sem Terra Brasil (2001–2014) Cadernos de Estudio, nº. 1–8. https://mst.org.br/
Porto-Gonçalves CW (2002) O desafioambiental. ABDR, Rio de Janeiro
Rodrigues AM (2004) Contribuição da AGB na construção da geografia brasileira – uma outra Geografia sempre é possível. Terra Livre, São Paulo 1(22):2004
Santos D (2002) Areinvençâo do espaço. Diálogos em torno da construção do significado de uma categoria. UNESP, Sâo Paulo
Santos M (1971) Les Villes du Tiers Monde. Genin, Paris
Santos M (1978) Por uma geografia nova. Hucitec, Sâo Paulo
Santos M (1979) Pobreza urbana (2ª Ediçâo). Hucitec, Sâo Paulo
Santos M (1996a) A natureza do espaço. Hucitec, Sâo Paulo
Santos M (1996b) De la totalidad al lugar. Oikos-Tau, Barcelona
SINGA-Simposio Nacional de Geografia Agrária (2019) IX SINGA. Para alem das cercas que nos cegam; as naturezas das r-existências na América Latina. Universidade Federal de Pernanbuco, Recife
Spósito ED (2016a) A pós-Graduaçâo em Geografia no Brasil: Avaliaçâo e tendências". en E. SavérioSpósitoet alt. (Orgs.) A diversidade da Geografia brasileira. Consequência, Rio de Janeiro
Spósito MEB (2001) Cidades médias: espaços em transição. Expressão Popular, Sâo Paulo
Spósito MEB (2016b) Espaços fechado e cidades-Insegurança urbana e fragmentaçâosocioespacial. Editora UNESP, Sâo Paulo
Suertegaray D (2017) (Re)Ligar a Geografia: natureza e sociedade. ComPasso Lugar-Cultura, Porto Alegre
Thomaz Junior A (2018) Geografia e Trabalho no século XXI. Centelha, Presidente Prudente

Part I
Theoretical Contributions and Challenges for Brazilian Geography

Chapter 2
Critical Geography: From the Office to the Streets

Alexandrina Luz Conceição and Sócrates Menezes

Abstract In the 1970s, with the approach of the theoretical contribution of the social sciences, Brazilian Geography consolidated the discourse of a critical geography, defending the renewal of an epistemological model based on Marxist approaches, emphasizing the concept of space as a central analytical geographical category, to give rise to a New Geography which revealed itself as a symptom of the crisis: the crisis of teaching and of research; the crisis of university; the sociopolitical crisis. The commitment to renewal arose in the context of the discontent of geographers who voiced criticisms/complaints against the hegemonic discourse of theoretical geography. We highlight that its theoretical foundations were solidified in a Marxist-structuralist perspective. In the writings of that time, we observe the absence of the fundamental principles of the ontology of social being as a structuring of the interpretation of the objective reality. In Brazil, almost 50 years after that movement, a chorus of voices rang out in defense of critical geography in a historical context of tension and the militarization of society. The crisis marked a period of confinement for bodies and minds. Reacting became a necessity, which materialized in the Association of Brazilian Geographers (AGB).This has become the focus of geography's discourse with society, which headed out onto the streets once and for all, even with the passing of the years, particularly in the nineties, with the clash of theoretical and epistemological differences, from the critics of the critics, and even from some of its formulators. This argument is supported by our analysis, because we understand that, in terms of mediation, space is the place and medium of social practice (of the reproduction of relationships of production), from which and within which geographers expect to unveil their contradictions and constitute strategies and political instruments, where the street is the concrete objective reality of knowing how to think/change and give meaning and significance to geography(ies).

A. L. Conceição (✉)
PPGEO, Universidade Federal de Sergipe, São Cristóvão, Brazil
e-mail: alexandrina.luzconceicao@gmail.com

S. Menezes
PPGEO, Universidade Estadual do Sudoeste de Bahía, Candeias, Brazil

R. C. Lois González and M. A. Mitidiero Junior (eds.), *Brazilian Geography*,
Advances in Geographical and Environmental Sciences,
https://doi.org/10.1007/978-981-19-3704-0_2

Keywords Critical Geography · Ontology of social being · Crisis · Brazilian geography

2.1 Introduction

From 1945 onward, after the stagnation that had persisted since 1929, a period of growth and prosperity based on production and consumption began, which favored a wave of expansion of the market system. Through the pact that brought together segments of the labor movement, the state, and capitalism, supported by John Maynard Keynes' theories of global demand, it was possible to guarantee a cumulative logic based on state intervention with the objective of reactivating production, consumption, and the profit rates of big capital. This process was essential for the hoarding of capital and the consequent beginning of the flow of fictitious capital with the financialization of the economy.

In this movement, the global economy experienced a long period of prosperity, between 1945 and the mid-1970s, marked by reconfigurations of labor productivity, conditioning a new distribution of income between workers and capitalists who were favored by the technological revolution. This period ensured a slight expansion of world trade and the creation of monopolies.

The transition from the thirty golden years of capitalism (1948–1977) to its financialized form was also marked by the exhaustion of Keynesian economics and the logic of the Welfare State, as well as the wave of deregulation in the labor market, capital markets, and banking systems that followed it. The crises resulting from the exhaustion of Taylorist/Fordist production, especially that observed during the deep recession of 1974–1975, provoked accentuated changes in the international division of labor, mainly with the insertion of the new pattern of technology and production developed in Japan and Germany. The archetypal form of regulation assumed by the state, once necessary for the functioning of the accumulative system in its golden years, is no longer necessary.

"Between 1974 and 1975, the international capitalist economy experienced its first widespread recession since World War II—the only one that, until that point, simultaneously affected all the great imperialist powers" (Mandel 1990, p. 09, translated). The heavy impact of this crisis, and the group of other crises that followed it, was identified by Mészáros (2003) as the beginning of the structural crisis.

In 1973, the crisis was accelerated due to the Arab oil embargo, with the rise in oil prices causing a crash in the real estate market. The real effect of the rise in the price of oil on the economic situation accentuated the inflationary trend, "restricting average profit rates from industrial capital" and causing a consequent recession. However, as Mandel (1990) states, this situation was not a direct result of the rise in the price of oil—it was an already-ongoing amplification of the crisis of overproduction.

Until that point, the scale of recessions had been smaller, with particularities in cyclical movements. Since then, hundreds of financial crises around the world began. "Financial crises serve to rationalize the irrationalities of capitalism. Usually they lead to reconfigurations, new development models. New fields of investment and new forms of class power" (Harvey 2007, p. 18, translated). The crisis spread and accelerated after 1986, resulting in a geographically uneven financial architecture.

The crises are part of the contradictory dynamics of the expanded reproduction of capital, are inherent to it, and are produced by the internal contradictions of its accumulative system, via processes that are converted into depreciation or devaluation. The crises serve to counterbalance the tendency of profit rates to fall by forcing the system itself to new, even more developed production patterns. The law of the tendency of the rate of profit to fall, as in Karl Marx's terms, is the result of the contradiction between the development and productivity of capital, through the proportional increase of its constant part in relation to its variable part. This process represents the very development of the organic composition of capital, empirically observed via the increase in productivity, which takes place through the adoption of production technology, resulting in a decrease in live labor crystallized into goods and, consequently, in the reduction of rates of surplus value.

The competition process accelerates this trend to the point of centralization and concentration of capital. When capital grows through the accumulation of surplus value, concentration occurs, while centralization results from the absorption of smaller capital by larger capital. A process of monopolization becomes prominent at the same time that the market itself tends to freeze. The crises occur at times when the capitalist system needs to restructure itself in order to restore the most compensatory general rate of profit to the new templates for capital accumulation, increasingly centralized in large monopolies. Productive restructuring, in turn, is the dominant productive mechanism in which the strongest capital, having survived the selectivity of the market in crisis, manages to carry out the extortion of surplus value more efficiently, thereby bringing about an increasingly explosive scenario, with economic, social, and even cultural confrontations.

According to Mészáros (2002), we can identify "three major social confrontations" that signaled the historical phase that ended in the 1970s and evidenced the "deepening of the general structural crisis of capital" (IBID, p. 1069, translated).

1. The Vietnam War;
2. May 1968—demonstrating the fragility and emptiness of the system's announced achievements;
3. The suppression of reform in Czechoslovakia and Poland uncovering the contradictions in the societies of "real socialism".

The general scenario being set up—considered by Mészáros (2002) as a "structural crisis"—stirred up the founding bases of capitalism demarcated by the supposed stability of the Fordist/Taylorist period. The critical dynamics and the procedural nature of restructuring that emerged in the wake of the destructive effects of the crisis reinvigorated protest movements across all dimensions: social, economic, political, cultural, and scientific. It was not only in the central countries of the capitalist economy, but also in their peripheries, that the crisis called for criticism in the rethinking of a socioeconomic model in the process of being dismantled. Faced with the limits of the template for accumulation and the logic of the Welfare State, their ideological bases, strongly supported by the scientific justifications of this previous context, also collapsed. In this context, science itself entered into crisis.

2.2 "The Crisis of and in Geography"

From the point of view of Geography, it was in the 1960s and 1970s that the defense of a "New Geography" was consolidated in Brazil, claiming a character of scientificity over the methodological exclusivity of the exact sciences. The theoretical scope held within it the denial of history as well as the option of mathematics as a universal language. In this way, the theoretical and epistemological body of geography was based on quantitative research, referenced in laws based on physical and mathematical theory.

The strong focus of the commitment to the interests of the state's planning models (and, consequently, their functionality with respect to the reproduction of capital) compromised the visibility of quantitative geographers, allowing for a great clash between these and those that opposed the use of geographic studies which contained no critique of the concentration of property, income and profit—in short, of the inherent condition of capitalist economic growth. The concern with the organization of space restricting the geographic reading to the studies on the differentiation of areas was deconstructed by the criticism of this geography, considered a utilitarian science. Within this context, it was through criticism that a "Crisis in and of Geography" was proclaimed.

In the intellectual atmosphere, there was a concern placed in the search for the need for "renewal" as a necessary—or better, decisive—condition in the conduct of science. We lived as if moved by the need to be "on the edge", "to be on time", "to do something new", a prerequisite for us to be recognized as intellectuals in the face of crises.

In the context of the crisis, the so-called Critical Geography broke through in the 1970s/80s, supported by a strong discourse in opposition to the geographers of New Geography. Within the dimension of Critical and Radical Geography, it was observed that inequality is inherent to the capitalist mode of production, with it soon being understood that defending emancipatory policies in the middle of the capitalist mode of production itself, without attacking the relationship of labor exploitation (the guarantor of surplus value), and without changing the mechanisms of capital reproduction, meant the continuation of conditions of poverty.

It is worth noting that, during that historical period in which the radical discourse of a militant geography broke through, the consideration and practice of a Critical Geography were not (as has been commonly identified) established within the context of a hegemonic force in geographic thought. It has always presented itself with different conceptions of the world, and therefore with different thoughts—supported, nonetheless, by two pillars that substantiate its discursive materiality: the strength of the Association of Brazilian Geographers (AGB), which guaranteed the institutionality of a geography that was opposed the positivist and neo-positivist method, based on the historical-dialectical method; and the demand for geographers committed to militant practice, managing a radical discourse, which implied not only the demand for a change in content, but especially in practice, since theory should be supported by praxis.

The philosophy of praxis has a signified to the reading of relationships as possibilities for historical determinants within the context of the totality of the relationships. Within this dimension, the category of space is seen as the main condition in geographic reading. Reflecting on the trajectory of geographic discourse and geographic praxis means reflecting on the forms of geographic thinking in their actions and interventions, in which spatial changes are implied in the making and contemplation of geography.

2.3 The Unity of Crisis-Criticism

The so-called Critical Geography would not be able to appear on the horizon of scientific possibilities if it was not for the critical framework presented above. The structural crisis of capital (Mészáros 2002) demonstrates the activation of its absolute limits in at least four elements which combine and reinforce each other:

1. The monopolization of capital resulting from globalized capitalist competition that increases the power and control of large corporations in relation to the market itself, impeding its minimally competitive operation;
2. The inability to maintain the expansionist pattern and the consequent inability to open "new markets", given the completeness of globalization through financialization across the globe, making its productive and expansionist system turn on itself through the destruction of its own production (decrease in the useful life of goods and the "environmental problem") and production relationships (structural unemployment and various forms of modern slavery);
3. The disintegration of minimally cooperative forms of sociability resulting from the flexibilization of production and the dismantling of the Welfare State and its public policies, in favor of the private appropriation of social space, through an ideology that denies the contradiction of classes in order to subjugate the emancipatory struggle for fragmentary and/or identity-based struggles;
4. The condition of work and its inability to insert itself into the system of production, which from a phenomenological point of view appears as massive unemployment of billions of people around the world, but represents the most serious contradiction of the capital crisis, as it means the radical separation of the means of production (extremely concentrated in the hands of a handful of global players) and the live force of labor that generates the value (hence, the very profit) of total real production.

These four elements that characterize the structural crisis that began in the 1970s, as emphasized by Mészáros, are not only consequences of the critical functioning of the system of capital accumulation, but are also the result of the attempt to overcome the same crisis. On the other hand, the restructuring of production that followed, based on the Toyotism-financialisation-deregulation tripod, was a response to the crisis which at the same time deepened its destructive effects.

In the central countries, these contradictions were exposed in quite intense social upheavals. In countries on the capitalist periphery, the same problems of massive unemployment, economic crisis, and fiscal crisis of the state were added to structural and explosive agrarian problems and problems of urban occupation and housing conditions, poverty, illiteracy, hunger, malnutrition, and infant mortality. In Latin America, and especially in Brazil, there was also the repression of the burgo-military dictatorships that made the political context considerably meaningful.

The movement for re-democratization, the founding of left-wing parties, and popular social movements in Brazil will be added to the general critical picture noted above as social-popular resistance. The capital-work relationship, formerly seen as too-distant abstraction that were too distant from each other, materialized in the cities via large mobilizations of workers, students, women, and peasants, in the struggle for rights. Taking to the streets is the real result of the history of resistance movements.

The crisis called for criticism not only because the criticality of the historical process needed a reasoned political counterpoint, but also because it exposed the founding categories of the historical reality of capital itself—a reality that needs to be understood and explained in order to be overcome (Menezes 2020). At this meeting point between socioeconomic crisis and the need for criticism, social theory ceased to be any kind of speculative formulation about reality in order to necessarily become its conscious interpretation. In these conditions in which reality is exposed by the opening of fissures that allow access to the contradictory internal structure of the accumulation-based system, previously veiled by the mechanisms of domination (ideology, economy, politics, etc.) in relative stability, it is also possible to glimpse its historical limits and, thus, its transitory character. History came into play. Social criticism, no longer a merely theoretical practice, was transformed into weapons of criticism.

The dimension of a reality in crisis showed itself as being eminently geographic. The logic of uneven development became observable in multiple scales of analysis: in the global dimension through US imperialism, in regional inequalities within Nation States, in the country–city relationship, and in the center-versus-periphery contradiction that is intertwined in all of them. This is the moment of exposure between the movements for the equalization and differentiation of space that Smith (1988) reveals to us. But it is also the time to understand the relationship between (concrete) space and the process of capitalist accumulation, as demonstrated by Harvey (2005). Theory proved to be an instrument of criticism.

Space gained social concreteness through the understanding of the crisis that disintegrated the capital accumulation model that had been in force since 1945. There was a qualitative leap from neo-positivist and mathematical abstraction, guided by the planning of capitalist modernization, to the understanding of space as a product and condition of man's work, his dwelling place, but also his prison (as Milton Santos would observe). The social space also became the space for political struggle, and geography the science of its foundation, as highlighted by Moraes (2005) regarding the main influences of this geography that purported to be militant:

> The point expressed by Lacoste clearly defines the objectives and posture of Critical Geography. This fully assumes an explicit political content, which appears fully in his statement, "Geography is a social practice in relation to the earth's surface", or in D. Harvey's, "the question of space cannot be a philosophical answer to philosophical problems, but a response based on social practice"; it also appears in the statement of M. Santos, "space is man's dwelling-place, but it can also be his prison". In terms of theory and practice, geographical renewal is seen as a revolutionary praxis, in the sense that explaining the world is not enough—it must be transformed. We can see how far this position reaches, in relation to the renovation undertaken through Pragmatic Geography (Moraes 2005; p. 123, translation).

The qualitative leap in the understanding of geography that was acknowledged beginning from the context of the crisis that began in the 1970s is important in terms of two main points: firstly, the inversion promoted by the concept of space—previously taken as an organized set of forms and objects provided for a certain purpose, and now understood as a product of sensible human practical activity, and therefore of work—where the subjective character is inserted into the externalized objectivity of the world; secondly, in the inversion promoted by the concept of time—previously conceived as an inevitability of the development/evolution of processes taken in themselves and now understood as historical processes resulting from the correlation of forces within the class struggle. Within these two instants, the centrality of social practice became evident, as an assumption and as an action. The geography of the streets became the object of the theory in the process of being constructed, at the same time that it became its political-finalistic objective.

In Brazil, the engagement and militancy of social movements, and geographers who were inserted within them and concerned with the critical framework of reality that is destructively assembled, were crucial for the movement of critical renewal. The crisis in geography, aligned to the context of a socioeconomic crisis, served to fuel a militancy that moved from the streets to the university and was later absorbed and inverted, from the universities to the streets. The indicative text by Carlos Walter Porto Gonçalves, A Geografia Está em Crise, Viva a Geografia! [Geography is in Crisis, Long Live Geography!], published in the midst of the boiling atmosphere of 1978, represented a step in that direction and at the same time called on geographers to rethink their science, shifting from the geography of crises to the crisis of geography.

Even though it was initially quite marked by its merely denunciatory character, this geography that took to the streets in Brazil was summoned by the very conditions of the practical reality stirred up by social movements and political collectives. The practice that resulted from militancy, in turn, gained theoretical breath through the effort made to understand and exercise praxis. The construction of Critical Geography—theoretically formulated, scientifically grounded, and epistemologically structured, as designed by Milton Santos in the thought-provoking Por uma Geografia Nova [For a New Geography], published in 1978, could not be consolidated without this dimension of the concrete reality exposed by the crisis. In the end, as Antônio Carlos Robert Moraes and WanderleyMessias da Costa would say, in Geografia Crítica: A Valorização do Espaço [Critical Geography: The Appreciation of Space], published in 1984, this indicated the Marxian admission of theory-making: that only from the perspective of transforming society are we able to understand it.

2.4 Geography Comes to the Streets

It was in 1968 that Lefebvre drafted the book Direito à Cidade [Right to the City], a work/manifesto that announces a "Movement oriented towards confronting and overcoming the mismatches of reality in relation to the possible, the contradictions between society and civilisation [...]" (Lefebvre 2008a, 11, translated).[1] This is a movement of "reappropriating the time-spaces of life swallowed up by the vortex of the commodity-based world in favor of the (ir)rationality that governs industrialisation" (2008a, p. 12, translated). According to Henri Lefebvre's thinking, based on the reproduction of urban space, it is possible to abstract the crisis and the capitalist production of and in space. In order to do so, "The urban must become threatening" (2008a, p. 20, translated). "Space is an intentionally-manipulated political instrument, even if the intention is disguised under the coherent appearances of the spatial figure" (p. 44, translated). Space is political, ideological, and strategic. The way that space is appropriated, dominated, and produced reveals the society that produced it. The production of space involves the materialization of a society in space, and this implies the production of goods, but also of ideologies, laws, culture, ways of life, and the social space itself (Lefebvre 2008a).

The idea of the right to the city arose from the streets (Harvey 2014), whose occupation (referred to as "occupying the streets") symbolized the democratization of public space. "The street? It is the place (topia) of meeting, without which there are no other possible meetings [...]. In it, movement and mixing take place, without which there is no urban life, but instead separation, stipulated and immobilized segregation" (Lefebvre 2008b, p. 27, translated).

Democracy is contingent on the exercising of citizenship carried out in the public space. Thus, the public space is the place of manifestation for the right to the city, but it is also the object of the claim itself, as it is in it that urban social life is manifested. The public space is simultaneously the place where problems present themselves, take shape, gain a public dimension, and at the same time can be resolved.

The antithesis of public space is conceived space. In their offices, designers, urban planners, geographers, and sociologists project a society without contradictions. Space, including as a crystallized accommodation—"neutral", "objective", "fixed"—hides the real subject acting in the struggle for domination and social control; the state promotes and appears only to assume, its disappearance, because in this way it does not reveal itself as a representative of the interests of classes or fractions of the ruling class. As abstractions, the urban space and the political subject of its ordering—the state—are placed in apparent conflict, while their determinations are already given. Thus, the "users" are also abstracted as impotent subjects—or, in the

[1] Foreword by the translator Sérgio Martins (p. 6–16).

words of Lefebvre (2006, p. 17, translated): "It is also forgotten that there is a total object, the absolute political space, the strategic space that seeks to impose itself as reality, while it is just an abstraction, albeit endowed with enormous powers because it is a place and a means of Power".

The limit of office-based science, and of urban planners and geographers involved with the multiple forms of spatial planning, is not in the failure of their undertaking, nor in the inefficiency of the state's logic and its policy. The problem lies in not admitting the contradiction between this abstract, neutral space, and its concrete dimension, of spatial practice, through which everyday life flows and the reality of the streets challenges and constrains the dominant power in the production of space. It is this practical dimension, of the geography of the streets, that comes into play in the context of the crisis and in the struggle for the right to the city.

Official Geography is also constrained within its Academic Offices. As an extension of the offices of the state, and through a presumptuous theoretical—and therefore ideological—monopoly, geography for capital finds itself divided between naturalistic empiricism and instrumental quantitative. None of these correspond in the slightest with the critical pulse that was coming from the streets. Until the 1970s, the National Association of Brazilian Geographers (AGB), for example, represented an institutional structure that was bureaucratic and averse to social agendas in the student forums (da França Antunes and Souza Neto 2008) and to the militancy of geography professors in schools and primary education courses (Moreira 2000). This institutional framework was inverted, making it the main space for the debate and publication of Critical Geography.

2.5 A Pause for Reflection: At Coffee Break

Geography that was meant to be critical, invoked by the destructive effects of the 1970 crisis (which even disintegrated the Welfare State), could not fail to objectify its internal criticism of that same Academic Office-based Geography. As Corrêa (1982) observes, the accusation directed at traditional geography, for being "naive" and removed from real/social issues, was also directed at instrumental geography, which was also accused of being both alienated and alienating.

In this first instance of critical activity, the criticism that took place in the street was one that demands the abandonment of the Academic Office. The concept of practice reached the theoretical construction that was also carried out at the university. Between militant street practice and the necessary theoretical practice, praxis was discovered and engaged on both of these fronts.

The theoretical construction forged through praxis, and intended to be critical, would become objectified in relation to two instances: social criticism and epistemological self-criticism. As noted, the crisis of geography could only be recognized through recognition of the socioeconomic crisis. Likewise, Critical Geography could not be conceived without the construction of a geographical criticism that was fueled by the real crisis, as represented by the subtitle of the main book of this movement,

"Por Uma Geografia Nova: Da Crítica da Geografia, a Uma Geografia Crítica" [For a New Geography: From the Criticism of Geography to a Critical Geography], by Santos (2002). The theory-practice unity preserved within the constitution of Critical Geography would also signify an epistemological restructuring of the conceptual and categorical bases of the discipline.

Therefore, the actual crisis is internalized as a crisis of geography and the epistemological race for a conceptual structure that best articulated the "new" notions of space, territory, place, etc., then began to dominate the theoretical efforts, especially between the 1980s and 1990. The academic space reclaimed its importance as the place for the construction of social criticism, while historical events anticipated questioning the foundations of Marxist theory, accepted as a matrix for Critical Geography. The end of the Soviet Union, the fall of the Berlin Wall and the so-called crisis of Marxism, also promoted by the advancement of neoliberalism and postmodernism, gave direction to geography. As Ana Fani A. Carlos assessed (2011, p. 17, translated), this movement "allowed many geographers to abandon the legacy of Marx with a 'certain ease' and the 'postmodern' method facilitated this behavior, relieving consciences insofar as it allowed the "mixing of various methods" in an a-critical way".

The standardization of science by the councils regulating research in the country—the National Council for Scientific and Technological Development (Conselho Nacional de DesenvolvimentoCientífico e Tecnológico—CNPq) and the Coordination for the Improvement of Higher-level Personnel (Coordenação de Aperfeiçoamento de Pessoal de Nível Superior—CAPES)—imposed new temporalities and rhythms of production for intellectuals. The difference between the productivity of research imposed by neoliberal logic and the temporality required in order to establish a coherent reflection of the world in crisis led toward a new rupture between theory and practice. The advancement and strengthening of postgraduate courses and the consequent "officialization" of Critical Geography within this academic dimension reinforced this trend, while intensifying a new process of institutionalization.

Academic Offices were full of geographers guided by a Critical Geography that is both a project of unity and the result of the merging of different theoretical and methodological orientations. It therefore became possible to find a certain Karl Marx alongside a certain La Blache; a certain Max Sorre alongside a certain Henri Lefebvre; a certain Milton Santos alongside a certain Martin Heidegger, all in favor of a certain theoretical unity that preserved the discipline's tradition, some reference from social criticism, and a contemporary representativeness. Without proper self-criticism, because the formal requirement around epistemological structuring appeared as a priority for the construction of the unity of Critical Geography, theoretical systems were forged from systems of theoretical fusions (Menezes 2016).

The unity that Critical Geography aspired to only became possible by disregarding the internal unities of the respective theories that were borrowed. But the decisive point was the concept of totality that was instrumentalized as a solely theoretical-methodological feature.

Admitted through postmodern "intellectual impostures", totality came to be identified more as a geographer's ability to relate different processes and readings on the world, rather than with the true ability to connect with this same world through praxis. The tendency to become imprisoned within an Academic Office by this geography—which was meant to be critical—finally occurred due to its distance from the practical geography of the streets.

As Carlos (2011, p. 24, translated) observes, about the to the need to renew geographical criticism, the starting point was to conceive the production of space as a reference for the constitution of man and his world, both under construction, starting with his objective and subjective conditions arising from its practice, "a practice that dramatically reveals crises".

Obviously, the theoretical practice aimed for by the critical geographer, the one that "dramatically reveals crises", could only become effectively emancipatory through analysis corresponding to the real and concrete criticality of this world. This required a conceptual and categorical formulation that corresponded to the revealed crisis and the emancipatory effectiveness of the theory. However, in order to understand space by means of its production, as observed above, in order to overcome the process of alienation promoted by capital, Carlos (2011; p. 24, translated) still believes that this "obliges us to shift the analysis from the field of epistemology to that of human reality", or ontology. And it is here, again, that the streets call for Praxis.

2.6 Geography in the Streets: The AGB—"A Necessary Breath of Fresh Air"

Reflecting on the 2013 Movement in Brazil[2], and bringing along Lefebvre's writings on the "Right to the City", Carlos (2017, p. 43, translated) points out that, In Lefebvre's reading, the "actions of the subjects are carried out in the spaces-times of daily life, of the lived experience in the urban setting, in the plane of the place, illuminating the plane of lived experience". Public space is the place of meeting—in the political plane, it is the place for resistance. Protests revitalize its use and reveal the oppression and the monopoly of violence of the state and market logic.

"Such a policy of space would not simply proceed with the accumulation of constraints; it would seek to bring together the appropriation of time and space by the users—individuals and groups. It would seek to unite this appropriation of space with socioeconomic organization on a broader scale, taking into account a factor of capital importance left aside by prospectivists, namely the complexification of society, the fact that society is becoming increasingly complex and diversified. In my view, this would be the project or the program of a left that would, finally, take care of these problems. What I am saying is perfectly utopian [...]" (Lefebvre 2008a,

[2] A mass movement that took to the streets of cities all over Brazil between June and October 2013, initially protesting against urban transport fare hikes and against the violence used to repress the protests, and then for the overall improvement of public and social policies in the country.

p. 73, translated). Through this space, as Lefebvre (2008b) states, a group manifests itself and appropriates places in a suitable time–space, the urban.

This was also the case with the 1968 movement, initiated in France, which represented the unleashing of a process of radicalization, reaching past the walls of the university and taking over the streets in basically all central (and some peripheral) countries. In Brazil, in this period, protests against the military government were added to the student agendas. However, beginning with Decree-Law 477 – AI-5, from 1969 to 1973, the dictatorship that was installed in the country imposed a deflating of the struggles in the streets through intense repression. Until the 1980s, the struggles took the form of micro-resistances, with specific indications marked by acts of confrontation, such as in 1977 with the return of students to the streets in the fight against repression, imprisonments, deaths and torture, for the end of AI-5, for a political amnesty, and for the reconstruction of the National Students' Union (UNE) and the Central Directory of Students (DCEs). This also happened in the university strikes of 1974 and 1975 against cuts in funding and in favor of public and free education and the boycotting of payment of annuities.

At the end of the 1980s, Brazil saw the most intense return of the student movement calling for the recovery of rights taken away during the period of military dictatorship, which reached its limit with the aggravating bureaucratization of academic centers that became strongly linked to the structures of power. It is understood that their institutionalization favored political neutralization and was sewn into a superstructure of relationships of dependency. From a strong behaviorist foundation, the contents of curricula were coated in an institutionalized a uncritical discourse, without breaking away from authoritarian relations, creating a bureaucratic superstructure that would strengthen power relations. A model of bankrupt education, assembled on power-control schematics, the elite university began to interact with distinct social classes. Within this context, human sciences assumed the profile "of the poor classes", brushing aside their contradictions while announcing their conflicts.

Within the specific scope of geography, such pressure favored the struggle for the disruption of Academic Geography, designated here as the Geography of Academic Offices and the release of the struggle in the streets, embodied in the practices and actions of the Association of Brazilian Geographers—the AGB.[3]

These practices and actions were mainly possible via working groups (WGs), which are thematic forums for debate, articulation, and political intervention belonging to the AGB itself. The WGs have been the great weapon of direct actions for the involvement of geographers outside of Academic Offices, despite reduced participations, albeit with oscillations in growth. The directly participatory actions

---

[3] The Association of Brazilian Geographers (AGB) is a civil entity with an autonomous, technical, scientific and cultural character – a non-profit founded in 1934 in the city of São Paulo. Organised at the national level and via local sections, it aims to bring members together, particularly within the same municipality. Its goal is to promote studies in geography, researching and disseminating new knowledge that aims to deepen geographic theory, as well as reinforcing joint actions to improve the lives of Brazil's population, being able to express itself publicly on the national/local reality in relation to social, economic and political problems.

with social movements, especially the Landless Rural Workers Movement (Movimento dos TrabalhadoresRurais Sem Terra—MST), began in the late 1980s. It was observed that, in this period, the MST was at the beginning of constructing its history, with a strong pursuit of political power.

As expressed in the AGB's Statute, working groups (WGs) are of fundamental importance to the entity. They develop numerous actions in the Local Sections of the Association, scattered among several universities across the country, providing aid to rural and urban communities, specifically when they are involved in conflicts, through political commitment to the geographical community and to society in general. The WGs also act in various spaces of debate and decision making, such as in forums with other entities, in institutional councils and in various negotiations, both at a local and national level, and in established spaces, in line with the entity's principle in the search for a transformation toward a fair society.

- In activities (of dissemination, raising awareness, or questioning, such as round tables, lectures, or courses on topics, almost always coordinating the agents involved—whether social movements, teachers, academics, etc.);
- In the strengthening of allies (such as social movements or trade unions, among others);
- In institutional surveys (with public bodies or private entities, depending on the situation);
- In lawsuits (with the Public Prosecutor's Office, for example);
- In the drafting of documents (such as letters, booklets, manifestos, compilations of texts, etc.).

The entity's principle is the decentralization of actions—hence the importance of working groups (WGs) in its structure. The functioning of the entity is ensured by the presence of Working Groups and Working Commissions. It is the Working Groups which guarantee the resistance and strengthening of the Association of Geographers in their direct action, construction of concrete measures and achievements for the improvement of society. The Urban, Agrarian, Environmental, and Education and Teaching Groups eventuate the foundation work, including through events. The emergence of the "Speaking Teacher" and the "Meeting of Teaching Practices"[4] is emphasized in discussions with unions, social movements, and the participation of primary education teachers.

Within the Urban WG, demonstrations also take place through the participation of Forums, as reported by Professor Dr. ArleteMoysés Rodrigues,[5] our important representative during some national board management meetings. According to Professor Rodrigues, from 1988 to 2010 many actions were taken—demonstrations, debates, and confrentations were taken—based on the principles of the right to the city as

---

[4] Periodic events organised by AGB in order to discuss the theme of the Education & Teaching Working Group.

[5] Professor Arlete Moysés Rodrigues is a lecturer at the State University of Campinas, deposition made in 2010 for the National Board of the AGB (2008-2010 Management meeting). It should be noted that, since the Temer government took over after the impeachment of President Dilma Rousseff (2016), the Cities Council was disrupted.

a collective right, and the right to housing, to environmental sanitation, to decent public transportation, and to accessibility as human rights. The slogans presented in the Popular Amendment for Urban Reform were partially incorporated into the Brazilian Constitution of 1982 (Articles 182 and 183), defining the social function of the city and property. However, the constituent parts included the Master Plan as a guiding element to ensure the fulfillment of the social function of municipalities of more than 20 thousand inhabitants.

The AGB has participated in the Forum for Urban Reform since its establishment. During 1988, the period of the Constitutional Congress, the action of urban social movements, construction workers, engineers, architects, geographers, and other professionals was fundamental so that the daily struggles confined within cities would be recognized as essential for the structural transformation of society, and that they should be escalated as a collective. As Rodrigues states, "We have made it clear that we at the AGB are committed to transformation and not to accommodation."

The Forum has no institutional status or characteristics. It is governed by principles in defense of the rights to housing, to good public transport, to a healthy environment, and to environmental sanitation, among others that may arise according to the specificity of each local section. As the author affirms, the important thing is to remain in the struggle for the attainment of the Right to the City.

Within this context, the Association of Geographers leads Geography to the Streets. In particular, the holding of the National Meeting of Geographers (Encontro Nacional de Geógrafos—ENG) on the Street (the "ENG ON THE STREET") consists of carrying out protests and public actions alongside social movements, trade unions, and civil society. The activity has the political character of carrying the agendas debated at the meeting onto the streets. This activity emanated from the management of the AGB meetings and began to constitute a common practice in events—in general those related to geography—such as the Regional and National Students' Meetings (EncontrosRegionais e Nacionais dos Estudantes—ENEG), or in general in the International and National Agrarian Geography Symposium (SimpósioInternacional e Nacional de Geografia Agrária—SINGA).

In the eighth International and National Agrarian Geography Symposium in 2017, during a public session, in the city center of Curitiba (Paraná), the SINGA on the street held debates with the general public. A public session, along with members from social movements, was held on issues related to the current political situation. In 2019, the 14th National Meeting of Geography Students (ENEG) was aimed at reflecting on "Subversive geographies: breaking barriers and building ties" in the city of Teresina (Piauí), or at the 18th National Meeting of Geographers (ENG)[6] in João Pessoa, taking to the streets in a cultural act, in the historical context of the crisis, realizing the hope of having "Everyone together and mixed", with the Groups "Coco das Manas" and "As Kalungas".

[6] The largest Geography event in Brazil marked by the intense participation of teachers and students (about 5,000 per event), held periodically every two years, which accompanies and historically drives the mobilisation of geographers within the country.

According to Harvey (2008, translated), the creation of new common urban spaces, of a public sphere with active democratic participation, requires backtracking the enormous wave of privatization that has been the mantra of destructive neoliberalism. […] Living under capitalism involves accepting or subjecting oneself to a set of rights that are necessary for the unlimited accumulation of capital. In this direction, insofar as we bring the debate on Critical Geography to the fore, we plead for our ability to change the foundations of transformation of the concrete reality. Within this movement, we position our Final Reflection.

2.7 Final Reflection

In dialectical logic, the relationship between theory and practice is an interdependent one, in which one does not exist without the other. Theory is a reflection that is made from the concrete context, from the researcher's experience with the reality in which they are inserted. In the object of analysis, for critical theorists there is no pure reason—neither a practical nor an aesthetic reason; theory and practice are dialectic and contradictory pairs, which allows a reading of reality and what is recorded about it, in order to be able to transform it. It is the act of transformation that defines the human condition. The unity between practice and theory is the guarantee of the dialectical unity of the objective and the subjective—it is the condition of overcoming, of revolutionary praxis (Conceição 2020).

In a lecture at the Federal University of Minas Gerais (UFMG) on October 23, 2007, Brazilian Philosopher MarilenaChauí stated that, inasmuch as politics are subjected to the logic of consumption-based marketing and spectacle, the notion of public space is being lost. According to her, what is now called public opinion is nothing more than a manifestation of individual tastes, preferences, and feelings, in place of an expression and discussion of the interests of the community.

Further, according to Chauí, the phenomenon is clearly seen in the mass media, where the figures of opinion makers and "people authorized" to speak on a certain subject have emerged. "The media have made the categories of truth and falsehood irrelevant, replaced by notions of credibility and plausibility". As onlookers, people are invited to say what they feel and not what they think: "This is one of the ways of destroying the public space" (Chauí 2008, translated).

For Losurdo (2014), the vogue for the anti-political is related to disillusionment and loss of hope "[…] in the face of the first difficulties and the first contradictions […] the naive enthusiasm turns into disillusionment and disgust […]. Now, the political world, the world as such, proves to be hopelessly mediocre and vulgar. This is the hypochondria of anti-politics" (Ibid., p. 13, translated). Further, according to the author:

The acceptance of a "truth" not submitted to critical research is replaced by the inability to recognize oneself in any truth other than the subject's own truth. The absence of criticism toward the object is thus replaced by the absence of criticism toward the subject—the dogmatism of the object replaced by the dogmatism of the subject (Losurdo 2014, p. 15, translated).

As Coutinho (2010) emphasizes, the misery of reason is the deformed and deforming expression of the capitalist world—capitalism at its extreme level of emptying reason deforms the operations of thinking, reaching the limit of deformation of the human in the animal condition. In this sense, with regard to geographical criticism that is alive, as Carlos (2016, p. 150, translated) summarizes, "Radicality requires the construction of a project for new society, founded in a renewed science [...]".

References

Carlos AFA (2017) A Privação do Urbano e o "Direito à Cidade" em Henri Lefebvre, in Justiça Espacial e o Direito à Cidade. Orgs. Ana Fani Alessandri Carlos; Glória Alves; Rafael Faleiros de Padua – São Paulo: Contexto, pp 33–62

Carlos AFA (2016) A Condição Espacial. Editora Contexto, São Paulo

Carlos AFA (2011) Crise e superação no âmbito da geografia crítica: construindo a metageografia. GEOUSP - Espaço e Tempo, São Paulo, 30

Chaui M (2021) Palestra na UFMG on 23 October 2008. Accessed July 2021. https://www.ufmg.br/online/arquivos/010174.shtml

Conceiçao AL (2020) A Epistemologia da Geografia na Formação dos Grupos de Pesquisa de Geografia Agrária. In: de Souza Mendonça Menezes S, Siqueira Pinto JES (Orgs) Geografias e Geograficidades: Escolhas, Trajetórias e Reflexões. Editora UFS, São Cristóvão

Corrêa (1982) Roberto Lobato, Geografia Brasileira: Crise e Renovação. In: Moreira R (org) Geografia: Teoria e Crítica – Os Saberes Posto em Questão. Rio de Janeiro, Vozes

Coutinho CN (2010) O Estruturalismo e a Miséria da Razão, Posfácio: José Paulo Netto. Editora Expressão Popular, São Paulo

da França Antunes C, Sousa Neto MF (2008) Os Estudantes, a Associação dos Geógrafos Brasileiros (AGB) e o Movimento de Renovação. In: Boletim Paulista de Geografia: 1978 + 30: AGB em Movimento, Specialedition, N° 88

Harvey D (2005) A Produção Capitalista do Espaço. Annablume, São Paulo

Harvey D (2007) A compressão do Tempo-espaço e a condição pós-moderna, in: Condição Pós-moderna: Uma Pesquisa sobre as Origens da Mudança Cultura. Translation: Adail Ubirajara Sobral & Maria Stela Gonçalves. Edições Loyola, São Paulo, pp 257–289

Harvey D (2008) El derecho a la ciudad, Translation for www.sinpermiso.info. Gerardo Pisarell, writtenon 05/10/2008

Harvey D (2014) Espaços de Esperança, translation: Adail Ubirajara Sobral & Maria Stela Gonçalves. Editora Loyola, São Paulo

Lefebvre H (2006) A produção do espaço. Trad. Grupo "As (im)possibilidades do urbano na metrópole contemporânea", do Núcleo de Geografia Urbana da UFMG, (fromthe original: La production de l'espace. 4th edition. Paris: Éditions Anthropos, 2000). First version: start - February 2006

Lefebvre H (2008a) Espaço e Política. Translation: Margarida Maria de Andrade & Sérgio Martins, Belo Horizonte. UFMG

Lefebvre H (2008b) A Revolução Urbana. Translation: Sérgio Martins, 3rd edition, Belo Horizonte: Editora UFMG

Losurdo D (2014) A Hipocondria da Antipolítica: História e atualidade na análise de Hegel. Translation: Jaime Classen. Editora Revan, Rio de Janeiro

Mandel E (1990) A crise do capital; os fatos e a sua interpretação marxista. TranslationbyJuarez Guimarães, João Machado Borges. Editora da Universidade Estadual de Campinas, São Paulo: Ensaio; Campinas SP
Menezes S (2020) A dialética do trabalho e a autocrítica geográfica. In: Trabalho, mobilidade e produção do espaço na relação campo-cidade/organizadores, Suzane Tosta Souza, Jânio Roberto Diniz dos Santos, Sócrates Oliveira Menezes. Vitória da Conquista: Ed. Uesb
Menezes S (2016) O Trabalho nas Fissuras da Crítica Geográfica. Thesis (Doctorate in Geography; final version) – Faculdade de Filosofia, Letras e Ciências Humanas, USP, São Paulo
Mészáros I (2002) Para Além do Capital rumo a teoria da transição. Translation: Carlos Cesar Castanheira & Sérgio Lessa, São Paulo: Ed. Boitempo
Mészáros I (2003) O Século XXI – Socialismo ou Barbárie. São Paulo, 2003. Editora: Boitempo
Moraes ACR (2005) Geografia, pequena história crítica. HUCITEC, São Paulo
Moreira R (2000) E Assim se Passaram Dez Anos. In: GEOgraphia – Year II – No 3
Santos M (2002) Por uma Geografia Nova. ED. USP, São Paulo
Smith N (1988) Desenvolvimento Desigual. Bertrand, Rio de Janeiro

Chapter 3
Far Beyond the 'Natural Environment': Geography at the Crossroads of the Capitalocene

Marcelo Lopes de Souza

Abstract The 'Anthropocene' has become a fad: it is difficult these days, in the academic milieu, not to come across allusions to this new geological era in books, papers and conferences. However, as several authors have already acknowledged, it would be unfair and ideological to hold humanity as a whole (i.e. the *anthropos*) generally responsible for the ecological-social evils that the planet is increasingly suffering. It is necessary to recognise that capitalism, due to its very economic dynamics, is behind the 'Great Acceleration' in terms of consumption, the use of resources and environmental degradation that can be observed since the middle of the twentieth century. This is the reason why '*Capitalocene*' has been suggested as a terminological alternative. As far as the crucial debate about the 'Capitalocene' and its challenges is concerned, geographers are still more 'supporting actors' than 'protagonists.' To a large extent, this has to do with the 'epistemological purification' strategy that many of us have pursued since the 1970s, in the wake of the discipline's radical turn. The weakening of the discipline's identity nucleus with regard to a commitment to the integration of natural and social knowledge led to a decrease in the ability to propose several types of reflections and research related to the '(ecological-)social metabolism,' the planetary dilemmas and the contemporary ecosocial conflicts, as well as a whole series of specific issues and problems—all things that require, to a greater or lesser extent, the construction of *hybrid epistemic objects*. In the last fifteen years, the emergence of a certain *environmental geography* has increased the hope that many radical geographers have finally begun to understand that in order to preserve a high level of sophistication and coherence from the point of view of social research, it is not necessary to give up a genuine interest in geobiophysical knowledge.

Keywords Anthropocene · Capitalocene · Capitalist world-system · Environmental geography

M. L. de Souza (✉)
Universidade Federal do Rio de Janeiro, Rio de Janeiro, Brazil
e-mail: mlopesdesouza@terra.com.br

R. C. Lois González and M. A. Mitidiero Junior (eds.), *Brazilian Geography*, Advances in Geographical and Environmental Sciences,
https://doi.org/10.1007/978-981-19-3704-0_3

3.1 'Anthropocene,' 'Capitalocene'... and the Geographers

The 'Anthropocene' has become a fad: it is difficult these days, in the academic milieu, to take two or three steps without coming across references to this new geological era. Coined by biologist Eugene Stoermer, but popularised by chemist Paul Crutzen from the 2000s on, the term designates a time that due to human's impact on Earth—and on the basis of stratigraphic evidence—it would be justifiably distinguished from the Holocene (which began less than 14,000 years ago and which constituted, until the debate on the 'Anthropocene,' the last epoch of the Quaternary period). Just search 'Anthropocene' in Google, and you will immediately have almost six million results ('*Antropoceno*,' in Portuguese, brings us to just over half a million); there is a huge proliferation of articles, books, congresses, symposia, etc. on the subject. From a strictly stratigraphic point of view, the discussion remains open as to whether it would be entirely appropriate to propose a new geological epoch. Nonetheless, in addition to the evidence in this regard (see, e.g. Zalasiewicz et al. 2017), both in the journalistic world and in the broader scientific debate (also involving the social sciences, the 'environmental humanities,' philosophy, and so on), the question seems already settled: we have entered, without a doubt, the 'Anthropocene.'

The fact that something turns into an intellectual fashion, however, is not enough to discredit the thing in question in itself: after all, the content behind the label 'Anthropocene' is arguably relevant. Nevertheless, a question needs to be asked, regardless of stratigraphic controversies and quarrels: is the term 'Anthropocene,' as such, really the most suitable to name this new era?

For an increasing number of scholars from outside the natural sciences, 'Anthropocene' amounts to a *very generic* and *poor* qualification. Consequently, in the wake of this dissatisfaction, it is not just the word 'Anthropocene' that has experienced an exponential multiplication of mentions in recent years, as there has also been a proliferation of alternative terms: 'Capitalocene,' 'Plantationocene,' 'Chthulucene'.... With the proliferation of words, unfortunately, the possibility of a deeper social criticism gradually fades. Symptomatically, the Anthropocene often seems even for intellectuals and researchers who raise objections to this term, to consolidate itself as a 'consensus,' because it has been so widely accepted and because it supposedly serves well the dialogue between natural sciences and humanities (even a critic such as Haraway (2016, 45 *et seq.*) seems to capitulate a little in the face of this huge influence). In the end, the chance to establish a truly strong alternative gets lost.

In any case, compared to other designations, 'Capitalocene' seems to be particularly persuasive. The term was proposed, as Jason Moore informs us right at the beginning of the acknowledgements section of a book he edited (Moore 2016), by Andreas Malm (then, a doctoral student in Lund, Sweden) in 2009. Moore himself has been, since then, one of the most enthusiastic disseminators of this alternative term, whose rationale is as follows: it would be unfair and grossly ideological to hold humanity or *Homo sapiens* (i.e. *anthropos*) generally responsible for the ecological (or ecological-social) evils under which the planet has increasingly suffered; hence, it is necessary to admit that a mode of production in particular—capitalism, due to its very economic dynamics, is behind the 'Great Acceleration' in terms of consumption, use of resources and environmental degradation that can be observed since the middle of the twentieth century. As much as this 'Great Acceleration' offers the immediate background that has justified, since the end of the last century, the feeling that led to the proposal of the 'Anthropocene,' the real rupture, historically, would refer to a change that occurred about two and a half centuries ago, with the gradual consolidation of industrial capitalism—or even longer, if we take into account the slow process of the emergence of this mode of production, from agrarianism to mercantilism to proto-industrialisation. Here is a type of objection that, without denying the kind of concern and discussion that gave rise to the debate around the Anthropocene, truly refines this debate from a socially critical and historically better informed point of view.

Not all scholars who have raised reservations about the Anthropocene from a broader political-philosophical point of view have adopted the term 'Capitalocene.' At least two other alternatives have been proposed so far, launching interesting debates: 'Plantationocene' and 'Chthulucene' (see e.g. Haraway 2016). However, it is easy to see how some observers who raise reservations about the term 'Anthropocene' have little concern for a comprehensive social critique. Donna Haraway, for example, admits that 'Capitalocene' is a powerful term, but she keeps it at some distance as she sees it as a product of 'a fundamentalist Marxism' (Haraway 2016, 50)—as if the critical outlook behind the idea of a 'Capitalocene' could or should be restricted to one of the currents of critical thinking, or even as if radical criticism was 'fundamentalist' in itself… Even more worrying is the fact that the suggestion of a 'Chthulucene' (derived from Cthulhu, the name of a cosmic entity, with the appearance of a tentacle monster, risen from the pages of sci-fi horror books) and a 'Plantationocene' (based on the historical-economic-geographical notion of plantation) reveals, at least at first glance, high sensitivity to ecological issues, but little social sensitivity: the symptomatic and shocking fact that the term 'plantation'—corresponding, as everybody knows, to the monoculture system so viscerally linked to slavery and colonialism in the Americas—awakens in 'post-humanist' thinkers like Haraway herself (although she no longer claims this label) first and above all the association with the destinies of non-humans, and not with the historical tragedy of slavery, is an attitude that has deserved clear criticism by anti-racist scholars (Davis et al. 2019, 5). This in itself shows the ideological limits of 'ecological postmodernism.'

I think it is a matter of intellectual honesty to concede that the term *Capitalocene* does not necessarily include, in the works of researchers who are less sensitive to types of oppression and structural asymmetries other than class exploitation and the contradiction between capital and labour, the whole range of negative aspects that need to be highlighted; racism and sexism, in particular, are continuously at risk of remaining in the shadows. However, I think it is highly doubtful that a word like 'Plantationocene,' even when used in a historically and socially denser way (as it is the case in Davis et al. 2019), has sufficient coverage to account for more than a very partial aspect of reality. For this reason, as long as it is not used in a narrow way, but rather to evidence or suggest the connections between capitalism (and associated phenomena, such as the *Conquista*, colonialism and neocolonialism, imperialism) and the multiple connections among class, race and gender, Capitalocene seems unsurpassable: it is the most powerful conceptual alternative to a too generic 'Anthropocene.' Moreover, nothing prevents us from considering, besides the problem of social justice and its environmental face—environmental justice—also *ecological justice*, illuminating the ethical issues ('speciesism,' animal rights, etc.) that arise in close connection with the 'dominion over nature' mentality that accompanies westernisation and capitalism.

As for this Anthropocene/Capitalocene/etc. debate as a whole, geographers have been more 'supporting actors' than protagonists. To a large extent, this has to do with the strategy of 'epistemological purification' that many of us have pursued since the 1970s, in the wake of the 'radical turn' (strictly speaking, basically a *Marxist* turn) of the discipline. The dissolution or weakening of the discipline's identity core in terms of a commitment to the integration of natural and social knowledge (a goal that was never completely abandoned on the surface and that still gives rise to a lot of lip service, but without significant support in research practices and in the daily lives of scholars and institutions) led to a decrease in the capacity to propose, from a geographical perspective, research agendas, dense reflections and in-depth studies on what Karl Marx called the 'social metabolism' (*gesellschaftlicher Stoffwechsel*), today's planetary dilemmas and contemporary ecosocial conflicts. As a result, a long list of more or less specific themes and problems has been neglected. The analysis of all these subjects requires to a greater or lesser extent the construction of *hybrid epistemic objects*, precisely what critical/radical human geography came to disdain in the 1970s and 1980s (as for physical geography, it has been 'diluted' amid the disciplinary fabric of natural sciences such as geology and meteorology, therefore placing itself in a secondary role). In this regard, unfortunately, Brazilian geography is no exception, on the contrary: in Brazil, the mutual distancing between human and physical geographers, in the wake of the 'radical turn' (on the part of human geographers) and the tendency towards hyper-specialisation (on the part of many physical geographers), was particularly virulent and traumatic. As a result of this, very few Brazilian geographers, especially between the 1980s and 2000s, dared to seek to build bridges over the intradisciplinary epistemological abyss.[1]

Despite the epistemological self-sabotage outlined in the previous paragraph, the story cannot be reduced to one of losses, setbacks and defeats. Several geographers have never ceased to be interested in the integration of geobiophysical processes with social relations, either within the conventional framework of 'land-change science' or under the shelter of the interdisciplinary and progressive field of political ecology. In the last fifteen years, the emergence of environmental geography (and of so-called critical physical geography even more recently) has raised the hope that geography

[1] Carlos Walter Porto-Gonçalves surely corresponds to an exception that proves the rule: apart from his several important contributions to what we might call a critical environmental geography (and at the same time to political ecology)—see e.g. Porto-Gonçalves 1984, 2006, 2013 and 2014–, few relevant works could be mentioned in this respect between the last decades of the 20th century and the beginning of the present century.

will finally stop throwing the baby out with the bathwater—that is, that radical geographers begin to understand that it is not necessary to give up a genuine interest in geobiophysical processes and factors in order to preserve a high level of sophistication and coherence from the point of view of social research. What is more, it increasingly seems that much of what geographers will have to say in terms of truly relevant contributions to the understanding and overcoming of social problems and challenges, henceforth, will depend on their ability to revalue, on a new basis (i.e. freed from the empiricism and functionalism of classical geography), the integration of different types of knowledge. In other words, on their ability to practise an *intra*disciplinary epistemic transversality. The present chapter intends to be a contribution to achieve more clarity in relation to geography's tasks, possibilities and conceptual, theoretical and methodological potentialities and limitations regarding the analysis of the Capitalocene and its challenges—which seem to represent a crossroads for both humanity and geography itself.

3.2 From the 'Natural Environment' to the Concrete Environment to the Environment-Territory-Place Approach

According to the *Webster's* dictionary, the word *environment* refers to the '(…) circumstances, objects, or conditions by which one is surrounded,' as well as to the '(…) complex of physical, chemical, and biotic factors (such as climate, soil, and living things) that act upon an organism or an ecological community and ultimately determine its form and survival.' The *Oxford* dictionary does not depart from that spirit: environment would be '[t]he natural world, as a whole or in a particular geographical area, especially as affected by human activity.'

We see that the term 'environment' is above all associated to a 'natural' facet of reality, or to 'natural' phaenomena and factors; that is, it is commonly taken as synonymous with 'natural environment.' The same is valid in relation to equivalent words in French (*environnement*) and German (*Umwelt*). In the case of the Romance languages of the Iberian Peninsula and Latin America, the situation is even worse: although the terms *ambiente* (Spanish/Portuguese/Galician) and *ambient* (Catalan) are used more or less frequently also with other, non-naturalising meanings ('social environment,' 'political environment,' 'built environment,' etc.), the mere mention of the words *ambiente*/*ambient* usually makes everyone—from journalistic discourse to academic circles—think primarily of '*medio ambiente*' (Spanish and Galician), '*meio ambiente*' (Portuguese) and '*medi ambient*' (Catalan). The trouble with the expressions '*medio ambiente*,' '*meio ambiente*' and '*medi ambient*' lies in the fact that they are very peculiar ones, because while they correspond to a redundancy (*medio*/*meio*/*medi* and *ambiente*/*ambient* cover approximately the same universe of meanings), when the words *medio*/*meio*/*medi* and *ambiente*/*ambient* are combined in this Iberoamerican linguistic tradition, the meaning is usually first and foremost that of *natural* environment, and therefore a rather restricted one. Interestingly, *two* words are used to commonly end up with *half* the meaning—a veritable semantic waste. Within such a tradition, the ambiguity of the termini environment, *environnement* and *Umwelt*—a cold comfort for sure, but comparatively less disadvantageous—gets frequently lost, since '*meio ambiente*' and its equivalents in Spanish, Galician and Catalan almost always unequivocally refer to 'nature,' and not (or not primarily) to society as well.

The very reduction of the 'environment' to a kind of 'wrapper' (i.e. to something that merely surrounds us) has dramatic consequences: the idea (largely inherited from biology, but already present in the etymology itself[2]) of the 'environment-as-surroundings' has the serious drawback of, in the case of those environmental debates that involve human needs and problems, reinforcing an impression according to which human beings and their social organisations are not part of the environment. This facilitates the previously mentioned reductionism that consists of conceiving the environment as a pure geobiophysical reality, while society is treated as a mere factor among others (in the best of all cases…).

Once we are interested in a non-mutilated environment—that is, in a geobiophysical-anthroposocial reality that is not reducible to a non-human or pre-human 'natural environment,' we are forced to talk about '*hybrids.*' This kind of conversation has been greatly stimulated by philosopher and anthropologist Bruno Latour since the 1990s, beginning with the publication of *We Have Never Been Modern* (Latour 1994). Nowadays, without necessarily agreeing with Latour on all aspects of his approach, this encouragement to question the strategies of epistemological purification (precisely what many human geographers embraced in the 1970s and 1980s) seems unavoidable—unless one wants to pay the price of an increasing gap in relation to the contemporary needs in terms of theoretical reflection and socio-political resistance. Everywhere we look, we will find hybrids, as Latour pointed out three decades ago. Currently, and as time goes by, the hybrid character—neither purely natural nor just social—of the problems that afflict and threaten us is increasingly becoming more obvious: from the greenhouse effect and global warming to the sea level rise to acid rain to daily landslides and floods to the risks represented by transgenics, epidemics and pandemics to the acidification and pollution of the oceans. Each of these phenomena, and even more so the synergy resulting from their interaction at various levels (from microlocal to planetary/global), is both ecological and social. This means that it is inevitable that the way to try to elucidate them and face them implies a cooperation of efforts and a combination of intellectual expertise unrelated to the comfort zones sanctioned by the positivist division of academic

[2] The uncomfortable connotation of 'surroundings' that is carried by the word 'environment' has accompanied the term from the very beginning, as attested by the etymology: *environment* derives from the verb '(to) environ,' which is the same as to surround, to encircle, to encompass, originating in the old French *environer*. It is essential to face the uncomfortable ambiguity that arises from this original element, otherwise we will contribute to perpetuate a source of confusion and limitation: human beings are not only influenced by their environment, at the same time as they increasingly influence it; they are *part of it*, at all scale levels.

labour. 'Nature' and 'society' do not and cannot have, today, the meaning that they still had in the nineteenth century, or even in the first half of the twentieth century.

On the other hand, it is advisable to be careful with a discourse that praises hybrids while inducing us to renounce *any* distinction between 'nature' and 'society.' This praise of indistinction is in vogue; but if this fashionable reading looks recent, its implications are not. From that kind of simplistic indistinction to the suggestion of adopting a single 'scientific method' inspired by natural science (as positivists have traditionally put it) or—completely reversing the reasoning—the neo-Marxist over-simplification according to which 'everything is society' and nature is nothing more than a kind of cognitive illusion, the way is never very long. Therefore, in addition to the hybrids, and without giving up their crucial relevance, it is necessary, at the same time, to refute the mere rejection a priori of the differentiation between what the ancient Greeks called *physis* (the realm of things and processes that are not properly created by human hands) and *nomos* (literally 'law,' but broadly referring to the realm of social creation), or what German philosophers such as Schelling and Hegel called between *erste Natur* (first nature) and *zweite Natur* (second nature). This is to say, the epistemic-political virtues of *unity in diversity* and *ontological pluralism* (from Cornelius Castoriadis (see, e.g. Castoriadis 1975, 1999) to Markus Gabriel [see Gabriel 2015]) need to be preserved, albeit at a level of complexity that has nothing to do with a *Cartesian* separation between 'nature' and 'society.' Inspired by the 'magmatic logic' (*logique des magmas*) explored by Cornelius Castoriadis—an especially broad and radical modality of non-classical logic, opposed to the Cartesianism of what he called 'ensemblistic-identitary logic' (*logique ensembliste-identitaire*) (see Castoriadis 1975, 1986a, b, 1997, 1999), we could talk about the imperative of challenging a *dualistic* view that simplistically separates 'nature' and 'society' in favour of a *dialectical* view that accepts distinctions without hypostatising or reifying them, and instead constantly relativising them. Not, however, a closed and rationalist 'dialectic'—that either preserves as 'synthesis' much of what had to be overcome (as in Hegel's case), or excessively dilutes the specificities of the original terms (as in Marx's case), but rather a 'magmatic' dialectic, in which the relationship between the determined (*bestimmt*) and the indeterminate (*unbestimmt*), or between what is conserved and what is overcome, is radically redefined. The graphic expression of this difference can be seen in Fig. 3.1: in 3.1.A, humans 'establish a relationship' with their '(natural) environment,' understood as a geobiophysical reality that surrounds them; in 3.1.B, in contrast, the integral environment encompasses the set of complex interactions (nonlinear, not univocal, whose limits are not trivially determinable) between the various dimensions of non-human geobiophysical reality (*physis*, *erste-Natur*) and the social relations, as well as the materiality of social products (beginning with the material aspect of social space) and, finally, the human beings themselves in their corporeality.

At this juncture, before moving on to the next section, it is convenient to supplement the discussion with the help of two other concepts, namely *territory* and *place*. It is true that the concept of environment, in its non-mutilated version (i.e. conceptually not prisoner of the so-called natural environment), seems to make a reinforcement of the dimensions of power (in the form of territory/territoriality)

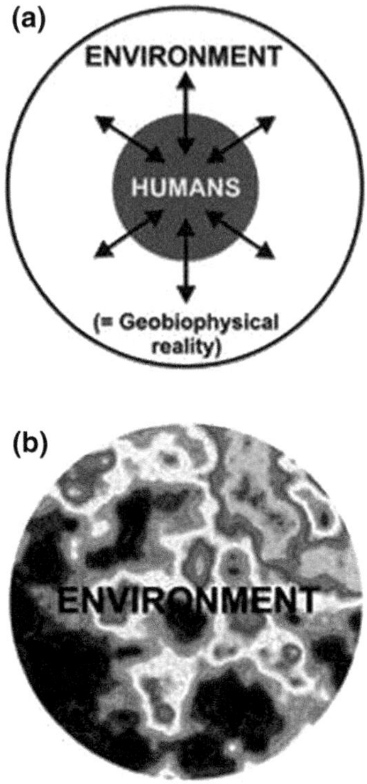

Fig. 3.1 Two approaches to the notion of *environment* (and to the links between 'society' and 'nature' in this context)

and culture/symbolism/identities (in the form of place) superfluous, as these dimensions of spatiality are already somehow implicitly contemplated by a holistic, robust concept of environment. Even so, it cannot be emphasised enough that facet of the geobiophysical-anthroposocial reality to which any notion of environment intuitively will always refer to in the first place—the *(ecological-)social metabolism*—can never be adequately examined without taking into account the power relations as well as culture/symbolism/identities. In other words, it can never be investigated without considering the territories—that is, the spaces defined and delimited by power relations, or, more precisely, the spatial projections of power—and the places—that is, the spaces of the 'lifeworld' (*Lebenswelt*), or, more broadly, the lived spaces, in relation to which feelings of belonging and emotional ties develop.

Just adding the concepts of territory and place to our analysis, however, seems to be insufficient. The analytical procedure of scrutinising the dimensions of geographical space has led to an extraordinary diversification of the conceptual arsenal available to researchers in the last decades; nevertheless, a movement in the opposite direction is also needed one that restores a sense of 'totality' to reality, in the wake of a process of synthesis. It is imbued with this spirit that it becomes reasonable to speak of *environment-territory-place* (or, for convenience, simply ETP), referring to a notion that promotes by means of a single expression the explicit interconnection of several fundamental ideas. The material transformation of spaces through human labour never ceases to be, at the same time, a process that cannot be simplistically detached from the struggles for the maintenance of territories—or for the seizure of another's territory, i.e. a *deterritorialisation*; this maintenance (or this seizure), for its part, will almost always be linked, be it directly or indirectly (and more or less explicitly), to the defence of a way of life and of values symbolised by place and somehow 'inscribed' ('*geo-grafada*' ['geo-graphed'], as Brazilian geographer Carlos Walter Porto-Gonçalves [2002, 2006, 2014] suggested) on the material space, with its landscapes, landforms and spatial processes (valleys and mountains, rivers and forests, fields, pastures, buildings, etc.). In short, the idea of environment-territory-place concerns a reality understood through the prism of the inseparability between ecological-social metabolism, projections of power onto space, and sense of place.

At first glance, this composite 'concept' seems useful, above all, as a kind of *reminder*: one that reminds us that doing justice to the complexity of reality, although it does not reject *differentiations*, certainly repels the epistemological one-way streets that worship *disjunctions* (in the sense of systematically distinguishing without reconnecting what actually does not exist in isolation). However, the theoretical power of what is being proposed here precisely derives from realising that ETP is rather an *approach*—or a methodological tool—than a concept. Firstly, because it is not the case to imagine a substantive conceptual innovation behind the combination of environment, territory and place in the same expression; secondly, because we must avoid the risk of falling into the type of reification criticised in the 1970s by the German and French geographers who raised objections, respectively, against the traditional concepts of *Landschaft* and *région*, which were intended to be holistic, but which ended up being instrumental for a functionalist view of reality that forced harmonies and convergences of spatial types and dynamics, hence neglecting conflicts and 'mismatches.' I do not assume that concrete environments, territories and places exist in perfect accord with each other; multiple environments, territories and places will interact with each other in a complex way—sometimes converging, sometimes not, for many historical and political reasons. What matters, and this has eminently operational value, is to emphasise the need to, when looking at reality and its discursive representations, never lose sight of the strategic relevance of a 'conversation' (albeit with tensions) among environment, territory and place.

3.3 Better Late Than Never: The Emergence of Environmental Geography

In the last fifteen or twenty years, the strategy of 'epistemological purification' uninhibitedly adopted by most critical geographers in the 1970s and 1980s has been—often explicitly—challenged by the emergence of *environmental geography*.

Although the environmental geography propagated by Anglophone geographers (see, for example, Castree et al. 2009) constitutes, given the scientific hegemony of the Anglo-American linguistic-cultural milieu, the major driving factor behind this intellectual project aiming at a partial re-appraisal of geography's disciplinary profile and self-image, similar efforts have been occurring in many places and academic environments (from France to Germany to Brazil). Furthermore, the gap created by the predominance of the 'purifying crusade' of the 1970s and 1980s was never absolute: even if some geographers were forced to 'seek refuge' in the interdisciplinary field of political ecology, they did not fail to oppose some resistance from inside geography itself, even at the height of the prestige of 'epistemological purification' as an ideal to be pursued.

Despite having ancient and deep roots—given the fact that the interface between natural and social research has been at the very heart of the disciplinary identity of geography for generations, contemporary environmental geography does *not* imply any nostalgic 'return to the past,' insensitive to the political-philosophical, epistemological and theoretical advances of the last decades. These notably include the overcoming of a typical disinterest for theorising, the overcoming of naive functionalism, a decisive objection to naturalism/positivism and the ultimate rejection of environmental determinism. Ancient roots therefore coexist with boldness and theoretical/methodological sophistication (which are, after all, crucial fruits of the 'radical turn').

The need for an environmental geography has not, on the other hand, excluded a sincere modesty in the way how hybrid epistemic objects have been explicitly and strongly advocated. No one believes nowadays—in contrast to what was common sense among geographers during the classic period of the discipline—that some kind of *environmental* geography must or could be synonymous with geography *tout court*. It is willingly admitted—or, in any case, it should be willingly admitted—that epistemic objects essentially designed to fit the boundaries of social research (such as studies into electoral geography, the formation of socio-spatial identities, urban networks or locational patterns of industrial or tertiary activities) or the boundaries of natural research (such as investigations on and modelling of the evolution of landforms or atmospheric dynamics) can legitimately and fruitfully capture the attention of geographers. Strictly speaking, this is not a mere question of 'possibility': a huge accumulation of knowledge and interesting debates has taken place under these conditions. The only thing that is defended here is that hybrid epistemic objects *also need to have their right of existence acknowledged*, which presupposes respect for them, instead of decreeing the 'epistemological purification' as an absolute goal.

Considered from this perspective, the kind of environmental geography that is currently gradually emerging corresponds to a 'rebirth' of the project of intradisciplinary epistemic transversality, but *on a new basis*, namely on one that combines a certain *epistemological modesty* with a remarkable *theoretical-methodological ambition*. The geographers who undertake this endeavour must seek to benefit with the caution of those who embark on a non-trivial adventure, but at the same time free from the hesitations and complexes which plague those who have been indoctrinated for decades in the dubious art of confusing the difficult with the impossible and the undesirable—from the synergy of two intersecting approaches: that of *ecogeography* (a term that, borrowed from Jean Tricart, expresses more accurately and critically than the old label 'physical geography' a type of research that is fundamentally polarised by the epistemological pole of geobiophysical inquiring, with its own methods) and that of *social geography* (an expression we owe to Élisée Reclus, which is much more precise than the traditional label 'human geography').[3] Once we reject the 'purifying' strategies of both the search for a 'pure physical geography' (which reduces society to a simple, vague, internally undifferentiated 'anthropic factor') and for a 'pure human geography' (implying a partial dematerialisation of reality, the denial of the biological dimension of the species and, more broadly, the ignorance of geobiophysical knowledge), *dialogue* becomes possible, no matter how tense it (inevitably) is. Dualism is abandoned in favour of a dialectic, without the search for better *integration* falling or degenerating into an obsession for *unifying* theoretically and methodologically macro-fields of knowledge (geobiophysical processes, on the one hand, and social relations, on the other), which resist a homogenisation of treatment.

Throughout this effort to accept the challenge of opening up our minds to hybrid epistemic objects, geographers identified with environmental geography are not and have not been alone. Be it in the form of the interdisciplinary field of political ecology—to which geographers have alongside anthropologists decisively contributed since the 1980s, be it in the form of a so-called critical physical geography, recently advocated by Anglophone geographers (Lave et al. 2014, 2019), a not insignificant number of geographers have tried to compensate for lost opportunities. Environmental geography, however, seems to be especially tailored to this task, as both political ecology and 'critical physical geography' have some limitations: in the case of political ecology, because it is an interdisciplinary field that was born as an initiative that primarily came from the social sciences and humanities (to the point that, at a given moment, a certain malaise emerged and a debate was established around the growing scarcity of *ecological* analysis in political ecology…)[4]; as for 'critical physical geography,' in spite of a somewhat excessive ambition, it consists of knowledge in which proficiency about geobiophysical processes clearly and ultimately overcomes familiarity with social relations and their study. The limits

[3] The objections against these two labels have already been explained in previous works, but it can be useful to repeat them here: 'human geography' is a designation that highlights an adjective—human—that refers to the epistemological realm of the biological species, not to the realm of society, which would be much more powerful and fruitful. As for expression 'physical geography,' it should be admitted that the adjective 'physical sounds rather restrictive and poor today; it does not possess the same potentiality of the term *ecological* (or of the prefix *eco*), which refers, in a more direct way, to the totality of biotic and abiotic factors. (When Alexander von Humboldt subtitled his masterpiece, *Kosmos*, the formulation *Outline of a physical description of the world* [*Entwurf einer physischen Weltbeschreibung*], the meaning of the adjective 'physical' was very broad then; however, it is necessary to recognise that since the consolidation of the discipline called 'physics,' its scope has shrinked.)

[4] See, on this, Walker (2005).

of political ecology and 'critical physical geography' do not in themselves constitute an insurmountable obstacle, nor should the inevitably overlapping efforts be seen as a major problem; the only real problem lies in overestimating the strengths and underestimating the weaknesses of each epistemic-discursive niche, with its trajectory and specificities. In the case of 'critical physical geography,' in particular, it is praiseworthy that—although late—English-speaking, anti-positivist ecogeographers seek to swell the ranks of those who support and defend intradisciplinary epistemic transversality; but it is frustrating to note a certain megalomania, along with a residual Eurocentric bias as well as the partly unfair assessment of the achievements of political ecology.[5]

With environmental geography, it is evident that a constructive overcoming of the 'human geography as a purely social science' and the 'physical geography as a purely natural science' approaches (or worse: geomorphology, climatology, etc. as independent fields, in the wake of a hypothetically irreversible fragmentation of 'physical geography' itself) has been attempted. In short, a 'new synthesis' or a new 'epistemological contract'—this time more modest but also more ambitious in comparison with classical geography, because it is theoretically and methodologically more sophisticated—has been defended and advanced. It is on this basis that geographers could be promisingly able to prepare themselves to contribute to elucidate the processes and dynamics of the Capitalocene, at several scale levels. This is exactly what the next section will deal with (even if in a purely introductory way), whose locus of discursive production and analytical focus will be the (semi-) periphery of the capitalist world-system.

3.4 The Capitalocene at the (Semi-)periphery of World Capitalism

In a famous and much-cited passage from the preface to the first edition of *Capital*, Marx asserts that 'the country that is more developed industrially only shows, to the less developed, the image of its own future'[6] (Marx 1962: 12). What seemed, at least until a few decades ago—when there was still significant optimism and the influence of historicist thinking was still considerable—to be something quite logical and credible, is difficult to sustain today. The idea according to which the state/stage of 'development' achieved by the most industrialised and 'advanced' countries is something that *will* be (=factual judgement, albeit related to a scenario) and *should* be (=normative judgement) reached by the least industrialised and less 'advanced' ones, makes us assume that this is not just a matter of inevitability, but also of

[5] We could also mention the insistence on maintaining a label ('physical geography') that geographers like Carl Troll, with his landscape ecology/geoecology (*Landschaftsökologie/Geoökologie*), and Jean Tricart, with his ecogeography (*éco-géographie*) had already intuitively perceived as weak.

[6] In the German original: "Das industriell entwickeltere Land zeigt dem minder entwickelten nur das Bild der eignen Zukunft".

desirability. However, both aspects are worth challenging. But incredible as it may seem, this old-fashioned approach remains alive in the work of many modern neo-Marxists, whose Eurocentrism is evident—even in those cases where the authors are proud about their supposedly 'post-colonial' or 'decolonial' approach to socio-spatial problems.

The peculiarities of the periphery and semi-periphery (i.e. those more or less industrialised countries of the 'Global South' such as Brazil, Mexico, Argentina and South Africa) of the capitalist world-system tend to be ignored or underestimated by the scholars based in the 'Global North,' but it is precisely there that the system shows its most cruel and anti-ecological face (respectively, persistence of exploitation on the basis of renewed forms of extraction of absolute surplus value and 'import'/absorption of eco-stress from core economies); it is also there that certain innovations (related to repression, social control, etc.) are tested, and certain processes, which will have reflexes or will be (partially) replicated in the 'Global North,' will be especially felt. Indeed, the (semi-)periphery has always been a laboratory for the geoeconomic and geopolitical global core; it is often a vanguard, and sometimes a testing ground, be it in relation to the testing of strategies and methods of repression and social control (see, for example, Graham 2010), be it in relation to trends of increasing inequality and conflict. For the purposes of this chapter, even more impressive is the intensity with which *the periphery* (and the semi-periphery) shows *to the core*, at least to some extent, its own future with regard to social issues; this is the rationale behind the suggestion made by sociologist Ulrich Beck twenty years ago by means of the expressions 'Brazilianization of the West' (Beck 2000, 1–9) and 'Brazilianization of Europe' (Beck 2000, 92–109), where 'Brazilianization' means an increase in inequalities in the wake of a growing precariousness in terms of labour and life conditions of poor workers.

At the periphery and semi-periphery, the colonial devastation that took place from the sixteenth to the nineteenth century was replaced by new forms of degradation in the twentieth and twenty-first century: transfer of polluting industries to the semi-periphery, 'import' of toxic waste by the periphery and semi-periphery, etc. The novelty, so to speak, lies in the reversal of prevailing expectations in comparison with the 1970s (not only among conservatives, but mostly on the part of the left itself): after the so-called lost decade of the great foreign debt crisis (the 1980s), the 1990s and 2000s witnessed an increasingly clear trend towards *deindustrialisation* in Brazil—loss of relative and sometimes even absolute importance in the manufacturing industry, along with an incessant increase in dependence of the country's export basket on agricultural and mineral commodities. Despite some conceptual and methodological divergences (and hence also in terms of interpretation) in relation to the degree of deindustrialisation in Brazil,[7] there is a notable consensus around the existence of such a process since the 1990s or 2000s, being the *reprimarização* ('reprimarisation') of the unquestionable export basket. All of this occurs within the framework of the emergence of neoextractivism as a 'model' for Latin America in the twenty-first century (Svampa 2019), which confronts us with the return, even if

[7] Compare e.g. Cano (2012) and Bonelli et al. (2013).

partially sophisticated or modernised, of typically (neo)colonial situations. It can be seen, therefore, that history is very far from unfolding in a linear manner.

What Argentina had already taught as a crucial lesson—given that deindustrialisation there began as early as the 1970s, catalysed by the adoption of a neoliberal economic policy—is now exemplified by the Brazilian experience: the '*desarrollo dependiente-associado*'/'*desenvolvimento dependente-associado*' ('associated-dependent development'[sic]) discussed by Fernando Henrique Cardoso and Enzo Faletto in the 1960s (Cardoso and Faletto 1979) is indeed possible to some extent, as it is viable to industrialise even though economic dependence usually establishes clear limits for this; *however*, there is no guarantee that there will be no setbacks, let alone the fact that the hypothetical rise to the position of semi-periphery will not necessarily open the door to a subsequent promotion to the category of a core country. As time goes by, the relative optimism of Cardoso and Faletto's interpretation has been showing its failures better and better. Not mentioning the fact that almost an entire continent, Africa, has been gradually forgotten, as it is considered by many as a 'hopeless case' (in spite of the Chinese investments, which have had some parts of Africa as a significant destination), a case like the Brazilian one is challenging enough: seen by many for generations as a 'land of the future,' Brazil increasingly resembles a 'land of the past,' both economically and politically (the increasing presence of right-wing extremism and the rising of neo-fascist nostalgia being to a large extent the result of economic failure).

It is somewhat surprising that a thesis such as that of 'planetary urbanisation,' propagated by Brenner (2014) and others, which is a direct legacy from the Lefebvrian theses of a '(complete) urbanisation of society' and the rise of an 'urban society,' becomes so popular in presenting the urbanisation of the world not only as an *inexorable* trend but also implicitly as a *positive* one, clearly paying tribute to Eurocentrism and its fixation on Western-style 'modernisation.' The ways of life of peasants, traditional populations and native peoples are treated as relics or something residual, which indirectly implies seeing their struggles and demands as of minor importance or anachronistic. At the same time, the concrete consequences of urbanisation in the 'Global South,' especially for the poor, are underestimated or obscured: deterritorialisation leading to *favelas* and peripheries; replacement of the harshness of agricultural work with urban hyperprecarious work; co-optation by religious fundamentalism or organised crime; and so on. Indeed, the dialectic between 'order' (which on its part constantly generates 'disorder') and 'disorder' (which gives birth to 'new orders' all the time), between legal and illegal, and between formal and informal, is one of the main characteristics of Latin American urbanisation, as well as in other parts of the 'Global South': the ubiquitous presence of crime and 'corruption' in the wake of the complex intertwining between exogenous and endogenous factors, at all scale levels. This demonstrates the complexity of the challenges many societies must face, including in terms of struggle and resistance against an increasingly promiscuous, criminal (and not only criminogenic!) capitalism. When, as in Brazil (with Rio de Janeiro as a 'lighthouse' or 'model'), the retail drug trade itself is increasingly replaced by paramilitary 'militias,' which extort firms and residents and whose activities are no longer limited to segregated spaces, it is the general conditions of anti-capitalist resistance themselves that are seriously threatened nowadays.

Definitely, the (semi-)periphery is not for beginners. And since it corresponds to most of the planet and its population, it must be understood correctly. Denouncing the (Eurocentric) superficiality of extemporaneous theses and theories aiming at 'modernisation' does not necessarily or automatically correspond to romanticising country life or idealising pre-capitalist societies (as 'anarcho-primitivists' and other people uncommitted to the reality principle do); similarly, pointing out the absurdity that—from an ecological-social perspective—consists of imagining the globalisation of the US-American consumption pattern and underestimating the global ecological footprint of such a pattern does not mean that one is fond of 'ecofascism' at all—i.e. insensitive to the needs and wishes of three-quarters of humanity.[8] Both our interpretive schemes and our projects of society need to take this into account.

The Capitalocene had several moments or 'phases,' on any considered level (planetary/global, hemispheric, national, regional, local…). This can be observed both in the core and (with obvious particularities) at the periphery of capitalism, in the wake of the constitution of the capitalist world-system or capitalist world-economy, to use Immanuel Wallerstein's well-known terminology (a world-*economy* that is, by the way, also a 'world-*ecology*,' as proposed by Moore [2011]). The current period is one of reversal and lowering of expectations: pessimism has replaced the more or less moderate optimism that prevailed until the 1970s and 1980s. If Thomas Piketty is right, then the high rates of annual economic growth (GDP per capita) of the twentieth century were an anomaly, and they will not be repeated in the twenty-first century (Piketty 2014, 93 *et seq.*). Simplifying the scenarios built by him, it is reasonable to imagine that the rates of the countries of the 'Global North' will be low (around 1% per year, at most, in the second half of the century), whereas in the 'Global South,' although higher, they will nevertheless remain at an unimpressive level of 4%, or even more modest rates after 2050. However, as Piketty himself points out, these assumptions are comparatively optimistic: other economists predict considerably lower rates. Demographic growth, for its part, will be even lower: in the second half of the twenty-first century, it will be close to zero or even negative in most continents, except for Africa (Piketty 2014, 79).

From an ecological and planetary point of view, this seems to represent, at first glance, a certain relief in the long run. But at what social cost? Taking into account the fact that capitalism is a mode of production inherently dependent on economic growth (due to the imperative of capital accumulation), some joker might well ask:

[8] The picture is surely complex, as we are not always facing 'ecofascism' in a strong sense: it suffices to bear in mind Haraway's (2016) 'soft' neo-Malthusianism, expressed by the acceptance of the slogan 'Make Kin Not Babies'—in which 'kin' comprise humans and non-humans alike—, to see how certain academic circles in the 'Global North' (and those in the 'Global South' which import their agendas) are distant, both intellectually and psychologically, from the lives and the everyday life of most of the planet's population. The fact that such a position is seen, often without reservations, as 'progressive,' is a proof of the political decay and the lowering of intellectual standards that characterise our time, despite the advances here and there (such as the willingness to refuse Eurocentrism, racism and sexism).

'what will end first, capitalism or the planet?' It is obvious, however, that 'the planet' will not end, and probably neither will capitalism. What can demise, and probably will, are our illusions, especially those related to the 'development' of the 'Global South': conflicts at various scale levels—but mainly in cities, regions and countries in Latin America, Africa and parts of Asia—will bury capitalism's 'civilising' promises for good.

What we will have ahead of us, more and more, is a kind of low-growth capitalism, politically manageable only by means of a combination of varying degrees of populism with neo-fascist barbarism (which are two sides of the same coin). Meanwhile, even if carbon dioxide emissions were reduced to zero in the medium-term—a totally unrealistic hypothesis, of course—the immense quantities already present in the atmosphere would by inertia account for the continuation of global warming, with all its cortege of effects. These effects include both the geobiophysical ones-—uch as the expansion of ocean water and the melting of continental glaciers and sea ice, the sea level rise, sharp changes in precipitation patterns as well as of salinity and acidity of the oceans—and the social ones—i.e. economic-spatial changes (mainly in terms of the global spatial patterns of agriculture), increasing frequency of disasters and impacts of extreme weather events on crops, worsening situations of socio-spatial inequality in coastal cities, increasing health problems, etc. Geographers can and should be among those best able to warn about the dangers of underestimating the ecological-social challenges and threats posed by a capitalist 'civilisation' that cannot survive in the long run—at least not in the form as we know it.

3.5 Conclusion: Escaping the Twentieth Century

What do geographers want today?... Does it not make sense—at last!—to re-interpret the 'burden' (so perceived by many) represented by the intradisciplinary epistemic transversality, valuing it on a new basis (i.e. with classical geography's weaknesses in mind) as an extraordinary *asset*? This is the hope that lies behind the epistemological project called 'environmental geography.'

Taking the Capitalocene's challenge seriously means that we urgently need to get out intellectually of the twentieth century, in which we were (and somehow still are) prisoners in a twofold sense: on the one hand, from an objectivist and, to varying degrees, either from an empiricist (positivism) or from a rationalist (Marxism) 'thesis,' in the dialectical sense of this term; on the other hand, from a relativistic, subjectivist 'antithesis' which tends, albeit with variations, to nihilism or, at least, to political (semi-)paralysis (examples being postmodernism, extreme social-constructionism, 'Actor-Network Theory,' and the like). Often relevant insights, such as pointing out the contradictions and dead ends of the enlightenment and the exaggerations of a 'provincial anthropocentrism,'[9] have been wasted and diluted into

[9] It has become fashionable, in academic and activist circles that discuss environmental ethics problems, to treat the term 'anthropocentrism' as an insult, as if *all* anthropocentrism shares the old

'deconstructionist' and relativist extremisms that have often made the postmodern 'antithesis' appear as not less nefarious than the modernist 'thesis,' at least from an emancipatory point of view. It suffices to have a look at contemporary intellectual trends and fashions such as 'anti-humanism,' 'post-humanism,' 'trans-humanism' and last but not least the various sorts of misanthropic ecologism and socially insensitive forms of animal rights advocacy. Even in the case of many so-called progressive people, ecological and animal rights activism (undoubtedly relevant and legitimate as such!) is often practised in a way that undermines social criticism, as for instance the controversial idea of proposing the complete equalisation of intraspecific and interspecific empathy and solidarity, hence denouncing everything that does not fit the radically 'biocentric,' 'abolitionist,' etc. outlook as 'speciesism,' which is simplistically compared to racism and human slavery... Incidentally, such 'progressive' intellectuals and activists are often unable to disguise their intolerance towards traditional cultures and practices that are regarded by them as incompatible with the ideals of veganism and 'animal liberation'... From my point of view, in contrast, the 'more-than-human' epistemic-ethical–political perspective, introduced and increasingly invoked by philosophers and researchers with a postmodern background, will only carry a genuine emancipatory dimension if the 'more-than-human' formula is not translated into social and intercultural insensitivity, pedantry and elitism.

Two events that took place during the Royal Geographical Society's annual meeting (with the Institute of British Geographers) in London in 2019 illustrate these concerns very well. **Scene No. 1**: a well-known British geographer remarked, during the debates after my presentation on environmental justice, that the term 'environmental justice' should be altogether abandoned in favour of *ecological justice*, given its 'anthropocentric' character. (With such a statement, he, with astonishing and arrogant self-confidence, disqualified and threw into the trash can nothing less than four decades of mobilisations and activism by poor and racially stigmatised workers from various countries, starting with the United States. By the way, I'm *not* suggesting that our vocabulary, theoretical approaches and even the existing, concrete political practices can *never* be subjected to *constructive, culturally and socially sensitive* debate.) **Scene No. 2**: during the same event I coordinated, together with three other colleagues (from the United Kingdom, France and Canada), three sessions on 'Environmental and Ecological Justice: Anarchist Contributions and Perspectives.' Together, the three sessions had an expressive number of presentations and many listeners. Both presenters and audience (the latter during the questions and comments made after the presentations), with rare exceptions, entirely neglected the

---

ideals (typical of the old Christian mentality and the Enlightenment) of a 'dominion over nature' and a relative insensitivity to the suffering of animals and a devaluation of non-human life forms. It is not seen that between capitalist/instrumentalist anthropocentrism (which I have termed 'provincial anthropocentrism') and deep green 'ecocentrism' or 'biocentrism,' there are many shades of gray. For example, to assume the premise that the solidarity and empathy of *Homo sapiens sapiens* does not need and should not be in principle the same as that our own species devotes to other species (in fact, with gradations depending on the level of sentience and other factors) does not automatically imply that the fate, suffering and rights of other forms of life and entire ecosystems cannot be taken into account by humans!

theme of environmental justice, focusing instead on discussions on animal rights, animal liberation and ('anarcha'-)veganism. Therefore, the possibility of comparisons and analytical cross-fertilisation was largely compromised, since the subject of environmental justice (and human needs and suffering) was treated, beforehand, as not that relevant (or so it seemed).

Today, more than ever, not everything that glitters is gold. Anyone willing to fight openly or implicitly against conservative positions needs to pay attention not only to the myths and limitations of previous centuries (undisputed validity of science, overoptimism reflected in ideas such as 'progress' and 'development,' absolute emancipatory centrality of the class struggle, ecologically imprudent productivism, etc.), but equally to the new myths and limitations nourished by a 'progressivity' immersed in an ocean of contradictions and traps (detachment between the agendas of 'identity politics' and the question of labour exploitation and class struggle, reinforcement of neo-Malthusianism in the name of 'ecology,' economic ignorance that uses the denunciation of economism as an alibi, petty-bourgeois/Eurocentric/academic/urban bias that leads to a disregard for the real living conditions of most of humanity, and so on).

We need—in fact we have needed it for a long time—a *new 'synthesis'* (again in the dialectical sense of this term). There have been efforts and steps in this direction, some of them brilliant, starting with the abstract level of the theory of knowledge, as illustrated by Markus Gabriel. However, much of the essence in this regard had already been anticipated by Cornelius Castoriadis several decades ago, with the advantage of establishing a strong connection of the rather abstract sphere of ontological/logical/epistemological reflection with the sphere of political-philosophical debate.

In spite of this, not few people still seem to favour the postmodern 'antithesis,' delighting themselves with its seemingly infinite variations—which are becoming more and more extravagant and therefore more and more elitist, since they are distant from the everyday life, the needs and language of the immense majority of the people on our planet, be it in the cities or in the countryside. The twenty-first century, despite appearances, only prolong the last decades of the previous century to a large extent. At least in some respects, the twentieth century was not exactly 'short' (1914–1991), as suggested by Eric Hobsbawm, but instead it has been too long, because it still casts its shadows (resurgence of fascisms, sclerosis of the left, dissolution of anti-capitalist criticism) in the third decade of the twenty-first century.

References

Beck U (2000 [1999]) The Brave New World of Work. Polity Press, Cambridge

Bonelli R et al (2013) Desindustrialização no Brasil: fatos e interpretação. In: Bacha E, De Bolle MB (eds) O futuro da indústria no Brasil: Desindustrialização em debate. Civilização Brasileira, Rio de Janeiro

Brenner N (ed) (2014) Implosions/Explosions: Towards a Study of Planetary Urbanization. Jovis, Berlin

Cano W (2012) A desindustrialização no Brasil. Economia e Sociedade, v. 21, número especial, 831–851
Cardoso FH, Faletto E (1979 [1969, 1976]) Dependency and Development in Latin America. University of California Press, Berkeley, Los Angeles and London
Castoriadis C (1975) L'institution imaginaire de la société. Seuil, Paris
Castoriadis C (1986a [1981/1984]) L'imaginaire: la création dans le domaine social-historique. In: Domaines de l'homme- Les carrefours du labyrinthe II. Seuil, Paris
Castoriadis C (1986b [1983]) La logique des magmas et la question de l'autonomie. In: Domaines de l'homme- Les carrefours du labyrinthe II. Seuil, Paris
Castoriadis C (1997 [1993]) Complexité, magmas, histoire. In: Fait et à faire - Les carrefours du labyrinthe V. Seuil, Paris
Castoriadis C (1999 [1991]) Mode d'être et problèmes de connaissance du social-historique. In: Figures du pensable- Les carrefours du labyrinthe VI. Seuil, Paris
Davis J et al (2019) Anthropocene, Capitalocene, ... Plantationocene?: A manifesto for ecological justice in an age of global crises. Geography Compass 13(5). https://doi.org/10.1111/gec3.12438
Gabriel M (2015) Fields of Sense: A New Realist Ontology. Edinburgh University Press, Edinburgh
Graham S (2010) Cities under Siege: the New Military Urbanism. Verso, London
Haraway D (2016) Staying with the Trouble: Making Kin in the Chthulucene. Duke University Press, Duke and London
Latour B (1994 [1991]) Jamais fomos modernos: Ensaio de Antropologia simétrica. Editora 34, Rio de Janeiro
Lave R et al (2014) Critical Physical Geography. The Canadian Geographer/LeGéographe canadien 58(1):1–10
Lave R et al (orgs.) (2019) The Palgrave Handbook of Critical Physical Geography. Palgrave, London
Marx K (1962 [1867, 1890]) Das Kapital: Kritik der politischen Ökonomie (Erster Band, Buch 1: Der Produktionsprozeß des Kapitals, "Vorwort zur ersten Auflage"; Marx-Engels-Werke, Band 23). Dietz Verlag, Berlin (based on the 4th edn, edited por Friedrich Engels)
Moore J (ed) (2016) Anthropocene or Capitalocene? Nature, History, and the Crisis of Capitalism. PM Press, Oakland, CA
Moore J (2011) Ecology, capital, and the nature of our times: accumulation & crisis in the capitalist world-ecology. J World Syst Res 17(1):107–146
Piketty T (2014 [2013]) Capital in the twenty-first century. The Belknap Press of Harvard University Press, Cambridge, MA, London
Porto-Gonçalves CW (1984) Paixão da Terra: Ensaios críticos de ecologia e Geografia. Rocco e Socii, Rio de Janeiro
Porto-Gonçalves CW (2002) Da geografia às geo-grafias: Um mundo em busca de novas territorialidades. In: Ceceña AE, Sader E (eds) La Guerra Infinita: Hegemonía y terror mundial. CLACSO, Buenos aires, pp 217–256
Porto-Gonçalves CW (2006) A globalização da natureza e a natureza da globalização. Civilização Brasileira, Rio de Janeiro
Porto-Gonçalves CW (2013 [2004]) O desafio ambiental, 4th ed. Record, Rio de Janeiro e São Paulo
Porto-Gonçalves CW (2014 [1989]) Os (des)caminhos do meio ambiente, 15th ed. Contexto, São Paulo
Svampa M (2019) Las fronteras del neoextractivismo en América Latina: Conflictos socioambientales, giro ecoterritorial y nuevas dependencias. Bielefeld University Press, Bielefeld
Steffen W et al (2015) The trajectory of the Anthropocene: the great acceleration. Anthropocene Rev 2(1):81–98
Walker PA (2005) Political ecology: where is the ecology? Progress Hum Geogr 29(1):73–82
Zalasiewicz J et al (2017) Making the case for a formal Anthropocene epoch: an analysis of ongoing critiques. Newsletters Stratigr 50(2):205–226

Chapter 4
Brazilian Geography and the Study of Territorial Formation

Manoel Fernandes de Sousa Neto

Abstract The formation of Brazilian society is inextricably linked to the territorial formation of Brazil, recognising the logic of Iberian colonisation and its subordinate placement in the international division of labour which, since as far back as the sixteenth century, has resulted in systematic colonisation which expropriated the original populations, transferred and subjected African ethnic groups to slave labour and established itself as an exporter of agricultural and mineral products. The process of territorial formation has led to an expansion towards the interior, with immense destruction of forests, ethnocides and wars against the indigenous peoples. Brazilian Geography was established as an important aspect in supporting the processes of colonial domination by producing knowledge about the territory, the native populations and nature. This vast knowledge of the territory was achieved through scientific expeditions, geographical societies and military institutions between the sixteenth and nineteenth centuries and, subsequently, by university institutions and research and planning institutes. This article seeks to unravel the role of geography as a body of knowledge and as a scientific discipline in the territorial and social formation of Brazil, setting out the contradictions in the process by proposing that the violence was not only material, but also symbolic and epistemological, expressed in a range of social conflicts as a result of the colonial logic of territorial appropriation. Finally, it seeks to elucidate the social conflicts in the formation of the territory and covered by geographical discourses and ideologies present in a range of types of literature, such as the school geography literature which treated Brazil as a peaceful, idyllic and paradisiacal country.

Keywords Territorial formation · Systematic colonisation · Historical geography

---

M. F. de Sousa Neto (✉)
Universidade de São Paulo, São Paulo, Brazil
e-mail: manoelfernandes@usp.br

R. C. Lois González and M. A. Mitidiero Junior (eds.), *Brazilian Geography*,
Advances in Geographical and Environmental Sciences,
https://doi.org/10.1007/978-981-19-3704-0_4

4.1 The Presence of the Past: Immanent Permanence

While leafing through the digital pages of the Universo Online (UOL) Portal, on that afternoon of 5 March 2021, I came across a piece written by the journalist Leonardo Sakamoto in which one can see a photo by Italvar Medina, an employee of the Public Labour Prosecution Office (MPT), underneath text that announces the release of 66 people subjected to work conditions analogous to slavery in a charcoal plant in the State of Minas Gerais where books were used as toilet paper.

Italvar Medina's photo is an evidentiary document, showing educational books piled on the shelves of an uninhabitable cabin in ruins. There are books for maths, Portuguese language and history, but the book *História, Sociedade e Cidadania* [*History, Society and Citizenship*] by Alfredo Boulos Junior grabs our attention. Alfredo Boulos's book, among those books used by human beings to clean up their excrement while living deprived of liberty under subhuman conditions, reveals a long historical tradition of the social and territorial formation of Brazil.

The existence of charcoal plants at a time when there are other less polluting and destructive energy sources; working conditions analogous to slavery well into the twenty-first century; the use of educational books that are ripped up to be used as toilet paper: all these elements together expose not a specific, isolated moment, but a shameful tradition in the processes of territorial appropriation since the spatial slicing that became a reference point in the Iberian dominion over Portuguese America (Moraes 2000; Sousa Neto 2020).

The territorial formation of Brazil, as a process of valorisation and devaluation of space (de Souza Martins 1986; Godoy 2018), and the creation and recreation of territorial funds (Moraes 2000, 2002) occurred with a certain educated cynicism that publishes books about citizenship and closes its eyes to contemporary slavery.

The territorial formation of Brazil, on one hand, could not have happened without an imaginary construction of this same country based on geographic (Moraes 1991), historical, anthropological and sociological ideologies that allow exhausting repetitions of certain apparent truths (Ribeiro 1979) that have always been shameless discursive constructions of social domination, legal justification of stolen properties and subordination of labour to all sorts of violent expropriations.

From the construction of the image of Brazil as an island (Kantor 2007; Magnoli 1997) to the Edenic conceptions of an earthly paradise (Diegues 1996; Galetti 2000), passing through that idea that we were a society where peaceful people lived in a beautiful racial democracy (Freyre 2003), we can say that it is impossible to read academic history books from the same point of view as contemporary enslaved people, a good portion of whom are possibly still illiterate today (Franco 1983).

Understanding this *territory in permanent formation in Brazil* requires an analysis of how contemporary events are connected over the long term (Braudel 1969). That long term was constituted by a world that sought to universalise the waters of the Mediterranean as an expression of one same, single, true capitalist civilisation that, for all its universality, is made up of a set of resuscitated singularities yet to be unravelled. Moreover, we must understand colonisation as a systematic process of modernisations that celebrate subjection, expropriation, plundering and destruction, based in archaism as a project (Prado Jr. 1976, 1980; Fragoso 1990; Fragoso and Florentino 1993).

4.2 The Case of the Colony that Was the Seat of the Metropole

Brazil is the only case in the historiography of the process that shaped capitalism in which the Colony was, for a time, the seat of the metropole. As we know, the Portuguese Court of Dom João VI, prodded by the Napoleonic armies that occupied Lisbon, quickly moved with their libraries, maps and debts to the city of Rio de Janeiro in 1808 and remained there until 1820. During this interval, the business of the Lusitanian Empire in many of its possessions around the world was administered from the tropics. This status as a colonial seat of the metropole, however, is only one of the many singularities in the process of the territorial and social formation of Brazil.

The gentry classes (Mattos 1990) in Brazil were managing a vast Portuguese territorial legacy (Moraes 2000) when D. João VI decided to return to Portugal, and they created an Empire surrounded by republics on all sides and in the sadness of the tropics they extended slavery and the forced labour of populations coming from Africa until the end of the nineteenth century (Martins 1986, 1994).

It is not a coincidence that many accept the theory that the territorial unity inherited from Portugal was maintained because it was in that territory that slavery was practised under a legal, judicial charter with broad social legitimation by, for the most part, those who owned the lands and enslaved people (Machado 1989; Carvalho 1980).

The antecedents of that territorial legacy, however, must be sought in the various possibilities of territorial dissolution (Holanda 1993), as exemplified by the Towns of the Seven Missions, the Dutch occupation, the experience of the Palmares Quilombo, the existence of a single Iberian Crown between 1580 and 1640, as well as by two States coexisting in the same Lusitanian America as the State of Grão Pará e Maranhão and the State of Brazil. Understanding that particular phenomenon typical of late capitalism almost totally South of the Equator reveals how systematic capitalist colonisation creates separate processes of colonisation to achieve the same ends.

The theme for understanding Lusitanian colonisation in Brazil comes from the geographer Moraes (2000), who proposes that colonisation in Brazil had been a sort of mixed colonisation, which was public and private at the same time. It used two "pure" colonial experiences: the Spanish one, characterised by being completely State run and the Dutch one, being completely private. Moraes proposes that in the case of Portuguese America, the plantation system and the economy based on the production and exportation of sugar, with an advanced "industrial" technology, resulted from a logic of mixed colonisation: the lands donated by the King were to be the subject of private endeavours that would establish large productive units of agricultural-mercantile slave monocultures (Novais 2011; Mattos 1990).

That logic has made Brazil, since its first primal mixed colonisation, a place of primary goods production, since it alternates between the economies of mineral extraction and those of agricultural production, in both cases giving us enslaved labour merchandise as another great branch of the colonial business of men of great stature (Fragoso 2000). Even if there are, from time to time, investments of capital, the gentry classes are always the ones seeking to maintain a primary economy, a phenomenon that leads to waves of modernisation that are always incomplete, failed, subordinate and dependent on external investments, investments that transfer riches produced in the form of debts, exchange rate changes and draconian commercial agreements (Fernandes 1987).

The formation of the territory of Brazil was, then, being shaped in two big ways as a constant advance from the coastal shores bathed by the Atlantic towards the places of expansion: in its march to the west when it was about agricultural products and to the borders with Hispanic America for the exploitation of mineral products, specifically gold, when there were certain mining outbursts like those in the State of Mato Grosso (Galetti 2000).

These two great vectors of the territorial formation of Brazil, however, must be understood as a more complex process of intense mobilisation, starting during the colonial period, of various internal commercial routes and paths. Many of these were controlled by the Portuguese tax State, but an infinity of troop routes, gold routes and cattle routes created processes of integration and networks of farms, settlements and villages connected to that dynamic (Goulart 1961).

The advancing march to the west always presupposed enormous processes of environmental destruction, like those that devastated the Atlantic Forest (Dean 1996), linked to the very institution of enslaved work, as we are correctly told by Pádua (2002), who defends the theory that we cannot understand environmental destruction in Brazil during the colonial and imperial periods without understanding its inextricable relationship with the institution of slavery.

The key to this understanding appears in works like that of Darcy Ribeiro, who tells us about the difference between models like the American *farmer*, that logic of being *colonies of occupation* and that one applied to Brazil, which he classified as *colonies of exploitation*. These were processes that ended up shaping an expansion into Brazil (Mattos 2005), but dictated slowly by the appropriation of indigenous lands through genocide, violence and squatting during the formation of estates aimed at agricultural production for the external market. In other words, instead of occupying the whole territory and creating systems of internal integration, they kept these same territories as a sort of future savings, creating vast territorial funds (Moraes 2000) which would be integrated insofar as those gentry classes could incorporate them into that agrarian-mercantile-slaveholder logic (Sousa Neto 2012).

This is the reason that leads the State in Brazil, after emancipation in 1822, to constitute itself much more as a Territorial State than as a Nation State, given that the territory was occupied by those gentry classes that owned lands and enslaved people, but not a nation in the classic sense, given that the idea of a citizen in that historical present only included white men with incomes and the ability to elect and to be elected.

Finally, but no less importantly, at the end of the colonial period, Brazil created itself as an Empire and the role of an Empire is its constant expansion. Thus, that monarchy surrounded by republics on all sides had almost no territorial losses, but rather it advanced its border until the permanent incorporation of Acre in 1904, under the long tradition of diplomatic activity from Alexandre Gusmão to the Baron of Rio Branco (Moraes 2015), and thus, it increased its territory.

4.3 Historical Geographies, Histories of Geography and Territorial Formation in Brazil

The effort to understand this process of the territorial formation of Brazil found different explanatory keys in various geographers, taking on distinct theoretical approaches, different scales of focus and innumerable thematic approaches aimed at achieving a view based on the processes of historical appropriation of the geographical space during the shaping of its territory.

In their book *O Brasil: território e sociedade no início do século XXI* [*Brazil: territory and society at the beginning of the twentieth century*] (2002), Milton Santos and Laura Silveira lay out a history of the territory of Brazil in the interval from the 1500s to 2000, trying to understand, based on processes of periodising space, the passage from the natural environment to the technical environment, from the technical environment to the technical-scientific environment and from the technical-scientific environment to the technical-scientific-informational environment. This perspective is to help understand territorial formation based on the creation of technical systems in the territory of Brazil considering technical changes that arose on a planetary scale.

de Moraes (2002), in turn, taking as his theoretical perspective the understanding that the formation of the territory entailed processes of social appropriation of the space under a logic of continuous valorisation, suggests that we use the Braudelian longue durée as the key to exploring Portuguese colonial territorial legacies and how the appropriation of territory is crucial for understanding Brazilian social formation.

For Ruy Moreira (2017), who works with the concept of Brazilian social formation, there are five phases in this process that run from the foundational vectors to the institution of a complex social formation. The theoretical line of investigation is centred on the understanding of the social division of labour at a world level and how, based on internal territorial arrangements, we can understand the production of the Brazilian territory.

These three perspectives do not exhaust the vast plethora of analyses proposed to understand the territorial formation of Brazil, but they are important in the Brazilian literature on the topic because they have in common the long periodisation that ranges from the colonial period to current times, having as their scale the territory of Brazil and as their theoretical framework a reading inherited from a certain Marxist tradition that tries to understand the singularity of social formations in the universality of the shaping of capitalism as a world system.

There are monumental and important investigations into the understanding of the territorial formation of Brazil on other scales. It is worth highlighting, for example, on the scale of urban formations, the seminal research conducted by the geographers de Almeida Abreu (2010) about the city of Rio de Janeiro and Pedro Vasconcelos about the city of Salvador (2016). Based on their guidance and following the perspective of solid work with primary documentation, much research has appeared over the decades analysing the formation of networks of cities, colonial pathways and the resulting materialisation of the implementation of railroad infrastructure, just to give a few examples.

The other important issue to remember when thinking about the territorial formation of Brazil has to do with thinking about how the history of geography dialogues with historical geography in the Brazilian geographic community, based on a set of research that deals with specific thematic, scalar and temporal slices and encompasses in its broad scope a set of elements from various spatial periods, creating quite a complex picture (Sousa Neto 2020).

Studies about institutions, individuals, Brazilian social thought, travellers, scholarly geographies, cartographies, original peoples and events such as world expositions shed various lights on how to interpret the territorial formation of Brazil. In the range of studies about scientific institutions that produce geographic knowledge, we have studies about the Brazilian Geographic and Historic Institute (IHGB) (Araújo 2001; Silva 1989), about Geographic Societies like the Geographic Society of Rio de Janeiro (SGRJ) (Pereira 2002; Cardoso 2013) and the Geographic Society of Lisbon (SGL) (Pessanha 2010); the Brazilian Geographic Association (AGB) (Antunes 2008); the Brazilian Institute of Geography and Statistics (IBGE) (Lamego 2014) and universities (Machado 2009).

Regarding investigations that dealt with individuals and their insertion into Brazilian social thought, so important for understanding the territorial formation of Brazil, we have myriad excellent works (Sousa Neto 2015; de Lira 2021; Duarte 2011). These studies that cover geographic intellectuals from various regions and distinct temporalities have a decisive role in the understanding of those vestiges that signal that knowledge of the territory is part of their production.

When dealing with works about travellers, naturalists and explorers, we have innumerable investigations that neatly clarify the way that the territory of Brazil in formation was being inventoried from the different perspectives of biodiversity and social diversity, a fundamental condition for the establishment of logics of colonisation and State action (Pádua 2002; Lahuerta 2006; França 2012).

Studies about academic geographies that have crossed the centuries since the 1800s demonstrate the role of scholarly knowledge in the effort to create a feeling of national belonging and formation of an imagined community, amidst the process of renewed indigenous genocides and the institution of slavery of black people. By shedding light on school books, scholarly institutions, curricula and educational policies, these studies allow us to understand how local, provincial and regional identities were being amalgamated to create a representation of the nation (Albuquerque 2011; Rocha 1996; Silva 2015; Vlach 1988; Gáudio and Braga 2007).

Meanwhile, studies about cartography have shown the way that the territory was being scrutinised by engineers, the military, cartographers and in coordination with the desires of the logic of Portuguese colonisation or of territorial policies of the State since the colonial period (Duarte 2018; Bueno 2011). Studies about original peoples (Corrêa 2021) have demonstrated how the territory was shaped by bloody conflicts and not at all linked to the idea of a peaceful process as some historiography has often tried to demonstrate, leading to erasures, silencing and forgetting.

Finally, studies about events in which Brazil participated, such as World Expos, are decisive for understanding how the territory of Brazil was being represented to the logics of permanent capitalist modernisation since the nineteenth century (Assis 2016). When crossing the history of geography with the processes of territorial formation, it is crucial to understand how practices are not separate from discourses, making the territory into an important document of the history of Brazil itself.

4.4 Eternal Returns in the Territorial Formation of Brazil

From Meyer (1993) comes the idea that Brazil is being eternally rediscovered, and that theme allows us to say that the territorial formation of Brazil is a process continuously linked to the logic of keeping it always on the map of the international division of labour as a producer "colony" of primary agricultural and mineral goods.

The national history narrative told by members of the dominant class was always that of talking about Brazil as a nation of the future, making its condition of immeasurable greatness explicit and proposing its mineral-agriculture mission, *natura naturans*, to carry out, with its systematic violence, the assassination of original peoples, the expropriation of indigenous territories and the enslavement of several non-white nations throughout the centuries of territorial formation.

The permanence of this eternal return makes explicit not only the sense of an unfinished democracy, conservative modernisations, repeated coups d'état, but also the practice of slavery and the genocide of which the original peoples are the victims.

As this small and incomplete overview finishes, we are under the aegis of a trial in the Federal Supreme Court (STF) that aims to deny the indigenous people of Brazil the right to the lands that they have occupied since long before the arrival of the Europeans to what today we call Brazil, a continent of contradictions where they live, in full process of formation, where green-and-yellow fascisms, structural racisms and colonised analyses are not left out. Instead of the future, Brazil is always positioned, by the large landholding classes and defenders of all forms of labour slavery, as a country that must open itself to stock markets with its feet firmly planted in a tradition grounded in the old colonial system.

References

Albuquerque MAM (2011) Dois momentos na história da geografia escolar: a geografia clássica e as contribuições de Delgado de Carvalho. Revista Brasileira de Educação em Geografia, Rio de Janeiro 1(2):19–51

Araújo R (2001) A Formação da Memória Territorial Brasileira (1838–1860). Doctoralthesis, FFLCH/USP, São Paulo

Braudel F (1969) Histoire et sciences sociales. La longue durée. In: Écrits sur l'histoire. Flammarion, Paris. p 76 (1st edn. – Annales E. S. C., no 4, October–December 1958, Débats et Combats, pp 725–753

Bueno BPS (352) Desenho e Desígnio: o Brasil dos Engenheiros Militares (1500–1822). São Paulo, EDUSP (352 p)

Cardoso LPC (2013) O lugar da Geografia Brasileira: a Sociedade de Geografia do Rio de Janeiro entre 1883 e 1945. São Paulo, Annablume (236 p)

Correa JA (2021) A resistência guaranítica na formação territorial do Brasil: o massacre das Coroas Ibéricas contra os Sete Povos das Missões (1753–1756). Master'sdissertation, USP

Dean W (1996) A Ferro e Fogo. A História e a devastação da Mata Atlântica brasileira. Companhia das Letras, São Paulo, 484 p

Del Gáudio RS, Braga RB (2007) A geografia, a educação e a construção da ideologia nacional. In: Livre T (ed) Presidente Prudente, vol 1, no. 28, pp 177–196

da França Antunes C(2008) A Associação dos Geógrafos Brasileiros (AGB) - origens, ideias e transformações: notas de uma história. Doctoralthesis. Advisor Ruy Moreira. Universidade Federal Fluminense, Rio de Janeiro (307 p)

da Rocha GOR (1996) A Trajetória da Disciplina Geografia no Currículo Escolar Brasileiro (1837–1942). Master'sDissertation – PUC, São Paulo

de Assis RJS (2016) A iminência da subordinação aos Estados Unidos: A afirmação do Brasil como periferia do capitalismo na exposição universal de Chicago. Doctoralthesis, 369 p. Faculdade de Filosofia, Letras e Ciências Humanas, Universidade de São Paulo. https://teses.usp.br/teses/disponiveis/8/8136/tde-08052017-121250/publico/2016_RaimundoJucierSousaDeAssis_VOrig.pdf

de Almeida Abreu M (2010) Geografica Histórica do Rio de Janeiro (1502–1700), vol 2. Rio de Janeiro, Andrea Jakobson

de Almeida Vasconcelos P (2016) Salvador: transformações e permanências (1549–1999), 2nd edn. Salvador, EDUFBA (538 p)

de Carvalho JM (1980) A Construção da Ordem: a elite política imperial. Campus, Rio de Janeiro

de Fragoso LF (2000) O Trato dos Viventes: formação do Brasil no Atlântico Sul. Companhia das Letras, São Paulo

de Holanda SB (1993) A Herança Colonial – sua desagregação. In: de Holanda SB (ed) História Geral da Civilização Brasileira - O Brasil Monárquico, Tomo II, 1o Volume – O Processo de Emancipação. Bertrand, Rio de Janeiro

de Lira LA (2021) Pierre Monbeig e a Formação da Geografia no Brasil (1925–1956). São Paulo, Alameda

da Silva JLW (1989) A Sociedade Auxiliadora da Indústria Nacional, matriz do Instituto Histórico. In: Wehling A (coord.) Origem do Instituto Histórico e Geográfico Brasileiro: ideias filosóficas e sociais e estrutura de poder no Segundo Reinado. Rio de Janeiro, IHGH, pp 11–20

da Silva MEF (2015) A Produção e Circulação do Saber Escolar: o Nordeste enquanto conteúdo escolar nos livros didáticos de Geografia. Revista Brasileira de Educação Em Geografia 4(8):71–87. Recovered at https://www.revistaedugeo.com.br/ojs/index.php/revistaedugeo/article/view/240

da Silva Guedes Galetti L (2000) Nos Confins da Civilização: sertão, fronteira e identidade nas representações sobre Mato Grosso. Doctoralthesis. FFLCH/USP, São Paulo

de Souza Martins J (1986) O Cativeiro da Terra. Hucitec, São Paulo

de Souza Martins J (1994) O Poder do Atraso. Hucitec, São Paulo

Diegues ACSA (1996) O Mito Moderno da Natureza Intocada. Hucitec, São Paulo
Duarte RB (2018) Cartografias Capitais: Os projetos do Mapa Internacional do Mundo e da Carta do Brasil ao Milionésimo (1891–1930). Doctoralthesis, 240 p. Faculdade de Filosofia, Letras e Ciências Humanas, Universidade de São Paulo
Fragoso J, Florentino M (1993) O Arcaísmo Como Projeto. Mercado atlântico, Sociedade Agrária e Elite Mercantil no Rio de Janeiro, c. 1790 - c. 1840. Diadorim, Rio de Janeiro
Fragoso J (1990) Economia Brasileira no Século XIX: mais do que uma plantation escravista-exportadora. In: Linhares MY (Org.) História Geral do Brasil. 5th edn. Campus, Rio de Janeiro, pp 131–176
França JMC (2012) A Construção do Brasil na Literatura de Viagem dos Séculos XVI, XVII e XVIII: antologia de textos (1591–1808). José Olímpio, Edunesp; Rio de Janeiro, São Paulo (614 p)
Franco MSC (1983) Homens Livres na Ordem Escravocrata, 3rdn ed. Kairós, São Paulo
Fernandes F (1987) A Revolução Burguesa no Brasil: ensaio de interpretação sociológica, 3rd edn. Guanabara, Rio de Janeiro
Freyre G (2003) Casa Grande e Senzala: formação da família brasileira sob o regime da economia patriarcal, 48th edn. Global Editora, São Paulo
Godoy PRT (2018) Categorias marxistas e análise dos processo de valorização capitalista do espaço. Revista Formação (online) 25(45):35–50
Goulart JA (1961) Tropas e Tropeiros na Formação do Brasil. Conquista, Rio de Janeiro
Kantor I (2007) Usos diplomáticos da Ilha-Brasil: polêmicas cartográficas e historiográficas. Varia História, Belo Horizonte 23(37):70–80
Lahuerta FM (2006) Viajantes e a Construção de uma ideia de Brasil no ocaso da colonização (1808–1822). Revista Electrónica de Geografía y Ciencias Sociales, Scripta Nova, vol X, no 218 (64)
Lamego M (2014) O IBGE e a geografia quantitativa brasileira, Terra Brasilis (Nova Série) [Online], 3 | 2014, placed online on 30 June 2014, consulted 16 January 2022. http://journals.openedition.org/terrabrasilis/1015
Magnóli D (1997) O Corpo da Pátria: imaginação geográfica e política externa no Brasil (1808–1912). Moderna/Unesp, São Paulo
Machado LO (1989) Artificio Político en los origenes de la unidad territorial de Brésil. In: Capel H (Org.) EspaciosAcotados. Geografia y dominación social. Barcanova, Barcelona, pp 213–237
Machado MS (2009) A Construção das Geografia Universitária no Rio de Janeiro. APICURI, Rio de Janeiro
Mattos IR (1990) O Tempo Saquarema: a formação do Estado Imperial. Hucitec, São Paulo
Mattos IR (2005) Construtores e Herdeiros: a trama dos interesses na construção da unidade política. Almanack Brasiliense, no. 1., São Paulo, pp 8–26
Meyer M (1993) Caminhos do Imaginário no Brasil. Edusp, São Paulo
Moraes ACR (2000) Bases da Formação do Brasil: o território colonial brasileiro no "longo" século XVI. Hucitec, São Paulo
Moraes ACR (1988) Ideologias Geográficas. Hucitec, São Paulo
Moraes ACR (2002) Território e História no Brasil. annablume\hucitec, São Paulo
Moraes ACR (1991) Notas sobre a identidade nacional e institucionalização da Geografia no Brasil. Estudos Históricos, Rio de Janeiro 4(8):166–176
Moraes ACR (2022) O Barão do Rio Branco e a Geografia. Terra Brasilis (Nova Série) [Online], 5 |2015, placed online on 17 December 2015, consulted 15 January 2022. http://journals.openedition.org/terrabrasilis/1546. https://doi.org/10.4000/terrabrasilis.1546
Moreira R (2021) SOCIEDADE E ESPAÇO NO BRASIL (AS FASES DA FORMAÇÃO ESPACIAL BRASILEIRA: HEGEMONIAS E CONFLITOS). Boletim Paulista de Geografia, [S. l.] 83:7–32. Available at: https://publicacoes.agb.org.br/index.php/boletim-paulista/article/view/760. Accessed on 1 Sept 2021
Neto S, Fernandes M (2018) Senador Pompeu: um geógrafo do poder no Império do Brasil. Consequência, Rio de Janeiro, p 160

Novais F (2011) Portugal e Brasil na crise do antigo sistema colonial (1777–1808). 9th edn. Hucitec, São Paulo
Pádua JA (2002) Um Sopro de Destruição: pensamento político e crítica ambiental no Brasil escravista (1776–1888). Rio de Janeiro, Zahar (318 p)
Prado Junior C (1976) História Econômica do Brasil, 37th edn. Brasiliense, São Paulo
Prado Junior C (1980) Evolução Política do Brasil, 12th edn. Brasiliense, São Paulo
Pereira SN (2002) Sociedade de Geografia do Rio de Janeiro: origens, obsessões e conflitos (1883–1944). Doctoralthesis. Universidade de São Paulo, 192 p
Santos M, Silveira ML (2002) O Brasil: território e sociedade no início do século XXI, 4th edn. Rio de Janeiro/São Paulo, Record
Schwartz LM (1993) O Espetáculo das Raças. Companhia das Letras, São Paulo
Sousa Neto MF (2020) Narrativas em História da Geografia na Geografia Histórica do Capitalismo. In: de Lira LA, Duarte RB, Sousa Neto MF (eds) Geografia das Ciências, dos Saberes e da História da Geografia. São Paulo, Editora Alameda, pp 129–134
Sousa Neto MF (2012) Planos Para o Império: os planos de viação do Segundo Reinado (1869–1889). São Paulo, Alameda
Sousa Neto MF (2015) A Ciência Geográfica e a Construção do Brasil. Revista Terra Livre 15:09–20
Vlach VRF (1988) A Propósito do Ensino de Geografia: em questão o nacionalismo patriótico. Master's Dissertation – USP, São Paulo

Chapter 5
Man is His Being in the World. Geography and Geographicity

Ruy Moreira

Abstract Man is an autopoietic being. A self-made being. However, he does not do it based on will, but because self-production is necessary. He does it himself, but within the possibilities of history. And this requires will, but also constant action and the capacity to materially organise his act of will. Here is where spatiotemporal conditionality comes in. Man does not construct himself without an act of will, but will is not sufficient if the objective reality does not allow for the possibility. This makes him the subject and the object of himself. To be this, you must construct yourself. And to construct, you need the means of construction. The supposition is the man-nature relationship. This is what location imposes. There, man makes space and space makes man. Self-construction is a geographical construction.

Keywords Geography · Geographicity · Socio-spatial relationship

5.1 Geographicity

Human construction is a combination of a spatially defined specific location on the earth's surface—every man is born in a place—and a universalist abstractivity—every man is born a human being—which leads the human species to historically transform into the human genus, the complete man, the integral man in the world, according to Lefebvre (1971, 1981). This means determining a form of man-nature relationship transformed into a defined form of socio-spatial relationship, strictly speaking man-spatial-nature, through the norms and rules of societal and sociability relationships that inform them and have in space their elevating link. The singularity of the man-nature relationship and the universality of the socio-spatial relationship materialised therein in the concrete particularity of the place. The human construction is this dialectic of the singularity and the universality realised in the particularity, in Lukács' (1970) thetic version, which Armando Correa da Silva refers to as geosociability (da Silva 1991), thus giving birth to man as the being, the real existence, which the

R. Moreira (✉)
Universidade Federal Fluminense, Niterói, Brazil
e-mail: ruymoreira@uol.com.br

R. C. Lois González and M. A. Mitidiero Junior (eds.), *Brazilian Geography*,
Advances in Geographical and Environmental Sciences,
https://doi.org/10.1007/978-981-19-3704-0_5

geographical construction translates into being, the being of essence defined by its spatial mode of existence, according to Sartre (2002) and Lukács (2010, 2012), of each real concrete form of geographical construction. It is a topic that is so dear to geography today (Martins 2007; de Oliveira Biteti 2014; Lima 2019).

Geographicity is this structure of reciprocal transfiguration of the man-nature and socio-spatial relationships, organised by their sociability relationships and regulated by their societal relationships of control and reproduction. In other words, the totalised structure of the man-spatial-nature relationship oriented to the regulatory mediation of the societal relationship and structuring sociability relationship, as Jean Brunhes saw for the Mediterranean irrigation societies (Brunhes 1962) and Karl Wittfogel for the hydraulic societies of the Asian world (Wittfogel 1966).

A concrete form of society, therefore, not a mode of production, a social formation, a spatial formation or a way of life, a pure categorical link, but a concrete way of inhabiting and living. Geographicity is a spatial mode of being, the spatial mode of the historical-produced, located and locating, induced and inducing, regulated and regulating, produced and producing, determined and determining space, which Lefebvre's creative infinity also saw as the reproducibility condition of modern society (Lefebvre 1973, 2013) and Milton Santos brought to the theoretical particularity of geography (Santos 1978). Space is empirically local and universally global at the same time, which gives man the uniqueness of being born in a place and, however, humanity's act of doing.

This is what we learn from the classics, when Reclus describes geography as history seen in space and history as geography seen in time (Reclus, undated), Vidal de La Blache, space as the cohabitation that unites man and things as a totality (Vidal de La Blache 1954), Brunhes, the habitat as his mode of dwelling (Brunhes 1962), Sorre, society as the totality that becomes more socially complex in time as the spatial fabric becomes denser (Sorre 1967), George, history as temporality concretised in space as an existing reality (George 1968), and historicity and geographicity expressing the multifaceted plasticity of autopoiesis, even when they did not see it clearly as so (Moreira 2015, 2019).

5.2 The Forms of Geographicity

A concrete geographical form of the construction of man in each time and each corner, geographicity varies in space–time according to the way in which the societal and sociability relationships order the man-nature and socio-spatial interaction, with the whole combining to the command of its mode of production. Thus, a community geographicity, an ancient geographicity, a Germanic geographicity, a Slavic geographicity, an Asian geographicity, a slave-owning geographicity, a feudal geographicity, and a capitalist geographicity, all emerge with as many forms of geographicity as there are modes of production and life that each one in each corner and each period has at its centre. Hence, present and past geographicities are distinguished and contrasted, due to their interlinking mode of man-nature and socio-spatial relationships, their fabrics of sociability and societal normativity, their forms of transition from one society in history to another, as in a relationship of communicating vessels (Hindess and Hirst 1976).

The geographicity of the present is the way of organising the reciprocal transposition of the man-nature and socio-spatial relationship regulated in the societal and sociability relationships of conflict. The way of being in expropriation (degrounding), expulsion (deterritorialisation), and purging (despatialisation), resulting from the separation between labour and the means of labour, on the one hand, and the concentration of money in increasingly fewer hands, on the other, that uproots and proletarianises the serf peasantry, its characteristic structural marks of configuration (Moreira 2007a, b). The unequal, stratified, and fragmentary way in which feudalism's sociability relationships give way to capitalist sociability relationships with their societal norms and rules of control, regulation, and reproduction in history (Moreira 2016).

Pre-capitalist geographicity is in all senses its opposite. In this geographicity, the man-nature relationship is a relationship of co-belonging, and involvement of identity that is consequently transported to the societal and sociability relationships, and this structure, in these terms, is the socio-spatial relationship. The relationship from which primitive accumulation will set man free, which is impracticable in the egalitarian structure of community society in the face of the absence of private property, enshrined in the mercantile privatism of the slave-owning society but hindered in the expansive stalemate of the productive forces, and lastly embryonic in the peculiarity of the classistlypersonalised land-territory relationship of feudal society, where, for this very reason, in the words of Massimo Quaini, capitalism, after all, is born (Quaini 1979).

Community geographicity, therefore, is the egalitarian mode of cohabitation that the societal and sociability relationships transport from the lower plane of the man-nature relationship to the upper plane of the socio-spatial relationship, regulating and organising the reciprocity of one relationship with the other, in order to maintain the whole of the man-nature relationship and the socio-spatial relationship in an all-encompassing solidary mode. The land and everything that men take from it is a common good. A general good that from the access to land to the distribution of means, the community relationship reaffirms itself as a way of life. It is a community of gentile conformation, where men and women are linked by consanguinity and kinship relations, differing in their interior by being grouped into nuclear families, in a great state of equilibrium. And it is these nuclear families that are responsible for the production, the community guiding the distribution and common use of land in a rotational occupation, the unequal productions of families being later brought together for collective distribution, according to the families' needs, a community relationship that is then maintained and reproduced by this unique societal mechanism of rotation. This consists of the annual exchange of use of parts of the territory, so that an equal distribution is agreed among the families, since, due to being distinct from each other in terms of their natural fertility conditions of the land, the parcels tend to lead the families to different results of production, some tending to produce more and others less, with the rotation of space fulfilling the role of remembering the temporary nature of the landscape and the perception of the benefit of keeping the distribution even regardless of moments and circumstances. It is an ordering mechanism of the socio-spatial relationship over the man-nature relationship that is typified as the societal and sociability relationships, guaranteeing the permanent maintenance of its community tradition for the entire geographicity.

The development of the productive forces tends, however, to put this permanence at risk. And the opportunity comes when this happens to the point of causing a surplus to emerge, that is community surpluses on a constant basis, and with this surplus the threat of breaking the existing communitarianism. Until then, the productive forces are restricted to natural and demographical elements (natural resources and population size), forcing the entire population to have to participate in the productive task. However, the rise in the level of development of the productive forces will allow the community to free part of its members to carry out non-productive tasks, creating the administrative, priestly, and military functions as permanent functions, in a social division of labour beyond the then existing natural division by sex and working age, bringing with it the risk of a managerial stratification within the gentile order. Production remains a function of nuclear families, but part of the production must now be transferred to support the fraction of the population involved in the social function of labour, in a formal structural differentiation of management that can progress to a more profound form of social differentiation. The spatial rotation must now be reinforced, to now also prevent and force the managerial stratum to remain integrated in the community regulation of life. However, where this does not occur, community relationships begin their dismantling, subsequently transforming into the stratified relationships of the ancient, Germanic, Slavic, Asiatic, slave-owing, and, later, feudal, modes of production, with their respective unequal forms of man-nature, socio-spatial, societal, and sociability relationships, and generating the cartographic map of pre-capitalist geographicities that transcend the capitalist transition itself (Hobsbawn 1977; Moreira 2007b; Réclus, undated).

The geographicity of Asiatism is an example of a hybrid case in this framework. Spatial regulation becomes dual. The social functions arising from the social division of labour assert their privileges, using them to define two distinct and combined ways of spatial ordering: on the one hand, the family rotation of the gentile community, maintained as a way to guarantee community production and transfer to it the surplus; on the other, the obligation of collective labour of the communities, then numerically multiplied, for carrying out the major works to regularise rivers and marginal lands subject to periodic flooding. A relationship of unequal combination of nuclear families that must continue to fulfil their obligation of surpluses and now also provide for those among their members who will regularly implement the works of rectification, canalisation, and preservation of rivers and areas subject to flooding that, in the end, enhance the obligatory task of surpluses. It is a relationship of extraction in which the managing layer, transformed into a superior community, enjoys the benefit, in a land (village communities) and territory (superior community) relationship of a multiplied list of communities now transformed into village communities, of increasing extraction, either in the form of taxes or in the form of goods, the community surplus. It is the geographicity of a societal relationship that regulates, within the organic whole of the axial man-spatial-nature relationship, the ordering of the contradictory interaction of an egalitarian man-nature relationship and a non-egalitarian socio-spatial relationship, at the service of a sociability of domain of castes.

Germanic, Slavic, and ancient geographicities differ from it, as they are geographicities of a simpler man-spatial-nature configuration, with the managing layer standing out little in the face of a community dismantled into nuclear families interconnected by a spatial rotation made inconsistent and fragile. This is how it is with the Germanic and Slavic societal and sociability relationships, which are very close forms of geographicity, where a military caste, above all, controls and regulates their domain through a territorial-productive ordering of a socio-spatial relationship loyal to it. They are geographicities in which the hallmark of the co-belonging of the man-nature relationship, although stratified here and there, changes little or nothing. It is also the same with the geographicity of the old mode of production, a form of man-spatial-nature relationship that arises from the migration and establishment of nuclear families in areas far from their origin, with sociability establishing itself in a socially regulated man-nature and socio-spatial relationship of heavily blurred exchanges.

Slave-owning geographicity is the extreme form of these past modes of geographicity. It is the geographicity that comes from the direct subordination of certain nuclear families to others, either by dependence or by subjection, to which time adds the imprisonment and conversion of prisoners of war into slaves through purchase and sale, in a society turned mercantile that soon becomes militarist and imperialist, expanding its domains across a vastness of peoples and territories. The former community land becomes state property arising from the emergence of conflicts in the now stratified society, which cedes it for private use to the lordly class in large estates, cohabiting with a myriad of small pieces of family-use land scattered either throughout the territory of the former areas of rotation and or throughout the territory taken from the surrounding communities of the ancient, Germanic, and Slavic modes of production. Here, slave labour predominates, as it considered by the societal and lordly sociability relationships as a physical thing equal to other things in nature, but at the same time a subject legally responsible for its acts, with the slave-lord relationship determining the private form of reciprocal transfiguration of the man-nature relationship and the socio-spatial relationship in a territorial whole administratively managed by a central slave-owning military state from the big cities.

Lastly, feudal geographicity is the socio-spatial relationship in which land is a relationship of ownership of the serf and the fiefdom is a relationship of domain of the lord, in a combination of land and territory regulated by the relationship of obligation in which, according to Leo Huberman, the serf provides for himself and his lord, with the serf supporting the lord with surpluses and the lord giving him cover and protection from other lords in return (Huberman 1969). It is a kind of intertwining of different orders of geographicity, here in the slave-owning society, and there in the societies still with community components, all of them present in a form of spatial arrangement in rings arranged around the lord's castle fortress that represent each one of them (Weber 1968). The lord's fortress is the centre of the arrangement, in a kind of mediaeval town-countryside relationship. It is surrounded by the ring of lots of urban inhabitants with their gardens and farms. Then, there is the ring of parcels owned by the serfs with their cultures associated with creation, with as many parcels as there are families. Followed by that is the ring of common use pastures, with the free raising of cattle. Lastly, closing the circle is the forest reserve ring, used for collecting firewood and feeding the community's pigs. Each ring expresses the form of man-nature relationship and regulation of socio-spatial relationship it expresses, with the fiefdom expressing the entirety of the slave-owning geographicity, in the guise of large estates, the plots of farmland for cattle the structure of nuclear families from the invading communities which, reacting to the former military-imperial domain, generate feudal order, the pasture and forest, the habit of common use by communities cohabiting with the slave-owing spatiality or communities of invading peoples, the castle fortress the form of defence against invasions with which feudalism is engendered. It is the historical origin that the lordship maintains through the constant wars waged among themselves, in the argument justifying the relationship of obligation and reciprocal servitude of the serfs. It is the mixed arrangement of community and private social relationships that explains for Quaini why capitalism came from the decomposition of feudalism, while it was hindered in the other geographicities of pre-capitalism.

5.3 Primitive Accumulation and the Framework of Modern Geographicity

Primitive accumulation is the way in which this transition takes place. The historical-structural form centred, on the one hand, on the separation between labour and the means of labour, and on the other, on the concentration of money in increasingly fewer hands. They are two processes that meet as two pillars in a single structure, pillars which have their master key in the expropriation, expulsion, and purging of the serf from the land. And their sewing stitch in the valorisation of money for the purchase and sale of labour.

The land is the centre of gravity of feudal geographicity. It is through the relationship with the land that the serf has his simultaneous relationship with nature, territory, and space. Land here means the contextual whole of nature in whose territorial spatial contour the serf sees agricultural soil, waters, rivers, forests, the diversity of the environment, and other men, as one and the same totality. The totality that Quaini calls the ecological-territorial structure (Quaini 1979). The rupture of the relationship with the land, thus, means the rupture and breaking of this totality. In a single act. An act that immediately expropriates man from nature and expropriates nature from man. Man and nature thus linking themselves in a relationship of radical externality. Breaking the link of co-belonging—the co-belonging that comes from community times—is still present in the rings, which primitive accumulation inherits and exorcises.

Community sociability is a man-nature relationship oriented in the space–time of the labour rites of the socio-spatial relationship. Each period and each area are the mark of a calendar of festivities that periodise and categorise the communion of life codified in the set of symbols and signs whose sense of community traverses time and arrives at feudal sociability. The breakdown of the community relationship by the forms of geographicity that succeed it in time brings with it the need for new signs and symbols, whose mark of transition in the West is the birth of Christianity.

Christian symbology is the expression of the separation of body and spirit that announces the changes in the form of co-belonging between man and nature of the community relationship, body and spirit separated through the separation of music and dance, until then an organic unity in the ritualistic moments of the festivity of labour, where music and dance are guided by the rhythm of percussion as the corporeal rhythm of the labour's movement. This is the case of the trampling of grapes to make wine. Christianity undoes this relationship, separating music, henceforth an attribute of the spirit, and dance, henceforth an attribute of the body, separating music and dance, spirit and body, divinity and worldliness, nature and man, excluded from/subordinated to the rite of charisma, in a spatiality of religious subjection. The process is the breaking of the music in two, the percussion music and the wind music, the former a body music and the latter an enchanting music (Andrade 1977). It is the way of symbolising the breaking of the man-nature relationship that then occurs, man and nature separated on the subjective plane of the religious representation of the world, but kept unified on the plane of the daily objectivity of labour.

Primitive accumulation radicalises this separation, via the expropriation that degrounds, the expulsion that deterritorialises, and the purging that despatialises, denaturising man and dehominising nature, breaking the man-territory-spatial bond both subjectively and objectively, creating a man with no land, no territory, no space, a man who no longer recognises himself, and is dissociated, abandoned, and uprooted from the estrangement from the world. The relationship between labour and money, labour purchased for the pure valorisation of money, is the aim and purpose of the relationship of estrangement. The degrounding, deterritorialisation, and despatialisation are the means. They are geographical strands through which primitive accumulation intervenes, producing capitalist geographicity via three corresponding processes of ordering: the emergence of the land market, the emergence of the labour market, and the emergence of the money market. Relationships through which primitive accumulation erects the labour-money-labour exchange relationship as the societal and sociability foundation of the socio-spatial relationship mode we know (Moreira 2016).

The starting point is the emergence of the land market, the legal and economic process that converts land from immovable property to movable property, and thus capital. The process that, at the same time, degrounds, deterritorialises, and despatialises man, and immediately breaks and transforms nature into a plethora of use values, thrown into the world as different modes of exchange value. A movement that is made in parts, from the rearrangement of the rings of sociability of the fiefdom, subjecting the reciprocal transfiguration of the man-nature and socio-spatial relationship of each ring to the same societal relationship of sociability of the market. First, it is the agricultural land (land in the sense that has since been current), dissolving and fragmenting the ring of crops and livestock arranged into multiple sections of privately owned areas. This is followed by the ring of community pasture and then the outer ring of forest, divided and fenced equally in multiple parts. Then comes the centre of the fortress and lots of farms and gardens, consumed and reinscribed in the vortex of urbanisation that the market brings with it. A whole new ordering of space organisation and regulation is carried out in this way, with the market rule at the same time dividing and uniting everything at the ends of the exchange relationship.

The labour force market is the next step. Expropriated and expelled from the land, the serf proletarianises himself, migrates between places seeking to sell the only form of property he has left, the labour force, with the land market unfolding into the labour force market. That is, labour separated from the means of labour, thereby made free labour, wage labour, labour offered for sale at a value measured by a corresponding amount of money. Labour bought and sold, and, like land, converted into capital. Labour that from the outset crosses and moves in dual paths, facing the opposite path of money. There is the serf who becomes bourgeois. And there is the serf who proletarianises himself. Dual paths, but that here and there cross each other the more the money market prevails and the land market and the labour market merge with it, eliminating from the list of paths the motive and purpose of the transition from feudalism to capitalism.

Lastly, and along the paths, comes the money market. With the peculiarity of money being the great helmsman on the path of progression and fusion with itself of the land and labour market, benefitting from the common origin in the emergence of cash income. And being able, in this way, to give, in time, the stamp and content of the societal and sociability relationships that primitive accumulation raises as the substance and foundation of the new form of geographicity, since the roadmap of cash income is its roadmap. The flight and abandonment of the fiefdom by the serfs in light of the overexploitation of income from labour and income from production that occurred during the course of the tenth century is the source of origin of cash income, as the flight and abandonment forced the lordly class to change the obligation of free labour (income from labour, called corvée) and income from production (income paid with a second or third of the peasant's production) to payment in a corresponding amount of money. A change that means leading the serf to previously having to sell the products on the market, where he used to go, in a goods—money—goods (G-M-G) exchange circuit, to only exchange his consumption surpluses for the surpluses of other serfs, thus obtaining and paying the cash obligation. A change that will guide the trajectory of the serf himself, along two paths, depending on the place and the seasonality of the moment, in a way anticipating the relationship between labour and money that little by little the course of exchanges will determine: that of capitalisation of the countryside and bourgeoisification of one part of the serfs and the proletarianisation and impoverishment of another, in a vertical differentiation. Wherever the situation favours, the serf can, after having resolved the dispute over the obligation of servitude, manage to keep a sufficient accumulation of resources in cash, which he uses to buy and revert to property the portion that previously belonged to him as possession, becoming the owner of the land and becoming a modern peasant, with the possibility of, now acting as owner, and venturing on a new purchase or lease, returning his production to the market, to become bourgeois and a member of the new class, the agrarian bourgeoisie, which thereby forms in the countryside. However, wherever the situation disfavours, the serf accumulates debt, is dispossessed of the land, and his cutback increases to the point that he will only be left with his labour to sell, an amount that grows the more the commodification of the land and agricultural production increases, often becoming expelled or employed as a wage earner by the former serf himself, now a rich peasant, from the fiefdom's coexistence, thus deepening and fuelling the creation of the growing labour force market. This movement of vertical differentiation fuels the joint expansion of the land market, the labour market, and the money market, and thus the monetisation that thrives on the side of change, putting money at the centre of everything (Dobb 1965).

However, the great transformation comes with the inversion of the order of money circulation, before a mere means of exchange in the G-M-G circuit, in which it is the goods that circulate, with money only mediating their movement, now becoming the centre itself with the transformation of the G-M-G circuit into the money—goods—more money (M-G-M') circuit, the circuit created in the influx of cash income and in the expansive impulse of monetisation, which comes from it, a circuit in which it is the money that now circulates, and the goods become its mere means of movement and expansion. The M-G-M' circuit in which the goods are no longer the surplus that is exchanged for other surpluses between the serfs, as before in the market, but the product that they now generate from the cradle as a product-for-sale, pushed by the demand for obtainment of cash income, and the capitalisation of land and agricultural production elevates as a project and presupposes the purchase and sale of labour force, the special goods created and expanded by this very capitalisation. Hence, the necessary prior commodification of land, without which there is no way to form a market for the labour force, the power of labour turned into a special form of goods, free labour, free for the wage-earning purchase and sale contract, and carry it over to the M-G-M' circuit to forge the labour-money exchange relationship that will continually valorise new money, the money valorised through labour, and make the proletarianised serf accept submission. It is impossible to imagine a sociability of serfs and peasants.

The inversion of the G-M-G circuit into the M-G-M' circuit, which comes from the emergence and development of mercantile intermediation, the result of the thought of a class then obscures in the dark theatre of feudal exploitation, which sees in the overlapping of the circulation of exchanges to the dispersiveness of peasant production, the opportunity to become enriched and emerge in the city as a new class alongside the agrarian bourgeoisie emerging in the countryside is the path of transformation. A lower class of inhabitants of the burgs, the bourgeoisie, the mercantile bourgeoisie sees in the growth of cash income, the driving force of this great leap from obscurity to the luminosity of dominance, through establishment of the overlapping of the sphere of circulation with respect to the deconcentrated sphere of peasant production, generating the embryo of sociability that will encase the reciprocal transfiguration of the man-nature relationship and the socio-spatial relationship in the societal rule of the market, the implicit rule, precisely, in the transformation of the land market, the labour market, and the money market, unfolded in the exchange of goods market.

A rule that leads to the growing expansion of precisely the centrality of the money circuit (Arrighi 1996; Hilferding 1985; Luxemburgo 1970).

And an inversion that puts serfs and market, income from production and cash income, lordship, and the consumer market beforehand, along with, on the one hand, the grounding, territorialisation, and spatialisation of the ascending money, and on the other, the degrounding, deterritorialisation, and despatialisation of the labour, destined to make money the key subject of primitive accumulation and labour the supporting subject that will amplify it as added value (the money M' of the relationship of the M-G-M' circuit that starts with money M).

The birth of industry, however, the darling child of mercantile intermediation, is the actual moment of this leap. Intermediation is the act of the bourgeois merchant circulating between the areas of peasant production, an artisanal production, buying their products in a certain area for a low price and reselling them to peasant craftsmen from other areas for a higher price, profiting from the difference, thus substituting, with his intermediation, the act of the peasantry exchanging income from production for cash income, fulfilling the act of buying and selling in their place, profiting in the exchange mediation, and taking the profit back to the bourg. The result, on the one hand, is the peasant craftsmen coming to dedicate themselves to the exclusive task of production, and, on the other, the merchant intermediary organising the exchanges, and the mercantile bourgeoisie submitting the production to its domain, centralising the relationship of buying and selling on a scale of circulation that only it can do. Soon, however, the bourgeois merchant realises that he can take care of the production himself, renting and gathering in a large shed the necessary craftsmen-produced mass collected from handicraft locations, becoming a merchant and an industrialist, and over time taking the sphere of production and overlapping the sphere of trade. It is the stage that internalises the labour-money relationship in the sphere of its own domain, that of the industrial valorisation of money, shaping and starting the very phase of the sociability of money capital (Moreira 2012b, 2015c).

5.4 The Adventures of Autopoiesis

"One of the presuppositions of wage labour and one of the historical conditions of capital is free labour and the exchange of free labour for money, with the aim of reproducing money and valorising it; the labour is consumed by money not as a use value for enjoyment, but as a use value for money", says Marx (1977), summarising the character and socio-geographical content of the sociability of money. Its presupposition, however, is the fragmentation of man. The fragmentation that begins with the intervention with which primitive accumulation dichotomises, fragments, and commodifies nature in the same act with which it reciprocally expropriates nature from man and expropriates man from nature, changing the man-nature relationship and the socio-spatial relationship, thus creating the basis for now fragmenting man's relationship with himself. It is something that is only possible in the industrial stage of the sociability of money capital. Indefinite in the dehominisation that leaves dehominised nature indefinite, the conceptual indefiniteness now completes itself in the dehominisation of man himself.

Degrounded, deterritorialised, and despatialised in the relationship with nature, man now degrounds, deterritorialises, and despatialises himself in the relationship with himself, whereby the radicality of the symbolic objectivity-subjectivity of Christianity that had completely distanced him from nature, repeats itself in the radicality of the objectivity-subjectivity of the sociability of money that now distances him from himself. And in the same externalist perspective. The perspective of the sociability of the market that defines nature as the inorganic body of man, based on the conceptualisation that defines man as the inorganic body of other men themselves. Presuppositions of nature and man needed to define the reciprocity of the man-nature and socio-spatial transfiguration as a reciprocity of use values and exchange values that substantiate man and nature in the value concept of the labour-money relationship.

The universe of industry is the field for this disintegration. If for nature, the reference is the diversity of the natural landscape of the environment, and the reference for man would have to be the diversity of the landscape in the world of labour. Nature could of course be divided into as many use values and exchange values as there are visual bodies in the natural landscape. Man asked for a technical and cultural criterion of objectivity. The landscape for nature. The labour for man. And so it is the case. Hence, the gap of more than a century between one fragmentation and another, the immediacy of nature and the mediacy of man. The crumbling space comes as a result.

Thus, as with nature, labour is first separated into productive and unproductive labour. The production and realisation of value are the distinguishing reference. Productive labour is labour that produces added value, the raw material with which labour valorises money. Unproductive labour is labour that realises the valorisation. In the same way, men's labour and women's labour are separated, the former productive, destined to the functions of labour in the factory, and the latter, unproductive, destined to the domestic functions of reproduction of the productive labour of men, the man inserting himself in a market economy and the woman in a (natural) domestic economy, separating the work space and the dwelling space, both set up in the same capitalist society and sociability, but unequally and as hierarchical as the labour-money relationship that this productive-unproductive separation of labour embodies. Ultimately, it is a way, according to Silvia Federici, of concealing that it is the woman's domestic labour that, in the end, is responsible for the reproduction of the man's industrial-productive labour (Federici 2017) and that, according to Francisco de Oliveira, gives the parameter for the very reproduction and replacement of the entire population as a population available to capital, replacing and reproducing the system itself as a whole (Oliveira 1977). Then, intellectual labour and manual labour are separated on one plane, and on another, managerial labour and executive labour, in a fragmentation of space, now the factory, with managerial labour and intellectual labour forming the government and manual labour and executive labour forming the ground floor, in a labour-money hierarchy structuring the factory itself

(Weil 1979). Lastly, labour is pulverised in the diversity of categories of professions, the sectoralised fragment of specialisations that breaks up and dissolves labour in the presence of the state of awareness of the reciprocal transfiguration of the man-nature and socio-spatial relationship, the labour that places man at the very centre of the world as a world built by labour (Braverman 1977).

A form of sociability that reproduces itself simultaneously on a world scale. Primitive accumulation has from the outset been a historical process on a local and global scale, while at the same time organising, since then, but under different forms of geographicity, the planet under different modes of reciprocal transfiguration of the man-nature relationship and the socio-spatial relationship, allowing capitalist sociability, according to Rosa Luxemburgo, to lead its reproducibility to take place in a relationship connected to the reproduction of non-capitalist geographicities, in a reproductive integrality of money capital that from the outset has assembled the world map as a fragmentary cartography of metropolises and colonies (Luxemburgo 1970).

The basic way is to take the act of degrounding, deterritorialisation, and despatialisation of men and women of European extraction to reproduce itself in a repetition of degrounding, deterritorialisation, and despatialisation of men and women from other parts of the world, arranging planetary space in the same relationship of separation of labour and means of labour and the concentration in increasingly fewer hands of money from the original primitive accumulation, to, in this way, form a world of metropolises and colonies under the same general rules of sociability. Here, in the metropolitan context, in the relationship of expropriation, expulsion, and purging that transforms serfs into proletarians, and there, in the colonial context, in the relationship of expropriation, expulsion, and purging that transforms men and women from the communities into slaves. An act that is completed in the ideologisation of the imaginary that divides the world into countries of rationality and countries lacking rationality, the Hegelian vision of the territorial transit of reason around the world, taken as the substance of the distinction between the presence of sociability in the metropolis and the absence of sociability in the colony, the metropolitan world of reason and enlightenment and the colonial world of darkness and a lack of light distinguishing between themselves in a relationship of civilisation and barbarism (Schüler 1995). Civilisation and barbarism are in a kind of reproduction, on a planetary scale, of the spirit and body dichotomy, music and dance, wind instruments and percussion instruments, and the enchantment and worldliness of the birth of Christianity.

5.5 The Spatial Malaise of Modern Society

Contemporary geographicity is that of sociability of money mounted on the side of primitive accumulation, taken to the scale of its globalisation. The sociability of what David Harvey calls accumulation by dispossession (Harvey 2004), today taken to every corner of the globe. The sociability of money transformed into the single subject

(Simmel 2015; Santos 2002). Sociability of the degrounded, deterritorialised, and despatialised man on an international scale, against which he moves in a tenacious struggle for counterspace (Moreira 2012a, b). The counterspace of a geographicity of space lived in pieces, a crumbling space, in the words of Lacoste (1988). The malaise denounced by Freud (1997a, b, c). And the culmination of a dispossessed man that begins with the reciprocal expropriation of man and nature, the source of the lack of space that is at the heart of the problem. The problem of rooting that space, from the rotational system of community sociability, is the key category.

Man is, therefore, the product-producer of the spatial context in which he lives, man producing geographicity and geographicity producing man, until primitive accumulation came to change this rule, no longer putting the man-spatial-nature relationship, but the labour-money relationship at the centre of reference. Thus, modern geographicity emerges in which man produces money, but money does not produce man. But, its own reproduction expanded through man. And that has its key piece in the rule of reciprocal transfiguration of the man-nature and socio-spatial relationship carried out through the metabolic exchange of commodified labour. The relationship of transfiguration whose hominisation money capital inverts, to convert it into its monetisation.

Geographicity, however, was to be the whole of sociability that weaves the socio-spatial relationship from the societal rule of organisation of the man-nature relationship in the sense of man as his own subject and object (Moreira 2020). The man who creates himself and makes himself an autopoietic being through it. The man who sees and finds his own project in the societal relationship of self-production. And so to become the self-made being, not out of pure desire, but out of the need for self-production. And it has in the historicity of the man-spatial-nature relationship—a relationship in which space organises, orders, regulates, controls, reproduces, and transforms the man of natural history into the man of social history—the axis that makes from the human species (the man of natural nature, the first nature) and the human genus (the man of the second nature, the socialised nature) the total man. The man at the same time the presence and the being that is the substance—the being-there (the presence) and the being-in-the-world (the being), by Heidegger; the social place (the presence) and the geosociability (the being), by Silva; the being-in-itself and the being-for-itself (the presence) and the being-for-the-other (the being), by Sartre; the telluric man, from the geography of plants, by Humboldt; the territorial social justice, by Harvey—of every form of geographicity. And there is in space—the worldhood of the world (the world-around-us), by Heidegger (1978), the practico-inert, the existence that becomes essence, by Sartre (2002), the presence of the man who makes himself be social, by Lukács (2010, 2012), the topos, by Lefebvre (1976), the territorialisation-deterritorialisation-reterritorialisation (TDR), by Raffestin (1993), the relational praxis, by George (1973), and the territorial machine, by Deleuze and Guattari (1976)—self's way of being. The real place and condition of existence.

It is through space, and as space, therefore, that primitive accumulation became a labour-money relationship. Man must be uprooted, denaturising, degrounding, deterritorialising, and thus despatialising him, forcing him to submit to the dictates of the sociability of money capital. And in this submission, leading him to hand over his potentiality as the potentiality of money, giving money the status of autopoiesis that is an attribute of man. Breaking the ecological and territorial aspect that integrates him in the world as a world-in-man (Quaini 1979). Subjecting him to the crumbling space (Lacoste 1974, 1977). Reducing him to corporeality and accusing him of madness (Foucault 1979, 1977). And, thus, leading him to schizophrenia (Deleuze and Guattari 1976). Making him a replicant, a Blade Runner lost in his own here-now.

References

Andrade M (1977) Pequena história da música. Martins Fontes, São Paulo

Arrighi G (1996) O longo século XX. Contraponto/Editora da UNESP, São Paulo

Braverman H (1977) Trabalho e capital monopolista: a degradação do trabalho no século XX. Jorge Zahar Editora, Rio de Janeiro

Brunhes J (1962) Geografia humana. Fundo de Cultura, Rio de Janeiro

Deleuzhe J, Guattari F (1976) O anti-édipo. Capitalismo e esquizofrenia. Imago, Rio de Janeiro

Dobb M (1965) A evolução do capitalismo. Editora Zahar, Rio de Janeiro

De Oliveira F (1977) A produção dos homens: notas sobre a reprodução da população sob o capital. In: de Oliveira F (ed) A economia da dependência imperfeita. Edições Graal, Rio de Janeiro

De Oliveira Biteti M (2014) O em-si-para-o-outro-para-si: o ôntico e o ontológico como dimensões do ser geográfico. PostgraduateProgramme in Geography, Niterói

Da Silva AC (1991) Geografia e lugar geossocial. Editora Contexto, São Paulo

Federici S (2017) Calibã e a bruxa. Mulheres, corpo e acumulação primitiva. Elefante, São Paulo

Foucault M (1977) Vigiar e punir. História da violência nas prisões. Editora Vozes, Rio de Janeiro

Foucault M (1979) Microfísica do poder. Edições Graal, Rio de Janeiro

Freud S (1997a) O mal-estar da civilização. Imago Editora, Rio de Janeiro

Freud S (1997b) O futuro de uma ilusão. Imago Editora, Rio de Janeiro

Freud S (1997c) Moisés e o monoteísmo. Imago Editora, Rio de Janeiro

George P (1968) A ação do homem. Difel, São Paulo

George P (1973) Problemas, doutrina e método. In: George P (ed) Geografia ativa. Difel, São Paulo

Harvey D (2004) O novo imperialismo. Edições Loyola, São Paulo

Hilferding R (1985) O capital financeiro. Nova Cultural, São Paulo

Heidegger M (1978) Ser e tempo, vols I & II. Editora Vozes, Rio de Janeiro

Hindess BH, Paul Q (1976) Modos de produção pré-capitalistas. J. Zahar Editor, Rio de Janeiro

Hobsbawm E (1977) Introdução a formações econômicas pré-capitalistas de Marx. In: Marx K (ed) Formações econômicas pré-capitalistas, 2nd edn. Editora Paz e Terra, Rio de Janeiro

Huberman L (1969) A história da riqueza do homem, 4th edn. Editora Zahar, Rio de Janeiro

Lacoste Y (1974) A geografia. In: Chatelet F (org.) História da filosofia: ideias, doutrinas. A filosofia das ciências sociais – de 1860 aos nossos dias, vol 7. Editora Zahar, Rio de Janeiro

Lacoste Y (1988) A geografia – isso serve, em primeiro lugar, para fazer a guerra. PapirusEditora, São Paulo

Lefebvre H (1971) O fim da história. Publicações Dom Quixote, Porto

Lefebvre H (1973) A re-produção das relações de produção. Publicações Escorpião, Lisboa

Lefebvre H (1976) Conversaciones con los radicales. Actuel, Madrid

Lefebvre H (1981) Para compreender o pensamento de Marx. Edições 70, Lisboa

Lefebvre H (2013) La producción del espacio. Capitán Swing Libros, Madrid

Lima EL (2019) Encruzilhadas geográficas. Notas sobre a compreensão do sujeito na teoria social crítica. Editora Consequência, Rio de Janeiro

Lukács G (1970) Introdução a uma estética marxista. Sobre a categoria da particularidade. Editora Civilização Brasileira, Rio de Janeiro

Lukács G (2010) Prolegômenos para uma ontologia do ser social. Boitempo, São Paulo
Lukács G (2012) Para uma ontologia do ser social, vols. I & II. Boitempo Editorial, São Paulo
Luxemburgo R (1970) A acumulação do capital (estudo sobre a interpretação imperialista do capitalismo). Editora Zahar, Rio de Janeiro
Martins ER (2007) Geografia e ontologia: o fundamento geográfico do ser. Revista Geousp, espaço e tempo, n. 21. DepartmentofGeography, São Paulo
Marx K (1977) Formações econômicas pré-capitalistas. Editora Paz e Terra, Rio de Janeiro
Moreira R (2015) O pensamento geográfico brasileiro. As matrizes clássicas originárias, vol 1. Editora Contexto, São Paulo
Moreira R (2016) O capítulo 24 e o segredo da atualidade de O Capital de Marx. In: Moreira R (ed) A geografia do espaço-mundo. Conflitos e superações no espaço do capital. Editora Consequência, Rio de Janeiro
Moreira R (2007a) O mal-estar espacial no fim do século XX. In: Moreira R (ed) Pensar e ser em geografia. Ensaios de história, epistemologia e ontologia. Editora Contexto, São Paulo
Moreira R (2007b) Sociabilidade e espaço: as sociedades na era da terceira revolução industrial. In: Moreira R (ed) Pensar e ser em geografia. Ensaios de história, epistemologia e ontologia. Editora Contexto, São Paulo
Moreira R (2007c) A sociedade e suas formas de espaço no tempo. In: Moreira R (ed) Pensar e ser em geografia. Ensaios de história, epistemologia e ontologia. Editora Contexto, São Paulo
Moreira R (2012a) O espaço e o contraespaço: tensões e conflitos na ordem espacial burguesa. In: Moreira R (ed) Geografia e práxis. A presença do espaço na teoria e na prática geográficas. Editora Contexto, São Paulo
Moreira R (2012b) O paradigma e a norma: a sociedade capitalista e o seu espaço. In: Moreira R (ed) Geografia e práxis. A presença do espaço na teoria e na prática geográficas. Editora Contexto, São Paulo
Moreira R (2019) Espaço, corpo do tempo. A construção geográfica da sociedade. Editora Consequência, Rio de Janeiro
Moreira R (2020) A torre, o palimpsesto e a expropriação. Olhando Tricart, Aziz Ab' Sáber e Quaini pelos olhos da totalidade homem-meio. In: Limonad E, Barbosa JL (eds) Editora Max Limonad, São Paulo
Quaini M (1979) Marxismo e geografia. Editora Paz e Terra, São Paulo
Raffestin C (1993) Por uma geografia do poder. Editora Ática, São Paulo
Réclus E, El hombre y la tierra, 6 vol. Maucci, Barcelona (undated)
Santos M (1978) Por uma geografia nova. Da crítica da geografia a uma geografia crítica. Editora Hucitec, São Paulo
Santos M (2002) O dinheiro e o território. In: Posgeouff: Território, territórios. Ensaios sobre o ordenamento territorial. DP&A, Niterói
Sartre J-P (2002) Crítica da razão dialética. Teoria dos conjuntos práticos, vol I. DP&A Editora, Rio de Janeiro
Schüler D (1995) Do homem dicotômico ao homem híbrido. In: Bernd Z, De Grandis R (orgs) Imprevisíveis Américas: questões de hibridação cultural nas Américas. Porto Alegre, Sagra/Abecan
Simmel G (2015) Psicologia do dinheiro e outros estudos. Edições Texto & Traços, Lisboa
Sorre M (1967) El hombre en la tierra. Labor S/A, Barcelona
Vidal de La Blache P (1954) Princípios de geografia humana. Cosmos, Lisboa
Weber M (1968) História geral da economia. Editora Mestre Jou, Rio de Janeiro
Weil S (1979) A condição operária e outros estudos sobre a opressão. Editora Paz e Terra, Rio de Janeiro
Wittfogel KA (1966) Despotismo oriental: estudio comparativo del poder totalitario. Ediciones Guadarrama, Madrid

Chapter 6
Physical Geography and the Study of Environmental Problems: The Brazilian Contribution

Dirce Maria Antunes Suertegaray

Abstract This text aims to discuss research into physical geography in Brazil, taking as a reference the transformations of Brazilian geography, especially since the 1970s. Since then, important aspects have been seen in the transformation of physical geography: the rift with geography from the classical viewpoint in some sectors and the adherence to systemic analysis—a branch of analysis related to the study of nature, in its specificities—for the environmental issue, an environmental analysis from the perspective of managing territory on different scales. In this context, there is a wide range of sub-fields and topics engaged in research related to physical geography, studies of nature in its specificity, studies of environmental impacts (understood as impacts on nature), and proposals related to diagnosis and monitoring, based on spatial analysis (GIS) and model building. Within the critical social approach, Porto Gonçalves (Porto-Gonçalves CW (1989) Os (des)caminhos do Meio Ambiente. Contexto, São Paulo, 147 p) stands out as the precursor to the discussion of the environment in Brazilian geography; followed by the analyses of (Mendonça, São Paulo 1:113–132, 2001) identified as socio-environmental geography; the proposed re-signification by (Suertegaray, Espaço Geográfico Uno e Múltiplo. Scripta Nova, Barcelona, n. 93, 2001) of the concept of environment; and the proposed new geographic approach from Souza, which can be identified as environmental geography. Within the environmental issue, the critical perspective highlights the analysis of social processes and the commercial exploitation of nature, especially through the demands of an economically globalised world. Sometimes, it takes the concept of "production of nature" as a reference. More recently, these analyses have incorporated a dialogue with territorial conflicts on their different scales. The new approaches are structured around an environmental concept associated with social (environmental) justice and the dispute for territory, especially between the major investors (agribusiness, among others) and native and traditional populations. This range of approaches broadly expresses the production of contemporary Brazilian geography.

Keywords Brazilian geography · Physical geography · Environment

D. M. A. Suertegaray (✉)
UFRGS. Universidade Federal da Paraíba, João Pessoa, Brazil
e-mail: dircesuerte@gmail.com

R. C. Lois González and M. A. Mitidiero Junior (eds.), *Brazilian Geography*,
Advances in Geographical and Environmental Sciences,
https://doi.org/10.1007/978-981-19-3704-0_6

6.1 Introduction

This text aims to discuss research in Physical Geography in Brazil, taking as references the transformations of Brazilian Geography from the 1970s onwards and, in particular, the production of Physical Geography and the studies associated with environmental issues. Initially, in order to collate such a vast amount of literature, it is important to indicate some central areas which aim to explain the main trends of Physical Geography in Brazil today. Firstly, however, in the topic labelled "Precursors" we highlight three renowned national geographers who, linked to their specific areas of activity, perform derivations for the analysis of environmental impacts, making a significant contribution to the continuity of geographic analysis in this area. In this sense, we have chosen to give a brief account of the portrayal, from 1960/1970 until the beginning of the 2000s, of the engagement of these geographers in environmental issues, particularly with regard to the preservation of the Amazon, the preservation of biodiversity, and the problematic environment of cities, expressed in the quality of the urban climate. As precursors, I have identified Orlando Valverde, Aziz Ab'Saber, and Carlos Augusto F. Monteiro, also remembering that Manuel Correia de Andrade (1994) expressed himself in his book *Uma Geografia para o Século XXI* (A Geography for the twenty-first century), indicating that Brazilian Geography: [...] could not fail to recognise that the world's tropical forests are being destroyed irrationally and that the interest in social transformations in the tropical world, naturally, should lead the geographer to reflect on problems catalogued, artificially, as Physical Geography [...].

Andrade (1994) wrote about the need for a vision of the whole, for the construction of the unity of the geographical vision, suggesting that geography should continue moving towards the unity of geographical thought, without abandoning specialisation. It should not be forgotten that, as early as the 1950s and 1960s, Manuel Correia de Andrade analysed the impacts of sugar mills and their waste materials on NE rivers.

Orlando Valverde (UFRJ) was a twentieth- and early twenty-first-century geographer who, according to Oliveira (2017), laid the foundations of agrarian geography in the country together with Andrade (professor at the Federal University of Pernambuco—UFPE) and Petrone (professor at the University of São Paulo—USP). The analysis and denunciation of the environmental implications of "national development"—above all, assuming the defence of the Brazilian Amazon—were derived from Agrarian Geography in the 1960s. His participation in the National Company for the Defence and Development of the Amazon (CNDDA) dates from the 1960s (from 1966), working with the aim of integrating and developing the Amazon, preventing its attack and depredation. He was, therefore, a precursor to the analysis and debate on policies for the Amazon, at that time and in subsequent decades. He was active in numerous projects, such as combating the implementation of the SUDAM (Superintendence for the Development of the Amazon) project in the region, the so-called yield forests; the sale of land in the Amazon to foreigners; the private appropriation of ore; SUDAM's agricultural projects; the silting up of

Lake Batata by the bauxite waste released by the Vale do Rio Doce Company; the presence of Eletronorte (CentraisElétricas do Norte do Brasil S/A) in the Amazon; and the fight against iron foundries (*guseiras*)—from the Carajás Project, to name a few, among so many projects analysed and opposed, as exposed in the analysis made by Oliveira (2017) of their political actions.

Aziz Ab'Saber (USP), a geographer dedicated to geomorphological studies, worked in the second half of the twentieth century and the beginning of the twenty-first century, making a significant contribution to Brazilian geomorphology, being the greatest exponent in this area of knowledge within the geographic context of the country. From the 1970s onwards, with the advent of environmental issues, he directed his analytical concerns towards the transformation of landscapes and their physiology, becoming involved in the debate with Brazilian society (through his participation in the Brazilian Society for the Progress of Science), in the Regional Planning of the Carajás Project area, in the Amazon, in the elaboration (together with the Institute of Advanced Studies at USP) of the FLORAN (Forest and Environment) project, proposing a social forestry plan on a national Brazilian scale. He participated extensively in the debate on the transposition of the São Francisco River, protesting against this work due to the social impacts resulting from it, having been active in the debates on the preservation of Brazilian biodiversity, against the approval of the Forest Code at the time of its reformulation by Law no 12.651/2012 (Brazil, 2012), and in the proposal of a Biodiversity Code as an alternative, as expressed in the work organised by Modenesi-Gautieri et al. (2010).

After a long path moving between different university environments, including the Júlio de Mesquita Filho State University of São Paulo (UNESP-Rio Claro), the Federal University of Santa Catarina (UFSC) and the Brazilian Institute of Geography and Statistics (IBGE), Carlos Augusto de Figueiredo Monteiro (2003) took on the role of research professor at the University of São Paulo (USP). His interest was in Climatology, but imbued with his desire to exercise a conjunctive analysis of the landscape, he began his teaching, focussing his analysis on Landscape Geography—in particular, on the physiology of the landscape, a conception expressed by Ab'Saber (1969) in his classic text *Um conceito de Geomorfologia a serviço das pesquisassobre o Quaternário* (A concept of Geomorphology at the service of research on the Quaternary). In this methodological text, through physiology Ab'Saber proposed the understanding of the landscape, including human activity. Following this, and in the search for articulation, Monteiro approached the systemic conception—initially through texts by Bertrand (1972), and later Berry (1972) and Sotchava (1977), among many other international geographers with whom he made contact via meetings of the International Geographical Union (UGI). His research, which was in part focussed on planning and linked to State institutions, experimented (under the most different themes) with the concept of Geosystem as an integrative alternative capable of unravelling environmental issues. His professorship thesis, *Teoria do climaurbano* (Theory of the Urban Climate) was a pioneering work on the climate of cities in Brazil, based on this analytical perspective. Unlike the systemic approach, widespread at the time, whose emphasis was on studies of functionality, prediction, and control of environmental imbalances, Monteiro's analysis goes beyond this perspective, insofar as it considers climate and environmental dynamics, including the temporal dimension, starting from the transformation of their pulsations, of their rhythms. The diffusion of his studies stimulated the continuity of the theme among Brazilian geographers and his contribution, as a researcher, in different institutional and multidisciplinary projects encouraged an integrated analysis in Physical Geography, in the areas of planning and geographic research.

6.2 Physical Geography Post-1970s

From the 1970s onwards, Brazilian Geography, particularly from the National Geography Meeting (ENG) held in Fortaleza in 1978, triggered an analysis of the production of geography in the country, promoting the criticism, on one hand, of the forms of doing Geography (influenced by the classics of this science, especially the French—the so-called Classic Geography), and on the other hand, of the emergence of Geography from a neopositivist perspective in Brazil, recognised as Quantitative Geography. The basis of this appreciation was founded on the so-called Critical Geography, centred on dialectical analysis—more precisely, on historical-dialectical Materialism, based on French and/or English-speaking geographers in the context of the radical and critical movement, which spread from this historical moment.

In Brazil, this moment corresponded to the advent of the environmental issue, the debate over which extends to Geography. However, at the beginning of this process, there was a rupture between the country's geographers, expanding the compartmentalisation between Physical Geography and Human Geography. Despite this, Brazilian geographers, especially those linked to the studies of nature, understood the debate on the environmental issue as significant, assuming this path as a possibility for studies in Geography. At this time, this debate was strengthened by the geographers who we call precursors—Valverde, Ab'Saber, and Monteiro—although each one of them specialised in Agrarian Geography, Geomorphology, and Climatology, respectively. Since the 1970s, Valverde and Ab'Saber were drawn to studies and the defence of nature and the environment, either through their research or via their actions in the context of Brazilian society. Monteiro, with his theory of urban climate, became a pioneer in the understanding of urban climate warming and in the studies of heat islands in cities.

Nevertheless, the expanded separation between physical geographers and human geographers occurred and promoted a rupture in which, thereafter, a significant disconnection was observed with these professionals from the National Association of Geographers, whose study was focussed on the constituents of nature. Along with this, associations were created over time, with the purpose of disseminating specific knowledge, such as the Symposium of Applied Physical Geography (SGFA), the Union of Brazilian Geomorphology (UGB), and later, the Brazilian Association of Climatology (ABCLIMA) and, more recently, the Brazilian Association of Biogeography (ABBIOGEO).

These associations would absorb the theme of environmental issues; if not in all of their publications—because, in their specificities, the works presented at their events were significant—in a significant part of them. Later, the field of humanities would enter the environmental discussion—and, post-2000s, this debate would be present within these two major fields of study in Geography.

6.3 From Studies of Nature to the Environment

In order to better illustrate this derivation in Geography—above all, the approach to environmental issues by Physical Geography—the survey on the production of Geography in Latin America and Brazil is exposed, from a comparison made between the years 2000/2001, and thereafter, data collected for Brazil at different times after the 2000s are presented.

The expansion of studies related to the environment is widely expressed, whether in Latin America or more specifically in Brazil. Observing the graph in Fig. 6.1, it is possible to verify the significant distinction between the number of works related to the environmental issue that was presented in two events, in comparison with the other themes grouped under the Physical Geography category. This is a small example of the interest in environmental issues, particularly on the part of physical geographers—whether at a continental level, or specifically at a national level.

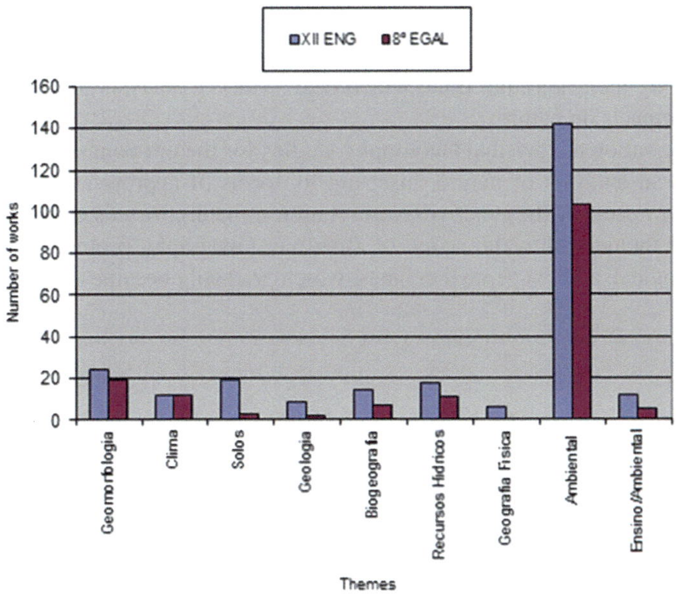

Fig. 6.1 Papers presented at the 12th national meeting of geographers (ENG) and at the 8th Latin American meeting of geographers (EGAL) (2000/2001). *Source* Suertegaray (2005)

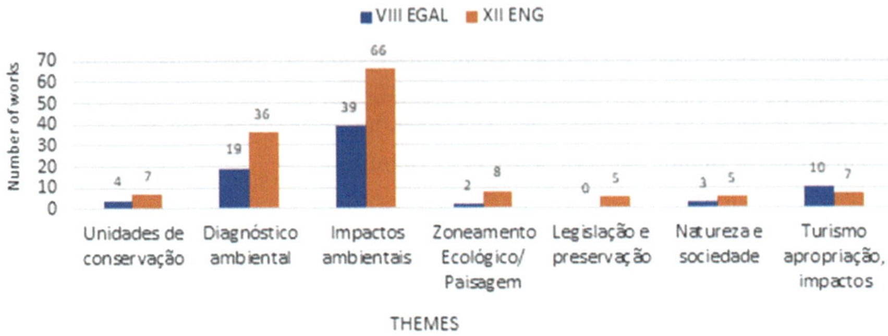

Fig. 6.2 Production of physical geography under the environmental theme at the national meeting of geographers (ENG) and the 8th Latin American meeting of geographers (EGAL). *Source* Suertegaray and Moretti (2014)

A closer look at the topics addressed on the environmental issue (Fig. 6.2) allows us to observe a range of topics, highlighting those shaped by the emerging themes of environmental regulations—demanded in the last 20 years of the twentieth century—whether in relation to Latin America or to Brazil.

In this figure, which expresses the theme raised in the annals of the National Geography Meetings, it can be seen that in the themes linked to the environmental issue, those most investigated by geographers concern the studies of environmental impacts. These environmental impacts are generally linked to aspects of nature, such as climate, terrain, vegetation or water. This trend is significant, whether in the data referring to Brazil or to Latin America. Studies on environmental impacts are followed by diagnostic and environmental planning studies. It is worth highlighting the emerging themes at that time, which dealt with the analysis of tourism and its possible impacts on nature.

The derivation of Physical Geography studies for the environmental perspective expressed an analysis of nature, inserting its forms of appropriation and emphasising the impacts of these activities in relation to nature. In this context, the environmental theme within the scope of Brazilian Geography promoted theoretical, methodological, and conceptual debates, which gradually became expressed in later years.

6.4 Environmental Studies, Topics Covered

We begin our exposition by indicating a trend towards studies in Physical Geography. This trend corresponds to the increase in studies from an environmental perspective, which leads us to think about the current production of Physical Geography.

In order to demonstrate this trend, data from different periods of these first 20 years of the twenty-first century have been retrieved under different information bases—either through the presence of these themes in the production of theses and dissertations, or through their publication in specialised journals.

Table 6.1 shows, based on pre-established categories, the number and percentage of doctoral theses and master's dissertations defended in different graduate programmes, distributed throughout Brazil, according to 2005 data. The universities that presented production in the themes of Physical Geography and Environmental Geography during the analysed period are indicated by category. Theses and dissertations in Physical Geography are taken as references. For this collection, production in this area of knowledge and the top positions of postgraduate courses at the time can be observed, namely: The Federal University of Rio de Janeiro

Table 6.1 Doctoral theses and master's dissertations in geography from 2000 to 2003, by topic

Programme	Physical geography		Human geography		Environment		Theory and method		Teaching	
	(no.)	(%)	(no.)	(%)	(no.)	(%)	(no.)	(%)	(no.)	(%)
UFRJ	*42*	*19.5*	*60*	*8.1*	18	5.8	7	9.2	2	3.0
USP-GF	*39*	*18.1*	0	0.0	*48*	*15.5*	6	7.9	9	13.4
UFU	*32*	*14.9*	*46*	*6.2*	11	3.5	1	1.3	14	20.9
UFSC	*24*	*11.2*	*47*	*6.4*	*25*	*8.1*	0	0.0	1	1.5
UNESP-RC	*21*	*9.8*	32	4.3	19	6.1	5	6.6	3	4.5
UFMG	13	6.0	34	4.6	*21*	*6.8*	8	10.5	9	13.4
UEM	12	5.6	14	1.9	*19*	*6.1*	2	2.6	5	7.5
PUC-MINAS	8	3.7	23	3.1	6	1.9	0	0.0	1	1.5
UFPE	6	2.8	27	3.6	4	1.3	1	1.3	1	1.5
UNESP-PP	3	1.4	*90*	*12.2*	*22*	*7.1*	2	2.6	6	9.0
UFGO	3	**1.4**	34	4.6	14	**4.5**	4	5.3	4	6.0
UFPR	3	1.4	15	2.0	*19*	*6.1*	1	1.3	1	1.5
UFBA	2	0.9	39	5.3	3	1.0	0	0.0	0	0.0
UnB	2	**0.9**	24	3.2	11	**3.5**	0	0.0	0	0.0
UFRGS	2	**0.9**	10	1.4	9	**2.9**	1	1.3	5	7.5
FUFMS	2	0.9	3	0.4	1	0.3	0	0.0	0	0.0
UEL	1	0.5	5	0.7	3	1.0	1	1.3	0	0.0
USP-GH	0	0.0	*173*	*23.4*	*38*	*12.3*	31	40.8	6	9.0
UFF	0	0.0	32	4.3	5	1.6	6	7.9	0	0.0
UFSE	0	0.0	26	3.5	3	1.0	0	0.0	0	0.0
UFRN	0	0.0	3	0.4	3	1.0	0	0.0	0	0.0
FUNECE	0	0.0	3	0.4	8	2.6	0	0.0	0	0.0
TOTAL	215	100.0	740	100.0	310	100.0	76	100.0	67	100.0

Source Suertegaray (2005), starting from data from CAPES (2005)

(UFRJ), the University of São Paulo—Physical Geography (USP-GF), the University of São Paulo—Human Geography (USP-GH), the University of Uberlândia (UFU), the University of Santa Catarina (UFSC), and the University of the State of São Paulo—Rio Claro (UNESP-RC). Regarding environmental analyses, there is a new order: The University of São Paulo (USP-SP), the University of Santa Catarina (UFSC), the University of Minas Gerais (UFMG), the State University of Maringá (UEM), and the Federal University of Paraná (UFPR). It is important to consider that, in the 1990s, the process of expanding postgraduate courses throughout the national territory began, and at this moment, the greatest expression in the environmental theme was centred on already-consolidated universities, since the institutionalisation of master's and doctorate programmes dates from the 1970s and/or 1980s, as in the Geography courses at the University of São Paulo (USP), the Federal University of Rio de Janeiro (UFRJ), and the Federal University of Santa Catarina (UFSC).

On the other hand, a distribution of studies on these themes among the set of programmes can also be observed, with higher numbers in the indicators of thesis defences and dissertations with these approaches. In some courses, at this time there was a greater number of productions in the environmental field than in Physical Geography, such as at the Federal University of Goiás (UFG), the University of Brasília (UnB), and the Federal University of Rio Grande do Sul (UFRGS).

In a second period of surveying this theme (Fig. 6.3), more specifically analysing the works presented at the national meetings of the Association of Brazilian Geographers (AGB) in 2008, 2010, and 2012, Suertegaray and Moretti (2014) promote an analysis of the production published in the annals of the events of this association, during the period indicated, with the objective of evaluating the production related to studies of nature and the environment. The quantitative analysis of these works allows us to identify the significant growth in studies on the environmental theme in 2012 (436) compared to 2008 (252), in addition to a quantitative similarity in relation to studies of nature during the same period. It is also possible to see the presence of works linked to the use of Geographic Information Systems (GIS), largely linked to the analysis of natural constituents and/or the environment.

When analysing the themes present in this production, we can see the greater prominence of studies on environmental policies and on watershed diagnostics

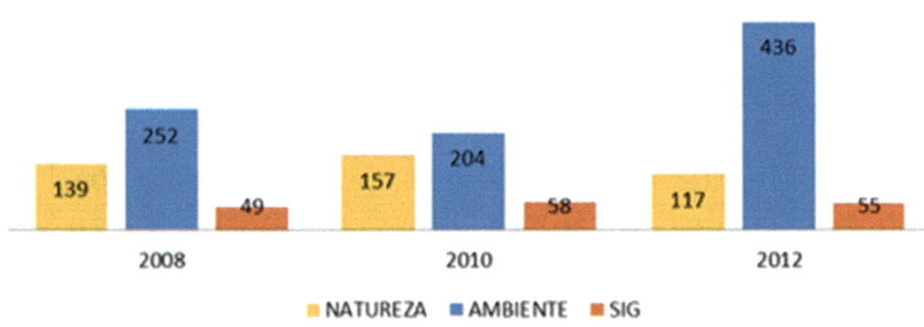

Fig. 6.3 Quantification of the production of studies on nature and the environment in the national meetings of the association of Brazilian geographers—AGB, between 2008 and 2012. *Source* Suertegaray and Moretti (2014)

(Fig. 6.4), in addition to works referring to environmental impacts, whose expansion is significant, jumping from 26 in 2008 to 120 in 2012.

The significant number of studies on the theme of the environment, in relation to the subdivisions of studies on nature, also indicates a reduction in the participation of physical geographers in national events promoted by the AGB, due to the constitution of new spaces for debate, particularly in relation to studies of Physical Geography. A survey of the production of geographic knowledge was carried out over a broader time frame, from 2000 to 2017, aiming to construct an article on the diversity of Brazilian Geography:

Given the complexity of this analysis, we followed the following path: we decided to analyse the production during this period, taking the articles produced in national magazines as a reference. In recent years, there have been many magazines for the dissemination of geographic knowledge. The choice was made based on their scope and the variability of themes. The following were chosen: the journals Geographia (UFF), Geografares (UFES), Mercator (UFCE), Terra Livre from the AGB (Association of Brazilian Geographers), Geousp (USP), and magazines with specific themes such as Cidades (Urban Studies Group), the Brazilian Journal of Geomorphology (UGB), and the Brazilian Journal of Climatology (ABCLIMA) […]. Although we are aware that this analysis does not cover the entire production, which is diverse, we believe that this flyover analysis, as mentioned by our colleague Marcelo L. de Souza, allows us to discuss the production of Brazilian Geography during the period (Suertegaray, 2016, p. 60). From this text, in order to compose this chapter we extract the specific analysis on environmental studies, indicating that:

The environmental theme as a theme of interface is not new—it is linked to the emergence of the discussion of the environment since the 1970s. Since then, the themes show a certain constancy. Although the concept of environment has been the subject of discussion and expansion, in the geographical context, in relation to the original concept of Biology (Environment), the issue of the environment is present in the different sub-fields and interfaces in Geography. Different perspectives and

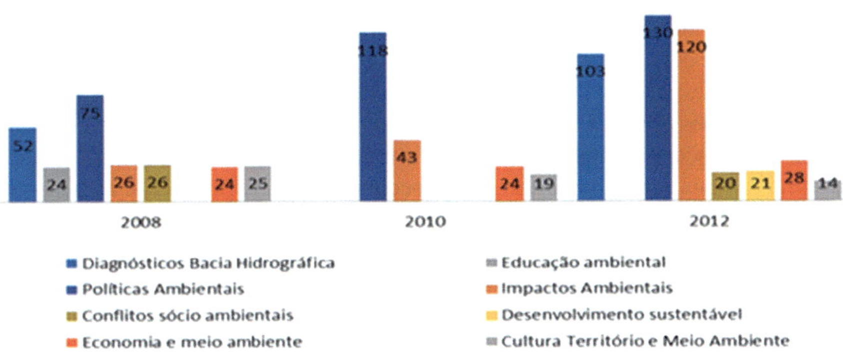

Fig. 6.4 Studies in the categories of nature and the environment in the ENG of 2008, 2010, and 2012. *Source* Suertegaray and Moretti (2014)

Table 6.2 Topics covered in geography and environment

Amazon—deforestation	Environmental diagnostics	Environmental quality
Socio-environmental analysis	Eco-dynamics	Water quality
Biodiversity	Environmental fragility	Recycling
Drought combat/monitoring	Environmental functions and services	Water resources
Environmental conflicts	Environmental impacts	Resilience
Environmental comfort	Economic-ecological modelling	Environmental risks
Ecological-distributive conflicts	Climatic changes	Geological hazards
Thermal comfort	Environmental perception	Conservation units
	Air pollution	Environmental vulnerability

Source Suertegaray (2016)

their connections can be observed in the environmental analysis—sometimes with Economics, sometimes with Biology. This can be seen when we analyse the concepts that are disseminated in environmental analyses, such as environmental functions and services, resilience or economic-ecological modelling, or even distributive ecological conflicts [...]. An analysis of these themes allows us to say that the environmental issue is sometimes linked to planning, and sometimes the defence of the preservation of nature, as well as the explanation of environmental conflicts. In the context of the environmental discussion, the emergence of new concepts stands out: these, in turn, derive from the perspective of the social sciences in relation to the transfigurations of nature—that is, the socio-environmental concept, or the re-signified concept of environment in the geographical context (Suertegaray 2016, p. 76, translated) (Table 6.2).

The survey carried out in these early years of the twenty-first century addresses some considerations: the derivation of studies from Physical Geography to environmental studies; the predominance of environmental analyses, more directly linked to environmental planning and the deciphering of impacts of "human actions" in relation to nature; the progressive expansion of the environmental theme, expressed in research on the economy, on politics, on culture, on territorial conflicts over demand for resources (especially in relation to the advance of the appropriation of nature), and on territories of native and/or traditional populations. This movement is accompanied by a conceptual analysis of the environment and the methods used to decipher this issue. In this sense, the different perspectives being discussed can be observed, in the context of Brazilian Geography, with regard to the concepts, methodologies, and social actions involved.

6.5 The Debate and Paths of Environmental Analysis in Brazil

Since the 1970s, with the derivation of studies from Physical Geography for environmental analysis, theoretical-conceptual discussions, and analytical procedures have expanded among Brazilian geographers, with a view to achieving a better understanding of the environmental issue, knowingly a matter of interface between nature and society. In this sense, we list some dimensions of the debate: the compartmentalisation of studies of nature, strengthened throughout the twentieth century, and the contemporary need, whether in the environmental issue or beyond it, for connection/mediation between nature and society.

This moment corresponds to the criticism of Geography, with a predominance of classificatory studies and the description of forms or of the understanding of the climate through its average pattern, as well as the search for ways to overcome the criticism. In this context, two trends can be seen: one values the study of processes in the search for the understanding of forms, emphasising functionality; and another promotes the appreciation of studies of forms, processes, and their transformations, considering the movement or dynamics of nature over time.

The perspective that emphasises the study of functionality finds, in the systemic analysis brought from the international literature—Bertrand (1972), Sotchava (1977), and Chorley and Haggett (1975), to cite the most widespread—and from the national literature—Christofoletti (1995)—the way to study the dynamics of nature and environmental impacts, both linked to "human action". In the context of (complex) systemic analysis, the investigation of nature and/or the environment can be observed, considering the combined movement of nature and society in time, indicating its transformation and derivation, according to Monteiro (2000, 2003).

Assuming a dialectical stance, another reading is made in the environmental analysis of Brazilian Geography. This, in a different way, expresses an environmental reading, which discusses the need to understand the problem, overcoming the concept of environment from a naturalistic perspective and promoting a discussion in which the analysis of environmental issues is approached from its close links with the social relations of production, unfolding in the appropriation, in the exploitation of nature and in its consequent degradation, extinction. This approach will be highlighted in this text.

From here on, in a synthetic way, different approaches to the environment are explained, without considering that they encompass the diversity of Brazilian production on the issue of the environment. The perspective most commonly seen in the works disseminated in Brazil is linked to an analysis of the environment as technical knowledge. Under this logic, analyses of environmental indicators are built, with objective parameters, in order to uncover impacts on nature, and technical-scientific works are developed, in part, based on systemic analysis, the intention of which is to subsidise environmental planning actions and/or project implementation. The concept of environment and/or environmental is sometimes conceived as synonymous with nature. It is important to mention the emphasis on studies of landscape, eco-dynamics, and eco-geography that is present in the works of Netto (1992, 2005) and Ross (1990, 2006).

Under another understanding, Suertegaray (1987, 2001, 2019) indicates the need to re-frame the biological/ecological concept of environment within the scope of Geography, questioning the use of the concept of environment in geographic studies, considering the criticisms—already widely made—about the naturalist conception of this notion. Its conception is centred on the notions of nature, understood as nature transfigured throughout its process of socialisation, transforming "natural-nature" into second nature—without, however, ceasing to be present in this transformation. This transfiguration, in turn, stems from the way in which capitalist society conceives, appropriates, and exploits "natural resources" and/or, more broadly, sullies/pollutes/degrades this nature. This expanded transfiguration of nature produces the environment, which is not only the primary nature transformed by the (generic) action of human beings on the surface of the planet, but also the product and materialisation of the relationship between nature and society, resulting from the ways in which humans organise themselves socially in order to produce their existence in and with nature. From the author's point of view, an environmental analysis implies the recognition of the economic, political, and cultural dimensions in its unveiling.

Humans promote the transfiguration of nature, through its appropriation and transformation. This transfiguration facilitates changes in natural forms and processes and, at the same time, in human natures, individually and socially. The impacts on the environment, analysed by Geography, are impacts on nature transforming it and transfiguring it, and demand an analysis of the connection between society and nature, aiming to unravel these implications in human individual and social lives, dialectically speaking.

Mendonça (2001) builds an environmental analysis, based on socio-environmental logic, explaining this understanding as an analysis of the social processes involved in the destruction of nature and their consequences for the human and social dimensions. This concept expresses an analytical perspective that attributes a fundamental and necessary importance to the social dimension in the analysis and interpretation of the environmental issue—which is where its conceptual construction comes from—which associates society with the (socio-environmental) environment. In his proposition, the author seeks to overcome the concept of environment and/or ambience, with a naturalistic and technical-scientific hallmark.

More recently, Souza joins the environmental debate, proposing an adjudication of environmental studies in Geography, suggesting the use of Environmental Geography. As he writes, this expression was considered redundant for some time, insofar as the environmental issue or environmentalism already constituted an interpretation concerning the studies of classical geographers and, in view of this, for some geographers today would constitute the centrality of Geography. Seeking a broader reading of the environmental in Geography, the author at present entrusts the environmental adjudication to Geography, stating that the Geography that proposes to be environmental (Environmental Geography) would not correspond to a subfield of Geography, but to an approach—that is, in looking at this science in the differentiated construction of the object of analysis, incorporating ethical and political values, I am dialoguing with Political Ecology.

Porto-Gonçalves is also a precursor, especially through his widely disseminated works *Paixão da Terra* (1985), *Os descaminhos do meio ambiente* (1989), *Amazonas Amazonias* (2001), and *A globalização da natureza e a natureza de globalização* (2006), among many others. The first two works cited introduce a new perspective in the environmental discussion of Brazilian Geography, thereafter expressing the link, in their analyses, from the environmental issue to the territorial expansion of capital and the consequent conflict with local populations, from an eco-socialist perspective. More recently, his approach expresses a connection with de-colonial themes (from Latin America), widely debated in this continent. It is noteworthy that his concerns about the conception of nature and his involvement with the struggle of native and/or traditional peoples were already present since the 1980s/90s.

The de-colonial perspective has, in its centrality, a proposition of defence of indigenous territorialities and/or traditional communities, starting from their cosmovisions, confronted in the struggle for territory and resources. In this approach, culture is rescued as a differentiated constituent of these communities, in relation to the western conception of nature. Bringing culture into the relationship with the economy, such an approach strengthens the struggle to maintain original and traditional territories and respect for nature and its historical uses. Regarding the de-colonial, there is an added value in including the cultural, neutralising the trend that is attributed to the predetermined importance of the economy or ecology, depending on the researcher's taste. In other words, economic crises are ecological crises and cultural crises. It is important not to separate these three areas, and to let them intermingle with each other (Escobar 2009, p. 75, translated).

6.6 Final Notes

As exposed in this text, whose period of analysis begins in the 1970s (a period considered to be a renewal in Brazilian Geography), relevant aspects in the transformation of Physical Geography are visualised: the rupture with Geography from a classical perspective, centred on the classification and description of nature's constituents, in some sectors, and adherence to systemic analysis. However, we should not disregard the fact that a part of nature studies, until that decade, was built from landscape studies, which promoted an integrated analysis of physical constituents and demanded explanations of the transformation of these landscapes in their process of change.

One of the most significant consequences of the social and academic contexts of the time was the derivation of part of the geographers dedicated to the studies of nature, in their specificities, for environmental themes, which were initially built with a view to subsidising planning in the perspective of management of the territory, at different scales. At present, there is a wide range of sub-areas and sub-themes involved in Physical Geography research, such as studies of nature in its specificity, studies on environmental impacts, proposals for the elaboration of diagnoses and monitoring based on spatial analysis (GISs), and the constitution of models, as in the approaches of Coelho Netto (2005) and Ross (2006), along with their extensive works.

Standing out as precursors to the discussion on environmental issues in Brazilian Geography, within the critical social approach, are Porto-Gonçalves (1984, 1989, 2000, 2001, 2004, 2006, 2016); from studies and the re-signification of the concept of environment, Suertegaray (1987, 2001, 2013, 2014, 2017, 2021a, b), Suertegaray and de Paula (2020); from the analyses identified as Socio-environmental Geography, Mendonça (1993, 2001, 2004, 2009, 2019); Mendonça and LIma (2020); Mendoça and Monteiro (2003); and from the proposition of a new geographical focus, to be identified as Environmental Geography, Souza (2018a, b, 2020).

In environmental issues, the critical perspective emphasises the analysis of social processes and the commercial exploitation of nature, above all, through the demands of an economically globalised world. And it sometimes takes the concept of production of nature as a reference (Smith 1988). More recently, such analyses have incorporated a dialogue with territorial conflicts, in their different scales. Therefore, new approaches are structured based on conceptions of the environment, associated with social (environmental) justice and the dispute over territory, especially among large investors, such as agribusiness, miners, ranchers, and large-scale timber explorers, demanding territories and resources from indigenous domains and traditional communities, among others. This diversity of approaches expresses, in a synthesised way, a reading of the production of contemporary Brazilian Geography. Finally, it is emphasised that the geographic production on environmental problems is vast, approaching the most diverse themes and incorporating different conceptions of nature, of the environment—which, in a synthetic way, we seek to express in this text, at the risk of overlooking other perspectives.

References

Ab'saber AN (1969) Um conceito de geomorfologia a serviço das pesquisas sobre o Quaternário. Geomorfologia, São Paulo, IG-USP, no 18, pp 1–23

Andrade MC (1994) Uma Geografia para o Século XX. Papirus, Campinas

Berry BJL (1972) Deliberate change in spatial systems: goals, strategies, another evolution. South African Geogr J 54:31–42

Bertrand G (1972) Paisagem e Geografia Física Global: um esboço metodológico. Cadernos de Ciências da Terra, São Paulo, IG-USP, n 13

CAPES (2005) https://www.gov.br

Chorley RJ, Hagget P (1975) Modelos físicos e de informação em Geografia. LTC; São Paulo: Ed. USP, Rio de Janeiro, 270 p

Christofoletti A (1995) A Geografia Física no estudo das mudanças ambientais. In: Becker BK, Christofoletti A, Davidovich FR, Geiger PP Geografia e Meio Ambiente no Brasil. São Paulo: Hucitec; Rio de Janeiro: CN-UGI, pp 334–345

Coelho Netto AL (1992) O Geoecossistema da Floresta da Tijuca-RJ. In: Abreu MA (Org.) Natureza e Sociedade no Rio de Janeiro. Coleção Biblioteca Carioca, Rio de Janeiro, vol 1, pp 104–142

Coelho Netto AL (2005) A Interface florestal-urbana e os desastres naturais relacionados à água no Maciço da Tijuca: desafios ao planejamento urbano numa perspectivasócio-ambiental. Revista do Departamento de Geografia (USP), São Paulo, vol 16, n 3, pp 46–60

Escobar A (2009). Ecología Política de la globalidad y la diferencia. In: Alimonda H (Coord.) La Naturaleza colonizada. Ecología política y minería en América Latina. [S.l.]: CICCUS/CLACSO (Working Groups Collection)

Mendonça FA (1993) Geografia e Meio Ambiente, 2nd ed. São Paulo: Contexto, 95 p

Mendonça FA (2001) Geografia socioambiental. Terra Livre, São Paulo 1(16):113–132

Mendonça FA (2004) Cidade, ambiente e desenvolvimento—Abordagem interdisciplinar de problemáticas socioambientais urbanas de Curitiba e RMC. Curitiba: Ed. UFPR/Imprensa Oficial do Estado do Paraná, vol 1, 276 p

Mendonça FA (2009) Geografia, Geografia Física e Meio Ambiente: uma reflexão a partir da problemática socioambiental urbana. Revista Da ANPEGE 5:123–134

Mendonça FA, Dias MA (2019) Meio Ambiente e Sustentabilidade, 1st ed, vol 1. Intersaberes, Curitiba, 296 p

Mendonça FA, Lima MV (2020) A cidade sob o enfoque socioambiental: Curitiba e região metropolitana como lócus de uma abordagem interdisciplinar da urbanização em vista da relação sociedade-natureza. In: Mendonça FA, Del Vechio M (Org.) A cidade e os problemas socioambientais urbanos—Uma perspectiva interdisciplinar, 1st ed. Curitiba: Ed. UFPR, vol 1, pp 11–27

Mendonça FA, Monteiro CAF (2003) Clima Urbano, 1st ed, vol 1. Contexto, São Paulo, 192 p

Monteiro CAF (2000) Geossistemas: a história de uma procura. Contexto, São Paulo, 127 p

Monteiro CAF (2003) A questão Ambiental na Geografia do Brasil. A propósito da "validade", "especialização" e "pesquisa universitária". Cadernos de Geografia, Florianópolis, CFH/UFSC, n. 5

Modenesi-Gauterri MC et al (2010) A obra de Aziz Ab'Saber. BecaBAL, São Paulo, 588 p

Oliveira AU (2017) A Geografia Agrária na Obra de Orlando Valverde. In: Suertegaray DMA, De Paula CQ, Pires CLZ, Silva CAO Orlando Valverde o geógrafo e sua obra. Porto Alegre: IGEO-UFRGS, pp 65–122

Porto-Gonçalves CW (1984) Paixão da Terra: ensaios críticos de Ecologia e Geografia, 1st ed. Socii, Rio de Janeiro, vol 1, 160 p

Porto-Gonçalves CW (1989) Os (des)caminhos do Meio Ambiente. Contexto, São Paulo, 147 p

Porto-Gonçalves CW (2000) Natureza e Sociedade: elementos para uma ética da sustentabilidade. Coleção Meio Ambiente, vol 3, pp 49–76

Porto-Gonçalves CW (2001) Amazônia, Amazônias, vol 1. Contexto, São Paulo, 178 p

Porto-Gonçalves CW (2004) O Desafio Ambiental, 1st ed., vol 1. Record, Rio de Janeiro/São Paulo, 179 p

Porto-Gonçalves CW (2006) A Globalização da natureza e a natureza da globalização, 1 ed, vol 1. Civilização Brasileira, Rio de Janeiro, 461 p

Porto-Gonçalves CW (2016) Lucha por la tierra: ruptura metabólica y reapropiación social de la naturaleza. POLIS, Santiago, Impresa, vol 45, pp 1–18

Ross JLS (1990) Geomorfologia, Ambiente e Planejamento, 1st ed, vol 1. Contexto, São Paulo, 80 p

Ross JLS (2006) Ecogeografia do Brasil, 1 ed, vol 1. Oficina de Textos, São Paulo, 208 p

Smith N (1988) Desenvolvimento desigual. Bertrand Brasil, Rio de Janeiro, p 242

Sotchava VB (1977) O Estudo de Geossistemas. São Paulo: IG-USP, 51 p (Métodos em Questão, no 16)

Souza ML (2018a) O que é a Geografia Ambiental? AMBIENTES—Revista De Geografia e Ecologia Política 1:14–37

Souza ML (2018b) Quando o trunfo se revela um fardo: reexaminando os percalços de um campo disciplinar que se pretendeu uma ponte entre o conhecimento da natureza e o da sociedade. GEOUSP—Espaço e Tempo 22:274–308

Souza ML (2020) Articulando ambiente, território e lugar: a luta por justiça ambiental e suas lições para a epistemologia e a teoria geográficas. AMBIENTES—Revista De Geografia e Ecologia Política 2:16–64
Suertegaray DMA (1987) A Trajetória da Natureza: um estudo geomorfológico sobre os areais de Quarai-RS, 1987, 243f. Thesis (Doctorate in Geography)—Faculdade de Filosofia, Letras e Ciências Humanas, Departamento de Geografia, Universidade de São Paulo, São Paulo
Suertegaray DMA (2001) Espaço Geográfico Uno e Múltiplo. Scripta Nova, Barcelona, n. 93
Suertegaray DMA (2005) O atual e as tendências do ensino e da pesquisa em Geografia no Brasil. Revista Do Departamento De Geografia Da USP, São Paulo 16:38–45
Suertegaray DMA (2013) Naturezas: epistemes inscritas nos conflitos sociais. Terra Livre 41:17–30
Suertegaray DMA (2014) Geografia e Ambiente: desafios e novos olhares. Revista Mato-Grossense De Geografia 17(1):3–14
Suertegaray DMA (2016) Conhecimento Geográfico no Brasil: diversidade temática e metodológica. In: Spósito ES, Sant'anna Neto JL, Melazzo ES (Org.) A diversidade da Geografia Brasileira escalas e dimensões da análise e da ação, 1st ed, vol 1. Consequência, Rio de Janeiro, pp 57–82
Suertegaray DMA (2017) (Re)Ligar a Geografia: natureza e sociedade. ComPasso Lugar-Cultura, Porto Alegre, 179 p
Suertegaray DMA (2019) Geografia, ambiente e território. Revista da Casa da Geografia de Sobral, Sobral, vol 17, no 3, pp 128–144, Dec. 2015. ISSN 2316-8056. Available at: http://uvanet.br/rcgs. Accessed on 11 Aug
Suertegaray DMA (2021a) Geografia e Análise Ambiental: um ano de pandemia. Humboldt—Revista de Geografia Física e Meio Ambiente, Rio de Janeiro, UERJ, vol 2, p 116. https://www.epublicacoes.uerj.br/index.php/humboldt/article/view/57373/. Accessed on 27 May 2021a
Suertegaray DMA (2021b) Meio, Ambiente e Geografia. Porto Alegre: Com Passo Lugar-cultura, 146 p
Suertegaray DMA, De Paula CQ (2020) Geografia e questão ambiental, da teoria à práxis. AMBIENTES. Revista de Geografia e Ecologia Política, vol 1, no 1, pp 79–102, 2019. ISSN: 2674-6816. Available at: http://erevista.unioeste.br/index.php/ambientes/article/view/22686. Accessed on 17 Mar
Suertegaray DMA, Moretti EC (2014) Considerações sobre o eixo temático Natureza no contexto dos Encontros Nacionais de Geógrafos (Brasil) 2008–2012. Terra Livre, vol 42, pp 83–100

Chapter 7
The Study of Cities in Brazilian Geography

Pedro de Almeida Vasconcelos

Abstract Urban geography is undoubtedly one of the most important subdisciplines of Brazilian geography: first, because it is a pioneer on the topic: a text on urban geography was found in a publication by an author from Bahia in 1913! second, because of the volume of production: in the surveys conducted by Mauricio Abreu in 1989, the author managed to count some 1000 titles; and third, because of the prominence of the authors who have been involved in the topic, from Josué de Castro in the 1940s, Milton Santos since the 1950s, Pedro Geiger in the 1960s, and Aldo Paviani, Maria Adélia de Souza, Mauricio Abreu and Roberto Lobato Correa in the 1980s. It is important to stress the continuity of the publications of Milton Santos abroad, which had an international impact. 1989 saw the start of the biennial National Symposia on Urban Geography (Simpurbs), which in addition to bringing together researchers from the entire country with an interest in the topic, gave rise to the publication of annals, journals, and books, which comprise a record of the main production over the course 30 years. At the same time, the Urban Studies Group was created, which began to publish the Cidades journal in 2004, as well as two books. With the annual meeting of the Urban Geography Commission of the IGU, held in Salvador in 2017, a process of expanding the participation in international events began. The production of this chapter will be supported by Mauricio Abreu's 1989 article, my book Dois séculos de pensamento sobre a cidade (1999, 2012) and an examination of the most recent publications, both individual and collective, placing special emphasis on the collections resulting from the Simpurb events. The main present-day topics of the subdiscipline can be proposed by studying the Working Groups of the mentioned biennial events.

Keywords Urban geography · Brazilian geography · Simpurb

P. de Almeida Vasconcelos (✉)
Universidade Federal da Bahía, Salvador, Brazil
e-mail: pavascon@uol.com.br

Catholic University of Salvador de Bahia, Salvador, Brazil

R. C. Lois González and M. A. Mitidiero Junior (eds.), *Brazilian Geography*, Advances in Geographical and Environmental Sciences,
https://doi.org/10.1007/978-981-19-3704-0_7

7.1 Introduction

Urban geography is undoubtedly one of the most important subdisciplines of Brazilian geography. First, because it is a pioneer on the topic: a text on urban geography was recently found in a publication by an author from Bahia in 1913. Second, it should also be noted that among the disciplines dealing with cities in Brazil, geography is the oldest and has the largest number of published works (Valladares et al. 1991). Third, because of the volume of production: in the surveys conducted by Mauricio Abreu in 1989, the author managed to count 931 titles. Fourth, because of the prominence of the authors who have studied specific cities such as Josué de Castro in the 1940s; Milton Santos, since the 1950s; Aldo Paviani, Maria Adélia de Souza, and Mauricio Abreu since the 1980s and Ana FaniAlessandri Carlos from the 1990s. It is important to stress the continuity of Milton Santos' publications in Brazil and abroad, which had an international impact.

1989 saw the start of the biennial National Symposia on Urban Geography—(Simpurbs), which in addition to bringing together researchers from the entire country with an interest in the topic, gave rise to the publication of books, journals, and annals, which comprise a record of the main specialised production over the course of 30 years. At the same time, the Urban Studies Group was created, which began to publish the Cidades journal in 2004, as well as two books in 2011 and 2013. With the annual meeting of the Urban Geography Commission of the IGU, held in Salvador in 2017, a process of expanding the participation in international events began.

The production of this chapter will be supported by Mauricio Abreu's 1989 article; my book from 1999 and 2012; the article by Dantas and Silva (2018), in which the most cited works from 2005 to 2016 are listed; the most recent books published on websites, both individual and collective; as well as the collections resulting from the Simpurb events and the issues of the Cidades journal.

7.2 Pioneers in Brazilian Urban Geography

Recently, young researchers from Bahia, led by André Nunes de Souza, found the text "A Geografia das Cidades", published in 1913, by Bernardino J. Souza, as a chapter of the book Por mares e terras (leiturasgeográficas), with several references to German authors, such as F. Ratzel, A. Hettner, K. Hassert et al., analysed by Wendel Henrique Baumgartner in the book authored by Souza and Vaz (2019).

After the publication of the text by Bernardino Souza, in addition to specific texts on Brazilian cities, commented below, other general studies on the urban topic, prepared by Brazilian geographers, can be highlighted, such as the book by Nice Lecocq Müller, O Fato Urbano na Bacia do Rio Paraíba, in (1960); Pedro P. Geiger's Evolução da Rede Urbana Brasileira, in (1963); and, above all, the series of books published by Milton Santos, in Brazil and France, such as A Cidade nos Paises Subdesenvolvidos, in (1965); Aspects de la géographie et de l'économie urbaine des pays sous-développés, in 19,694; Dix essais sur les villes des pays sous-développés, in (1970); Les villes du Tiers Monde, in (1971); L´Espace partagé. Les deux citcuits de l´économie urbaine des pays sousdéveloppés, in (1975); Pobreza Urbana, in (1978); A urbanização desigual in (1980); Manual de Geografia Urbana, in (1981); Ensaios sobre a urbanização latino-americana in (1982) and A urbanização brasileira, in (1993). Likewise, the book authored by Speridião Faissol Tendências atuais na geografia urbano-regional: teorização e quantificação, in (1978), and the book Governo Urbano by Maria Adélia de Souza, in (1988), were also published. Therefore, in addition to the pioneering text by Bernardino de Souza, the 10 books published by Milton Santos on the city and urbanisation, in Portuguese and French, in addition to his doctoral thesis on Salvador in (1959), should be highlighted.

7.3 Brazilian Cities Analysed by Urban Geographers

Another possibility in terms of examining the production of Brazilian urban geography is to select important texts on the most studied cities (or metropolises).

São Paulo appears as the city with the highest number of publications and, at the same time, has the oldest text on a Brazilian city, written by historian and geographer Caio Prado Jr., with his article "O Fator Geográfico na Formaçâo e no Desenvolvimento da Cidade de São Paulo", published in 1935, followed by Delgado de Carvalho's text "A Cidade de São Paulo", in (1954). Followed by that is the set of four volumes authored by Aroldo de Azevedo, A cidade de S. Paulo; Estudos de Geografia Urbana, written to commemorate the city's fourth centenary and published in (1958). In 1963, Pasquale Petrone published Pinheiros: aspectos geográficos de um Bairro Paulistano, and in 1971, the book A Estruturação da Grande São Paulo was published by Juergen R. Langenbuch. In 1980, the book O Centro da Metrópole Paulistana, by Helena K. Cordeiro, was published. In 1989, the books A Identidade da Metrópole. A verticalização de São Paulo, by Maria Adélia A. de Souza and Metrópole Corporativa Fragmentada: o caso de São Paulo, by Milton Santos, were published. In 1994, the same author published the book Por uma economia política da cidade: o caso de São Paulo. In 1998, Maria Adélia published the book São Paulo. Ville mondiale et urbanisme français sous les tropiques in France. 2004 saw the publication of the two volumes authored by Ana Fani A. Carlos and Ariovaldo Umbelino de Oliveira, entitled Geografias de São Paulo: representação e crise da metrópole, which celebrates, at the time, the city's 450th anniversary. Lastly, São Paulo appears in a comparative analysis by Silvia Aparecida Guarnieiri Ortigoza entitled Paisagens do consumo: São Paulo, Lisboa, Dubai e Seul, in (2010).

Brasília appears as the second city with the highest number of publications, especially considering the collections authored by Aldo Paviani: Brasília, Ideologia e Realidade: espaço urbano em questão (1985, 2010); Urbanização e Metropolização—A gestão dos conflitos em Brasília (1987); A Conquista da Cidade: movimentos populares em Brasília (1991, 1998, 2010); Brasília: moradia e exclusão (1996); Brasília—Gestão Urbana: conflitos e cidadania (1999, 2008), as well as together with other authors: Brasília: controvérsias ambientais (2003); Brasília: dimensões da violência urbana (2005, 2010); Brasília 50 anos: da capital a metrópole (2010), in addition to the author's own book Brasília: a metrópole em crise—Ensaios sobre urbanização (1989, 2010). The book by José W. Vesentini A Capital da Geopolítica, which traverses between urban geography and political geography, was published in (1986).

Rio de Janeiro was the third most studied city. The text by Lysia M. C. Bernardes "Importância da posição como fator do desenvolvimento do Rio de Janeiro", published in 1957–1958, initially stands out. It is followed by the book by Aluizio Capdeville Duarte A Área Central da Cidade do Rio de Janeiro, published in (1967). In 1987, Mauricio Abreu published the book Evolução Urbana do Rio de Janeiro, followed by the books by the same author: Natureza e Sociedade no Rio de Janeiro, in (1992), and Rio de Janeiro: formas, movimentos, representações, in (2005). Lastly, in 2010 Mauricio Abreu published his magnum opus Geografia Histórica do Rio de Janeiro (1502–1700), in two volumes with more than 900 pages.

With regard to Salvador, there are three texts from the 1950s, Aroldo de Azevedo's "Salvador et le Reconcavo de Bahia", from 1951, Aziz N. Ab'Saber's "A cidade de Salvador: Fotografias e Comentários", from 1952, and the doctoral thesis by Milton Santos, published in (1959), O Centro da Cidade de Salvador. In 1961, Milton Santos authored the book Cidade do Salvador, which included the texts by Ab'Saber "O Sítio da Cidade do Salvador" and by Aroldo de Azevedo, "Fisionomia da Cidade do Salvador". In 2002, the book Salvador: transformações e permanências (1549–1999), by Pedro de A. Vasconcelos, was published, with a 2nd expanded edition in (2016) and in (2005) Salvador de Bahia (Brésil) transformations et permanences (1549–2004) was published in France. Lastly, in (2014), the book by Jânio Santos A cidade poli(multi) nucleada: a reestruturação do espaço urbano de Salvador was published.

Recife was the subject of Josué de Castro's thesis A Cidade de Recife: Ensaio de Geografia Urbana, published in 1948. In 1978, the book Metropolização e Sub Desenvolvimento. O caso de Recife, by Mário Lacerda de Melo, was published, and the following years was Recife: Problemática de uma Metrópole em uma Região Subdesenvolvida by Manuel Correa de Andrade. More recently, in (2007), the book by Edvânia T. A. Gomes Recortes de paisagens na cidade do Recife was published.

The city of Belém was studied by Antônio Penteado who in (1968) published Belém Estudo de Geografia Urbana, in two volumes. In (1997), Saint-Clair C. Trindade Jr. published the book Produção do Espaço e Uso do Solo Urbano de Belém and, in (2019), together with Veloso dos Santos, he authored the book O urbano e o metropolitano em Belém: (re) configurações socioespaciais e estratégias de planejamento e gestão.

Fortaleza was analysed by José Borzacchiello da Silva with the book Os incomodados não se retiram. Fortaleza em questão, from 1992 and by Eustógio Wanderley Dantas with the book Mar à Vista: estudo de maritimidade em Fortaleza, was published in 2002 and 2011.

With regard to Aracaju, there is the book Transformações do Espaço Urbano: o Caso de Aracaju by Neuza Maria G. Ribeiro, from 1989 and the book Aracaju:

estado and metropolização, by Vera Lúcia A. França, published in (1999). Studies on Goiânia were published in the book authored by Lana de Souza Cavalcanti entitled Geografia da Cidade: a produção do espaço urbano de Goiânia, in (2001). Lastly, Manaus was historically analysed in the book by José Aldemir de Oliveira entitled Manaus de 1920–1967. A cidade doce e dura em excesso was published in (2003).

We can summarise this part by identifying a relationship between the sizes of the metropolises and the number of publications made by geographers. The scale would therefore go from São Paulo to Aracaju. Although São Paulo and Rio de Janeiro are among the most studied metropolises, Brasília, thanks mainly to the texts published by Paviani, places the national capital in second place. On the other hand, the metropolises of the Northeast (Salvador, Recife, and Fortaleza) and the North (Belém) form a second block with a very similar number of publications. Lastly, the three smallest metropolises (Aracaju, Goiânia and Manaus) had a more limited number of studies.

7.4 The Production of Urban Geography Over the Last 30 years (1989–2021)

Given the enormous production in the subarea, especially considering the growth of postgraduate programmes, it is difficult to select the most outstanding books or the most productive colleagues with quality production. Priority was given to books, a tradition of the discipline, which had a national impact, despite the existence of articles of fundamental importance. Small books for dissemination of the subdiscipline or topical texts, such as housing and transport, will not be considered, in view of the text size.

The production of this more recent period was divided into three parts: individual production, co-production and collections, always following the chronological order: Ana Fani A. Carlos published the books A (Re) Produção do Espaço Urbano in (1994); followed by Espaço-Tempo na Metrópole, in 2001 and by O Espaço Urbano: novosescritos sobre a cidade, in 2004. Arlete M. Rodrigues published Produção e Consumo do e no Espaço. Problemática Ambiental Urbana, in (1998). Pedro de A. Vasconcelos published Dois Séculos de Pensamento sobre a Cidade (1999 and 2012) (in addition to two books regarding Salvador). Marcelo Lopes de Souza published the books O Desafio Metropolitano; um estudo sobre a problemática sócio-espacial nas metrópoles brasileiras, in (2000); Mudar a Cidade, in (2002); ABC do desenvolvimento urbano, in (2003); A Prisão e a Ágora: reflexões em torno da democratização do planejamento e gestão das cidades, in (2006) and Fobópole. O medo generalizado e a militarização da questão urbana, in 2008. José Aldemir de Oliveira published Cidades na Selva in (2000) (in addition to the book regarding Manaus). In 2002, Paulo César da Costa Gomes published the book A Condição Urbana; in (2005) Sidney G. Vieira published A Cidade Fragmentada and Roberto Lobato Corrêa the book Estudos sobre a Rede Urbana in (2006) (in addition to the books published in

1989). In 2007, Angelo Serpa published O espaço público na cidade contemporânea and Cidade Popular—Trama de relações sócio-espaciais (in addition to the book authored with Ana Fani A. Carlos regarding the Simpurb in Salvador, in 2018). In the same year, Álvaro H. de S. Ferreira published A Cidade no Século XXI and Maria Encarnação Beltrão Sposito Cidades Medias—Espaços em transição. In 2008, the books by Eliseu S. Sposito, Redes e Cidades and Marcio M. Valencia Cidade (I)Legal were published. Wendel Henrique published O Direito à Natureza na Cidade in 2015. Lastly, Sandra Lencioni published Metrópole, Metropolização e Regionalização in (2017).

With regard to co-authored books, in 2006 Heloisa S. de M. Costa, G. M. Costa, J. G. de Mendonça and R. L. de M. Monte-Mór published the book Novas Periferias Metropolitanas. In 2012, Eliseu Sposito published with Paulo F. J. Silva the book Cidades pequenas: perspectivas teóricas e transformações socioespaciais and in 2016 together with Maria Encarnação B. Sposito and Oscar Sobarzo the book Cidades médias. In 2008, Saint-Clair da Trindade Jr. and M. G. C. Tavares published Cidades ribeirinhas na Amazônia (in addition to a book regarding Belém and another co-authored book regarding the same city). Aldo Paviani wrote with Suely F. N. Gonzales and Jorge G. Francisconi Planejamento e Urbanismo na actualidade brasileira: objeto, teoria, prática in (2013) (in addition to the written book and collections regarding Brasília). In the same year, Maria Encarnaçâo Beltrâo Spósito, together with Eda Maria Gomes, published the book Espaços fechados e cidades.

In terms of the collections, also a tradition in geography, many were the results of events, such as the Simpurbs. Ana Fani authored the book Crise Urbana in 2015; and together with Carles Carreras authored Urbanização e mundializaçãi estudos sobre a metrópole in (2005); with Ariovaldo U. Oliveira, the book Geografias das metrópoles in 2006 (and another book regarding São Paulo in 2004); with Marcelo L. de Souza and Maria Encarnação B. Sposito, A produção do espaço urbano: agentes, processos, escalas e desafios in 2011; with Danilo Volochko and Isabel P. Alvarez, A cidade como negócio in 2015; with Glória Alves and Rafael F. de Padua, Justiça espacial e o direito à cidade in 2017; and with César S. Santos and Isabel P. Alvarez, Geografia Urbana Crítica. Teoria e Método in 2018 (in addition to the authored books regarding the Simpurbs, the first in São Paulo in 1989; the second with Amália Inês G. Lemos in 2003; and the book regarding Salvador in 2018 together with Angelo Serpa). Jan Bitoun authored the book Tipologia das cidades brasileiras e políticas territoriais in 2009. Arlete M. Rodrigues authored the book Estudos Urbanos in 2011. Pedro de A. Vasconcelos, together with Roberto L. Corrêa and Silvana M. Pintaudi, authored the book A cidade contemporânea: segregação espacial in (2013) (in addition to the book authored with Sylvio C. Bandeira de Melo e Silva regarding the Simpurb in Salvador in 1999). Sylvio C. Bandeira de M. e Silva co-authored with I. M. Carvalho, A. M. G Souza and G. Pereira, the book Metrópoles na atualidade brasileira: transformações, tensões e desafios na Região Metropolitana de Salvador in 2014, and together with I. M. Carvalho and G. C. Pereira, the book Transformações metropolitanas no século XXI. Bahia, Brasil e América Latina in 2016 (in addition to the book authored with Pedro de A. Vasconcelos regarding the Simpurb in Salvador in 1999). Maria Encarnação B. Sposito authored with Arthur M. Whitacker the book Cidade e Campo: relações e contradições entre urbano e rural in 2006, and with William R. da Silva Perspectivas da Urbanização. Reestruturação urbana e das cidades in 2017 (in addition to the authored book regarding the Simpurb in Presidente Prudente in 2001). Lastly, Álvaro H. de S. Ferreira authored with João Rua and Regina C. de Mattos the books Espaço e metropolização. Cotidiano e ação in (2017) and Metropolização do Espaço. Gestão territorial e relações urbano-rurais and Desafios da metropolização do espaço in 2019.

In the production of the last 30 years, Ana Fani A. Carlos can be highlighted for her high production of individual books, for her co-written books, and, above all, for those authored by her colleague, including the one regarding the first Simpurb. Marcelo Lopes de Souza presents a solid production of individual books. The collections were, to a large extent, the results of the Simpurbs, which are very representative of local leaders in urban geography.

7.5 The National Symposia on Urban Geography (Simpurbs)

The National Symposia on Urban Geography—Simpurb started in 1989 and continued biennially until 2019. The event scheduled for this year 2021 has been postponed to 2022 as a result of the current Pandemic. The dissemination of the events through the publication of 11 books, three journals and two annals should be highlighted. From the titles of the lectures and the round tables, it is possible to examine the topics of greatest interest in Brazilian urban geography over the last 30 years, as well as highlight the geographers who were invited to hold the lectures or those who were honoured throughout the 16 biennial events.

The I National Symposium on Urban Geography was held at the University of São Paulo—USP, in November 1989. Only in 1994 was the book Os caminhos da Reflexão sobre a Cidade e o Urbano, authored by Ana Fani A. Carlos, with the main results of the event are published. The book, with 390 pages, was not authored by round tables and has 14 chapters: we can highlight the 1st, which corresponds to the author's presentation, "Rethinking Urban Geography: the Balance of a Symposium"; the 2nd, the text of the opening lecture by Milton Santos, entitled "Trends of Brazilian Urbanisation at the End of the XX Century"; and the 13th, with Mauricio Abreu's extensive chapter "The Geographical Study of the City in Brazil: Evolution and Evaluation", with a total of 123 pages and a bibliography with 931 texts. The book concluded with two statements: one by Pedro Geiger and one by Lysia Bernardes.

The II Symposium was held in October 1991 at the São Paulo State University (UNESP Rio Claro), São Paulo, and the main results were published, in the same year, in issue 42 of the Boletim de Geografia Teorética (Journal of Theoretical Geography), with all 262 pages dedicated to the event, and whose topic was "Rethinking the City and the Urban". The journal begins with a tribute to the couple Lysia and Nilo Bernardes. The inaugural lecture was held by Roberto Lobato Corrêa, entitled

"New geographical dimensions of the urban area in Brazil". The results of six round tables were also published (Cities and Social Movements; The Daily Life of the Metropolis; The Urban Environment; Urban Geography: Theoretical and Methodological Perspectives; Space and Modernity and City—Space and Time) in 130 pages. The texts of the talks were published on the remaining 113 pages.

The 3rd Symposium was held at the Federal University of Rio de Janeiro, in September 1993, whose main results were published in the respective Anais, with a total of 225 pages. They contain the results of five round tables (Matrices of Urban Geography; Networks, Flows and Territories; Geopolitics of the Urban Environment; Time and Space in Urban Daily Life and Urban Crisis and Urban Reform), in a total of 146 pages, while another 70 pages were dedicated to the five talk sessions. The lecture held by Fany Davidovich was not published in the event's annals.

The 4th Symposium took place at the Federal University of Ceará, in Fortaleza, in October 1995, and resulted in the publication of the book A Cidade e o Urbano, authored by José Borzacchiello da Silva et al., in (1997) with 317 pages. The book begins with the text of the lecture by Pedro Geiger, "Geography and regional and urban development in Brazil" and is followed by the publication of texts presented in five round tables (Epistemologies of cities; Cities, metropolises, and megalopolises; Urban environment; Urban daily life and Urban development policies). The lectures by José Borzacchiello da Silva and Armando Correia da Silva, as well as the scientific talks, were not published in that book.

The 5th Symposium was held in Salvador, in October 1997, had about 250 participants and resulted in the publication of the book authored by Pedro de Almeida Vasconcelos and Sylvio Bandeira de Mello e Silva entitled Novos Estudos de Geografia Urbana Brasileira, in (1999), with 271 pages. The book begins with the text of the lecture by Maria Adélia de Souza "City: Place and Geography of Existence" and is followed by the texts of the five round tables (Matrices of urban geography; Geographical networks; Space, power and culture; Space and time in the urban and Socio-spatial dynamics and forms of intervention). With regard to the talks, only the texts of round tables with special talks were published.

The 6th Symposium, held at UNESP Presidente Prudente, São Paulo, in October 1999, resulted in the book authored by Maria Encarnação Beltrão Sposito, entitled Urbanização e Cidades: perspective as geográficas, published in (2001), with 643 pages. The seven parts of the book correspond to five round tables (The city in Brazil: transformations and permanence; Dilemmas of method: urbanisation and fragmentation; Social production of space and urban environmental issues; Cities and public policies: from planning to management and The urban network and its recent dynamics); to two open debate sessions on "The meaning of the city at the end of the XX century" and the last part was on "Medium-sized cities". There was no lecture at the event, but Lea Goldstein was honoured.

The city of São Paulo and the University of São Paulo (USP) once again hosted the 7th Symposium in October 2001, which resulted in the publication of the book authored by Ana Fani Alessandri Carlos and Amália Inês Geraides Lemos, entitled Dilemas urbanos: novas abordagens sobre a cidade, published in 2003, with a total of 430 pages. The book is divided into seven chapters that correspond to the seven debate sessions (The problems of the city, the metropolis and the way of life, segregation, conflicts, strategies; Urban spatialities and temporalities; Culture and the city; Work in the city; Space and State: urban policies under discussion; The clash between environmental and social issues in the urban; The urban in the world of merchandise: introduction). The lecture held by French geographer Marcel Roncayolo was not published in the aforementioned book.

The 8th Symposium was held in Recife, at the Federal University of Pernambuco, in November 2003. The event featured the opening lecture by Ana Fani Alessandri Carlos; three discussion tables (Urban network; Urbanism/geography interface; and Management instruments—master plans—new features), seven round tables (Small- and medium-sized cities in different regional contexts; New forms of management: proximity in question; City centres: interventions, balances and perspectives; Urban peripheries, types and dynamics of expansion, agents and their transformations; Metropolises: between the global, the national and the local; Urban environmental risks; and Time, culture and space) and the closing lecture by Manuel Correia de Andrade. The V. I Anais were published in (2004), with a total of 228 pages, and did not follow the event's structure, with 19 texts being chosen for publication.

In Manaus, the 9th Symposium was held at the Federal University of Amazonas, in 2005, and resulted in the publication of the book authored by José Aldemir de Oliveira entitled Cidades brasileiras: territorialidades, sustentabilidade e demandas sociais, in two volumes published in 2009 and 2010, totalling 930 pages. Texts presented in the round tables and coordinated talks were selected. In the first volume, with 359 pages, there are three parts: Part 1 (Central areas) includes the opening lecture given by Portuguese geographer Tereza Barata Salgueiro, "Mobility, new social demands and urban sustainability"; and is followed by Part 2 (Small cities) and Part 3 (City, space and time). The second volume, with 571 pages, is also divided into three parts: Part 1 (Management of the City); Part 2 (New and old topics of the city and the urban) and Part 3 (The space of metropolises). The text of the final lecture by French geographer Martine Droulers "Urbanisation, sustainable development in the Amazon" was not published.

The 10th Symposium was held at the Federal University of Santa Catarina in Florianópolis, in November 2007, and resulted in the publication in 2011 of the book As cidades e a urbanização no Brasil: passado, presente e futuro, with 447 pages, authored by Elson Manoel Pereira and Leila Christina Duarte Dias. The book is divided into three parts: Part 1 (Future and present of cities and urbanisation in Brazil); Part 2 (Similarities and differences in the urbanisation process in Southern Brazil); and Part 3 (Trajectories of urban geography in Brazil), which bring together the discussions held in six round tables. Roberto Lobato Corrêa was honoured at the event.

The next symposium, the 11th, took place at the University of Brasília in September 2009, and its results were published in three issues of the Cidades journal, edited by Maria Encarnação Beltrão Sposito. Issue 10, "The city and the urban: a conceptual pursuit", included the publication of the lecture by Ana Fani Alessandri Carlos entitled "The illusion of the transparency of space and blind faith in urban planning: the challenges of a critical urban geography", as well as the texts presented at the round table "Current issues regarding the city and the urban: limits and possibilities in urban geography". Issue 11, "Spatial forms and urban policy(ies)", includes texts from the round tables entitled "Urban policy(ies): state versus social movements" and "New spatial forms and new urban roles". In issue 12, "Urban Brazil: challenges and agendas", the texts of the round tables entitled "Urban Brazil: challenges and agendas" and "The dynamics of administrative fragmentation of territory and urban land use laws: contradictions and conflicts" were published. Aldo Paviani was honoured at the event.

The 12th Symposium took place in Belo Horizonte, in November 2011, at the Federal University of Minas Gerais, with the title "Science and Utopia: for a geography of the possible". The opening table featured a presentation by Sérgio Martins. The topics covered were as follows: Avatars from the world of merchandise; Urban operations aimed at mega sporting events; Right to the city as a utopia or just as "improving the city"; Reproduction of real estate capital in capitalist urbanisation; Urban and metropolitan restructuring of a reformist and circumstantial nature; and Culture and environmental issues in the city. The closing lecture was given by Amélia Damiani.

The 13th Symposium was held at the Rio de Janeiro State University in November 2013. It had 871 participants and resulted in the publication in 2014 of the book Geografia urbana: ciência e ação política, with 404 pages, authored by Floriano Godinho de Oliveira et al. The book, divided into five parts, begins with the publication of the lecture by Arlete Moysés Rodrigues "Science and political action: a critical approach", followed by the closing lecture held by British geographer David Harvey. Part II, "The production of the urban: contemporary trends, conflicts and resistance in the urban space", has five texts; Part III, "Urban economy, social inequalities and public policies" (also five texts); Part IV, "Information and communication in political and cultural practices in the production of the city" (seven texts), and Part V with the talk "Reflections on Geography and the Urban: a decalogue", by the honouree at the event Pedro Geiger.

The 14th Symposium was again held in Fortaleza, at the Federal University of Ceará in September 2015, with the title of "Perspectives and Approaches to Urban Geography in the XXI Century". The event began with an interview with two research group coordinators, Maria Encarnação Beltrão Sposito and Luiz Cesar de Queiroz Ribeiro. 12 texts presented at the event's round tables were published in the special issue (Vol. 14, No. 4) of the bilingual journal Mercator, edited by EustógioWanderley Correia Dantas, also in 2015. Jan Bitoun, José Borzacchiello da Silva, and Pedro de Almeida Vasconcelos were honoured at the event.

Salvador returned to host the 15th Symposium at the Federal University of Bahia in November 2017, which resulted in the publication in 2018 of the book, authored by Angelo Serpa and Ana Fani Alessandri Carlos, entitled Geografia urbana: desafios teóricos contemporâneos, with 511 pages. The book is divided into seven parts, with the first six corresponding to the round tables: Thinking about the city and the urban today: interdisciplinary dialogue in the field of human and social sciences; The future of the city: urban policy or utopian project?; Production of the urban and the city in the era of financialisation; Specificities of the city and the urban in Brazil; The spatiotemporal scales of the urban; and Which theories to understand the city and the urban in this century? (Contribution of geography) and a special table: Theoretical contributions to urban research: Milton Santos, Maurício Abreu, Neil Smith, and Edward Soja.

The 16th Symposium was held in Vitória, at the Federal University of Espírito Santo, in November 2019. The results of the event were published in 2020 in the book Geografia Urbana: cidades, revoluções e injustiças: entre espaços privados, públicos, direito à cidade e comuns urbanos, authored by Ana Maria L. de Barros, Claudio L. Zanotelli and Vivian Albani. The seven parts of the book correspond to the event's seven round tables: Cities and revolutions. Public spaces, urban common areas and social networks: interdisciplinary dialogue in the field of urban and social studies; Practices, utopias, dystopias and heterotopias in and of cities in the global North and South: what similarities, what differences?; Scales, cultures and decoloniality: new perspectives in urban geography; Fragmentation, segregation, violence and urbicide in Brazilian cities; Production of space, the city and the urban and the new international financial order: unity of social and environmental issues; Urban and regional economy and migrations; and Urban Geopolitics in Contemporary Latin America: dispute of the common areas.

Lastly, the XVII National Symposium on Urban Geography, entitled "The production of the urban and the urgency of the transforming praxis: theories, practices and utopias in the midst of a convulsed world", which was scheduled to be held in October of this year at the Federal University of Paraná, in Curitiba, was postponed to 2022 due to the current COVID-19 Pandemic.

The event titles themselves show us the dominant trends in each period: the first event, in 1989, in São Paulo, dealt with "Paths of reflection" on both the "city" and the "urban", indicating a significant theoretical concern, while the 2019 event, "Cities, revolutions and injustices", indicates a strong political concern. With regard to the titles of the round tables, since the book of the first event was not authored by tables, in the second event, in 1991, in Rio Claro, we found tables dealing with "Social Movements", "Daily Life" and "Urban Environment", while in the event in the city of Vitória in 2019, new topics appear such as "deconiality", "urbicide" and "common areas", confirming the topics that have come to dominate the debate in Brazilian urban geography.

7.6 The Cidades Journal

The Cidades journal was founded by the Urban Studies Group (GEU), which initially comprised geographers Ana Fani Alessandri Carlos, Jan Bitoun, Maria Encarnaçâo Spósito (1st editor), Mauricio de Almeida Abreu, Pedro de Almeida Vasconcelos, Roberto Lobato Corrêa and Silvana Maria Pintaudi (2nd editor). Issue 1 dates from 2004. Up to issue 15 (2012) the journal was published on paper. Then, it began to be published only virtually. The last published issue was issue 22, dated 2016.

The topical issues from 2007 onwards which inform of the topics of interest in the period should be highlighted: Differentiation and inequalities (No. 6, 2007); Images of the city (No. 7, 2008); The Urbanisation of society (No. 8, 2008); Social activism and urban space (No. 9, 2009); The city and the urban: a conceptual pursuit (No. 10, 2009); Spatial forms and urban policy(ies) (No. 11, 2010); Urban Brazil: challenges and agendas (No. 12, 2010); The city and the party (No. 13, 2011)17; Mauricio de Almeida Abreu (No. 14, 2011)18; Libertarian thought and praxis and the City (No. 15, 2012); The Production of Urban Space. The Meaning of Public Policies (No. 16, 2012); Microterritorialities in cities (No. 17, 2013); Trade and consumption in the contemporary city (No. 18, 2013); Extreme processes in the formation of the city [from crisis to the emergence of contemporary spaces] (No. 19, 2014); Readings on the City (No. 20, 2015); Diffuse Urbanisation and the Disperse City (No. 21, 2015) and Justice and Law: a debate on the urban (No. 22, 2016). The topics also begin dealing with issues of differentiation and inequalities, and end in the last issue, with issues of Justice and Law, which have a certain affinity.

7.7 Conclusions

Brazilian urban geography, in addition to its importance in relation to other subdisciplines, has a long tradition in Brazil, with emphasis being placed on the pioneering text by Bernardino José de Souza on the topic in 1913, greatly anticipating the production of the first professional geographers in the 1930s. Individually, we can highlight the production, in the initial period, of Milton Santos, the Brazilian geographer with the greatest international impact, and, in the current period, the extensive production of the geographer Ana Fani A. Carlos.

Both the National Symposia on Urban Geography—the Simpurbs over the last 30 years, and the Cidades journal—have allowed us to follow the topics highlighted over time in the subdiscipline. The most recurrent topics in the Simpurbs' round tables were as follows: issues of time and space (nine tables), followed by environmental issues (seven tables), urban policies and management and theoretical and methodological issues (six tables each). The openness to international issues in the last event in 2017 can also be highlighted, such as the discussion on cities in the North and South, as well as those in Latin America and the debate on decoloniality. In a shorter period (2004/2016), the Cidades journal brought specific topics not highlighted in the Symposia, such as the city and the party, microterritorialities, commerce and consumption, and diffuse urbanisation. Lastly, we detected topics that have been relatively under-addressed in the events and beforehand, such as the urban economy and migrations, which were re-addressed in the last event of 2019.

As a whole, geographers initially studied the main Brazilian cities empirically and then moved on to an eminently theoretical discussion, on very diverse topics, and there is, therefore, no hegemonic thinking in current times.

References

Abreu MA (1994) O estudo geográfico da cidade no Brasil: evolução e avaliação. In: Carlos AFA (ed) Os Caminhos da Reflexão sobre a Cidade e o Urbano. Edusp, São Paulo, pp 199–322

Dantas EWC, Silva JB (2018) A geografia urbana brasileira: de uma análise diacrônica às obras mais citadas no último decênio. Confins 38. http://journals.openedition.org/confins/16331. https://doi.org/10.4000/confins.16331. Accessed on April 2021

Souza EAL (2012) Ciência, Cidade e Utopia: XII Simpósio Nacional de Geografia Urbana (SIMPURB), Belo Horizonte, 16 to 19 November 2011, Geousp, 32, 222–224

Valladares LP, Sant'Anna MJC, Cailleux AML (1991) 1001 teses sobre o Brasil Urbano: catálogo bibliográfico (1940–1989). Rio de Janeiro: ANPUR, IUPERJ

Vasconcelos PA (1999, 2012) Dois séculos de pensamento sobre a cidade. Salvador: Ilhéus: Editus; Salvador: Edufba, Ilhéus: Editus

Cited Texts

Abreu MA (1987) Evolução Urbana do Rio de Janeiro. IPLANRIO/Jorge Zahar, Rio de Janeiro

Abreu MA (ed) (1992) Natureza e Sociedade no Rio de Janeiro. S.M.C.T.E, Rio de Janeiro

Abreu MA (ed) (2005) Rio de Janeiro: formas, movimentos, representações. Da Fonseca Comunicação, Rio de Janeiro

Abreu MA (2010) Geografia Histórica do Rio de Janeiro (1502–1700). Andrea Jakobson; Prefeitura Municipal do Rio de Janeiro, Rio de Janeiro, 2 v

Ab'Saber AN (1952) A cidade de Salvador: fotografias e comentários. Boletim Paulista de Geografia 11:61–78

Ab'Saber AN (1961) Sítio da Cidade do Salvador In: Cidade do Salvador, ed. Milton Santos. Salvador: Imprensa Oficial, pp 11–12

Andrade MC (1979) Recife: problemática de uma Metrópole em uma Região Subdesenvolvida. E.U.; UFPE, Recife

Azevedo AA (1951) Salvador et le Reconcavo de Bahia (Brésil du Nord-Est), Les Cahiers d'Outre-Mer 15:189–203

Azevedo AA (1961) Fisionomia da cidade do Salvador. In: Cidade do Salvador, ed. Milton Santos. Salvador: Imprensa Oficial, pp 5–54.

Azevedo AA (ed) (1958) A cidade de São Paulo: estudos de geografia urbana. Nacional, São Paulo, 4 v

Barros AML, Zanotelli CL, Albani V (ed) (2020) Geografia Urbana: cidades, revoluções e injustiças: entre espaços privados, públicos, direito à cidade e comuns urbanos. Consequência, Rio de Janeiro

Baumgartner WH (2019) A geografia das cidades em Bernardino José de Souza: influência alemã e conhecimento local nas monografias descritivas do Estado da Bahia. In: André N, Caroline BNV (eds) A Geografia no alvorecer da República: contribuições à história da ciência geográfica no Brasil. Edufba, Salvador, pp 93–112

Bernardes LMC (1957–1958) Importância da posição como fator do desenvolvimento do Rio de Janeiro. Anais da AGB 11(1):175–196

Bitoun J (ed) (2009) Tipologia das cidades brasileiras e políticas territoriais. Recife: FASE/Observatório das Metrópoles

Carlos AFA (1994) A (Re) Produção do Espaço Urbano. Edusp, São Paulo

Carlos AFA (ed) (1994) Os Caminhos da Reflexão sobre a Cidade e o Urbano. Edusp, São Paulo

Carlos AFA (2001) Espaço-Tempo na Metrópole. Contexto, São Paulo

Carlos AFA, Lemos AIG (ed) (2003) Dilemas urbanos: novas abordagens sobre a cidade. Contexto, São Paulo

Carlos AFA (2004) O Espaço Urbano: novos escritos sobre a cidade. Contexto, São Paulo

Carlos AFA, Oliveira AU (ed) (2004) Geografias de São Paulo: representação e crise da metrópole. Contexto, São Paulo, 2 v
Carlos AFA, Carreras C (ed) (2005) Urbanização e mundialização: estudos sobre a metrópole. Contexto, São Paulo
Carlos AFA, Oliveira AU (ed) (2006) Geografias das metrópoles. Contexto, São Paulo
Carlos AFA, Souza ML, Sposito MEB (ed) (2011) A produção do espaço urbano: agentes, processos, escalas e desafios. Contexto, São Paulo
Carlos AFA (ed) (2015) Crise Urbana. Contexto, São Paulo
Carlos AFA, Volochko D, Alvarez IP (ed) (2015) A cidade como negócio. Contexto, São Paulo
Carlos AFA, Alves G, Padua RF (ed) (2017) Justiça espacial e o direito à cidade. Contexto, São Paulo
Carlos AFA, Santos CS, Alvarez IP (ed) (2018) Geografia Urbana Crítica. Teoria e Método. Contexto, São Paulo
Carvalho IM et al (ed) (2014) Metrópoles na atualidade brasileira: transformações, tensões e desafios na Região Metropolitana de Salvador. Salvador: Edufba19
Castro J (1969) [1948] A cidade de Recife: ensaio de geografia urbana. In: Ensaios de geografia humana. Brasiliense, São Paulo, pp 159–236
Cavalcanti LS (ed) (2001) Geografia da Cidade: a produção do espaço urbano de Goiânia. Editora Alternativa, Goiânia
Cordeiro HK (1980) O Centro da Metrópole Paulistana: expansão recente. São Paulo USP/IG
Corrêa RL (2006) Estudos sobre a Rede Urbana. Bertrand Brasil, Rio de Janeiro
Costa HSM et al (2006) Novas Periferias Metropolitanas. Belo Horizonte: Editora C/Arte
Dantas EW (2002, 2011) Mar à Vista: estudo de maritimidade em Fortaleza. Fortaleza: Museu do Ceará/SECUL; Fortaleza: Edições UFC
Delgado de Carvalho CM (1954) A Cidade de São Paulo. In: Aspectos Geográficos da Terra Bandeirante, ed. Conselho Nacional de Geografia. IBGE, Rio de Janeiro, pp 289–295
Duarte AC (1967) A Área Central da Cidade do Rio de Janeiro. In: A Área Central da Cidade do Rio de Janeiro. IBGE, Rio de Janeiro, pp 11–42
Faissol S (ed) (1978) Tendências atuais na geografia urbano-regional: teorização e quantificação. IBGE, Rio de Janeiro
Ferreira AHS (2013) A Cidade no Século XXI. Consequência, Rio de Janeiro
Ferreira AHS, Rua, João, Mattos RC (ed) (2017) Espaço e metropolização. Cotidiano e ação. Consequência, Rio de Janeiro
Ferreira AHS, Rua, João, Mattos RC (ed) (2019a) Metropolização do Espaço. Gestão territorial e relações urbano-rurais. Consequência, Rio de Janeiro
Ferreira AHS, Rua, João, Mattos RC (ed) (2019b) Desafios da metropolização do espaço. Consequência, Rio de Janeiro
França VLA (1999) Aracaju: estado & metropolização. São Cristóvão: Editora UFS; Aracaju: Fundação Oviêdo Teixeira
Geiger PP (1963) Evolução da Rede Urbana Brasileira. Centro Brasileiro de Pesquisas Educacionais, Rio de Janeiro
Gomes ET (2007) A. Recortes de paisagens na cidade do Recife. Recife: FJN; Editora Massangana
Gomes PCCG (2002) A Condição Urbana. Bertrand Brasil, Rio de Janeiro
Henrique W (2015) O Direito à Natureza na Cidade. Salvador: SciELO; Edufba
Langenbuch JR (1971) A estruturação da grande São Paulo: estudo de geografia urbana. IBGE, Rio de Janeiro
Lencioni S (2017) Metrópole, metropolização e regionalização. Consequência, Rio de Janeiro
Melo ML (1978) Metropolização e Subdesenvolvimento: O caso de Recife. Recife: UFPE
Müller NL (1960) O Fato Urbano na Bacia do Rio Paraíba. IBGE, Rio de Janeiro
Oliveira FJG et al (eds) (2014) Geografia urbana: ciência e ação política. Consequência, Rio de Janeiro
Oliveira JA (2000) Cidades na Selva. Editora Valer, Manaus

Oliveira JA (2003) Manaus de 1920–1967. A cidade doce e dura em excesso. Manaus: Editora Valer; Governo do Estado do Amazonas; Edufa
Oliveira JA (ed) (2009, 2010) Cidades brasileiras: territorialidades, sustentabilidade e demandas sociais. Edua, Manaus, 2 v
Ortigoza SA, Guarnieiri (2010) Paisagens do consumo: São Paulo, Lisboa, Dubai e Seul. SciELO; Editora UNESP, São Paulo
Paviani A (ed) (1985, 2010) Brasília, Ideologia e Realidade: Espaço Urbano em Questão. Editora UnB, Brasília
Paviani A (ed) (1987) Urbanização e Metropolização—A gestão dos conflitos em Brasília. Editora UnB, Brasília
Paviani A (1989, 2010) Brasília: a Metrópole em Crise—Ensaios sobre Urbanização. Editora UnB, Brasília
Paviani A (ed) (1991, 1998, 2010) A Conquista da Cidade: Movimentos Populares em Brasília. Editora UnB, Brasília
Paviani A (ed) (1996) Brasília: Moradia e Exclusão. Editora UnB, Brasília
Paviani A (ed) (1999, 2008) Brasília—Gestão Urbana: Conflitos e Cidadania. Editora UnB, Brasília
Paviani A, Gouvêa LAC (ed) (2003) Brasília: Controvérsias Ambientais. Editora UnB, Brasília
Paviani A et al (ed) (2005, 2010) Brasília: Dimensões da Violência Urbana. Editora UnB, Brasília
Paviani A et al (ed) (2010) Brasília 50 anos: da Capital a Metrópole. Editora UnB, Brasília
Paviani A, Gonzales SFN, Francisconi JG (2013) Planejamento e Urbanismo na atualidade brasileira: objeto, teoria, prática. Livre Expressão, São Paulo
Peluso VA (1991) Estudos de Geografia Urbana de Santa Catarina. UFSC, Florianópolis
Penteado A (1968) Belém. Estudo de Geografia Urbana. Belém, UFPA
Pereira EM, Dias LCD (eds) (2011) As cidades e a urbanização no Brasil: passado, presente e futuro. Insular, Florianópolis
Petrone P (1963) Pinheiros: aspectos geográficos de um bairro paulistano. Edusp, São Paulo
Prado C (1983, 1935) O Fator Geográfico na Formação e no Desenvolvimento da Cidade de São Paulo. In: A cidade de São Paulo: geografia e história. Brasiliense, São Paulo, pp 7–37
Ribeiro NMG (1989) Transformações do Espaço Urbano: o Caso de Aracaju. Recife: FUNDAJ; Ed. Massangana
Rodrigues AM (1998) Produção e consumo do e no espaço: Problemática ambiental urbana. Hucitec, São Paulo
Rodrigues AM (ed) (2011) Estudos Urbanos. Editora Unicamp, Campinas
Santos J (2014) A cidade poli(multi)nucleada: a reestruturação do espaço urbano de Salvador. Edufba, Salvador
Santos M (1959) O Centro da Cidade de Salvador. Livraria Progresso, Salvador
Santos M (ed) (1961) Cidade do Salvador. Imprensa Oficial, Salvador
Santos M (1965) A Cidade nos Países Subdesenvolvidos. Civilização Brasileira, Rio de Janeiro
Santos M (1969) Aspects de la géographie et de l'économie urbaine des pays sous-développés. Centre de Documentation Universitaire, Paris
Santos M (1970) Dix essais sur les villes des pays sous-développés. Paris: Ed. Orphrys
Santos M (1971) Les villes du Tiers Monde. Editions Génin, Libraries Techniques, Paris
Santos M (1975) L'Espace partagé. Les deux circuits de l'économie urbaine des pays sous-développés. LibrariesTechniques, Paris
Santos M (1978) Pobreza Urbana. Hucitec, São Paulo
Santos M (1980) A urbanização desigual. Vozes, Petrópolis
Santos M (1981) Manual de Geografia Urbana. Hucitec, São Paulo
Santos M (1982) Ensaios sobre a urbanização latino-americana. Hucitec, São Paulo
Santos M (1990) Metrópole corporativa fragmentada: o caso de São Paulo. Nobel, São Paulo
Santos M (1993) A urbanização brasileira. Hucitec, São Paulo
Santos M (1994) Por uma economia política da cidade. São Paulo: Hucitec; Editora PUC-SP
Serpa A (2007a) O espaço público na cidade contemporânea. Contexto, São Paulo
Serpa A (2007b) Cidade Popular—Trama de relações sócio-espaciais. Edufba, Salvador

Serpa A, Carlos AFA (eds) (2018) Geografia urbana: desafios teóricos contemporâneos. Edufba, Salvador
Silva JB (1992) Os incomodados não se retiram. Fortaleza em questão. Multigraf Editora, Fortaleza
Silva JB, Costa MCL, Dantas EWC (eds) (1997) A Cidade e o urbano: temas para debates. Eufc, Fortaleza
Silva SBM, Carvalho IM, Pereira GC (eds) (2016) Transformações metropolitanas no século XXI. Bahia, Brasil e América Latina. Edufba, Salvador
Silva WR, Sposito MEB (eds) (2017) Perspectivas da Urbanização. Reestruturação urbana e das cidades. Consequência, Rio de Janeiro
Souza AN, Vaz CBN (eds) (2019) A Geografia no alvorecer da República: contribuições à história da ciência geográfica no Brasil. Edufba, Salvador
Souza BJ (1913) Por mares e terras (leituras geográficas). Romualdo dos Santos Livreiro Editor, Salvador
Souza ML (2000) O Desafio Metropolitano: um estudo sobre a problemática sócio-espacial nas metrópoles brasileiras. Bertrand Brasil, Rio de Janeiro
Souza ML (2002) Mudar a Cidade: uma introdução crítica ao planejamento e à gestão urbanos. Bertrand Brasil, Rio de Janeiro
Souza ML (2003) ABC do desenvolvimento urbano. Bertrand Brasil, Rio de Janeiro
Souza ML (2006) A Prisão e a Ágora: reflexões em torno da democratização do planejamento e gestão das cidades. Bertrand Brasil, Rio de Janeiro
Souza ML (2008) Fobópole. O medo generalizado e a militarização da questão urbana, Bertrand Brasil, Rio de Janeiro
Souza MAA (1988) Governo Urbano. Nobel, São Paulo
Souza MAA (1989) A Identidade da Metrópole. A verticalização de São Paulo. Hucitec; Edusp, São Paulo
Souza MAA (1998) São Paulo. Ville mondiale et urbanisme français sous les tropiques. L'Harmattan, Paris
Sposito Eliseu SS, Sposito MEBS, Sobarzo O (2006) Cidades médias. Expressão Popular, São Paulo
Sposito ES (2008) Redes e cidades. Editora UNESP, São Paulo
Sposito E, Silva PFJ (2012) Cidades pequenas: perspectivas teóricas e transformações socioespaciais. Paco Editorial, Jundiaí
Sposito MEB (ed) (2001) Urbanização e cidades: perspectivas geográficas. Presidente Prudente: [s. n.]
Sposito MEB, Whitacker AM (eds) (2006) Cidade e Campo: relações e contradições entre urbano e rural. Expressão Popular, São Paulo
Sposito MEB (2007) Cidades Medias—Espaços em transição. Expressão Popular, São Paulo
Sposito MEB, Gomes EM (2013) Espaços fechados e cidades. Editora UNESP, São Paulo
Trindade, Saint-Clair C (1997) Produção do Espaço e Uso do Solo Urbano de Belém. UFPA/Naea/Plades, Belém
Trindade, Saint-Clair C, Tavares MGC (2008) Cidades ribeirinhas na Amazônia. Edufpa, Belém
Trindade, Saint-Clair C, Santos V (eds) (2019) O urbano e o metropolitano em Belém: (re) configurações socioespaciais e estratégias de planejamento e gestão. Consequência, Rio de Janeiro
Valença MM (2008) Cidade (I)Legal. MAUAD, Rio de Janeiro
Vasconcelos PA, Silva SBM (eds) (1999) Novos estudos de geografia urbana brasileira. Edufba, Salvador
Vasconcelos PA (2002, 2016) Salvador: transformações e permanências (1549–1999). Ilhéus: Editus; Salvador: Edufba
Vasconcelos PA (2005) Salvador de Bahia (Brésil): transformations et permanences (1959–2004). L'Harmattan, Paris
Vasconcelos PA, Corrêa RL, Pintaudi SM (eds) (2013) A cidade contemporânea: segregação espacial. Contexto, São Paulo

Vessentini JW (1986) A Capital da Geopolítica. Ática, São Paulo
Vieira SG (2005) A Cidade Fragmentada. Pelotas: Ed. Univ. Federal de Pelotas

Annals

Anais 3° Simpósio Nacional de Geografia Urbana. Rio de Janeiro, 13–17 September 1993
Anais/VIII Simpósio Nacional de Geografia Urbana. Recife: Editora Universitária da UFPE (2004)

Journals

Boletim de Geografia Teorética, vol 21, no 42. Rio Claro—SP (1991)
Cidades: Revista Científica/Grupo de Estudos Urbanos, vol 7, no 10, Presidente Prudente—SP (2009)
Cidades: Revista Científica/Grupo de Estudos Urbanos, vol 7, no 11, Presidente Prudente—SP (2010a)
Cidades: Revista Científica/Grupo de Estudos Urbanos, vol 7, no 12, Presidente Prudente—SP (2010b)
Mercator, Revista de Geografia da Universidade Federal do Ceará (UFC), vol 14, no 4. Fortaleza, 2015

Pedros Almeida Vasconcelos is Emerit Full Professor of UFBA and Catholic University of Salvador de Bahia.

Chapter 8
The Production of Urban Space and "Critical Geography"

Ana Fani Alessandri Carlos

Abstract Knowledge is part of the process of reproducing reality, a need to capture its *novel* aspects. I am referring to the profound transformations of reality since the 1990s, noting that, tending towards the urban, they take place in a global space, under the aegis of financial capital. These metamorphoses need to be explained in their contradictions which can be interpreted, in this century, through the urban conflicts exposed by demonstrations in public spaces in major cities, demanding an effort to interpret them. This is based on a certain theory of reality that assumes an open and contradictory social totality. This revelation, which is based on the dialectical method, enables the elaboration of a transition from Geography towards a *metageography* which, going beyond the limits imposed by a specialisation, accentuates the need to address the foundations of explaining reality. This movement is performed through the centrality of the concept of a "social production of space", focussing on how the spatiality of social relationships takes places in the modern world. From this perspective, we define space as a "condition, medium and product of social reproduction", within a critical-radical geography about urban space in the present day.

Keywords Metageography · Social production of space · Urban space

In Brazil, the construction of geographic thought has occurred from various theoretical perspectives, indicating that there is not only one path, or a single way of producing geographic knowledge about the world, particularly in understanding the Brazilian reality in its historical determinations. One possible way to understand the world and explain reality, with its contradictions, is presented by the critical view in Geography. What is currently called Critical (Urban) Geography was created at the end of the 1970s, marked by a crisis in Geography. At that time, the response to superseding the critical moment was the need to build a Geography committed to society (Lacoste 1988), based on a profound critique of "the ways of thinking about

A. F. A. Carlos (✉)
Universidade de São Paulo, São Paulo, Brazil
e-mail: anafanic@usp.br

R. C. Lois González and M. A. Mitidiero Junior (eds.), *Brazilian Geography*, Advances in Geographical and Environmental Sciences,
https://doi.org/10.1007/978-981-19-3704-0_8

and doing Geography", anchored in positivism, with a new theoretical approach and method (Rodrigues 1998). This dispute was characterised as a breaking point.

Historical materialism emerged as fertile ground for this change. This approach, however, called into question the status of the discipline as a field that falls within the scope of Human Sciences. Born from the division of labour of investigations about the world, which tends to limit the field of explanations to problems, the state of the discipline was fragmented. In an attempt to resolve this issue towards understanding reality as a social whole, interdisciplinarity emerged on one side, while on the other, the dialectical method emerged as a way of addressing the whole, signalling the direction of this research. In the critical phase, new questions were presented in the debate and needed to be confronted. Geography was committed to building an understanding of the profoundly unequal Brazilian reality. Its starting point was space and Geography was a science of space with the argument that there is no such thing as an a-spatial society, as suggested by Milton Santos. This assertion challenged geographers to consider society in its relation with space.

This critical moment can be understood as an investigation of the world, the impasses experienced, the existence of inequality, and an observation of the underlying conflict in society as a whole. If historical materialism created the foundations to locate these processes in the heart of the course of history, dialectical materialism structured the knowledge base as an adaptation of thought and man, including the dimension of social practice as a component of what is real. Understanding space from the perspective of the movement of things and things in movement led to an understanding of the contradictory nature of the relations that produce space (Carlos 2011a, b).

Geography began to discuss the nature of space and explain the process of spatial production based on the production–reproduction of life. However, there was no consensus among the various perspectives of what the readings of Marx provided for the research outlook. Despite this, the understanding of praxis, such as that of the subject in real relationships, in the dialectical method, connecting theory and practice, has enabled us to consider the movement and the moment of action that transforms reality and gives it direction. It entails a critique of theory and practice. The outlook was built upon a critique of so-called traditional Geography. Geography needed to consider the world and be located within it. Theoretical concerns emerged in full force and, by characterising this period and the production process of geographical knowledge, new horizons opened up. The epistemological debate thus became central (Oliveira 1982; Correia 1980).

As a consequence, new categories of analysis began to enter the debate and components of this concept of space became central. As such, producing an intelligibility of the world indicated a process and act of learning, based on the criticism of a Geography separate from social problems. The first challenge was to surpass the level of phenomenon, and to understand reality as a dialectical connection between form/essence.

A word of Greek origin, critique means the capacity to judge, to make a rational examination without prejudice, engaging with theoretical productions as a stance imparted by philosophy. This construction lies in superseding categories of analysis as a necessary moment of understanding the world in movement. Based on the critique of the discipline and given the new direction in relation to theory (materialist), as well as method (dialectic), Geography recreated itself. The Marxist perspective, which has enabled us to consider the world to be a practice (a process of transforming itself in its uninterrupted movement) and the subject to be a society realised in a changing reality, has led geographers to critically review the concepts that constitute Geography as a discipline.

On the level of practice, it would require an understanding of the contradictions that lead to the alienation of man, finding the breaking points towards the "other thing", which meant assuming the fact that knowledge about the world in movement contemplates utopia. On the level of building a so-called critical Geography, the path of this construction started by reconsidering the man–nature relation indicated in the works of Marx. From the relation between man (a generic being) and nature (offered to man) as a starting point for geography, we are faced with the possibility of considering space as a product that emerged in the history of the civilisation process, of the capacity of man to transform the conditions found in nature and, based on it (and without leaving it), produce a social world (Carlos 2011a, b). At the same time, we can claim that in this relation, man produces himself.

Therefore, this critical moment led to building the foundations of the shift from the concept of the "organisation of space" (Andrade 1993) to that of the "production of space". In this sense, we can consider space in its real and concrete materiality as a world, an objectivity that reorganises/highlights the producing subjects of spatial reality and its meaning, and in its contradictions. The production of space is thus immanent to the production of man (Carlos 1979; Rodrigues 1998). In this movement, the producing subject of the built world is society and the content of produced space is the social. Social relations are realised concretely in space, producing a space. In other words, man becomes human, working in the world, assuming the role of a producing subject. Through the process of labour, society transformed nature into a social world. Therefore, it is an active condition, an action that includes a strategy in which an end supersedes supporting life itself, while it is open to the production of human history. Geography was no longer the discipline that describes and distributes activities by human groups over the earth's surface. As a result of the analysis by Marx, this path requires superseding the following:

Superseding the idea of geographic space as a stage for human action, reduced to the status of a physical framework where one can observe the distribution of activities by men/human groups, towards considering space as a social production and a moment immanent to the production of life;

Superseding the idea of human groups moving across space towards establishing a society, divided into classes, in its position as producing subject of space, the world and itself.

This critique produced an inventory of reformulated or superseded concepts, leading to a profound understanding of reality. This was a moment in building what was called "Critical Geography" (Moreira 2000; Silva 1983; Oliveira 1982).

It is important to recall from this time the reflection on space as an object of study, which in many cases indicated a way of understanding the man–nature relation as a society/space dialectical relation (one term is realised in and through the other) and the mediations between them (social labour), without, however, disregarding the importance of the location/distribution of the phenomena analysed. "*Produzir* (to produce) and *producir* (to produce) are two indissociable acts"[1], wrote Milton Santos (1978, p. 163).

The idea to consider society enabled us to not only question the path to understanding space as an absolute objectivity (Correia 1980), resulting from its consideration as a stage for human activity, but also the empiricist view that this objectivity guaranteed to Geography.

Over the course of its development, this new way of thinking about the geographical investigation of the world enabled us to think of space as a commodity (Carlos and Lencioni 1982), as a consequence of understanding it as a social production under capitalist logic, to the extent that one understands that each mode of production produces a space according to its rationale. As such, the focus on the social construction of space revealed the commercial side of spatial production and the production of a commodity space and, as a consequence, changed the focus of the analysis from spatial distinctions based on the distribution of activities to considering them a class issue as well. Along these lines, it also enabled the construction of an understanding of capitalist space as an alienated social production. In other words, the dialectic of the acquisitions of this moment in the production of geographical thinking enabled the theoretical leap of radically changing the terms of the debate. Highlighting the underlying conflicts of a society founded on inequality (a society ranked into classes), Geography, which was being outlined, directed the perspective towards the construction of a project of transforming the world as a moment of transforming space (Soja 1993).

Geography, as a way to think about reality, transforms with it and calls into question the idea of an absolute truth. The temporariness of these concepts emerges from the social foundation of reality, and, therefore, there are no eternal truths, but constantly changing thoughts that always demand new parameters to challenge old paradigms. This pattern indicates the relativisation of knowledge itself and questions the categories of analysis used by Geography, as well as its theoretical–methodological procedures, exposing the fragility of past arguments to explain the contemporary world. The real transformation of society and space demands a transformation of Geography in movement as well.

8.1 From Geography to Metageography

"Critical geography", which rests on the foundations of historical materialism, gradually gave rise to a diversified Geography, from the theoretical as well as methodological points of view, starting at the end of the 1980s. I can claim that today critical thinking, which was hegemonic in the 1970s, is residual today in Geography. The

[1] "Produzir e producir o espaço são dois atos indissociáveis."

speed imposed by the new productivist orientation relies on a competitiveness that undermines theoretical thought, given the demand for short-term results, which is certainly not specific to Geography.

The open perspective in the 1970s, which questioned the capitalist logic that guides the continued production of the world, lost its strength. The challenge of Geography, today, is dual in nature: (a) to supersede the divisions resulting from the intensification of its specialisations, the commitments to models of interpretation of the world and to the fragmentation of the world, constituting the specific fields of investigation; and (b) to escape the trap facing the social crisis, searching for quick solutions through public policies within the scope of Institutions and the State, without questioning their logic.

However, critical thinking still continues today[2], guiding investigations[3], research and education, proposing new interpretations (and categories of analysis) of the world in which we live, with the social totality as a perspective towards building another project of the world, as a negation of capital. This is because the tendency towards thinking of the totality includes the negation and the critical moment, and assumes a deeper understanding of what is real in that which makes it explicit and guides it (Carlos 2018).

Building a rationale towards metageography assumes the need to understand the radical critique of today's reality, going to the roots and finding the foundations of inequality that sustain capitalism, though renewed (as private property, and the contractual relations derived from it, class society, and abstract work). Therefore, the idea of metageography does not mean building another geography, but building a line of thought within Geography, leading to the need to consider the role of geography in the modern world permeated by the techno-bureaucratic rationale, given the ever-increasing needs of the process of accumulation, which sustains the concentration of wealth (and power).

The path of this critique involves putting categories of analysis in order, in a connected way using the path of the dialectical method, reviewing components of the main concepts of Geography, in a higher system of thinking and superseding that is focussed on the investigation of urban practice (today, society is urban at its foundation) and globalised in a profound metamorphosis during a time that is compressed and requires an explanation. A Geography with critical input informing the investigation aims to understand the contemporary transformation and the restricted time

[2] I cite GESP here, a group I have coordinated for the past 20 years. GESP is a forum engaged in organizing knowledge with methodological foundations that support an analysis of socio-spatial reproduction, which seeks to supersede formal analytical procedures such as systems, index and models through geographical analysis. This "Critical Geography", understood as a commitment to analyzing urban reality in its contradictory movements and focusing the analysis on the contents that lay bare the inequality concretely lived, has the objective of building a project of "another city", given that each historical moment opens up an unlimited array of possibilities that poses the need for a reflection that explains our time (http://www.gesp.fflch.usp.br).

[3] In numerous works by Brazilian authors, such as Sandra Lencioni, Arlete Moisés Rodrigues, Ariovaldo Umbelino de Oliveira, and Alexandrina Luz Conceição, which are important names in critical geography today.

for change, diving into the possibilities and limitations of this transformation seen in places and in everyday life, as a socio-spatial–temporal practice. Volochko (2018, p. 69) draws attention to the fact that:

> the scale of everyday life refers to the scale of the place, of its conflicting production and appropriation by social subjects. The socio-spatial practice revealed by everyday life harbours the dimensions of political and economic controls, among many others: subjective, cultural, symbolic, etc. Therefore, everyday life expresses chance, the ordinary, the residual, the irrational, as well as the wealth of possibilities for struggles and the transformation of reality[4].

Thus, assuming the indissociable union of theory–practice, this investigation dives into urban social practice, revealing the level of experiences in its distinctions and differences, finding that which is at its foundation. Understanding urban reality through the central notion of the "production of space" results from the assumption that the production of space is immanent to the production of life and, therefore, social relations would require the connection of space and time, given that they are necessarily realised in a determined space requiring a determined duration.

The construction of metageography with the central notion of the production of space, thus characterising urban practice as socio-spatial–temporal, means placing this production within the social whole of other productions which, under the aegis of capitalism, is located in the movement of valuation, which is the foundation of accumulation. The metamorphoses of space reveal the process of capitalist accumulation that is realised today, preferably, through the reproduction of metropolitan space mediated by public policies, revealing a fragmented social praxis in the form of segregation (Carlos 2013).

In capitalism, the production of space is based on and, in this process, replaces that of valuation, which is realised through the world of commodities. "Space and everyday life are taken as spheres to be dominated, controlled, and produced strategically for accumulation"[5] (Pádua 2018, p. 37). The contradiction between the social production of space and its private appropriation experienced in the segregated city presents the potential to challenge it. A life of deprivation leads to criticism. Knowledge of what is possible, using the explanatory foundations of the conditions that produce inequality, manifests in a life of deprivation.

In a historical regression, the terms of this condition rest on the private ownership of wealth (and its concentration) and on society divided into classes, thus associating wealth and power. These topics have remained removed from the analysis of the world.

In this sense, this critical moment can be understood as one of examining the world, the impasses experienced, the existence of inequality, the observation of the

[4] Original text: "a escala da vida cotidiana remete à escala do lugar, da sua produção e apropriação conflituosas pelos sujeitos sociais. A prática sócioespacial revelada pela vida cotidiana abriga as dimensões dos controles políticos e econômicos e, também, muitas outras: subjetivas, culturais, simbólicas etc., portanto, a vida cotidiana expressa o acaso, o banal, o residual, o irracional, expressa os conflitos, a miséria, o conformismo, a alienação, assim como a riqueza das possibilidades de luta e de transformação da realidade."

[5] Original text: "O espaço e a vida cotidiana são tomados como esferas a serem dominadas, controladas, produzidas estrategicamente para a acumulação."

underlying conflict in society considered as a whole, and renewing the conditions under which they promote alienation in the modern world. Based on the observation of intensifying inequality (as well as conditions of alienated life), this critique leads to the possibility of building a project for transforming this reality, to the extent that utopia is part of the learning process.

8.2 Repercussions of This Investigation

From this perspective, the investigation leaves the field of epistemology in order to focus on understanding reality dialectically, by linking theory and practice. On the level of theory, we begin with the thesis according to which social relations are realised as spatial–temporal relations, highlighting the need to discover society–space relations, focussing on the social production of space. On the level of urban reality, we begin with the thesis according to which the accumulation of capital today is realised, primarily, through the reproduction of urban space.

8.2.1 The Level of Social Relations Producing Space[6]

Society is established by producing a world that is unique to man. In other words, man becomes human, acting in the world, assuming the role of a producing subject. Therefore, this condition is active, an action that is a strategy and a purpose that supersedes the maintenance of life itself to the extent that it is a production of its own history. Therefore, at the start of the investigation, an importance is placed on the meaning and content of the concept of production, in order to derive the terms of reproduction. The production of space as a moment of social reproduction reveals the entirety of this production.

The active role of a society that builds an objective world in a practice that can be characterised as socio-spatial indicates the reproduction of social relations as a space–time relation, highlighting the body as an expression of social relations. This is because human beings are related to space through their body. This is the mediation necessary according to which we are related to the world and others, a relation with space-times defined in everyday life, by using these places of life. Therefore, the body travels through different scales, connecting them. First, there is the place, the home (in which the subject starts to build family relationships and primary references); then,

[6] The concept of production has a broader meaning than that given by the economy, since it is linked to the production of man and his humanity based on the transformation of nature into a social world. In the *lato sensu*, this concept links the production-reproduction of social relations. In its *stricto sensu*, it links the creation of things and commodities. Therefore, this concept demonstrates the perspective of understanding a whole that is not restricted to the level of the economic, enabling an understanding of society in its broader movement, which assumes a whole. This understanding changes the terms of urban analysis.

the street, in which one is seen in relation to others, establishing identity networks; then, there is the scale of the neighbourhood, which takes shape as a relation of various space-times mediated by social exchanges (of all kinds); and, connecting these scales, the body comes to the city, multifaceted and multiple, with simultaneous actions and images that seduce and influence. It outlines a juxtaposition of moments in life and spatial scales (Simoni-Santos 2018).

In other words, space is considered a production/product of the action of society, immersed in the whole of social reproduction. Social relations, in turn, take on a concreteness and materiality in a space that is produced on a level that transcends the classical idea of a location for phenomena and human activities, which involves an analysis of the objective space–time conditions that reveal and define this activity as a social practice. This thought process provides a glimpse of the role of the production of space in the reproduction of contemporary society. Considering that the consequence of the production of space is its reproduction, we are faced with the need to think about the movement of history, which makes this new moment of society explicit.

Our investigation has led us to define the production of space as a *condition, means and product of the reproduction of society*, highlighting a process/movement establishing society itself as an urban society, identifying the urban issue as a spatial problem (Carlos 1994; 2011a, b). This triad is built in the movement of the method that focuses on the concept of social production of space as a result of the subject–object relation, in its theoretical requirements (i.e. reviewing the content of concepts and discovering new ones). Space as a condition/means and product of society suggests new components upon broadening thinking to include the social whole, discovering the subjects and object of the action that transforms nature into a social world.

Space *as a condition* involves and supersedes, without excluding, the idea of the pure materiality of space. If human activities are distributed in space involving a set of actions that are carried out in everyday life (in a space–time dialectic) through the use of places for this purpose, there is a subjective awareness of this process of creating identities. Practical reality leads to the idea of reference to human beings. This is because the material dimension of space involves the movement of history where the present action is reproduced in confrontation with the dead labour accumulated through the constant transformation of nature into human space. This means that if the starting point of the production of the world is nature, its reproduction no longer starts with nature, but with its transformation. As such, the continued production (in its dialectic with the discontinued production) of space involves the dialectical relation between cyclical time (the time of nature) and linear time (fabricated in the process of history). The metamorphoses and the continuity that include accumulation, in space, of time in history summarise the "new". Therefore, production–reproduction is simultaneous, but there is also a scale to be considered in the connection between space–time scales. Taken together, all these qualities emerge as different/unequal places and, among these same places in time, differences are marked on the space–time scales. This spatial dynamic is realised in a structure, a form. It takes on individualised functions depending on the time. In other words, the production of space is part of the production of the objective material conditions of the production of human history.

Consequently, space as a condition for the realisation of life involves a whole that evades the material in order to incorporate the universe of producing subjects in their relation with this materiality that is subject to historical determinations. This means that life develops according to the logic of this society that imposes on everyone access to places mediated by the market, whether buying a house, accessing the city centre, accessing leisure, enjoying public spaces, etc.

In capitalist society, the use of places, as a condition of life, is defined by the space and time of the commodity imposed on everyday life. As a condition for the realisation of capitalist accumulation in this century, space takes on the form and function of fixed capital. From the political point of view, space transforms into a territory upon which the State is not structured.

Space defined *as a means* is that which underscores the mediation of the action that produces life, revealing society "active and in action". The relation with nature is not direct and requires mediations such as labour, involving knowledge, technique, its division, its representations, the norms and laws that govern it, which are defined on the level of the State, through public policies of budget distribution and orientation. Among these places, and in these places, we find circulation to be a mediation. Life is established as a set of relations based on commodities and social exchange, creating a network of places through the spatial division of labour, imposing networks of circulation, distribution and exchange. The social relations that are realised as spatial relations are mediated by defined places, without which social exchanges would not exist; here, circulation is imposed. Social life is established according to a centre (a fixed point in space) that brings together actions and enables gatherings that, subjectively, have a meaning of representation created by collectivity. This gathering requires the mobility of members of the group.

From the point of view of the cycle of capital accumulation, space is primarily circulation; the realisation of profit depends on it. That is, if the process of labour produces profit, its realisation depends on the exchange/consumption of the commodity. The larger the space–time shift of the commodity in space—from factory to consumer—the greater the cost of circulation, decreasing the value of the capital. From the political point of view, the space-territory is a means of domination.

Space as a product (Soja1993; Pádua 2018; Lencioni 2017) takes on the form of human labour, summarising the set of social relations in its historical determinations. This material dimension also involves planes and levels that are relativised. That is, it is not merely a built space (architectures and forms), though it includes this dimension. Space is the product of the mode of production that, in its development under capitalism, is in the form of a commodity, a product of society, whose social relations are defined by the logic of the valuation process. In this sense, the space that results from this process includes the double meaning of commodity: it is an exchange value at the same time in which it is, dialectically, a use value.

Space produced according to capitalist logic presents a brutal conflict that is founded in the contradiction between space as use (space appropriated as a condition and means for the realisation of life) and space produced as an exchange value, subject to the process of valuation, as a condition of the realisation of accumulation. In this context, space is fragmented and sold in pieces on the market, producing the circuit of reproduction of commodities (and thus, realising the private ownership of land). From the point of view of capital, space is consumed productively as an ever-increasing possibility for realising the capital cycle. Specifically, space as a production of capitalist society follows the *logic of commodities* and is the focus of public investments. This process of reproduction—in the form of commodities—is realised by co-opting almost all levels of human reality, from access to housing to the production of the imagination based on the establishment of the consumer society. In this movement, the contradiction between the needs for increased reproduction of capital and those of society as a whole intensifies, in which the private ownership of wealth is a central element that defines the contractual relations that permeate everyday life, thus determining it. This is because capital, in its development, created the world of commodities with its language, signs and representations.

The production of space, as a commodity, is constitutive of the explanatory problem of the modern world. The segregated city (a product of the logic of valuation and the concentration of wealth, manifests in the centralisation of power and in the private ownership of land) is the more complete expression of the commodification of space under the logic of capital. Under capitalism, inequality is expressed in the concentration of wealth in the hands of a class that, over the course of its development, materialised in space, according to the terms and direction of its social production, which defines each one's place according to their access to housing and everything that this implies (from the infrastructure that defines its location to its relations with other places, including use of the street). This inequality also defines the strategies that underlie the act of planning space as a class strategy. In this process, space assumes various forms/modes as a condition/means and product of the process that aims for the broadened reproduction of capital, realising it as an incessant movement of valuation, as a realisation of the world of commodities.

This process of the production–reproduction of social space indicates a *central contradiction*. Space produced by the whole of society is, by nature, a social result, but its appropriation by members of this society is private. This conflict cannot be avoided and is revealed in struggles for the space of life.

8.2.2 The Accumulation of Capital as a Moment of the Reproduction of Urban Space

We can place the urban issue today in the connection between the urbanisation of society and the reproduction of capital in a globalised space that culminates with the reproduction of urban space itself as a condition (and product) of the realisation of accumulation under the hegemony of finance capital. In this space, capitalism reproduces the city itself as a commodity, a moment in which the creation of new economic sectors, integrated into global capitalism, makes the city a source of its business. At this time, urban space has taken on the status of a real estate product and is consumed productively, as is the case of tourism and the production of modern services (Carlos 2020). Consequently, accumulation is asserted through the spatial extension, establishing functions specific to portions of space based on the development of productive forces, in their new modalities of production, including urbanisation (which is processed under the pressure of the global market).

This means that, in the broad process of accumulation, capitalism extends to the production of space (Lefebvre 1970) encompassing and connecting new areas given the need to expand to new productive spheres. This is because capital is limited by the cyclical process of its realisation, thus implying the need for time–space compression, making it so that accumulation extends in a broader way to the production of space as a need to expand and grow the cycle of reproduction. At this time, the production process not only creates classic commodities, but also the very production of the city itself is a commodity. The move from the classic type of production, that of the means of production or consumer goods, towards the production of space itself as fixed capital, highlights a new moment of accumulation. It is thus a new direction for urbanisation that has become a business in the process of expanding capital (Carlos et al. 2015).

The metropolis has become a place for realising a new phase of capital accumulation that is revealed in the changed meaning of space: *the production of the urban "as a commodity"* (as a source of profit) under the aegis of financial capital (Padua 2015). In this context, urban land itself has become a place for investments and a source for the realisation of surplus value. In line with this, the production of space highlights a contradiction. While there is a movement that constitutes the globalisation of urban society, the fragmentation of space (due to the development of private property) as well as individuals (manifested as a result of a break from the past, the fraying of social relations, and the development of mass hedonism) is exacerbated in a movement that, in its extension, ranks places by creating and recreating new functions as a process of constantly renewed accumulation of capital in the scope of the reproduction of the social relations of production" (Carlos 2015; Volochko 2015; Alvarez 2018).

In this century, this production takes on a new meaning, since the crisis of the productive process requires the development of new productive sectors, enabling the shift of capital to more productive spheres, which has been occurring with the abandonment/spatial redirecting of the classical type of production (that of the means of production or consumer goods) towards the production of urban space (portions of the city or of the entire city) as a moment of the realisation of accumulation. Therefore, beyond meaning the growth of a city, urbanisation reveals a profound process of transformation in the forms and components of the production of the urban, based on the central locations of the territory.

In summary, at the current time, the reproduction of urban space is realised as a condition of the reproduction of capital based on three fundamental elements: (a) the realisation of *financial capital* (through the real estate sector)[7] reveals the

[7] For Lefebvre, real estate would be a second sector of the economy at this time, in 1970 and 1974.

"new achievement of space". As an extension of capitalism, space is continuously reproduced, as an exchange value, which assumes its interchangeability and consequently, commodification realising the private ownership of urban land. Financial capital enters the real estate circuit by buying and selling space that emerges as an investment, when the industrial capital in the metropolis loses its hegemony; (b) the production of *spaces for leisure and those aimed at tourism* indicate the move from the production to the consumption of space. Here, space itself enters the commercial circuit, as a commodity to be sold-consumed based on its particular attributes. It is the production of specific characteristics in space, creating qualities that can be sold; (c) *the development of* drug trafficking, a new economic activity that is unique due to being an illegal commodity, assumes different strategies (Barbosa 2015). In other words, in order for this commerce to occur, it must first dominate space, producing a specific form of spatial segregation.

In this coming together of sectors, the process of valuation through the production of space is realised. From the finance perspective, accumulation results in the incorporation of space. Urbanisation in the world highlights the role and importance of space in the process of capitalist accumulation, whether as a productive force for capital, or as a business in the hands of those who have a monopoly on land (through the existence of private property). In this sense, the production of space emerges as a renewed possibility for realising reproduction, a process that leads to the generation of immense conflicts. Today, these three sectors make the fragmentation of space explicit by producing forms of segregation that ultimately reveal the conditions of the extension of commodities in the current phase of capitalism through the reproduction of space.

8.3 The Struggles for Space in Space

The accumulation of capitalism requires the incessant renewal of the process of valuation. Expanding the process of accumulation involves the time of what is reproducible and repetitive, and proposes an expansion of the world of commodities. The production of space in this process replaces the valuation that is realised through the world of commodities. In the city, the contradiction between use value and exchange value dominates life. Under capitalism, it is impossible to eliminate the monopoly over the distribution of social wealth generated, as well as that of property in its private form and class relations, where the weight of the economic results in power relations.

The exchange value evolves towards increasing abstraction since these spaces are strictly subject to the domination of exchange for speculation, for capital investment, while urban space is conversely, for those subject to the possibility of the realisation of life: a use that involves the use of time. Under the aegis of exchange value, the production of space is subject to the logic of property, which is established as a moment of the private appropriation of portions of the city (an expression of this production), which can be arranged as exclusive spheres of private discretion, excluding all other possibilities (which, in its various forms, is the foundation of wealth). As such, the urban reproduced in the modern world, guided by the exchange value, places the citizen in the city, limiting and determining their everyday life. This movement reveals urban segregation. Private property as a foundation of production under capitalism imposes the contractual structure upon which society rests. On this path, land ownership becomes abstract in its private form, the foundation of segregation, which delimits the possibilities for using places, while it creates the possibilities of questioning them through urban social movements, revealing the spatial morphology/social morphology juxtaposition.

The development of the real estate market giving a new role to space in the realisation of capital accentuates the barbarity that takes shape in the access of subjects to the city. These actions reveal that the process of the social production of space is realised as a process of valuation, where social space and time, dominated by exchange, become a space and time of the commodity imposed on everyday life. The public policies that promote urban removals reveal the glue between the economic and political, highlighting the power of class in the realisation of these alliances. This urbanity and urban life are subjugated to the process of valuation (real or fictitious).

In this dimension, space is founded on inequality, revealing social life in its entirety, in the contradiction between necessity (imposed by economic reproduction) and desire (imposed by the reproduction of social life). A violent conflict emerges between the production of space (in the city), guided by the logic of the realisation of the exchange value, and the needs of a society that uses the city as a moment necessary to reproduce urban life. In this context, the city as a social production (a space–time of the realisation of life) appears to be external to society, an alien force to the extent that practice is permeated by the ever-increasing possibilities of the realisation of accumulation, which imposes strong restrictions on the forms of appropriation of space with the goal of use. This movement renews alienation in the modern world by shifting the struggle, primarily, from the factory to the city. In other words, from the struggle regarding work, salaries, and the workday, to the struggle for life in its broader and more complex aspects.

Therefore, everyday life reveals an urban space saturated with conflicts reflecting the drama and the urban tragedy of our time amid the perception of insecurity, the "absence of identity" promoted by the loss of urban references (carriers of local history) and by the parameters of commodities that uphold a new urbanity, which ignores the other. The call for a world "beyond capital" on social forums means disentangling this society under capital, the logic behind inequality. The opposing social movements represent the contrary as a real and practical criticism of capital. I believe they are the product of this contradictory urban society, reclaiming their other. It exposes the class society in its deepest contradiction marked by the existence of the private ownership of social wealth produced in all of its forms, since reproduction under capital, with new forms, has replaced its foundations. This means that the socio-spatial practice reveals the objective condition of human existence in its alienations and possibilities. It also reveals beyond this objectivity, the subjectivity in the awareness that comes from and in practice, and with practice. It is a practice that dramatically reveals crises because there are divisions, confrontations, and decisions.

On the theoretical level, it entails an investigation of what is possible as a combination of what is real and knowledge exploring the developing whole. This development emerges within this line of thinking and is open to the global level. Here, there is the potential for utopian thought. In his analysis of the 1965 events in Los Angeles, regarding the demonstrations by blacks for civil rights, Debord writes that they highlighted "the existing problems in this act (…) and the role of a revolutionary publication is not only to give credence to the insurgents of Los Angeles, but to contribute to giving them credence, theoretically explaining the truth, the pursuit of which expresses this action-practice"[8] (2006, p. 14).

8.4 Conclusion

Building an understanding of the modern world as the future of geography indicates several possibilities. This chapter was based on a Marxist-Lefebvrian perspective to position the explanatory power of geography within the analysis of today's reality in the heart of an urban society that is tending globally to be under the aegis of finance capital. This turns the production of space into a moment of capital valuation. This process of accumulation intensifies the inequalities experienced and demands of Geography an explanation of the foundations of these inequalities according to Geography. Here, we outlined a movement that indicates the need to think theoretically and practically about reality which is open to a social totality mediated by the concept of the production–reproduction of social space.

References

Alvarez PI (2018) A produção e reprodução da cidade como negócio. In: Carlos et al (eds) A cidade como negócio. Contexto, São Paulo, pp 65–80

AndradeMC (1993) Uma Geografia para o seculo XXI. Hucitec, São Paulo

Barbosa JL (2015) A geografia dos negócios do narcotráfico.In: Carlos et al (eds) A cidade como negócio. Contexto, São Paulo, pp 247–266

Carlos AFA (1979) Reflexões sobre o espaço Geográfico. Dissertation, Universidade de São Paulo

Carlos AFA (1994) A reprodução do espaço urbano: o caso de Cotia. EDUSP, São Paulo

Carlos AFA (2011a) Da "organização" à "produção" do espaço no movimento do pensamento Geográfico. In: Carlos AFA et al (eds) A produção do espaço urbano: agentes e processos, escalas e desafios. Editora Contexto, São Paulo

Carlos AFA (2011b) A condição espacial. Editora Contexto, São Paulo

Carlos AFA (2018) Geografia critica-radical e a teoria social. In: Carlos AFA et al (eds) Geografia urbana crítica: teoria e método. ColeçãoMetageografia.Editora Contexto, São Paulo

Carlos AFA (2020) Segregação socioespacial e o "Direito à Cidade". GEOUSP Espaço e Tempo. https://doi.org/10.11606/issn.2179-0892.geousp.2020.177180

[8] Original text: "em ato os problemas existentes (...) e o papel de uma publicação revolucionária é não só dar razão aos insurgentes de Los Angeles, mas contribuir a dar-lhes razão, explicando teoricamente a verdade cuja busca expressa essa ação-prática".

Carlos AFA, Lencioni S (1982) Alguns elementos para discussão do espaço geográfico como mercadoria. BORRADOR—Teoria e Método Em Geografia 1:01–09
Carlos AFA, Lencioni S (1982) A categoria da formação econômica da sociedade na análise geográfica. BORRADOR—Teoria e Método em Geografia 1
Carlos AFA, Volochko D, Alvarez IP (eds) (2015) A cidade como negócio. Editora Contexto, São Paulo
Correa RL (1980) Geografia, teoria e crítica:o saber posto em questão. Editora Vozes, Rio de Janeiro
Conceição AL, Silva JD (eds) (2019) O fim do pensamento crítico reflexivo? A negação do humano e a banalizaçãoda teoria. Editora UFS, São Criatovaõ
DebordG, (2006) El planeta enfermo. Anagrama, Barcelona
Lacoste Y (1988) A geografia—issoserve, em primeiro lugar, para fazer a guerra. EditoraPapirus, São Paulo
Lencioni S (2017) Metrópole, metropolização e regionalização. EditoraConsequência, São Paulo
Moreira R (2000) E assim se passaram 10 anos. Geografia 2(3):27–49
Moraes ACR, Messias da Costa W (1984) Geografia crítica: avalorização do espaço. Hucitec, São Paulo
Oliveira AU (1982) Espaço e tempo: uma compreensão materialista e dialética. In: Santos M (ed) Novos rumos da geografia brasileira. Hucitec, São Paulo
Pádua RF (2018) Pensando a noção de prática socioespacial. In: Carlos AFA et al (eds) Geografia urbana crítica: teoria e método. ColeçãoMetageografia. Editora Contexto, São Paulo
Pádua RF (2015) Produção estratégica do espaço e os novos produtos imobiliários. In: Carlos AFA et al (eds) A cidade como negócio. Editora Contexto, São Paulo, pp 145–164
Rodrigues AM (1998) Produção e consumo do e no espaço: problemática ambiental urbana. Hucitec, São Paulo
Santos M (1978) Por uma geografia nova. Hucitec, São Paulo
Silva AC (1983) A renovação geográfica no Brasil—1976/1983 (as geografias crítica e radical na perspectivateórica. Boletim Paulista de Geografia 60:73–139
Simoni-Santos C (2018) Escalas geográficas: instrumento de observação ou objeto de investigação? In: Carlos AFA et al (eds) Geografia urbana crítica: teoria e método. Coleção Metageografia. Editora Contexto, São Paulo, pp 89–107
Soja E (1993) Geografias pós-modernas: a reafirmação do espaço na teoria crítica. Zahar, Rio de Janeiro
Volochko D (2015) A moradia como negócio e a valorização do espaço urbano metropolitano. In: Carlos AFA et al (eds) A cidade como negócio. Editora Contexto, São Paulo, pp 97–119
Volochko D (2018) Conflitos socioespaciais, socioambientais, e representações: dimensões da análise urbana. In: Carlos AFA et al (eds) Geografia urbana crítica: tfeoria e método. Coleção Metageografia. Editora Contexto, São Paulo

Chapter 9
Dialogues on Brazilian Political Geography and Its Perspectives in the Twenty-First Century

Adriana Dorfman and Lício Caetano do Rego Monteiro

Abstract The challenges faced by Brazilian political geography in the twenty-first century are (1) a reconsideration of its history, in view of its relationship with geopolitics; (2) the necessary dialogue (overcoming the idea of transposition) between academic and school geography, reflecting upon imagined geographies and stereotypes of global politics disseminated in school; (3) intradisciplinary exchanges with sub-categories of geography (such as urban, agrarian, environmental, economic and cultural geographies); (4) interdisciplinary exchanges, especially with political sciences, international relations and history; (5) the geopolitics of knowledge, translated into the relationship between Brazilian and Francophone political geographies, Anglophone geography and emerging Latin American decolonial geopolitics; (6) the 1990s "territorial turn" and the uses and abuses of the concept of territory in different areas of knowledge and social practice; (7) connections with the state, management of territory and public policies and the mutual interference in categories of analysis and in practices; (8) regional and global horizons of Brazilian political geography and geopolitics in the face of Brazil's international standing; (9) the incorporation of Brazilian territory and societal issues and finally, (10) the acknowledgement of actors beyond the state and of spaces beyond academia that produce political knowledge, which challenge and nourish scientific production in political geography.

Keywords Territory · Political geography · Brazil · Geopolitics · Knowledge production

A. Dorfman (✉)
Universidade Federal do Rio Grande do Sul, Porto Alegre, Brazil
e-mail: adriana.dorfman@ufrgs.br

L. C. do Rego Monteiro
Universidade Federal Fluminense, Niterói, Brazil
e-mail: liciocaetano@id.uff.br

R. C. Lois González and M. A. Mitidiero Junior (eds.), *Brazilian Geography*, Advances in Geographical and Environmental Sciences,
https://doi.org/10.1007/978-981-19-3704-0_9

9.1 Political Geography and Geopolitics in Brazil: Traditions and Changes

The first question raised about Brazilian political geography explores its interweaving with and its difference from geopolitics generated in Brazil. In the country, geopolitics studies in the 1920s and 1930s, with Everardo Backheuser (1926) and Mário Travassos (1935). Despite Backheuser's pioneering spirit, some authors highlight Travassos as a precursor of geopolitics in Brazil, given the greater influence of his work (Vlach 2003; Costa; Théry 2012). In the 1940s, interest in geopolitics intensified with publications in the military, diplomatic and academic fields, with the emergence of the first courses in geopolitics at the Rio Branco Institute, the Brazilian Cultural Institute, the Law Institute of the Pontifical Catholic University of Rio de Janeiro (PUC-RJ) and even with the creation of a Brazilian Institute of Geopolitics in 1949, which operated until 1961 (Miyamoto 1981, pp 79–81).

1949 also marked the creation of the War College (*Escola Superior de Guerra*), which became the main production centre for Brazilian geopolitics during the following decades, representing the conservative doctrinal inclination of Brazilian geopolitical thought and its association with the principles of the National Security Doctrine: the fight against communism, national integration, population and defence of the Brazilian territory. It is in this context of military prominence in the management of the territory during dictatorial times that a "Brazilian geopolitical school" was formed, with exponents such as Golbery do Couto e Silva (1981, for example), Carlos de Meira Mattos (1975, among others) and Therezinha de Castro (1982). From the Brazilian alignment on the western side of the Cold War to the geopolitics of the internal colonisation of the Amazonian space, through the transport network and the creation of colonies and military regiments on the borders, many decisions and orientations have been influenced by the geopolitical military imagination, with evident authoritarian traces.

Since the mid-1980s, the conservative geopolitical tradition has been challenged. With re-democratisation and the emergence of a geopolitics theory that is not identified with the military, Bertha Becker, José William Vesentini and WanderleyMessias da Costa, among others, published texts, taught courses lead the discussion on the relationship between politics and space in Geography (Vesentini 1986; Becker 1988; Costa 1991). From that moment on, different curricula in Brazil included political geography, still contrasting it with geopolitics.

The most common distinctions pointed to geopolitics being the practical and applied knowledge of geography for the political objectives of certain actors—the most evident actor was, of course, the state—with Political Geography being a scientific approach to the relations between space and politics, without commitment to any immediate application. Another view, very close to the first one, focused not on the aim of knowledge, but on differentiating ideological content, associated with geopolitics, from scientific knowledge, associated with political geography. There was also a frequent distinction linking geopolitics to the expanded scale of political phenomena, at the international arena, while political geography could be applied to any scale of analysis, creating an immediate association between foreign policy and geopolitics versus domestic politics and political geography.

Today these distinctions seem outdated and out of place in an expanding arena of studies shared by political geography and geopolitics, along with several other traditions.

9.2 Geopolitics of Knowledge

Another challenge posed to political geography refers to the geopolitics of knowledge—considered as a correlation between the place of enunciation and thought, within a colonial difference in the production of knowledge (Mignolo 2020). The geopolitics of knowledge preside over national and linguistic contexts, and in the case of Brazilian political geography, we observe the way in which it incorporates,ignores and/or dialogues with Francophone political geography, Anglophone critical geopolitics and the emerging Latin American decolonial geopolitics. Here, we move away from the widespread view, still present in certain histories of political geography, which classifies authors, works and theories into national schools. Although initially conditioned by the link between geopolitics/political geography and nationalism, reinforcing the links between contexts of production, theories and reception, the present is marked by the consolidation of internationalised publications and debates—albeit characterised by inequality in the circulation of knowledge. The idea of national schools today can be resized, taking account of linguistic elements, evidently highlighting the expanded effects of production in English.

In a previous work, we analysed the development of production involving political geography and geopolitics in German, English, French and Spanish throughout the twentieth century (Rego-Monteiro 2018). We found that interest in geopolitics grew in Germany in the 1920s and 1930s, and in the English-speaking world in the 1940s, with no appropriation of the term "Geopolitics" in the French-speaking world until that period. In Spanish-speaking countries, interest began in the 1940s and did not decrease, as in Germany and the USA—only starting to decline in the 1980s. The same thing happened in Brazil, with the adoption of geopolitics by military schools and in subjects taught as part of university courses. Meanwhile, political geography as an academic discipline has been consolidated in all of these contexts.

The presence of studies on Brazilian political geography varied slightly throughout the twentieth century, increasing at the turn of the century. The rehabilitation of geopolitics in the English-speaking (Hepple 1986) and French-speaking (Lacoste 1976) worlds in the 1970s and early 1980s had no parallel in Brazilian geography. This mismatch can be attributed to the association between geopolitical thought and conservatism in Brazil. It was only in the mid-1980s that studies in the field of political geography and geopolitics expanded, influenced by French-speaking authors such as Lacoste (1976), Claval (1979) and Raffestin (1980). The circulation of their main works was guaranteed by their translation into Portuguese and represents the continuity of an influence that dates back to the institutionalisation of geography in Brazil's universities, when French geographers were of seminal importance. Lacoste's book has been circulating in pirated versions since the late 1970s; and to this day, the Brazilian academic world appears to be one of those which most strongly incorporated the ideas of Raffestin (Amilhat-Szary 2019). Anglophone authors from critical geopolitics, such as Agnew (1998, among many), Tuathail and Dalby (1996, 1998, for example) were not translated into Portuguese—except for brief texts—and even today are rarely cited in our country.

The field of political geography in Brazil was also organised with the publication of important books. In 1991, Wanderley M. da Costa wrote the book *Geografia Política e Geopolítica: Discursos sobre o Território e o Poder*, the result of his doctoral thesis, which became the most complete textbook in this field, with an important historical assessment of the different schools of geopolitics and political geography in Germany, France, England, the USA and Brazil, culminating in a review of the renovations of the 1970s and 1980s and a deepening of the themes of states, borders, nations and nationalism.

In 2005, Iná Elias de Castro published a more up-to-date textbook called *Geografia e Política: territórios, escalas de ação e instituições*, focused on concepts and themes without an explicit concern for the historical reconstitution of the geopolitical and political geographical schools of thought. The book discusses the field of political geography, the relations between territory and conflict, the concept of power, political representation, the relations between state and territory in Brazil and the contemporary international system. In a way, both books represent the geopolitics of knowledge on a national scale, being written by professors working in São Paulo and Rio de Janeiro, in the main research centres, with the oldest and best-rated postgraduate programmes in Brazil (the University of São Paulo (USP) and the Federal University of Rio de Janeiro (UFRJ), respectively), in cities that have publishers of national scope. These conditions certainly contributed to the dissemination of these works.

Two other important textbooks by foreign authors in the field of political geography were published in Portuguese. *Por uma Geografia do Poder* (French title: *Pour une géographie du pouvoir*), by the French-Swiss Claude Raffestin, was first published in 1980, and later published in Brazil in 1993. *Geopolítica, Identidade e Globalização* (Spanish title:*Geopolítica, identidad y globalización*), by the Spanish authors Joan Nogué Font and Joan Vicente Rufí, was published [in Portuguese] in 2006. In addition to textbooks, several collections by Brazilian authors made up an overview of political geography produced here (Castro, Rodrigues, Ribeiro 2013; Silva 2016; Rückert et al. 2018; Costa, Vasconcelos 2019; Rodrigues, Rego-Monteiro 2020). Academic discussion has also advanced through periodicals published in Portuguese, such as the *Revista de Geopolitica*, and dossiers dedicated to political geography in journals such as volume 4, number 12 of the *Revista Geonorte* (2013), the 31st edition of the French magazine *Espace Politique* (Silva, Théry 2017), volume 63 of the Revista Brasileira de Geografia (Rodrigues, Rego-Monteiro 2018) and volume 16, number 30 of the Revista da Associação Nacional de Pós-Graduação em Geografia (ANPEGE) (Silva, Dorfman 2020).

Exchanges with Latin American production in the field of political geography remain a challenge, partly driven by the dialogues with Argentine, Mexican, Colombian and Venezuelan colleagues, among others, with whom we share contexts and conflicts. In addition to the discipline, in Latin America the intellectual and political environments gave rise to what was called the "decolonial turn" in the Human Sciences, which took place in the 2000s. At first, this movement had little impact on political geography, which remained concerned with the global North. It is possible to identify publications, already during the course of the 2000s and 2010s, which, without identifying themselves as part of the field of political geography and geopolitics, began to relate space, knowledge and power within a decolonial perspective (Cruz, Oliveira 2017).

The idea of a situated and positioned knowledge, present in decolonial thinking, was quickly incorporated by Brazilian geographers, as it was already being exercised through theories and methodologies based on the multi-scalarity of the territory and on critical geography. Haesbaert (2021) provides a synthesis of the decolonial and (multi-)territorial "turns", including the definition of the body as a political scale. On the other hand, decolonial methodologies are not always deepened, and certain concepts such as "geopolitics of knowledge", "frontier thinking" and "epistemicide", dear to decolonial thinking, have not yet been developed into a geopolitical key by Brazilian geographers.

9.3 Political Geography and Its Disciplinary Dialogues

Exploring the relationship of political geography with other disciplines, we start from the observation that, despite political geography being a field that is restricted to geography, geopolitics is present in several other fields of study, which is an opportunity for interaction between disciplines. This is what happens with political science, international relations, strategic studies, security and defence.

Political science shares, with geography, the very paternity of geopolitics, since the term appears in the work of the political scientist Rudolf Kjéllen, receiving the contribution (without using the term) of geographers such as Ratzel and Mackinder for the delineation from classical geopolitics.

Geopolitics appears in political science when it deals with internal relations within the state territory, until the point at which international relations (IR) became demarcated as a separate field of study. This division between the domains of internal and external sovereignty ended up becoming a basic presupposition for the differentiation between the two areas. According to RBJ Walker, this concept of spatiality (internal/external duality, exemplified by the borders of the state's territory) limited IR's understanding of the international system throughout the twentieth century, making it difficult to interpret world politics beyond the spatial frameworks defined by the modern (state) political imagination (Walker 1993). It would be more appropriate to consider that the problem lies in the way that these theories conceive the spatial dimension of politics, falling into what Agnew and Corbridge called the "territorial trap" when they identified three limiting geographic assumptions in this conception of space: (1) state territories are exhibited as fixed and sovereign units; (2) the domestic/foreign polarity obscures processes that operate at different scales and (3) state territory is depicted as existing before society, and defines its borders (ethnic, national) (1995, p. 84).

The expression "territorial trap" is the most widespread contribution of critical Anglophone geopolitics to debates in Brazil. Geographers John Agnew and Stuart Corbridge use it to point out that much of the IR literature implicitly assumes that the state is a fixed territorial entity which does not vary over time nor is conditioned by its place within the global geopolitical order (op. cit.). This expression is wrongly understood as equal with the idea of "methodological nationalism", defined by the sociologist Herminio Martins as "national pre-definitions of social realities", characteristic of the sociological analyses in which the national community appears as the unit that demarcates social phenomena and problems (1974, p. 276).

Another very relevant discussion with origins in both postcolonial theory and Anglo-Saxon geography is concerned with social representations. On the one hand, the reference is orientalism, a concept proposed by Edward Said, who also proposes an imaginative geography (1978), explaining the discourses which take space as a reference and which produce hierarchical geopolitical identities of the "self" and the "other". An analogous idea circulates through Anglo-Saxon geography under the name of "geographic imagination", again emphasising political aspects of the representation of territories and their inhabitants (Massey 2008). At present, the concern with geopolitical representations of the self and the other constitutes the mark of critical geopolitics in the English language, when examining the images circulating in the media, political institutions and academia (Tuathail 1996), explained in the expression "modern geopolitical imagination" (anchored in the scale of the state—masculinist and realist) proposed by Agnew (1998). Analyses that question social representations, imaginative geography and geographic imagination are quite frequent in Brazilian political geography and the absence of references to critical geopolitics is probably due to the existence of translations of Said and Massey (not identified as political geography theorists) and the aforementioned lack of translations of critical geopoliticians.

Thus, the critique of the personification of the state and the nation as a primordial scale of territorialised processes brings political geography closer to political science and sociology and opens up space for dialogues with anthropology, in a movement that seeks to undo the instrumentalisation of these sciences by the modern colonial and state-centric project.

9.4 State Management and Political Geography

The link between political geography and the state appears in the management of the territory and in Brazilian territorial public policies, when geographers are called upon to prepare studies and analyses to support state action. In Brazil, the profession of geographer is a recognised one, and there is recurring participation from geographers in state programmes such as zoning, territorial ordering, management plans or city master plans. Despite the risk of instrumentalising academia in order to legitimise policies that are not always accompanied by effective social control, the interchange between the practical and analytical uses of concepts and methodologies usually generates good insights.

Here, we can cite examples such as the Border Strip Development Program (PDFF) (MIN/RETIS 2004) and the Survey on Public Security at the Borders (2013–2014) (Brasil 2016) as projects with a national scale. The PDFF was developed based on methodological and conceptual subsidies developed within the framework of the Retis Research Group (UFRJ), coordinated by Lia Osorio Machado, who had previously published an *Atlas da Fronteira Continental do Brasil* (2001). The experience acquired in the scope of the research was then adapted in order to implement public policies in the Border Strip. The concepts used in this study (twin cities, arcs, border zones and sub-regions) entered the vernacular of politics and studies on Brazilian borders, and the maps produced achieved wide impact and circulation.

The Research on Public Security at Borders was coordinated by the Centre for the Study of Citizenship, Conflict and Urban Violence (NECVU/UFRJ) with the additional participation of the Retis Group. The survey was commissioned by the National Secretariat for Public Security (SENASP), and the geographers who participated in it were invited to carry out a socioeconomic and demographic diagnosis that draw a picture of security policies at the borders. As a result of this research, studies on violence and security gained momentum in the production of political geography in Brazil. In this case, the interaction between public power and academia originates in the demands of governance and bears fruit in academia, in search of instruments for analysing issues of public safety and "urban violence", often represented as territorialised disputes that mark contemporary Brazilian society.

9.5 Intradisciplinary Dialogues

The study of territorialities and the criticism of Brazilian state policies have for decades mobilised areas of geography that do not identify themselves with political geography. Analyses of urban polarities and segregation; on transport and service networks; of borders, fronts and settlement projects; of the economy and its legal regimes; of conflicts over land; the appraisal, exploration and conservation of natural resources in regions such as the Amazon and western Brazil; among other themes and problems, are carried out by different areas within geography, such as urban, agrarian, economic, cultural and environmental geography. Once again, the concept of "territory" allows us to identify the bridges that integrate these areas, in a cross-pollination that is still capable of bearing many fruits. The politicisation of the concepts of "place", "environment" (Souza 2020) and "landscape" represents the inflow towards political geography.

An important space for interaction in the field of geography is the Brazilian Congress of Political Geography, Geopolitics and Territorial Management (CONGEO), which has had three editions since 2014 and is the largest event in this field of studies in Brazil, bringing together around 600 researchers in its latest edition. The different core concepts of the congress indicate the central themes and interfaces with different areas of geography: (1) political and geopolitical geography: from classical approaches to contemporary renovations; (2) (geo)policies of the environment, resources and sustainability; (3) localisms, nationalisms, regionalisms and globalisms; (4) borders: a constant territorial challenge; (5) the management scales of territorial public policies; (6) regional integration and new spaces for cooperation and international conflict; (7) territory, politics, citizenship and democracy; and (8) political geography and teaching: thematic, conceptual and methodological innovations (Rodrigues, Rego-Monteiro 2020).

9.6 A Note on School Geography and Political Geography

In addition to intradisciplinary relations, in this list, we observe another relevant source of dissemination of political geography in Brazil: school geography. Political issues are present in the history of schooling in Brazil. Here, as in other processes for the construction of "national consciousness", one of the central tasks in the universalisation of primary schools was the strengthening of national projects, with nationalism framing the design of the curriculum for languages, literature, history and, also, geography, always tracing the differences between national communities (Dorfman et al. 2021). By including geography among the compulsory subjects in the elementary school curriculum, the large number of courses in Brazilian universities was also guaranteed, which allowed the deepening of the different disciplines into which geography unfolded—the geopolitics of Brazil in the authoritarian period and political geography after democratisation. But it was not simply a question of the number of students, professors and researchers in geography. Many authors of textbooks and didactic atlases, such as Carlos Delgado de Carvalho, Geraldo José Pauwels and, much more recently, José William Vesentini, have also dedicated themselves to political geography.

At present, the teaching of geography has paid attention to geographic imagination, imaginative geography and social representations—although, contradictorily, most textbooks insist on the distinction between geopolitics and political geography, perhaps due to the influence of Yves Lacoste's work. The starting point of this author's work is the distinction between the geography of the teachers and the geography of the Military Chiefs of Staff, stating that the first is responsible for purposely hiding the historical construction of the state, as well as its contradictions, leaving on display only the nationalist character, falsely naturalised from the territorial formation of nation states (Lacoste 1976, p. 26).

However, since the 1980s Brazilian school geography has been strongly guided by critical geography. In this way, the agenda became "the interpretation and questioning of spatial phenomena, based on the premise that the relationships between society, nature and work shape space" (Froehlich 2021, p. 29, translation). This trend is incorporated into the current Common National Curriculum Base in Brazil, into which themes of political geography unfold in the formation of territories and their conflicts, in the legal and political outlines that define nation states, in globalisation and economic blocks and in migrations and the issue of refugees, to name a few examples. In addition, teaching geography at school is seen as "a facilitator for the student to understand the nuances of the geographic space (world) at all levels, from local to global, so that this young person can be an active citizen" and one who is politically engaged (idem, p. 30, translation). This perspective of positioned teaching leads to the relationship with the communities in which the school is inserted—communities that are marked by conflicts that demand territorial explanations.

9.7 Territory in the Dialogues of Political Geography

A key element in bringing political geography closer to other fields is the territorial turn shared by social and human sciences, social movements and public policies. Territories and territorialities are key terms for political geography. Although they are not exclusive to this sub-discipline, it is remarkable that the increasingly recurrent use of a territorial approach to social phenomena has been accompanied by an appreciation of the field of political geography as a source and reference for the social sciences. Conversely, the use of territorial vocabulary by the social sciences fertilises debates within political geography.

In a traditional conception, territory is an attribute of the state, with its political-administrative divisions that are well-defined and marked from top to bottom. Historically, territories have been considered as objects of conquest, appropriations/expropriations (by state or private agents), and as a resource of power, as they provide access to various other resources (land, water, minerals, labour, etc.). Conceptually, territory was also seen, on the one hand, as a continuous, delimited and material space, a container; on the other, it was naturalised (and depoliticised) as mere soil, nature preceding political power, whose timeless myths reinforce a sense of belonging to a given group or nation. In all these senses, in this traditional conception, it is the territory of the state that prevails as the superior hierarchical level and orders the other forms of territoriality that exist within society (Souza 1995).

Several sources of dispute contributed to the redefinition of the concept of territory. Here, it is worth highlighting how the movements of struggle for land in Latin America started to assume territory as a key element in their political demands (Porto-Gonçalves 2015). This territorial turn happened when social movements realised the limitations of struggling for land without struggling for the preservation of community, autonomous forms of political organisation and regulation of territory, resources and social life, in addition to the proper ways of using and caring for nature. Territory is, thus, more than an object to be defended, it is the condition for the struggle itself. It is the locus of resistance and conflict, something that is given and needs to be recognised in its objectivity, but it is also a strategy of belonging and affirmation, a projected intention, a horizon of struggle—hence its subjective and intersubjective dimension.

It is possible to place this "territorial turn" in the 1990s, occurring firstly in the context of social movements in Latin America, which began to include among their demands the struggle for life, dignity and territory (Porto-Gonçalves 2015). The diversity of these territories could no longer use the territory of the state as a model. After all, one of the pillars of state power lays in activelly deny and delegitimise other forms of political territoriality, which are either subsumed as official state units or discarded and attacked. In the language of the movements and the intellectuality associated with them in different ways, these are described as autonomous, alternative, emerging territories, among other more specific formulations. These territories take different forms and names, as is the case with traditionally occupied lands (indigenous lands, *quilombos* (Brazilian hinterland settlements founded by people of African origin), *seringais* (rubber plantations), *faxinais* (community agriculture/farming sites), *babaçuais* (babaçu palm plantations), pasture lands, among others) (Almeida 2004). But territorial struggles today do not only take place in traditional territories—they also appear in "black territories" (Vieira 2021, translation) and in urban occupation movements (Cigolini and Lima 2021), for example.

This is a way of telling the story of the passage from state territory (as a basic unit) to multiple territories. In parallel to this re-signification of territories based on social struggles—also in geography, and more broadly in social thought—territories began to be seen in a different way, taking account of not only their material aspects, but also their symbolic ones; not only their formal boundaries, but also the multi-territorialities and discontinuities (in time and space) (Souza 1995; Haesbaert 2002). Territory came to mean not only the state but, mainly, power (Raffestin 1980), in a broader and more diffuse sense: "space defined and delimited by and through power relations" (Souza 1996, translation). The return to the territory, as Milton Santos (1994) said, was one of the solutions to counteract the ineluctable discourse of globalisation that permeated and encompassed everything. The idea that society would be increasingly de-territorialised, in an absolute sense, was dismantled by identifying the processes of de-territorialisation and re-territorialisation promoted by global market agents and by states (Haesbaert 2002).

The territorial turn in social movements and critical social thinking also had an effect on the way in which public policies themselves came to be oriented. Because of this, Rogério Haesbaert speaks of a (multi-)territorial turn in "Latin America", recognising at least two major approaches: a more "top-down" approach, based on numerous state policies labelled "territorial" and broad business strategies for land exploration (either via "agribusiness", or via neo-extractivism of "natural resources"), and another more "from the bottom up" in the forms of resistance—or rather, of r-existence—, which configure struggles for territory based on the organisation of subordinate groups (p. 60, 2021, translation).

Regional planning was also reoriented, as in the past, it was generally top-down planning on larger scales. In the words of Nunes and Fonseca, since the implementation of the 1988 Federal Constitution it has been possible to identify "political-territorial arrangements anchored in the premises of participation, decentralisation and cooperation" (2018, p. 301, translation). Initiatives such as participatory budgeting, municipal consortia and other forms of defining territorial policies based on consultation and establishment of guidelines by the communities directly involved have been claimed as strategies that are capable of reconciling plans originating from different scales of action.

9.8 Final Considerations

Brazilian political geography and geopolitics in the 1930s and 1940s walked very closely together, took different paths during the period of hegemony of military thought in geopolitics, and were even explicitly divergent in the 1980s. In the last three decades, we can observe a rapprochement between the two fields and a loss of emphasis on demarcation, rehearsed during the 1980s, which opposed political geography and geopolitics as strictly separate disciplines. Brazilian geopolitics at its outset, received German and French influxes in the 1930s and added an American influence after World War II. It ended the twentieth century with a relatively low influence from contemporary Anglophone geopolitics, a remote and disowned presence of German *Geopolitik* in its origins, an enduring fraternal affinity with the French—past and present—and a still low but growing interest in the production in Spanish.

The maturity of studies in political geography in Brazil can be evidenced by several facts: publications in the field, the consolidation of a congress at the national level, the critical review of its own history, and the originality in relation to different foreign contributions, in addition to a well-resolved relationship with geopolitics which allows a transition into areas beyond geography and academia.

Contemporary Brazilian political geography has a relevant expression within the scope of Latin America, where geopolitics is generally associated with disciplines other than geography. In spite of Brazilian foreign policy, which has promoted a distance between Brazil and its regional neighbours, with the absence of integrationist policies, there has been a rapprochement between Brazilian political geography and Latin American production, through decolonial and territorial studies beyond the state.

The current concept of territory brings into play the closest scales, local social controls and knowledge that comes from the practise of social agents and the tacit rules of local coexistence, to the point of creating an understanding of the fact that action in places and the capillarisation of welfare networks can be expressed as "territorialisation". In other words, the conceptualisation circulating in the human sciences and in societal organisations brings the concept of territory closer to local experiences, in which communities have a prominent participation, even if sometimes in dialogue with state agents. The centrality acquired by the territory demands a maturation of political geography, as its instruments are called upon in the ongoing territorial struggles and practices in Brazil.

References

Agnew J, Corbridge S (1995) Mastering space: hegemony, territory and international political economy. Routledge
Agnew J (1998) Geopolitics: re-visioning the world politics. Routledge, New York
Almeida AWB (2004) Terras tradicionalmente ocupadas: processos de territorialização e movimentos sociais. Revista Brasileira de Estudos Urbanos e Regionais, vol 6, no 1
Amilhat Szary A-L (2019) Préface In: Pour une géographie du pouvoir [en ligne]. Lyon: ENS Éditions. Available at http://books.openedition.org/enseditions/10361
Backheuser E (1926) A Estrutura Política do Brasil. I—Notas prévias. Rio de Janeiro, Mendonça, Machado & Cia
Becker BK (1988) A geografia e o resgate da geopolítica. Revista Brasileira de Geografia. 50, specialissue t. 2
Brasil: Ministério da Justiça (2016) Segurança Pública nas Fronteiras: Diagnóstico Socioeconômico e Demográfico. Brasília: MJ, SENASP
Castro IE (2005) Geografia e política: territórios, escalas de ação e instituições. Bertrand Brasil, Rio de Janeiro
Castro IE, Rodrigues JN, Ribeiro RW (Org.) (2013) Espaços da democracia. Para a agenda da geografia política contemporânea. Bertrand Brasil, Rio de Janeiro
Castro T (1982) O Brasil no mundo atual: posicionamento e diretrizes. Colégio Pedro II, Rio de Janeiro
Cigolini AA, Lima NR (2021) Organizações populares em Curitiba: Sujeitos e conflitos, organização e luta. Revista da ANPEGE, [S.l.], vol 16, no 30, pp 229–251
Claval P (1979) Espaço e poder. Zahar, Rio de Janeiro
Costa WM (2008) Geografia política e geopolítica. São Paulo: Edusp [1991]
Costa WM, Théry H Quatre-vingts ans de géopolitique au Brésil : de la géographie militaire à une pensée stratégique nationale. Hérodote 2012/3, no 146–147
Costa WM, Vasconcelos DB (Org.) (2019) Geografia e geopolítica da América do Sul: integrações e conflitos. FFLCH/USP, São Paulo
Couto e Silva G (1981) Geopolítica do Brasil. In: Couto e Silva G (ed) Conjuntura política nacional: o Poder Executivo & Geopolítica do Brasil. José Olympio Ed, Rio de Janeiro
Cruz VC, Oliveira D (Orgs.) (2017) Geografia e giro descolonial: experiências, ideias e horizontes de renovação do pensamento crítico. Letra Capital, Rio de Janeiro
Dorfman A, Filizola R, Félix JM (Orgs.) (2021) Ensinando Fronteiras. Porto Alegre: Diadorim Editora
Font JN, Rufí JV (2006) Geopolítica, identidade e globalização. Annablume, São Paulo
Froehlich CCP (2021) A fronteira no ensino básico de geografia: políticas nacionais e práticas situadas. Master'sDissertation, UFRGS
GRUPO RETIS (2001) Atlas continental da fronteira continental brasileira. UFRJ/CNPq, Rio de Janeiro
Haesbaert R (2002) O mito da desterritorialização: do "fim dos territórios" à multiterritorialidade. Bertrand Brasil, Rio de Janeiro
Haesbaert R (2021) Território e descolonialidade: sobre o giro (multi)territorial/de(s)colonial na "América Latina". CLACSO, Buenos Aires
Hepple L (1986) The revival of geopolitics. Political Geogr Quart Supplement to vol 5, n 4

Karol E (2013) Geografia política e geopolítica no Brasil (1982–2012). Doctoral Thesis, USP, São Paulo

Lacoste Y A geografia—isso serve, em primeiro lugar, para fazer a guerra. Papirus, Campinas 1988 [1976]

Martins H (1974) Time and history in sociology. In:Rex J (org) Approaches to sociology, an introduction to major trends in British sociology. Routledge &Keagan Paul

Massey D (2008) Pelo espaço: uma nova política da espacialidade. Translated by Hilda Pareto Maciel & Rogério Haesbaert. Bertand Brasil, Rio de Janeiro

Mattos MC (1975) Brasil: geopolítica e destino. Bibliex, Rio de Janeiro

Mignolo W (2020) A geopolítica do conhecimento e a diferença colonial Revista Lusófona de Educação, 48

Miyamoto S (1981) Os estudos geopolíticos no Brasil: uma contribuição para sua avaliação. Perspectivas, 4

Novaes ARA (2015) Geopolítica pelas Imagens, Terra Brasilis (Nova Série) [Online], 6. Available at http://journals.openedition.org/terrabrasilis/1722

Nunes J, Fonseca AAM (2018) Cooperação e coesão territorial no pacto federativo brasileiro: potencialidades e limites a partir de estratégias cooperativas intermunicipais nos estados da Bahia e do Rio de Janeiro. In: Rückert AA, Silva ACP, Silva GV (Orgs.) (eds) Geografia Política, Geopolítica e Gestão do Território: integração sul-americana e regiões periféricas. Editora Letra1, Porto Alegre, pp 287–304. https://doi.org/10.21507/9788563800367-18

Ó Tuathail G (1996) Critical geopolitics: the politics of writing global space. University of Minnesota Press, Minneapolis

Ó Tuathail G, Dalby S (1998) Rethinking geopolitics. Routledge

Porto-Gonçalves CW (2015) Pela vida, pela dignidade e pelo território: um novo léxico teórico político desde as lutas sociais na América Latina/AbyaYala/Quilombola, Polis [Online], 41. Available at http://journals.openedition.org/polis/11027

Raffestin C Por uma geografia do poder. Ática, São Paulo, 1993 [1980]

Rego Monteiro LC (2013) Linhas cruzadas no resgate da geopolítica pós-anos 1970. Revista Geonorte, Special Edition 3, v 7

Rego Monteiro LC (2014) Segurança na América do Sul: a construção regional e a experiência colombiana. Doctoral Thesis, PPGG/UFRJ

Rego Monteiro LC (2018) Linhas cruzadas: a geografía política e a geopolítica no século XX no Google NgramViewer e o debate no Brasil. In: Rückert A, Silva ACP, Silva GV A integração sul-americana e a inserção das regiões periféricas. Porto Alegre: Letra1

REVISTA GEONORTE, vol 4, no 12 (2013)

Rodrigues JN, Rego-Monteiro L (2018) III Congresso Brasileiro de Geografia Política, Geopolítica e Gestão do Território: interfaces, pluralidade e renovação de um campo. Revista Brasileira de Geografia, 63, v 2

Rodrigues JN, Rego-Monteiro LC (org.) (2020) Crise e reinvenção dos espaços da política. Consequência, Rio de Janeiro

Rückert A, Silva ACP, Silva GV (2018) A integração sul-americana e a inserção das regiões periféricas. Porto Alegre: Letra 1

Said E Orientalismo. O Oriente como invenção do ocidente. Companhia das Letras, São Paulo, 1990 [1978].

Santos M (2005) O retorno do territorio. In: OSAL: Observatorio Social de América Latina. Year 6 no 16. Buenos Aires: CLACSO, 2005 [1994]

Silva ACP (2016) Geografia Política, Geopolítica e Gestão do Território: racionalidades e práticas em múltiplas escalas, organizado por Augusto Cesar P. da Silva. Gramma, Rio de Janeiro

Silva G, Théry H (2017) Geografia Política e Geopolítica Brasileira no século XXI. L'Espace Politique, n 31

Silva M, Dorfman A (2020) Editorial: a Geografia Política se faz presente. Revista da Associação Nacional de Pós-Graduação em Geografia (ANPEGE), v 16, n 30

Souza ML (2020) Articulando ambiente, território e lugar: A luta por justiça ambiental e suas lições para a epistemologia e a teoria geográficas. Ambientes Revista de Geografia e Ecologia Política, vol 2, n 1
Souza ML (1995) O território: sobre espaço e poder, autonomia e desenvolvimento. In: Castro IE et al (org.) Geografia: conceitos e temas. Bertrand Brasil, Rio de Janeiro
Travassos M (1935) Projeção Continental do Brasil. Nacional, 2ª ed., São Paulo
Vesentini JW (1986) A capital da Geopolítica. Editora Ática, São Paulo
Vieira DM (2021) Territórios negros em Porto Alegre, RS (1800–1970) [e-book]: geografia histórica da presença negra no espaço. Belo Horizonte: Associação Nacional de Pós Graduação e Pesquisa em Planejamento Urbano e Regional—ANPUR
Vlach V (2003) Estudo preliminar acerca dos geopolíticos militares brasileiros, Terra Brasilis [Online], 4–5. Available at http://journals.openedition.org/terrabrasilis/359
Walker RBJ Inside/outside: relações internacionais como teoria política. Apicuri/PUC-Rio, Rio de Janeiro 2013 [1993]

Chapter 10
The Consensual Divorce of Geography. Adherence to Neoliberalism, the Cult of Freedom and the Overthrow of Democracy

Tadeu Alencar Arrais

Abstract The connection with the State has shaped, in a certain manner, the emergence of geography. This discipline has participated in different ways in the national development agenda. It has constructed a theoretical and empirical framework which has allowed us, since the transition from the national and regional scale, to interpret the nation. It has supported, from a technical and bureaucratic apparatus, the construction of development plans which have historically translated into authoritarian, hegemonic visions of national development. The fact is that the State needed geography and geography, as a scientific field of knowledge, needed investment from the State, especially in the field of training researchers. This history, however, is not linear and, above all in modern times, points to a rift, given that the historical circumstances are marked, on the one hand, by the unrestricted adherence to the neoliberal model, marked by the New Tax System (NRF), and, on the other, by the discrediting of scientific knowledge in general and the field of humanities in particular. The State has completely lost its ambition for a national development project and, due to the lack of this ambition, it has also abandoned geography.

Keywords Geography · State · Neoliberalism

10.1 Introduction

The connection with the State has shaped the history of academic geography. Geography has participated in different ways in the national development agenda. It has constructed a theoretical and empirical framework which has allowed us, since the transition of scales (national and regional), to interpret the territory. It has supported, from a technical and bureaucratic apparatus, the construction of development plans and territorial planning.

T. A. Arrais (✉)
Universidade Federal de Goiás, Goiânia, Brazil
e-mail: tadeuarraisufg@gmail.com

R. C. Lois González and M. A. Mitidiero Junior (eds.), *Brazilian Geography*,
Advances in Geographical and Environmental Sciences,
https://doi.org/10.1007/978-981-19-3704-0_10

Theoretical reflection on the State and instrumental action, via participation in state bureaucracy, have always been on geography's radar. This history, however, is not linear and, above all in modern times, points to a rift, given that the historical circumstances are marked, on the one hand, by the unrestricted adherence to neoliberalism, marked by the New Tax System (NRF), and, on the other, by the absence, in academic reflection, of a critical reflection on the nature of the Brazilian Welfare State.

Brazilian universities, like academic geography, are not a body detached from reality, which means that we have also adhered, under different circumstances, to the neoliberal project. This project is in different ways incompatible with democracy. The erosion of Welfare State policies and the predominance of fiscal adjustment policies, even in times of a pandemic, are symptoms of this moment. We advocate that in other times these topics would be a central concern of Brazilian academic geography.

10.2 The Vulgate of Liberal Freedom

The discourse on freedom has been the most efficient in the history of liberalism. Legal freedom against Divine Law. Freedom to sell the labour force which presupposed the end of the primacy of mediaeval serfdom. Its importance can only be measured when we consider the political, economic, and cultural ties of the Old Regime. The American Revolution, in 1776, and the French Revolution, in 1789, announced individual freedom and equality of opportunity as common projects. As described in Article 1 of the Declaration of the Rights of Man, published in 1789: "Men are born and remain free and equal in rights. Social distinctions may be founded only upon the general good" (UNICEF 2021, online).

In different ways, enlightenment-inspired movements placed the question of individual freedom and the defence of private property within the scope of political discussion. The philosopher John Locke (1632–1704), in the Second Treatise of Government (1963), admitted that the purpose of government was to "preserve property" (p. 77). The bourgeois State demanded the creation of a law that would preserve it. Thus, the State does not appear, therefore, as a natural evolution of society. As Marx and Engels (1986, p. 98) recalled: "Private law develops simultaneously with private property out of the disintegration of the natural community". This is the, shall we say, revolutionary character of the bourgeoisie, as well described by Marx and Engels (1997) in the metaphor "all that is solid melts into air". The bourgeoisie, as a revolutionary class, subverted all dimensions of life to its domain, and liberalism, as a doctrine, linked its destiny to that of men.

The rhetoric of freedom, under the State, finances, in the context of the Industrial Revolution, a substantial change in the labour regimes. Freedom is, above all, the freedom to sell one's labour force. The centrality that labour assumed as an analytical category in the interpretations of modernity lies in the fact explained by theorists that the wealth of a nation underwent industrial transformation and the consequent increase in productivity via the division of labour. It is the freedom to produce more and more pins, to recall Smith (2008). It is always good to point out that Smith (2008) praises liberalism and freedom in a colonialist nation. Polany (2000, p. 51), in just one question, sums up the meaning of the disarticulation of communal production in liberal England: “What ‘satanic mill’ ground men into masses?”.

The mass and whirlwind of ordinary people being pushed into a disciplinary economy is a common image in the eighteenth and nineteenth centuries. Industrial cities appear as a historically unprecedented form of centralisation of consumption and production of surpluses. Marx’s (1999) metaphor of capital as a vampire that constantly revives itself by sucking living labour is all too appropriate. The narrative that follows is of the worker as an abstract subject. For capital, age or gender does not matter, as after all, everyone is free to sell their labour force. The central issue is that bourgeois liberalism, which promised freedom, did not keep its promises. But the working class, consumed by machinery, did not inertly witness the transition. The unequivocal proof is the workers’ movements that multiplied in Europe and the USA in the nineteenth century.

The undeniable fact is that liberalism has fulfilled its generational role. It cultivated, in men, at the same time, the fear of a return to the Dark Ages and the hope that the overcoming of material scarcity would come from individual energies. It demonstrated, didactically, that the bourgeoisie found in the State the political means to defend its external interests, through colonialism, and its internal interests, through the sedimentation of the accumulation of national capital and the repression of workers’ movements. The most efficient dimension, however, came through the possibility of fostering a type of ideology that externally denied the presence of the State or any other regime of social class, party, or group solidarity, characterised as democratic. In addition to the Old Regime, they violently fought the ghosts that haunted Europe.

10.3 Neoliberalism Against Democracy

Thirty-five years separate the Military Coup that brought General Augusto Pinochet (1915–2006) to power in Chile, in 1973, from the Walter Lippmann Colloquium, held in Paris, in 1938. Nine years after the aforementioned colloquium, a society was founded on Mont Pèlerin that still bears the name of the Swiss rock formation today. Its founding statement anticipates the history of neoliberalism:

> The central values of civilisation are in danger. Over large stretches of the Earth’s surface the essential conditions of human dignity and freedom have already disappeared. In others, they are under constant menace from the development of current tendencies of policy. The position of the individual and the voluntary group are progressively undermined by extensions of arbitrary power. Even that most precious possession of Western Man, freedom of thought and expression, is threatened by the spread of creeds which, claiming the privilege of tolerance when in the position of a minority, seek only to establish a position of power in which they can suppress and obliterate all views but their own, etc. (Pelerin 1947, online).

At the founding meeting of the Mont Pèlerin Society, the economists Ludwig von Mises (1881–1973), Friedrich A. Von Hayek (1899–1992), and Milton Friedman (1912–2006) were present. This trio were protagonists of the economic and political debate in the twentieth century. If classical liberalism, in its essence, was revolutionary in terms of ideas, the same applies to neoliberalism, because, in 1947, when the Mont Pèlerin Society was founded, there was an understanding that State intervention in different fields of society, was necessary. After all, it was necessary to rebuild post-war Europe as much as to create mechanisms of political and economic regulation on a global scale. It is a period between 1945 and 1975 called The Glorious Thirty. The growth of employment and the emergence of the Welfare State with the consequent fall in wage inequality, especially due to the growth deficit in the preceding period, as far as the European continent is concerned, endorses this argument. The maturing, in the USA, of the Social Security Act, which proposed a set of federal benefits for the most vulnerable, in addition to unemployment insurance, and the Beveridge Plan (Beveridge 1943), in England, promoting a broad retirement and pension plan, in addition to other allowances, illustrate the State's primary concern with social vulnerability and worker integration.

Under the scorn of economists such as John K. Galbraith (1908–2006), as reported by Wapshott (2016), neoliberals elected opponents and cultivated a simple and efficient system of ideas. Historian Perry Anderson (1995), citing the Mont Pèlerin Society, writes: "Its purpose was to combat the reigning Keynesianism and solidarity and to prepare the foundations of another kind of capitalism, hard and free of rules for the future" (p. 10). The first intervention matrix, which was countercyclical, was represented by the British economist John Keynes (1883–1946) and the second also by the British economist William Beveridge (1879–1963). The battle of ideas was not limited to the economic dimension, as it also returned to the political and cultural realm. There was, so to speak, a materiality in the ideas that shaped the Welfare State and Social Democracy which should be attacked in essence. The movement to expand the State's social expenditure, whether in France or in the USA, could not be accepted.

The topic of freedom, unlike classical liberalism, would no longer support a discourse against the Old Regime. The industrial bourgeoisie had already solved this problem during the eighteenth and nineteenth centuries. As a winning class, so to speak, the bourgeoisie refunctionalises the discourse of freedom by targeting the political conjuncture of the first two quarters of the twentieth century. The First World War, the Russian Revolution, the Wall Street Crash of 1929, and the Second World War are events, from the point of view of causes, that are permeable. The discourse of freedom, as opposed to communism and socialism, arises in this environment. The narrative constructed by Hayek, in The Road to Serfdom, a book published in 1944 and dedicated, sarcastically, to "socialists of all parties", set the tone of the battle: "few are ready to admit that the rise of Nazism and Fascism was not a reaction against the socialist tendencies of the preceding period, but the result of those very tendencies". (Hayek 2010 [1944], p. 31).

The change that took place, beyond the rhetoric of threat to individual freedom worthy of trailers such as that of the 1983 film The Day After, took place, on the one hand, on the factory floor and, on the other, in the financing systems of public services and security that had been sustained in Europe and the USA by the expansion of the Fiscal State. The changes in the pattern of accumulation, described by Harvey (2008), affected not only the labour market, but also the way in which public policies are financed. What Harvey (2008) terms "accumulation by dispossession" is supported, at different times, by privatisation programmes, financialisation, crisis management and manipulation, and also redistribution via the State. This was the prescription, even before the so-called Washington Consensus, in 1989, experimented in Latin America.

The political response to the crisis of accumulation and its translation into unemployment and the increase in social inequality, from the perspective of neoliberals, began to expand corporate liberties that, ironically, had a highly mapped national origin. Freedom meant fighting, with all weapons, against the Welfare State. In Latin America, this is done by the dagger of generals, as warned by Galeano (2020). In Chile, General Augusto Pinochet (1915–2006) overthrew Salvador Allende (1908–1973). The cynical character of the neoliberal programme is expressed didactically in the way Milton Friedmann and F. Hayek fawned over the Chilean dictatorship. There is no incompatibility between neoliberalism and dictatorship. In an article published in 1976, criticising Friedmann's visit to Chile, as described by Duke and Montes (2016), the journalist Orlando Letelier clearly exposed the limits of economic freedom. The transcription of fragments of the article, although long, is timely, given that, shortly after its publication, Orlando Letelier was murdered:

> It is curious that the man who wrote the book Capitalism and Freedom, and developed the argument that only classical economic liberalism can support political democracy, can now so easily dissociate economics from politics, when the economic theories he defends coincide with an absolute restriction of all kinds of democratic freedom. It is logical to expect that if those who oppose private enterprise are held accountable for the effects of its measures in the political sphere, those who impose unlimited "economic freedom" should also be held accountable when the imposition of that policy is inevitably accompanied by massive repression, hunger, unemployment, and the subsistence of a brutal police state. (Letelier 1976, p. 1).

In Brazil, the criticism of neoliberalism appears more frequently in the 1990s, manifested by the economic opening and the consequent process of privatisation of state-owned companies. Biondi (1999) prepared a detailed account of the privatisation of Telesp (Telecomunicações de São Paulo), Vale do Rio Doce (Companhia Vale do Rio Doce), Light (Light Serviços de Eletricidade), Embratel (Empresa Brasileira de Telecomunicações), Usiminas (Usinas Siderúrgicas de Minas Gerais), Cemig (Companhia Energética de Minas Gerais), etc. However, as in the Chilean military regime (1973–1990), the fact that the genealogy of Brazilian neoliberalism finds itself in the 1964 Military Coup escapes criticism. João Goulart's (1919–1976) speech on 18 March 1964 announcing the so-called Basic Reforms demanded a harsh response from the internationalised economic system and the national bourgeoisie (Goulart 1964). In response to the threat of private property, the March of the Family with God for Liberty took place days later [1] The taxation of foreign capital, nationalisation of refineries, in addition to social policies associated with the resumption of the social function of land ownership, especially in areas bordering highways and public dams, set the tone for the political conjuncture of the time. The ghosts exorcised in Europe could not find shelter in Brazil. For those who have doubts about this genealogy of neoliberalism, it is enough to remember that the report of the US-Brazil Economic Mission, known as the Abbink Mission (Bulhoes 1950), pointed to the end of the employment stability policy and the need to exonerate international capital as predicates for growth of the Brazilian economy, a recipe promptly followed by the Military Regime.

The interpretation of neoliberalism only from the perspective of privatisation and the pursuit of productive efficiency minimises its political dimension, translated into the very capture of the State, as well as its subjective dimension, resulting in the acceptance of a narrative that there is no incompatibility between authoritarianism and development, something that is frequent in the speeches of the current president Jair Messias Bolsonaro (2019–) and naturalised by a portion of Brazilian society. The discourse that promises to fight the communist threat, even though it seems folkloric, is not at all naive and has already prompted, at various times throughout Brazilian history, the implementation of authoritarian regimes.

10.4 The Neoliberal Bureaucratic Symbiosis

Economist John Kenneth Galbraith (1908–2006), in Economics and the Public Purpose (1975), calls the continuous feedback between the market and the State a bureaucratic symbiosis, citing, among other examples, the relationship between the department of transportation and the North American automotive industry. The separation between State and market was not even seen from Smith's point of view (2008). This bureaucratic symbiosis was exactly the same in Brazil. The capitalisation of contractors during the Military Regime (1964–1985), as pointed out by Campos (2014), or even the fable of the Brazilian Miracle sustained by the intense exploitation of labour, are enough to demonstrate that this symbiosis has always been present in military and civil halls. This symbiosis happens, for example, through the regulation of the labour force, tax incentives, infrastructure financing, interest payments, and, especially, direct procurement. This understanding is fundamental, since the distorted understanding of this proof reinforces the idea about the inefficiency of the State. Neoliberalism is not against the State. There is not a struggle against the State, but a struggle for power to direct spending priorities, control natural resources, and

[1] The March of the Family with God for Liberty first took place in São Paulo, on 19 March. It is considered a response to João Goulart's rally on 13 March 1964. The march brought together segments of organised society, such as landowners, employers' unions, segments of the church, right-wing politicians and organised women's movements. The motto *God, Family and Liberty* well represents the conservative agenda that would be assumed, institutionally, from the 1964 Military Coup.

Table 10.1 Federal Government—implementation of expenditure by nature, 2020

Expenditure group	Value (r$)	% Relative to the grand total
Other current expenditures	1,637,305,500,695.39	47.91%
Debt amortisation/refinancing	1,036,043,438,509.74	30.31%
Debt interest and charges	346,683,477,200.01	10.14%
Personnel and social security contributions	263,672,279,200.67	7.72%
Financial investments	113,422,267,860.68	3.32%
Other	20,471,839,942.65	0.60%
Total	3,417,598,803,409.14	100.00

Source Portal da Transparência (2020)

intervene in strategic sectors of national economies. This is evident when we look at the contemporary reform agenda in Latin America. The Brazilian NRF, approved in 2016, provides a summary of this perspective: "the New Tax System is hereby instituted within the scope of the State Tax and Social Security Budgets, which will be in force for twenty financial years, under the terms of Arts. 107 to 114 of this Transitional Constitutional Provisions Act" (BRASIL 2016, online).

Tavares and Ramos (2018) argue that the PEC-95 constituted a threat to Brazilian democracy. Essentially, it compromises public investments in health, education, science, etc., for 20 years, leaving room only for financial commitments.[2] The fiscal austerity amendment is the backbone of contemporary neoliberalism. Table 10.1 shows that the amortisation and financing of public debt, in 2020, consumed 30.31% of the Federal Government's expenditures.

The priority for public debt interest and refinancing compromises the solvency of the State, stimulating fiscal adjustment policies that, in turn, aim to create fiscal surpluses to pour more resources into the financial market, in an endless spiral. If, on the one hand, the financial market drains public resources, on the other, part of the primary expenditure capitalises on the internal market. The State's struggle does not only find itself in the dispute over public assets or natural resources. Financialisation is sponsored by the indebted State in order to feed and drain resources that would be destined, for example, to social policies. The indebtedness aspect shaped what Streeck (2018), when analysing the crises of capitalism, called buying time.

When considering the Federal Government expenditures (excluding those destined for public debt amortisation and refinancing), Social Security and Welfare

[2] "The most severe contractionary fiscal policy in the recent history of capitalism, perhaps the biggest ever proposed in the history of humanity, takes shape in the Brazilian scenario in December 2016. Constitutional Amendment 95 legislates on public expenditures, preventing the growth of expenditures destined to primary expenditures for twenty years and excluding from the debate the restriction of the increase concerning the public debt interest and its rollover. In addition, this amendment does not cover non-dependent state-owned companies, responsible for increasing the State's liabilities without control of the main public budget maintenance bodies." (Tavares & Ramos, 2018, p. 1)

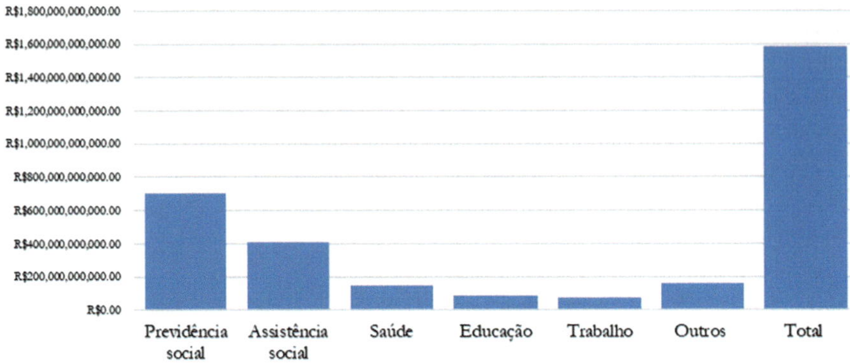

Fig. 10.1 Area of activity (function) of the Federal Government with the highest expenditure, 2020. *Source* Portal da Transparência (2020)

represented 70.14% of the total in 2020, which is equivalent to 1.11 trillion reais, compared to 1.37 trillion reais (Fig. 10.1) destined for the payment of public debt interest, amortisation, and financing. Health, in the year in which the COVID-19 pandemic began, represented 9.48% of the total Federal Government expenditures.

But the State also capitalises companies via a procurement system. In 2020, 34.8 billion reais were spent on procurement through contracts by the Federal Government. The main beneficiaries of the contracts are contractors, companies linked to the automotive industry, companies in the pharmaceutical and chemical sector, publishers, as well as service providers. This is a form of capitalisation, let us say, in retail, which is repeated at the state and municipal scales, in addition to the Federal District.

It is the budget framework, in a repeated discourse on the public sector deficit, that fuels the rhetoric of reforms. It is not by chance that the Labour Reform (BRASIL 2017), approved in 2017, the Social Security Reform, approved in 2019, and the Administrative Reform, currently being considered in the National Congress, arise so much interest. The first reform relieved capital of the burden of labour and stimulated precarious forms of labour at a time of intense unemployment, whilst also removing the obligation to pay union contributions. The result is that, unprecedentedly, there are more people unemployed and/or in informal jobs than formally employed people, that is, with a Work and Social Security Card. The total number of registered Individual Microentrepreneurs, as of 30 June 2021, reached 12,223,716 (SIMEI 2021). It is an army of entrepreneurs made up, mostly, of manicurists, hairdressers, maids, app delivery workers, drivers, etc. The second reform condemned a significant portion of the population, linked to Brazil's National Institute of Social Security (INSS), to contribute for 40 years, in an unstable and regionally differentiated labour market, with the possibility of retiring from a minimum wage. The discourse on the social security deficit, based on the idea of privileges, was one of the most accepted by the Brazilian population (Arrais and Viana 2019). In May 2021, 66.67% of the issued benefits reached the value of up to one minimum wage and 83.39% the value of up to two minimum wages (INSS 2021). In that month, 31,237,449 benefits were issued, which sets the tone for the importance, from the point of view of combatting income inequality, of Social Security for the 5568 Brazilian municipalities (Arrais 2019). The third reform, which is in progress, criminalises the civil service, spreading the myth of inefficiency. It is as if the expansion of public services did not depend on the expansion of the State's technical and bureaucratic framework.

These three structural reforms have in common the inscription of a subjectivity that fosters a false consensus, sponsored by the financial system, the hegemonic media, and the Federal Government, that the reforms constitute a sine qua non condition to overcome the crisis produced by the system itself. Neoliberal magic requires, for its effective functioning, the outsourcing of the crisis.

10.5 The State and the Mapping of Social Rights

The history of the Brazilian Welfare State is not disconnected from the secular challenges of a society that insists on concealing its slave-based condition. The demands, partial conquests and even the incomplete expansion of social rights are a direct result of the struggle of urban and rural workers. The Constitution of 1988 (BRASIL 1988), called the Citizens' Constitution, due to the environment of democratisation, translates, in two articles, the challenges of our young and, now, threatened democracy.

The challenge of democratisation was not only the resumption of representative democracy. It was necessary to universalise access to public services and guarantee welfare and social security benefits for a population dispersed over 8,510,345,538 km2 (IBGE 2021). The challenge was enormous, especially for a country whose poverty was urbanising with extreme speed. In 1991, Brazil reached the mark of 146,825,475 inhabitants. It is in this territory, which is socially unequal, differentially populated and ecologically distinct, that the State has expanded its social protection network through public services. In 1991, 85.8% of Brazilian municipalities were in the range known as the low range of municipal development, a percentage which was reduced to 41.8% in 2000 and 0.6% in 2010. The middle range of municipal development prevailed in 2010, covering 40.1% of the municipalities, which is much higher than the 0.8% recorded in the same range of development in 1991 (UNDP 2010). In 19 years, through public policies of income transfer, increase in the minimum wage, in addition to inflation control and the progressive universalisation of health and education services, the map of municipal development has completely changed its features. The path of universalising public services and protecting the most vulnerable was a fundamental step towards the "permanence" of that young democracy.

Expanding State action in the provision of public services, especially education and health, would not be efficient in combating inequality without Social Security. The expansion of the social security floor to a minimum wage and its extension to rural workers and the elderly and people with disabilities in poverty was a decisive factor in reducing income inequality. The minimum wage for the Rural Retirement Pension, for both men and women, changed the features of the Brazilian rural space (Arrais 2016). The Bolsa Família (Family Allowance) Programme, created in 2003, benefitted 6.5 million families that year, then reaching 13.7 million families in 2015 and, in May 2021, 14,695,025 families. In the midst of the COVID-19 pandemic, with the deepening income inequality and unemployment, the country recorded 29,847,849 low-income families (MDS 2021), which means that there was no expansion of the Bolsa Família coverage. It is essentially an urban, efficient, and cheap programme (Arrais 2016). But the protective tripod of the Welfare State (Rural Retirement Pension, Continuous Benefit and Bolsa Família) would not have a positive effect without the presence of the Unified Health System (SUS), whose principles of universality, integrality, and gratuitousness, give expectation, in each of the more than 5568 municipalities, of access to health.

The recipe for austerity, which precedes the pandemic, announces that the most vulnerable groups will not have a place in the Minimal State. The efficiency of the doctrine starts to affirm the individual's protagonist role in a meritocratic market, which is essential to block the construction of political environments outside the atmosphere of individual and corporate competition. This demonstrates that the enemy of neoliberalism is a regime marked by different degrees of solidarity, tensions, and political consequences, whether or not derived from the electoral arena we call democracy. These consequences are incompatible with authoritarian governments. Levitsky and Ziblatt (2018) point to the rejection of the rules of democracy, the denial of the legitimacy of opponents, the stimulation of violence and the propensity to restrict civil liberties as indicators of authoritarian behaviour. The recipe applies to the Brazilian case.

10.6 The Consensual Divorce of Geography

As of 30 October 2021, Brazil had recorded 607,694 deaths from COVID-19 (COVID-SAÚDE 2021). As of the same date, 21,804,090 people had been infected. The total number of after-effects, as in a war, will be difficult to count. Social inequality has unequivocally become a risk factor in a country that is inertly witnessing one of the most radical neoliberal experiences on the planet. The idea of socialising the risks of the pandemic, very early on, proved to be a fable (Arrais et al. 2020). The possible question in this period of the history of geography is: Do we have theoretical and conceptual instruments that can help in the interpretation of the conjuncture (political, economic, and social) and, at the same time, stimulate the necessary mobilisation for the political struggle in defence of democracy.

The pandemic arrives at a particular moment in our political, economic, and also epistemological history. Agendas for research and political intervention that were once central to geography, even if located in the umbrella of political geography and economic geography, have lost their centrality. Disciplines on theory and/or functioning of the State and even on public policies are not included in the curricular matrices of undergraduate courses in geography. The slightest hint of attachment to wholeness in the corridors can even be understood as orthodoxy. It is illustrative that the word State and the concept of public policy do not appear in any of the titles of the 73 Working Groups of the XIV National Meeting of Postgraduate Studies and Research in Geography (ENAPEGE), the main scientific event in Brazilian geography. Public policy, even from an instrumental perspective, seems to be off the radar of Brazilian academic geography. Research agendas increasingly seem to respond to individual appeals. The possessive pronoun (my research, my group, my laboratory, my students) has never been so used in academic corridors.

Is the hypothesis that the neoliberalism that erodes State policies is the same that surrounds the university and, therefore, geography, exaggerated? It is no secret that contemporary geography has made interpretation of the State secondary and also neglected politics. Ironically, for this very reason, one can be alarmed, in contemporary times, with the erosion of public policies in several dimensions. It seems that the radical austerity policies would not affect riverside communities, indigenous peoples, quilombolas, rural settlers or even housing, urban mobility, and public education funding policies.

Neoliberal pedagogical projects, with a modern discursive veneer, intend to reform the school, leaving aside the understanding that the primary school teacher is, above all, a worker. The issue of the precariousness of teaching work is, indeed, often infantilised in the teaching of geography. This geography seemed to be unaware that the Unified Health System (SUS) preceded the COVID-19 pandemic. The Unified Health System, created from the Constitution of 1988 (Brasil 1988), started from the principle of health as a right of all and a duty of the State. The SUS Tripod is composed of integrality, universality, and gratuitousness. The SUS is the main asset of Brazilian democracy.[3] This geography is equally surprising when realising that the Rural Retirement Pension, as well as the other Social Security benefits, is responsible for the capitalisation of municipal economies. But the worst thing is that this geography still seems to condemn, a priori, the State, by accepting the vulgate of endemic corruption when, from a financial point of view, it is the expenditures with interest and public debt amortisation that compromise the very existence of the State's bureaucratic structure. This a priori condemnation of the State was explored by Souza (2019). A part of this academic geography closed its eyes to the PEC-95 (BRASIL 2016), as after all, a constitutional amendment of a few hundred characters could not affect our research! The young professor and researcher, recently hired into the federal education system, busy producing high-impact articles, did not even realise that the dismantling of the retirement and pension system would strike at their jugular. But that did not seem like a problem, as after all, on their own, they are enough! They do not need unions and even less so class associations! The competitive

[3] The Brazilian Welfare State Observatory, on the channel Porque O Estado Importa, produced a series of informative videos labelled Geography of the SUS. In.: https://www.youtube.com/watch?v=D6z3lamGtA8&list=PLtScHQ6ZcbiHcQQQkBQRi5ZrQDCvUK4d8

sphere is, above all, individual. The classic film The Working Class Goes to Heaven, by Elio Petri, could, if updated, be called The Academic Class Goes to Heaven. The intellectual repeats, without the same revolt, the fate of Lula Massa, as they try to adapt themselves to the gears of productivity. The shift from collective subjectivity to individual subjectivity was a slow, gradual operation, capturing institutions, and people. The intellectual research agenda did not escape this determination, as it incorporated both the grammar of competition and the practice of competition. This is a dimension of what Dardot and Laval (2016, p. 16) called the new way of the world:

> For almost a third of a century, this rule of life has governed public policies, commanded world economic relations, transformed society, and remodelled subjectivity. The circumstances of this normative success have been described numerous times. Sometimes in its political aspect (the conquest of power by neoliberal forces), sometimes in its economic aspect (the rapid growth of globalised capitalism), sometimes in its social aspect (the individualisation of social relations at the expense of collective solidarities, the extreme polarisation between the rich and poor), and sometimes in its subjective aspect (the emergence of a new subject, the development of new psychological disorders).

The human sciences, and also geography, have not always neglected reflecting on the functioning and nature of the State. Max Weber (1864–1920) devoted energy to understanding state bureaucracy. For him, the power of the State was manifested through "bureaucratic management" (2001, p. 23). Marx (1818–1883), before this sociologist, observed that the "… State bundles, controls, regulates, supervises and keeps civil society under tutelage" (Marx 2008 [1974], p. 60). This occurs through bureaucracy. The State, therefore, since modernity, as Pikety (2014) taught, has added, to the traditional functions of security and war, other demands that resulted from social pressures. The State has built physical and abstract networks legitimised by legislation that populate national territories. Geography has interpreted the action of this State through the prism of territory and modernisations. We find this tradition in the works of Santos (1979, 2008) who considered the Nation State as a totality. Undoubtedly, it is one of the most competent ways to participate in the national political debate, as well as in the proposal of government public policies. The historic link to State planning, at various scales, is an unclaimed legacy. The strange thing is that we are living in times of a pandemic in which, every day, we demand the presence of the State or even revolt with the erosion of public investment in research, education, health, culture, etc. Unfortunately, policing in relation to the non-use of the gender neutral will not solve our demands for public policy or even our chronic problem of mobilisation. It is true that we experienced a kind of enchantment in the 2000s, under the aegis of the Lula government (2003–2013), which blocked the necessary distance for a critical assessment of the neoliberal project. Lazzarato (2019, p. 36) offers a point of view that deserves to be explored:

> Neoliberalism did not suddenly arrive at the end of Lula's terms, it was cultivated, favoured, fostered—irony of fate!—by the Workers' Party. Capital has an excellent relationship with the institutions of the labour movement, as financialisation would have been inconceivable without the "pension funds" of US wage earners (teachers, employees, labourers), and major institutional investors in the stock market.

Lazzarato (2017) says that indebtedness has become the driving force behind the neoliberal financialisation movement. This perspective is not highlighted, theoretically or empirically, in contemporary analyses of Brazilian neoliberalism. A discreet change, carried out in 2003, in Article 192 of the Federal Constitution of 1988, is enough to endorse Lazzarato's arguments (2017, 2019). Article 192 limited interest and commission rates to 12% per year, further conceptualising raising interest beyond this level as a crime of usury. A Constitutional Amendment, dated May 2003 (BRASIL 2003), at the beginning of the Lula government and under the presidency of the Federal Chamber of PT Deputy João Paulo Cunha, revoked the provisions that would regulate the charging of abusive interest. This change responded to the political commitment signed in the Letter to the Brazilian People (PT, 2002) in June 2002. The grammar, throughout the letter, adopts a moralistic and conciliatory tone. A less explicit type of amnesty. For the financial market, everything worked well, since it was guaranteed the draining of resources from the State and also from the consumer. The accumulated interest on revolving credit cards exceeded, in February 2021, 320% per year, and on overdrafts over 120% per year (BCB 2021). In February 2021, there were 61.6 million people in default and another 11.2 million companies in default (Serasa Experian 2021). Debt and interest have become the biggest input of neoliberalism, which is now institutionalised. Santos (1993), in a book published shortly after the country's democratisation, warned about this tendency of converting citizens into consumers. In an interview with the newspaper Folha de S. Paulo, on 22 September 2003, Francisco de Oliveira made the following diagnosis:

Folha—Why does the platypus conspire against politics?

Because the platypus is something that is truncated. It produces this "informal" army [of workers] and you can't do politics like that. Doing politics without something that objectively structures interests is a missionary's task. It is necessary to call on Fathers Nóbrega and Anchieta again to re-evangelise. Why did these class parties arise in the wake of unions? Because it is a material structure of interest that is capable of aggregating. They are interests that aggregate, on the side of capitalists and workers. When this material structure disintegrates, doing politics is a missionary's thing. The conditions that formed the PT no longer exist. There will certainly be other conditions, but which ones? Nobody knows. It will be done by social experimentation. And, it may come to nothing. Let us end this automatic belief in progress. It may come to nothing.

We can hypothesise that, concomitantly with the adherence provoked by the enchantment in the face of redistribution and credit policies for families, we have also advanced in the direction of epistemological relativism, which has distanced us from structural and political debates. Postmodernity has reaped its fruits in the soil of geography. The adopted relativism, whether coincidental or not, has kept us away from criticism. The conditions for consensual divorce were given. The provocative text by François Dosse, published in the 1970s, called The last minute guest: geography finally awakens to epistemology (Dosse 2007), could even be called, nowadays, The late guest, geography finally discovers particularisms. In being content with lateral interpretations of the world, geography has, in different ways, distanced itself from politics. It is no longer a matter, as in the 1980s, of the dichotomy between physical geography and human geography or even the tensions between bachelor's and teaching degrees. From today's perspective, those debates are naive. Geography has adopted particularism, which is the highest stage of fragmentation, and unknowingly flirts with neoliberalism on a daily basis. It tries to build, from epistemological gymnastics, a type of neo-exceptionalism that has been giving up on the ambition of a more globalised narrative regarding social phenomena.

The divorce of the State, which has taken place in geography, manifests itself in our difficulty in understanding that neoliberalism is not an abstract, moral discourse, but a practice that could easily be mapped, as is the case with the effort that has been undertaken by the Brazilian Welfare State Observatory. The purification of the grammar of freedom, the one most socialised by neoliberalism, and this has a history, is the most difficult to understand. It is not without reason that the most successful books of the neoliberal trinity, formed of economists awarded with the Nobel Prize, are very attractive to the general public. They are easily digestible political narratives that have often opted for blackmail and the spread of fear. Mises (2015) insists that the anti-capitalist mindset is the work of fanatics. Hayek (2010 [1944]), in comparing socialism, communism, and fascism, denounces serfdom. Friedman (1977), a man who frequented the White House Oval Office, praises the role of freedom in capitalism. The author, on the role of government, highlights: "Its major function must be to protect our freedom both from the enemies outside our gates and from our fellow citizens; to preserve law and order; to enforce private contracts; to foster competitive markets (p. 12)".

This consensual divorce took geography away from the discussion of a nation project. Geography is the heir to a competent tradition of territorial analysis. The challenge is to interpret the complexity of contemporary territory, as well as the technical and bureaucratic functions of the national State, using the inputs of geography. Many geographers have undertaken this task with competence and conceptual rigidity. Many understand that the Nation State, as well as the Welfare State, even if incomplete, is historical facts resulting from conflicts between social classes and groups. Many advocate its overcoming through more horizontal regimes of solidarity. This utopian horizon, especially in times of crisis, is essential, and this does not imply disregarding the materiality represented in the dispute over the budget, as well as the bureaucratic symbiosis. These disputes are not, however, only in the economic or fiscal dimension. They are investments, as Dardot and Laval (2016) point out, in the construction of a corporate, state, and individual subjectivity whose norm is competition and aim is blocking any regime of political solidarity. Our challenge is justified, in this sense, for three reasons. First, as geographers, obliged by our professional duty to unveil and denounce the impact and materiality of each neoliberal project in this continental territory. Second, as civil servants who, by our professional duty, are part of a bureaucracy that has been criminalised on a daily basis. Third, as citizens who, being interested in participating in the national debate, do not disregard the fate of the most vulnerable. Perhaps it is time, in the face of the violence that replaces politics and the fascism that surrounds democracy, to annul the divorce.

References

Anderson P (1995) Balanço do neoliberalismo. In: Sader E et al (eds) Pós-neoliberalismos: as políticas sociais e o Estado Democrático, 8 ed. Paz e Terra, Rio de Janeiro

Arrais TA (2016) O Bolsa Família e a tradução regional da questão social. Rev. Inst. Estud. Bras, vol 65. Available at: https://www.scielo.br/j/rieb/a/TStPNDQcMBfvg7zMvwWFXJB/?lang=pt. Accessed on 27 July 2021

Arrais TA (2019) Income inequality, public employment and income transfers in contemporary Brazil. Mercator, Fortaleza, vol 18, pp 1–18. Available at: http://www.mercator.ufc.br/mercator/article/view/e18008. Accessed on 24 July 2021

Arrais TA, de Oliveira AR, Viana JL, Alencar DP, Salgado TR, Neto JM, de Souza ME (2020) Celeiros da pobreza urbana: suplementação de renda e isolamento social em ambientes metropolitanos nos tempos pandêmicos. Vigilância Sanitária Em Debate: Sociedade, Ciência & Tecnologia (HealthSurveillanceunder Debate: Society, Science & Technology)—Visa Em Debate 8(3):11–25. Retrieved from https://visaemdebate.incqs.fiocruz.br/index.php/visaemdebate/article/view/1609

Arrais, T. A.; Viana, J. L. Pequeno atlas da tragédia previdenciária brasileira. Goiânia: UFG-IESA, 2019. Available at: http://obsestadosocial.com.br/literatura/pequeno-atlas-da-tragedia-previdenciaria-brasileira/. Accessed on 24 July 2021

BCB. Banco Central do Brasil (2021) Available at: https://www.bcb.gov.br/estatisticas/reporttxjuros?parametros=tipopessoa:1;modalidade:204;encargo:101. Accessed on 11 March 2021

Beveridge WHOPB (1943) Livraria José Olympio, Rio de Janeiro

Biondi AO (1999) Brasil privatizado. Perseu Abramo, São Paulo

Brasil. Constituição Federal de 1988. Available at: http://www.planalto.gov.br/ccivil_03/constituicao/constituicao.htm. Accessed on 12 Mar 2021

Brasil. Emenda Constitucional (2003). Available at: http://www.planalto.gov.br/ccivil_03/constituicao/Emendas/Emc/emc40.htm#art2 Accessed on 12 Mar 2021

Brasil. Emenda Constitucional número 95 (2016) Available at: http://www.planalto.gov.br/ccivil_03/constituicao/emendas/emc/emc95.htm. Accessed on 18 July 2021

Brasil. Lei no 13.467, de 13 de julho de 2017. Available at: http://www.planalto.gov.br/ccivil_03/_ato2015-2018/2017/lei/l13467.htm. Accessed on 24 July 2021

Bulhoes OG (1959) À margem de um relatório. Available at: https://bibliotecadigital.seade.gov.br/view/listarPublicacao.php?lista=0&opcao=5&busca=A%20Margem%20de%20um%20relatorio&tipoFiltro=pa.id_autor&filtro=70&descFiltro=BULH%C3%95ES,%20Octavio%20Gouv%C3%AAa%20de&listarConteudo=T%C3%ADtulo%20%20%C2%BB%20A%20Margem%20de%20um%20relatorio. Accessed on 22 July 2021

Campos, P. H. Pedreira. Estranhas catedrais: as empreiteiras brasileiras e a ditadura civil-militar, 1964–1988. Niterói: Eduff, 2014.

COVID-SAÚDE. Brasil: estatísticas. Available at: https://covid.saude.gov.br/. Accessed on 24 July 2021

Dardot P, Laval C (2016) A nova razão do mundo: ensaio sobre a sociedade neoliberal. Boitempo, São Paulo

Dosse FA convidada de última hora: a Geografia desperta para a epistemologia. In: Dosse F (ed) História do estruturalismo: o canto do cisne, de 1967 a nossos dias. Bauru: Edusc, 2007 [1992] (v. 2)

Duke BC, Montes L (2016) Friedrich Hayek y sus dos visitas a Chile. Available at: https://www.cepchile.cl/cep/site/docs/20160304/20160304101209/rev137_BCaldwell-LMontes.pdf. Accessed on 15 Dec 2020

Friedman M (1977) Capitalismo e liberdade. Artenova, Rio de Janeiro

Galbraith JK (1975) A economia e o objetivo público. Martins, São Paulo

Galeano E (2020) As veias abertas da América Latina. LM&PM Pocket, Porto Alegre

Goulart J (1964) Discurso na central do Brasil. Available at: https://memoria.ebc.com.br/cidadania/2014/03/discurso-de-jango-na-central-do-brasil-em-1964. Accessed on 20 July 2021

Harvey D (2008) O neoliberalismo: história e implicações. Loyola, Rio de Janeiro
Hayek FA O caminho da servidão. Instituto Ludwig von Mises Brasil, São Paulo 2010 [1944]
IBGE (2021) Organização do Território. Available at: https://www.ibge.gov.br/geociencias/organizacao-do-territorio/estrutura-territorial/15761-areas-dos-municipios.html?=&t=o-que-e. Accessedon 20 July 2021
Lazzarato M (2017) O governo do homem endividado. M1 Edições, São Paulo
Lazzarato M (2019) Fascismo ou revolução? M1 Edições, São Paulo
Letelier O (1976) The Chicago boys in Chile: Economic freedom's awful toll. Available at: https://www.tni.org/my/node/12111#:~:text=Deeply%20involved%20in%20the%20preparation,the%20intellectual%20assets%20it%20lacked. Accessed on 15 Dec 2020
Levitsky S, Ziblatt D (2018) Como as democracias morrem. Zahar, Rio de Janeiro
Locke J (1963) Segundo tratado sobre o governo. Ibrasa, São Paulo
Marx KO (1999) Capital: o processo de produção do capital. 17 ed. Civilização Brasileira, Rio de Janeiro, vol 1
Marx KO (2008) 18 de Brumário de Luís Bonaparte. Martin Claret, São Paulo
Marx K, Engels F (1986) A ideologia alemã, 5 ed. Hucitec, São Paulo
Marx K, Engels F (1997) Manifesto do Partido Comunista. Available at: https://www.marxists.org/portugues/marx/1848/ManifestoDoPartidoComunista/index.htm. Accessed on 20 June 2021
Mises LV (2015) A mentalidade anticapitalista, 2 ed. Vide Editorial, Campinas
MDS. Estatísticas do Cadastro Único e do Bolsa Família (2021). Available at: https://aplicacoes.mds.gov.br/sagirmps/bolsafamilia/painel.html. Accessed on 22 July 2021
MONT PÈLERIN. Statement of Aims (1947). Available at: https://www.montpelerin.org/statement-of-aims/. Accessed on 15 Dec 2020
Observatório Do Estado Social Brasileiro. Available at: http://obsestadosocial.com.br/. Accessed on 24 July 2021
Piketty T (2014) O capital no século XXI. Intrínseca, São Paulo
Polany K (2000) A grande transformação: as origens de nossa época. Campus, Rio de Janeiro
Santos M (1979) Espaço e sociedade. Vozes, Rio de Janeiro
Santos M (1993) O espaço do cidadão. Nobel, São Paulo
Santos M (2008) Da totalidade ao lugar. Edusp, São Paulo
SerasaExperian. Dados gerais (2021) Available at: https://www.serasaexperian.com.br/conteudos/indicadores-economicos/. Accessed on 12 March 2021
SIMEI (2021) Receita Federal. Estatística Microempreendedores. Available at: http://www8.receita.fazenda.gov.br/simplesnacional/aplicacoes/atbhe/estatisticassinac.app/EstatisticasOptantesPorDataMunicipio.aspx?tipoConsulta=2&optanteSimei=1&anoConsulta. Accessed on 22 July 2021
Smith A (2008) Riqueza das nações. Hemus, São Paulo
Souza J (2019) A elite do atraso. Estação Brasil, Rio de Janeiro
Streeck W (2018) Tempo comprado. Boitempo, São Paulo
Tavares FM, Ramos PVG (2018) O novo regime fiscal brasileiro e a democracia: interpretações sobre uma incompatibilidade. Boletim Goiano De Geografia 38(3):450–472
UNICEF. Declaração dos Direitos Humanos 1789. Available at: https://www.unicef.org/brazil/declaracao-universal-dos-direitos-humanos. Accessed on: 20 July 2021
Wapshott N (2016) Keynes X Hayek: as origens e as herança do maior duelo econômico da história. Record, Rio de Janeiro
Weber M (2001) A objetividade do conhecimento na ciência social e na ciência política. In: Weber M Metodologia das ciências sociais: parte I. Cortez, São Paulo

Chapter 11
Scientific Research and the Construction of the Field of Teaching of Geography in Schools: Trends and Challenges

Ângela Massumi Katuta and Maria Adailza Martins de Albuquerque

Abstract The aim of this chapter is to present the cutting edge Brazilian academic–scientific production focused on the area of Teaching of Geography, limited to the years 1987–2019. This limitation is justified due to the fact that it was the period in which there was an increase in the amount of work produced, national coverage, research topics and clashes of theoretical-methodological, epistemological and political guidelines which developed the debates on the concepts of Teaching of Geography, course plans, teaching practices, research and outreach, among much other production, which influenced the field of geography taught in schools and the initial and ongoing education of professors. To do so, we will cover the origins of the field of research, the transformation which it has undergone, as well as its current configuration and challenges, with the critical, libertarian and emancipating perspective of education as a horizon for action. We have conducted qualitative and quantitative studies of books, specialist journals, dissertations and theses produced on the issue in order to reflect on the path for constructing this field of knowledge and, above all, on its challenges with a view to strengthening Piagetian education, which together with other conservative and currently neoliberal policies, have strongly impacted and continue to impact the field with regard to the expansion of the number of jobs focused on the methodologies of teaching which help to alienate the political and transformative role of education in a country of broad social and territorial inequalities, such as Brazil.

Keywords Teaching of Geography · Scientific research · Field of knowledge · Education · Trends · Challenges

---

Â. M. Katuta (✉)
Universidade Federal do Paraná/Litoral, Matinhos, Brazil
e-mail: angela.katuta@gmail.com

M. A. M. de Albuquerque
Universidade Federal de Paraíba, João Pessoa, Brazil
e-mail: dadaufpb@hotmail.com

R. C. Lois González and M. A. Mitidiero Junior (eds.), *Brazilian Geography*, Advances in Geographical and Environmental Sciences,
https://doi.org/10.1007/978-981-19-3704-0_11

11.1 Introduction

This text aims to present a brief systematisation of the construction of the research field of the Teaching of Geography[1], taking academic studies as a reference and their relations with how the discipline is taught in basic education. We understand the limitations of this perspective to the extent that the Teaching of Geography is materialised at different levels in the confluence of public policies. At certain historical moments and in certain countries, these policies have created education systems that often preceded the creation of academic courses and the field of research in the area. The area will be gradually structured with the organisation of the initial and continuing training courses, departments, research, graduate courses (specialisations, master and doctoral studies, post-doctoral studies, among others), research groups, events, journals, etc. With this structure, the most diverse lines of research are created, which have for some time been evidence of the quantitative and qualitative expansion of themes, and the theoretical-methodological reference perspective within the production of this field.

In a country of a continental size, such as Brazil, the diversity and quantity of the existing academic studies require us to make temporal and methodological cut-offs. As this is a first approach, whose objective is to break down and present the field of research and the productions in the Teaching of Geography, we established a temporal cut-off, emphasising the period between 1987 and 2019. This cut-off is justified by two facts: 1987 was the first year in which the data on dissertations and theses were inserted into the Coordination System for the Improvement of Higher-Level Staff (CAPES), a foundation linked to the Ministry of Education (MEC); and the year 2019, because it was the last year the data were consolidated.

A brief search in the catalogue of theses and dissertations in CAPES using the term Teaching of Geography, without discriminating the areas of knowledge in which these studies were conducted shows a total of 1,280,486 dissertations and theses defended. However, as our intention is to present a picture of the production in the field of research with reference to the context of the Post-Graduate Courses in Geography (PGCG) as the reader will see later, we distinguish the works produced in them that are defended in programmes of other areas.

The first part of the historical analysis sees the educational processes of the field in question, in the period from 1994 to 2020, because we understand that, together with the theses and dissertations defended in the areas of Geography and Education, the most important constitute a locus and expression of growth and diversification of academic–scientific production, whose results are broken down into dissertations, theses, journal articles and other types of production. Because of the nature and time to elaborate this systematisation, we chose to address these productions, because we understand that they make it possible to give an idea of the transformations in this field of research. In the text, we also refer to the publications of books on topics of the field at hand. However, it has not been possible to systematically survey them

[1] When referring to the field of research, we use the expression Teaching Geography, with capital letters, which differentiates it from teaching the subject.

because of the growth and dispersion of this type of print over time and also because we do not yet have a national database to make an adequate survey possible.

We start from the proposition that the understanding of the research field of Teaching of Geography has an intimate relationship with the structuring of geography as a school discipline, with the formation of universities, teacher training courses, post-graduate courses, with the policies and educational programmes of basic education, with curricular reforms, among others. Therefore, in the text below, we present the constitution of the research field based on some data that characterise it. In the end, we pointed out some challenges based on the organic unity between initial and continuing education and the inseparability of teaching, research and outreach.

11.2 The Development of the Field of the Teaching of Geography

A field of scientific knowledge is usually built over a long period. In the case of this text, in order to analyse the research field in question, we have emphasised the most recent period of development, based on the justifications set out above. For this approach, we use two elements: The publications of academic research and research groups, elementary references to expand and spread the Teaching of Geography throughout the country, which also express the transformations that have taken place in the area, both from a qualitative and quantitative point of view.

First, we discuss the publication of books and the creation of specialised journals in the field and, finally, the research groups linked to the National Council for Scientific and Technological Development (CNPq), based in higher Education institutions (IES) or Basic Schools, whether or not they are linked to the PGCG, undergraduate and other courses and their respective productions of dissertations and theses, the basic source of the field publications (works in events, articles in magazines, chapters of books, complete works, among others). These axes of academic–scientific production selected are indeed interrelated, and the increase in their abundance indicates the increase in the production of the research coming from the field under analysis.

Before submitting the data, some considerations need to be made regarding this approach. The selection that we chose and that will be displayed in the text below is not the only possible route to an analysis for this purpose. The reflections expressed here are provisional in nature and should be further elaborated and re-elaborated at the appropriate time. With regard to book publications, we consider the first editions, but some have been republished in subsequent years, which are sometimes different from those indicated.

For geography students and teachers in Brazil, books are traditionally the most important means of scientific dissemination. Despite the convenience of accessing articles published in electronic journals created today, there is a tradition constituted in books. However, the second medium, particularly in view of its accessibility, has redefined its role in the dissemination of research and, little by little, it has been incorporated into the readings aimed at training professionals in the area at the various levels of teaching and in the guidelines of teaching work.

The publication of books intended for the field is difficult to quantify because up to the 1980s, it was restricted to three or four works; whereas today the number has grown exponentially. In this text, it is impossible to present precise data, since there is no archive and/or institutional repository in the country for cataloguing and disseminating them. For journals and research groups, there are official data from federal government institutions. There were hardly any books published between the 1920 and 1970. Those that were published mainly revolved around the methodology, and were linked to the pedagogical guidelines that were in vogue in each period. In general, they were intended for teachers already exercising their vocation, who needed specialised training due to the lack of higher courses in the country[2], and were characterised by criticism of the value of mnemonic geography[3], typical of past experiences, but bringing new propositions (Albuquerque 2011). Regarding the publication of journals in the period, some that circulated among teachers included articles that, in general, did not result from academic research. We highlight the journals *Revista do Ensino* published in several Brazilian states (Dias 2013, 2020), and *BoletimGeográfico* published by the Brazilian Institute of Geography and Statistics (IBGE), which provided specific sections and was intended for geography professors.

Publications of a methodological nature remained for a long time, because there was no core research or specific post-graduate courses in the field, and therefore, dissertations and theses that, as we will see later, caused qualitative and quantitative transformations in production. Geography professors and researchers collaborated for the first productions in the field, including indicating the need for research, as is the case with the text written by Pierre Monbeig (2010, p. 18 et seq.), by IBGE in 1956, entitled *Papel e valor do ensino da Geografia e de suapesquisa*, republished in 2006 by the journal *Tamoios*. This text, despite highlighting the author's concern with the Teaching of Geography, records a very common perspective of professionals at the time, which still stands in the departments of geography: the failure to recognise the need for specific research in the field Teaching of Geography. The following excerpt demonstrates the understanding of many geographers that research in geography referred to the themes it has classically addressed:

Specifically, if the geographer studies the complex relationships of natural resources and human groups in their location and as a function of their spatial

[2] The first courses aimed at training geography teachers in Brazil date back to the 1930s, despite having been offered free courses for this training in the previous decade (Albuquerque 2011).

[3] Mnemonic geography was and remains a complex problem for the Teaching of Geography in Brazilian elementary schools. In the 19th century, it served a school project that was developed based on the memorization of names of places, aspects of elements of nature, such as names of rivers, mountains, lakes, etc. emphasizing quantitative data to cover up colonial domination and territorial organization supported by large estates and slavery. Later, during the military dictatorship, it was used by conservative and authoritarian projects to hide the exploitation of the working classes by international and national capital and even today, even in the face of the development of analytical approaches based on the awareness and understanding of spatial processes, it still remains in part. significant from schools.

area, the geographer is, par excellence, the technician of the organisation of space. However, it is not enough to organise the space, it needs to be organised well. [...] As geographical knowledge is indispensable to the education of modern man, geographical research is indispensable to the good management of a great undertaking, the good administration of a modern state. Such is the multiplicity of problems that arise in the current medium, where the contribution of each technician risks falling into a narrow, sometimes sterilised specialisation. Due to the largesse of their views, the geographer completes the technicians. The problems that they can help solve are more evident in a country like Brazil, where there are so many areas to organise, where so many possibilities remain unexplored, where there is so much dynamism in so many contradictory directions.

In view of the above, we can state that this understanding shows that, at that time, there was already a concern with the Teaching of Geography, although it was not understood as a field of research, only as an action, which needed input from the point of view of the methodological prescription. It is important to point out that this concern was based on concrete demands that have arisen and accumulated since the Brazilian Empire, because according to some authors (Rocha 1996; Albuquerque 2014; Feliciano 2017), the teaching of this discipline was already contained in the literary texts of Jesuit schools. However, the origin of the discipline in the country is central to debates in the field of research, in which two theoretical proposals stand out. The first part of the historical analysis sees the educational processes from the core to the periphery (the provinces). In addition, because Colegio Pedro II[4] was a referential institution created to be followed by the secondary schools and provincial high schools, these researchers understood that what was established for the first one would be faithfully reproduced by schools throughout the country. Since this important institution created the discipline of geography in 1837, this group became the cornerstone of the discipline (Rocha 1996). A second theoretical perspective observes the historical fact from the provinces, because it questions the idea of curricular fidelity and proves the creation of schools with another organisational and curricular structure that already had the aforementioned discipline in a previous historical period. Thus, the Isolated Disciplines of Geography that were established in the provinces of Paraíba, Piauí and Rio Grande do Norte in 1831 (Albuquerque 2014) are seen as another landmark. According to Vlach (1988), the reason for creating this discipline was the result of the requirements for entry into higher education (courses in law and medicine), from 1831, when the preparatory examinations started to require this knowledge.

What has been observed in the most recent research (Silva 2012; Maia 2014) is the difficulty of demarcating an exclusive institution in which the discipline of geography was established for the first time, because this fact occurred spatially in

[4] Colégio Pedro II was the first public secondary education institution founded in 1837 in the city of Rio de Janeiro, then capital of the Empire of Brazil. This, over time, became a reference for secondary schools that were founded in Brazilian provinces. Their Teaching Programs, the name given to the curricula at the time, should be followed by the provincial Lyceums.

a diffuse and concomitant way, both in the city of the court and in the provinces, in distinct educational levels and in institutions located throughout the country.

Only a century later, and with growing social pressure on the education of teachers, among other things, the undergraduate course in geography was created in 1934 at the School of Philosophy, Sciences and Letters (FFCL) of the University of Sao Paulo (USP). The following year, a degree in geography was created at the School of Economics and Law of the University of the Federal District (UDF) in Rio de Janeiro. In the early years, these courses had contributions from geography teachers from the university missions established in an agreement between Brazil and France. They include Pierre Deffonteines (1894–1978) and Pierre Monbeig (1908–1987).

Based on the above, we can say that there was a gap of at least 100 years between the installation of geography as a school discipline (1831) and the first teacher training course in the area (1934) (Albuquerque 2011). Therefore, other professionals worked in the area, as noted.

Prior to FFCL/USP, there was no bachelor's degree programme in geography nor any professor with a degree in geography in Brazil. There were people who, from different colleges or even from normal schools, taught this discipline, as well as others. They were mainly professors of geography, lawyers, engineers, doctors and seminarians.

The material conditions that created the Teaching of Geography in Brazil, in the basic schools, together with the characteristics of the professionals who, first of all, acted in it and with the lack of courses in the education of geography teachers in the country. For more than a century, they created and consolidated a profile of demands of teaching geography centred on the geographic studies of the Brazilian territory as well as teaching methodologies and teaching studies. According to the above mentioned authors (*ibidem*, p. 46), initially, before the opening of the degree course at USP, the publications intended to teach the discipline were written by non-geographers. They expressed concepts of nineteenth-century science and were directed towards "[…] the enumeration of names of rivers, mountains, islands, cables, capitals, major cities, population totals of countries, cities, etc. Memory was the primary skill for the student to do well on tests." This was the significant part of the production in the period and was focused on the concrete demands of non-specialists in the area.

We can mention as examples of works aimed at meeting the demands listed in publications of the IBGE, entitled *Curso de Férias para aperfeiçoamento de professores de Geografia do Ensino Médio (Holiday Course for the Improvement of High School Geography Teachers)*, in the years 1967 and 1968[5]. Both of them provided several texts of different authorship, whose contents were directed to the geographic knowledge proper and didactic-pedagogical reflections, aiming at meeting the educational demands.

An interview granted by Lívia de Oliveira, during the preparation of Laubstein's dissertation (2008), highlights the lack of research in the field of Teaching of

[5] Available in: <https://biblioteca.ibge.gov.br/visualizacao/livros/liv13960.pdf>; <https://biblioteca.ibge.gov.br/visualizacao/livros/liv13959.pdf>.

Geography, even after the formation of degrees in universities. In his speech, Prof. Lívia advocated that his thesis, entitled *Contribution to the Teaching of Geography*, defended at the State University of Campinas (UNICAMP) in 1967, and guided by José Araujo de Oliveira Jr., was the first of its kind constituted as a field:

He (João Dias da Silveira) wanted me to come to work in the Didactics of Geography and for me to do research—he used the word well—he wanted me to start research in Teaching of Geography, not in Geomorphology. [...] He wanted me to do research in the Teaching of Geography! Not even USP had this kind of work! So my work was pioneering, and led the way because of Dr João's vision, because it was he who got me started in it (Gonçalves 2011, p. 11).

In summary, we can state that, from the installation of the Teaching of Geography in Brazilian schools (1831) until the late 1960s, academic research (dissertations and theses), books and magazines, in the field of the Teaching of Geography, were almost non-existent. This does not mean that there was no production directed towards the professionals who worked in Brazilian basic education. We have already demonstrated in this text that there were concerns about the educational demands surrounding the Teaching of Geography, shown by the productions that we set as an example. However, as Pontuschka et al. (2009) stated, the publications were made by professionals who, as a rule, did not have academic education in the area. This demonstrated that the gap caused by the lack of training courses was an important element to understand the academic production of this time, a production that was elaborated from criticisms of them. The text by Monbeig, to which we referred, is a didactic example of a narrative that criticised what was put in a mnemonic tradition by someone who was already working in the education of professionals in the area.

Since the 1960s, there have been many transformations in education that reverberated throughout academic–scientific production, including in the field of the Teaching of Geography. It began in this decade with the first productions founded on research, many elaborated in the context of the military dictatorship in the country, which lasted more than two decades, between 1964 and 1985. The 1968 University Reform was a milestone and didactic example in this sense, due to the technicist nature[6] that the Brazilian military state promoted in the education of teachers:

[...] since a new field of study was inaugurated in university faculties/education centres. At this point, the degrees, which until then were linked to the Faculty of Philosophy, Sciences and Letters, gained another configuration: specific departments of each area of knowledge; a space of pedagogical formation; an Elementary Education Teaching Certificate and a Teaching Degree were instituted, aimed at

[6] "Trend seen in the 70s, inspired by behaviorist theories of learning and the systemic approach to teaching, which defined a highly controlled and teacher-directed pedagogical practice with mechanical activities inserted in a rigid educational proposal that could be fully programmed in detail. According to educator José Mário Pires Azanha, what is valued in this perspective is not the teacher, but the technology, and the teacher becomes a mere specialist in the application of manuals and his creativity is within the possible and narrow limits of the technique. used. "This guidance was given to schools by official bodies during the 1960s and until today it persists in many courses with the presence of didactic manuals with a strictly technical and instrumental character"." Source: Menezes (2001).

training teachers who taught what was then called primary and secondary education, respectively.

For Pereira, the creation of central institutes or departments was important for the consolidation of practising research in Higher Education Institutions (IES), since the reform of the university structure made the indissociability of teaching and research possible (dos Santos and Mororó, 2019, p. 6, Grifonosso).

As we can see, after the University Reform, during the military government, the organisation of specific departments was established by areas of knowledge, also present in basic education. The Geography Departments started to work with the specific knowledge of the area, separated from the pedagogical training, which took place in the departments of Education, whose role in the training was and has been to work on the specific characteristics of pedagogical knowledge. This separation of teacher education has constituted and strengthened a famous and criticised model known as 3 + 1—a three-year training of specific knowledge in the area, as part of the bachelor degree and[7] one more year of pedagogical training. This model was the expression of the materialisation of the technicist tendency in education, much strengthened at the time of the military dictatorship. By means of Law 5692 of 23 August 1971, it eliminated the contents of History and Geography from basic education and inserted a discipline called Social Studies, instead of both. The objective was to "educate for adequacy, cultivate cooperation, discipline the spirit, and mirror itself in behaviours." These "[...] were the frontline elements of this authoritarian period" (Martins 2014, p. 48).

In this period, many degree courses in geography from public universities ceased to function. Others continued to resist military charges, and at the same time, many undergraduate programmes in social studies were opened in private colleges to meet the educational demand installed under the military regime. In addition, more disciplines of a civic-moral character were included in basic education, based on the precepts of the military dictatorship: Moral and Civic Education and Brazilian Social and Policy Organisation (OSPB). For the same author,

[...] there is a wide gap between academic curricular contents and school curriculum contents. It can even be said that, in this case, the school disciplines were taught by the demarcation of their existence in a curricular framework or grid, with specific space for classrooms. It is also important to remember that, in this period, given the need to train teachers who could act in the teaching of these disciplines. Elementary Education Teaching Certificates were created, as well as a bachelor's degree in Social Studies. These higher education courses were not included in epistemological, disciplinary, academic or scientific references. However, it is true that they educated many teachers who, attending these general courses, ended up teaching

[7] Unfortunately, in Brazil, even today, many misunderstand it as research-oriented training. Underlying this understanding, there is another that, little by little, advances, with the strengthening of postgraduate courses and, above all, after the creation of the Teaching Initiation Program—PIBID which, according to Freitas (2016), caused a transformation of understandings, summarized in the theme of his dissertation: "Training professors-researchers in a bachelor's school: the PIBID culture of geography at UNICAMP"—State University of Campinas/State of São Paulo/Brazil.

without teaching the strategies for producing the knowledge that guided the content they had to teach (Ibidem. Grifonosso).

The military dictatorship in the country negatively impacted research in Teaching of Geography, because we understand that the prohibition of its teaching and training also reverberated in its academic production. This topic must be delved deeper into another research. For now, let us remember the statement made by Prof. Lívia de Oliveira on the absence of studies in the field in 1967. Education, as it was organised later, revealed, in a way, the influence of technicism in the way it was thought of. In many cases, the education of teachers in the country is still thought of as a sum of parts that rarely converse with one another. There are universities where teacher training or the "pedagogical part" occurs in other buildings or campuses. This shows and marks the separation of the bachelor's degrees and teaching degrees, the latter of which are understood historically, in this view, as "the pedagogical educations", which lacked research training.

In addition to the dichotomised education between a teaching degree and a bachelor's degree in Brazil, as indicated by dos Santos and Mororó (2019, p. 7), in the University Reform (1968), Elementary Education Teaching Certificates were created, whose training programme was lighter: "[...] In this case, the student could complete the course in between a minimum of one and a half years and a maximum of three years". The degrees that required research were intended, respectively, for the primary and secondary levels in basic education.[8] These educations were created to meet the historical lack of trained teachers and benefited the private schools, where traditionally, and still today, there is no research.

This dichotomised understanding of teachers' education, which still exists in Brazil (bachelor's degree versus teaching degree), together with the creation and strengthening of a dual network of degrees in public and private institutions, in which, according to Gatti's research (2014), the second, which currently predominates, has migrated to Distance Education Courses (DEC). This has had and still has a great impact on the lack of academic–scientific production in the field of the Teaching of Geography in the relationship between the academic scope and basic education, since, at this level of teaching, there are still teaching demands for contents of the area and prescriptive methodologies.

It is in this context in which a significant part of the themes related to the field is now being researched in Education Departments in which, professionals from the specific areas are rarely hired, except in those of methodologies. In addition to this fact, Teachers in Geography Departments often guide students who want to research topics related to the Teaching of Geography to guide them towards professionals in the Education Department. Therefore, a significant part of the studies, theses and dissertations of the field under analysis began to be developed in the area of Education. However, the data we present in the following lines show the expansion of production in the Post-Graduate Courses in Geography, linked to the departments

[8] Denomination given to the two levels of schooling that made up the eight years of basic education, established by Law 5,692/71, which establishes Guidelines and Bases for teaching 1st and 2nd grades.

of the area, and are indicators of changes in the concepts in the same areas regarding the subject in question.

According to Katuta (2020, p. 140), with the creation of the Post-Graduate Courses in Physical and Human Geography at the University of Sao Paulo in 1971 and 1972, the post-graduate programme in geography at the Federal University of Rio de Janeiro (UFRJ), there are processes of strengthening and expanding research in various topics, including those linked to the Teaching of Geography and cartography. Thus, we can confirm that, although we have had production regarding the topic for more or less 90 years, as Archela pointed out, it was effectively strengthened and expanded starting in the 1970s, with the creation of graduate courses in geography in Brazil.

The creation of the first Post-Graduate Courses in Geography in the south-eastern region of Brazil in 1970, and its later capillarity in other capitals and regions of the country, expanded the production of research in the area of geography as a whole, including in the Teaching of Geography, as we will show later.

Since the 1980s, with the process of political opening in the country and its consequent re-democratisation forged by organised social movements and/or groups, a debate was inserted based on today's precepts. In Brazil, it is known as Critical Geography, an expression of the adoption of critical and Marxist theories in its various perspectives, with a strong political imprint and dialectical methodological approaches. This movement made a resounding criticism of the geography taught, regarding its aspects of conservative, mnemonic, and descriptive content. In this decade, we highlighted three books that circulated between students and teachers and were references to the formation of a generation that initiated critical debates in school geography: *Para ondevai o Ensino de Geografia?* (Oliveira 1989); *A Geografia do aluno trabalhador: caminhos para uma prática de ensino* (Resende 1989) and *Geografia e ensino: textos críticos* (Vesentini 1989). It is essential to link the publications of this decade with the context of political opening, with the end of the military dictatorship and with the organisations of social and popular movements through public, universal and high-quality education.

In the same period, unlike the previous decades, there were no specialised journals in the field. However, it is essential to highlight the role of some journals in the area of geography, published in print, which in their editions published articles resulting from research on teaching and texts, which expressed the political debates inherent to the curriculum reforms aimed at basic school and higher education, including the issue of mathematics, called dossiers. Among these journals we highlight *Terra Livre* and *BoletimPaulista de Geografia*, published by the Associação dos GeógrafosBrasileiros (Association of Brazilian Geographers) (AGB) and *Orientação*, published by FFLCH-USP.

In this interim, journals were created linked to the state sections of the AGB, which also published articles on the Teaching of Geography. In addition to the publications, in 1987, the AGB started to organise a specific event held every four years called *Fala, Professor!* (Speak up, Teachers!). The event held debates, encouraged and strengthened the dialogue between professors and researchers from the undergraduate courses in geography and those of Basic Education, which resulted in the publications of annals with the articles presented at the event.

Starting in the 1990s, with debates on professionalising teachers, the emergence of collectives aimed at developing new curricula for basic education and teacher training, the expansion and internalisation of post-graduate programmes in geography, the number of research that dealt with the Teaching of Geography was expanded, as was already demonstrated by Pinheiro (2003) and Cavalcanti (2016). In this decade, the field mobilised its studies from a great theoretical plurality, a result of clashes between the currents that Pinheiro (2003, p. 120) identifies as: "[…] "Traditional Geography", "Quantitative and Theoretical Geography", "Humanist and Perception Geography", and "Radical and Critical Geography" (Pinheiro 2003, p. 120).

The following figure shows that, from 1987 to 2019, there was a practically constant increase in the number of dissertations and theses produced on the topic of the Teaching of Geography, both in Post-Graduate Courses in Geography and in others. It also shows that, in almost three decades, the number of dissertations and theses in the post-graduate programmes in geography, jumped from 28 in 1987, to 562 in 2019, and from 14 to 334, respectively. It is noteworthy that most of the production of the field of research is still produced in post-graduate programmes outside the area of geography (Fig. 11.1).

As a result of this context of expanding master and doctoral research, in post-graduate courses in general, an editorial market focused on topics related to the Teaching of Geography was growing. Publishers who, until then, did not have books in their catalogues intended for this field saw market possibilities in disseminating the results of the research. It is at this time that publishers located in the south-eastern region of the country, such as Cortez, Loyola, Attica, Papirus, Contexto and Marco Zero, among others, invested in publishing studies in this field, expanding its catalogue. It began to feature titles aimed at an expanding and professionalising audience, due to the increase in vacancies in public and private universities, especially in undergraduate studies. In addition to this, the number of teachers working in the various teaching networks increased. This catalogue was composed of works written

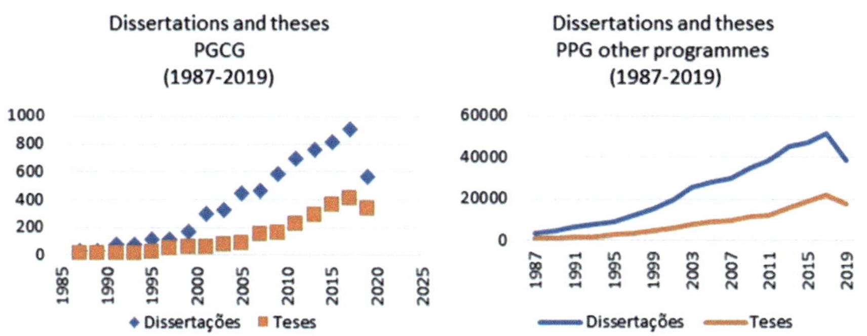

Fig. 11.1 Dissertations and theses on Teaching of Geography (1987–2019). *Source* CAPES database. Organised by: Ângela Massumi Katuta. Prepared by: Jéssica Cristina Lozovei

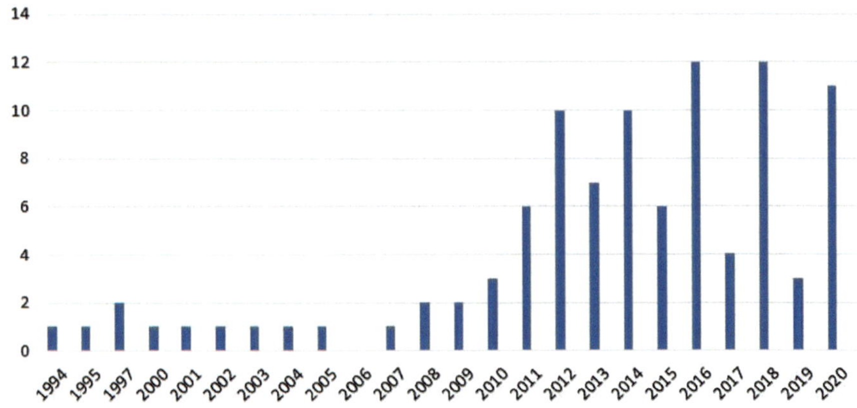

Fig. 11.2 Research groups specialised in Teaching of Geography by year of creation. *Source* CNPq database—organised and elaborated by Maria Adailza M. de Albuquerque

by a restricted group of specialists and "non-specialists"[9], in general, professors from the largest and oldest universities, located in the most developed regions of the country, linked with the few existing PGCG, still with a small number of vacancies for research in the field.

Regarding the production and circulation of journals, there was a continuous lack of specialised journals. However, those linked with the AGB continued to disseminate the academic and political debates regarding the Teaching of Geography, highlighting the relevant role of the entity to install, strengthen and deepen the guidelines related to this field. In this period, the group of researchers from the field was held in teaching laboratories, where debates took place, educational resources were produced and small publications and coordination of events were organised. Then, the research developed in PGCG was initiated in the post-graduate programmes in geography and projects aimed at continuing teacher education were implemented, stimulating the articulation between the school and the university. The culture of organising researchers in institutionalised research groups had no roots in this field. Of the one hundred research groups specialised in Teaching of Geography today, only four were established in the period, as shown in Fig. 11.2.

In the following decades, in the context of popular governments of Lula da Silva and Dilma Rousseff (2003–2016), the increase in education funds, the expansion of access to higher education for people of less favoured social classes, and the internalisation of post-graduate programmes by the country expanded access, permanence, diversity and democratisation of Brazilian education. Of course, there are many criticisms that can be made of these governments, but one cannot deny the social achievements that their policies and programmes have made possible.

[9] Here we call "non-specialists" authors who wrote about school geography, however, their research did not refer to topics related to the field.

Thus, the increase in the number of publications of books from specialised journals and the emergence of research groups was exponentially higher than in previous periods. This is attributed to several factors related to the policies and educational programmes established in this context, highlighted by the following: The establishment of the Curriculum Guidelines for Teacher Education (DCN) (Brazil 2002, 2015), which was responsible for separating the bachelor's and teaching degrees, and required professionals specialised in the Teaching of Geography to teach new disciplines established by the curricula and focused on the training of geography teachers; The need for specialised teachers, which triggered pressure for vacancies and the creation of Geography, Geographic Education, Teaching of Geography or PGCG School Geography, in view of the fact that in order for teachers to enter university, they were required to have this training, with rare exceptions; the establishment of guarantees that enabled basic school teachers to depart from their activities to study, while receiving stipends. This fact was of extreme importance to the field, since it enabled these subjects and the topics addressed by them from their experiences, to be objects of research, expanding the subject of the field; the creation of programmes intended for research and the teacher training, such as the Institutional Programme of Scholarships for Teacher Training (PIDIB)[10] and the Pedagogical Residency[11], which, stimulated and strengthened the culture of the professor-researcher, defended since 1980 by critical perspectives; the editorial publications of books financed by institutions to promote research; and the offering of scholarships in an adequate number to the demand for students, which enabled many subjects, who previously had no economic resources to study, to enter and stay at the university doing research. In addition to these factors, we highlight the regularity of national events: *Speak up, Teachers!*; School Cartography and National Meeting of the Practice of Teaching in

[10] “The program offers scholarships for teaching initiation to students from on-site courses who dedicate themselves to internships in public schools and who, when they graduate, are committed to teaching in the public network. The objective is to anticipate the link between future masters and public classrooms. With this initiative, Pibid articulates higher education (through licensure), the school and state and municipal systems. The program's intention is to unite state and municipal education departments and public universities in favor of improving teaching in public schools where the Basic Education Development Index (Ideb) is below the national average of 4.4. Among the Pibid proposals is the encouragement of a teaching career in the areas of basic education with the greatest shortage of teachers with specific training: Science and Mathematics, from the fifth to eighth grades of elementary school, and Physics, Chemistry, Biology and Mathematics for teaching medium” (Available at: <http://portal.mec.gov.br/pibid#:~:text=PIBID%20%2D%20Apresenta%C3%A7%C3%A3o&text=A%20inten%C3%A7%C3%A3o%20do%20programa%20%C3%A9,national%2C%20de%204%2C4> Because it is a program, it is changed according to the concepts that guide the actions of the Ministry of Education, so , we chose to present the characterization present on the website of the aforementioned ministry.

[11] “Initially, the Pedagogical Residency is linked to the formation of the subjects of the National Common Curricular Base. The program is one of the actions that are part of the National Teacher Training Policy. With the aim of improving practical training in undergraduate courses, it promotes the immersion of the licentiate in the basic education school from the second half of their course.” Available at: <http://portal.mec.gov.br/component/tags/tag/45681>. As it is a program, it is changed according to the concepts that guide the actions of the Ministry of Education, so we chose to present the characterization present on the website of the aforementioned ministry.

Geography (ENPEG), and regional meeting in the field, always with publications of annals. It is also important to highlight the role of the National Association for Research and Post-Graduate Studies in Geography (ANPEGE), by incorporating in its struggles and publications, such as the *Revista da ANPEGE* and in its biannual event, the National Meeting of Research and Post-Graduate Studies in Geography (ENANPEGE), the major debates in the field.

This context was responsible for extending the number of publications of books, initially on paper and, more recently, on digital media. There are collections from research groups; publications of research results from the PIBID and the Pedagogical Residency; books on partnerships with researchers from other areas of knowledge; books with results from master and doctoral research; other books with the texts of the roundtable discussions at events, among other organisational structures of the works. There is also a wide range of theoretical-methodological perspectives in these works, besides a balanced spatial distribution of publications by all states of the country, which differs from the 1980s, in which production in the south-eastern region predominated.

The publishing houses for publications in the IES, from their official publishers, have increased this number of publications. In addition, there has been an improvement in the accessibility to services of smaller publishers and editorial improvement of the works, with the creation of editorial boards and caring for the quality of the paper of the books, covers, images and a significant and more appropriate structure than there were in previous decades.

In parallel with the editorial expansion of books, the number of specialised journals also grew. Between 2001 and 2019, 19 thematic journals were launched, which are listed in IES or other institutions of basic education distributed by the country (Fig. 11.3), indicating the spatial decentralisation, also an expression of the phenomenon of the internalisation of the PGCG and the research groups.

The increase of books and journals demonstrates the expansion of the scientific field throughout Brazil, which is the result of a set of elements related to the internalisation of the post-graduate programmes. Another fact to be highlighted that expressed and at the same time, revealed the enlargement of the field, is the growing number of research groups linked with graduate courses, post-graduate courses and federal primary schools. According to Fig. 11.1, the most constant growth period began in 2008, one year after the first Public Notice of the PIBID was published, and continued to grow throughout the evaluated period. In this teacher training programme, the increase in the number of scholarships for students and teachers in basic schools and the IHE allowed groups to be created that had research as their purpose as a formative principle. As such, teacher training was consolidated as a research space. This stimulated the permanence of the students at the university, their association with these groups and their future entry into the PGCG. The dissertation of Freitas (2016) also shows that the PIBID transformed the school's culture of bachelor's degrees, to which we referred earlier, because research in the bachelor degrees gained strength, institutionalisation and funding.

Another programme that has distinctive principles that is currently linked with the dissemination of the precepts of the National Common Curriculum Base (BNCC) in

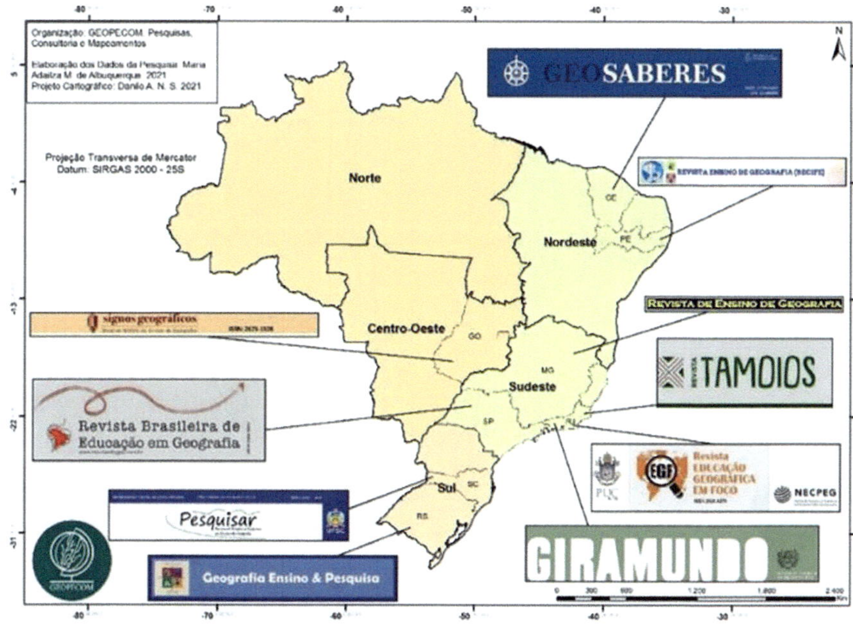

Fig. 11.3 Locations of magazines by federal states Organised and elaborated by Maria Adailza Martins de Albuquerque

higher Education, but which was launched in 2011 and started to operate in 2012, is the Pedagogical Residency. It has also encouraged students to stay in the universities and to be linked to research groups, since it offers scholarships for students and teachers of the projects.

The location of the groups on Fig. 11.4 demonstrates the capillarity of the field throughout the country and its link to the policies of fostering research to educate teachers and to interiorise the PGCG. Most of them are located in the most developed regions, where there are the oldest and greatest number of programmes. However, they can be found in all regions of the country, even where the PGCG does not exist. This confirms its vigour with regard to the expansion of the research field of Teaching of Geography, even when only linked with graduate courses.

This growth and encouragement in the field took a new direction, starting with the 2016 coup against President Dilma Rousseff, and the assumption of ultra-right governments. In this context, gaps and resources were lost, because education was and continues to be vilified by right-wing groups that see it as a critical space of their fascist and neoliberal ideology. Thus, the government of Jair Messias Bolsonaro cut off the promotion of research, making the greatest cut in its history, both for Higher Education Institutions (HEI) and for basic education. In addition, many research groups are being prevented from continuing a vigorous collective project, which was contributing to this fundamental field for Brazilian Geography. Also resulting from the current context of the prohibition of emancipatory pedagogy directed towards

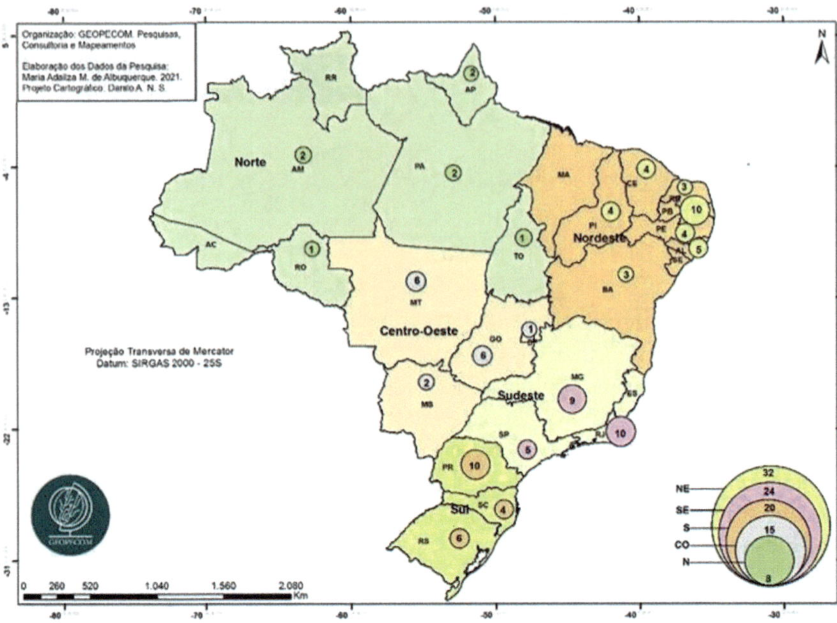

Fig. 11.4 Distribution of research groups specialised in Teaching of Geography by federal state Organised and elaborated by Maria Adailza Martins de Albuquerque

autonomy, the training of teachers connected with the research runs a serious risk of being interrupted due to the imposition of Resolution No. 2 of 2019, called the Common National Base Education (CNB Education). This resolution authoritatively prescribes a curriculum model for the undergraduate courses aimed at training teachers exclusively as CNB implementers and reproducers.

In the next topic, we have listed some challenges that, in our view, are still present in the field of research in question, because they are linked with the necessary integration between the initial and continuing training of teachers, based on the inseparability between teaching, research and outreach.

11.3 The Indissociability Between Initial and Continuing Teacher Education, Research and Outreach

In this brief reflection, we show the close relationship between the constitution and the development of the research field of Teaching of Geography with a series of elements, highlighted by the following: The demands produced by basic education schools, after their constitution as a school discipline from 1831 onwards; The creation of the Undergraduate Courses in Geography in 1934 (USP/SP) and 1935 (UDF/RJ), Post-Graduate Courses in Geography (USP 1971; UFRJ 1972); The creation of educational policies and programmes aimed at strengthening the relations between public HEI degrees and schools and teachers of basic education (PIBID, Pedagogical Residency), curricular reforms, among others. These elements helped to deepen and increase the research field of Teaching of Geography in Brazil, which, until the military government in the 1980s, was quite restricted.

We also observed that, in the process of political opening of the 1980s, the geography taught in Brazil was questioned, debated and criticised by educators and researchers who used distinct theoretical-methodological frameworks, establishing a profitable basis among what was generally called traditional, quantitative, theoretical, humanist and perception geography and radical or critical geography. The first books of the area written by professionals who worked in it date from this period, which gradually became involved in the field of research, due to the expansion of federal and state public policies that internalised undergraduate and post-graduate studies and, consequently, teaching, research and outreach. These actions resulted in an increase in the number of research groups, dissertations, theses and journals in this field.

It is important to note the performance of the Brazilian Geographers Association (AGB), which was fundamental in researching the Teaching of Geography, because inside it, there were working groups divided by subject area, among them, those linked to the Teaching of Geography in the various local sections in the country[12], whose militant action resulted in the first thematic dossiers for the area and at the event still held today, titled the National Teaching of Geography Meeting—Speak up, Teachers!, whose first version occurred in Brasilia, at the University of Brasilia Campus (UNB), in July 1987, which brought together two thousand people.[13] The entity also supported other events in the area that helped to expand, foster and consolidate the field of research and academic–scientific production: The National Teaching of Geography Meeting (1985) and the Colloquium of Cartography for Schoolchildren (1995), both initially held at the Universidade Estadual Paulista (UNESP), Rio Claro campus, and organised by Professor Dr. Lívia de Oliveira. It is also worth highlighting the role of ANPEGE, which has been incorporating the debates in the field for some years.

The following table presents a systematisation of what we address in this text: The formation of the research field of Teaching of Geography. We organised this so that readers can have a summary of this long process. Therefore, it presents a timeline that highlights historical facts and the educational milestones that reverberated in the field in question that help readers to situate themselves in the reflections elaborated here (Table 11.1).

Finally, despite the increase in academic–scientific production in the field of Teaching of Geography, expression of the increase in the number of research groups,

---

[12] In 2020, AGB organized a map of active local Sections that serve to elucidate the entity's capillarity, which, in the years 1980 to 1990, was even greater. Available at: <https://www.agb.org.br/mapa-das-secoes-locais-ativas-2020/?unapproved=853&moderation-hash=997eb3ca4a27ad7c869bc5441c07ebc5#comment-853>.

[13] To access the history of the event, narrated by Zeno Soares Crocetti, see: <http://geocrocetti.com/zeno/falahistgeo.htm>.

Table 11.1 School Geography in Brazil

Historical period	Educational milestones	School Geography in Brazil
Lands that were constituted as Brazil—1530	Education of native peoples	Spatial knowledge fundamental to life
Colonial Brazil 1500–1822	Jesuit education Pombaline education	Addressed in classical or religious texts aimed at the catechisation of the natives and the classical formation for the ruling classes
Imperial Brazil 1822–1890	Additional act—1834	• Geography remains as knowledge among the royal classes • Creation of the isolated disciplines of geography after 1831 • Geography in exams of preparatory courses for entry into higher education (1831) • Geography as a school discipline in secondary education in provincial high schools, Colégio Pedro II, military schools and private schools • Publications of textbooks on Geography, cosmography and the chorography of Brazil and the provinces
	Couto Ferraz reform—1854	
	Leoncio de Carvalho reform—1879	
Brazil Republic 1890–1937	1892—origin of school groups	• Publications of textbooks with regional references (1905–1913) • The discipline of social studies is established for primary education • Foundation for higher courses in the education of geography teachers (FFCL—USP in 1934 and the University of Brazil in 1935) • Insertion of modern school geography • Publications of articles intended for the training of teachers of the basic school (IBGE journal and pedagogical journals)
	1930—creation of the ministry of business, education and public health	
	1932—new school Manifesto	
	1934—Francisco Campos reform	
New state (Vargas dictatorship) 1937–1945	1942–1946—Capanemare forms	Strengthening of the Teaching of Geography in Brazil to disseminate nationalist ideals corroborated by the Vargas Government

(continued)

Table 11.1 (continued)

Historical period	Educational milestones	School Geography in Brazil
Brazil: re-democratisation 1945–1964	Brazilian education guidelines and bases act (1961)	• Education of Geography teachers (3 + 1): 3 years of specific training in geography plus 1 year of pedagogical training • First thesis on the Teaching of Geography (1967) • First Post-Graduate Courses in Geography (USP/UFRJ/UFPE/UNESP) (1970–1980) • Replacement of Geography and history academic disciplines with social studies (1971) • Replacement of a bachelor's degree in geography (4 years) with a bachelor's degree in social studies (1 year) in private colleges
Military civilian dictatorship 1964–1985	University reform—1968	
	Brazilian education guidelines and bases act 5692 (1971)	
Re-democratisation 1985—current days	1996—education guidelines and bases law 9394	• State and municipal geography curricula (1980s) • First congress of *Speak up, teachers!*—AGB (1987) • Publications of books with critical perspectives for the Teaching of Geography • National geography curriculum parameters (1996) • Expansion and internalisation of post-graduate education in geography • Creation of the lines of research in geographic education/Teaching of Geography (2000) • Start of the foundation of research groups on the Teaching of Geography • Separation of bachelor and undergraduate degree education in geography • Establishment of specific disciplines for the undergraduate degree, with the expansion of post-graduate vacancies in the area • Institutional teacher training programme—PIDIB (university residency and research on Teaching of Geography) (2007—present) • Pedagogical residency (2011-present)—teacher training policy linked with BNCC (2016) • Replacement of geography with the area of human and applied social sciences in high school education
	1998—national curriculum parameters—NCP	
	2001–2015—national curriculum guidelines—DCN	
	National common curriculum base (NCCB)—fundamental education (2016) High school reform (2017) National common curriculum base (BNCC)—teaching (2019) National curriculum base—teacher training (2019)	

dissertations, theses, magazines and events, as in all academic circles, there are challenges inherent to its constitution, its strengthening and consolidation to be collectively worked on. One of them refers to the necessary relationship between the field of research, the researchers linked with the HEI and the professionals who work in the Teaching of Geography in public schools of basic education, whether they are teachers, technicians and coordinators of the State and Municipal Secretaries of Education, which presupposes the strengthening of the inseparability of teaching, research and outreach. Although research on Teaching of Geography cannot be reduced to school geography, and each institution has autonomy—basic education schools and teacher education universities/colleges—we have found that research in this field helps and/or is a condition for the qualitative improvement of school geography, it also constitutes demands and helps to transform the field of research from the initial and continuing education into close dialectic interaction. Therefore, the construction of a place where there is a confluence and collaboration between basic education and academic–scientific production on the Teaching of Geography is of paramount importance. For this reason, it is essential to transform the PIBID and Pedagogical Residency programmes into a public policy focused on undergraduate courses, because, as indicated by Freitas' research (2016), until it was carried out, they helped strengthen a teacher training founded on research. Currently, they are being used by the current government as a tool for implementing the curriculum imposed on basic education.

Another challenge to the field of research is the question of methodological prescription, especially in the current context of neoliberal reforms, which advocate alienation, control and subalternation of teachers through prescriptive curricula, focused on active methodologies, skills and abilities aimed at getting an education for the labour market in the context of the flexible accumulation of capital and accumulation by spoliation. This reform has greatly impacted the demands of state and municipal education systems, including academic production. For some time, there has been an academic–scientific production that strengthens the prescriptive meaning of neoliberal educational policy. In other words, they prescribe how classes should be taught using active methodologies and suggest how to use certain resources and/or languages, strengthening processes that help crumble the teaching-pedagogical autonomy. In addition, they hide the problem related to epistemological questions and the method, which are fundamental in the approach of content, in general, understood as neutral in a neoliberal project that, by making them secondary, removed the school from its fundamental role: to serve as a key to the interpretation of reality through the scientific concepts produced in the academic sphere and resignified by the school or vice versa. In times of obscurantism and neoliberal disputes, contributing to the strengthening of an emancipatory school is fundamental and, in our understanding, this occurs through teaching, research and outreach. Academic–scientific actions can, to a large extent, assist the collaborative construction of teachers that bring together educators of different levels of teaching in order to strengthen, as Freire (1987, 1996) advocated, an education and an emancipatory and autonomous field of research.

References

Albuquerque MAM de (2011) Século de prática de ensino de geografia: permanências e mudanças. In: Rego N, Castrogiovanni AC, Kaercher NA (eds) Geografia: práticas pedagógicas para o ensino médio. Porto Alegre, Penso, pp 13–30

Albuquerque MAM de (2011) Século de prática de ensino de geografia: permanências e mudanças. In: Rego N, Castrogiovanni, Carlos A, Aercher NA (eds). Geografia: práticas pedagógicas para o ensino médio. Porto Alegre: Penso. pp 13–30

Albuquerque MAM de (2014) Um debate acerca da origem da geografia escolar no Brasil. Interf Científica Educação, Aracaju 2(2):13–23

Brazil (2002) Ministério da Educação. Resolução CNE/CP nº 1, 18 February 2002. Institui Diretrizes Curriculares Nacionais para a Formação de Professores. 2002. Diário Oficial [da] União, Brasília, 19 Jan 2002c, Section 1. Available at http://portal.mec.gov.br/seesp/arquivos/pdf/res1_2.pdf. Accessed on 26 May 2021

Brazil (2015) Ministério da Educação. Parecer CNE/CP nº 2, 9 June 2015. Diretrizes Curriculares Nacionais para a Formação Inicial e Continuada dos Profissionais do Magistério da Educação Básica. Diário Oficial [da] União, Brasília, 25 June 2015a, Section 1, p 13. Available at http://pronacampo.mec.gov.br/images/pdf/parecer_cne_cp_2_2015_aprovado_9_junho_2015. Accessed on 20 May 2021

Cavalcanti L de S, Goiânia BG de G (2016) Para onde estão indo as investigações sobre ensino de geografia no Brasil? Um olhar sobre elementos da pesquisa e do lugar que ela ocupa nesse campo 36(3):399–419

Dias AML (2013) Linguagens lúdicas como estratégias metodológicas na educação geográfica: a Revista do Ensino como referencial de análise (1925–1940). Dissertation (master's in geography), post-graduate programme in geography—PPGG, Universidade Federal da Paraíba. João Pessoa

Dias AML (2020) A Revista do Ensino e a Geografia Escolar (1932–1942): inovações educacionais na Paraíba. Doctoral thesis (Ph.D. in geography), post-graduate programme in geography—PPGG, Universidade Federal da Paraíba. João Pessoa

Feliciano LA dos (2017) O ensino de geografia no Brasil: do Colégio Pedro II à Universidade de São Paulo—1837 a 1934. Congresso Nacional de Educação EDUCERE, XIII, Curitiba. Anais. PUC/PR, Curitiba, pp 3920–3932

Freire P (1987) Pedagogia do oprimido. São Paulo, Paz e Terra

Freire P (1996) Pedagogia da autonomia: Saberes necessários à prática educativa. São Paulo, Paz e Terra

Freitas ASF de (2016) Formar professores-pesquisadores numa escola de bacharéis: a cultura do Pibid de Geografia da Unicamp. Dissertation (master's in geosciences), post-graduateprogramme in geosciences, Universidade Estadual de Campinas. Campinas, SP. Available at http://www.repositorio.unicamp.br/handle/REPOSIP/325050. Accessed on 15 June 2018

Gatti BA (2021) A formação inicial de professores para a educação básica: as licenciaturas. Revista USP 100:33–46. https://doi.org/10.11606/issn.2316-9036.v0i100p33-46. Available at https://www.revistas.usp.br/revusp/article/view/76164. Accessed on 22 June 2021

Gonçalves AR (2011) A geografia escolar como campo de investigação: história da disciplina e cultura escolar. In: Revista Bibliográfica de Geografía y Ciencias Sociales. Barcelona, vol XVI, no 905, s.p. Available at http://www.ub.edu/geocrit/b3w-905.htm. Accessed on 20 June 2021

Katuta ÂM (2020) Reformas educacionais: retrocessos e resistências na atual conjuntura brasileira. In: Caderno Prudentino de Geografia, Presidente Prudente, Dossiê "Conjuntura no Brasil: retrocessos sociais e ações de resistência, vol 4, no 42, pp 14–44, Dec 2020

Laubstein GSP (2008) História da disciplina geografia: contribuições da memória de uma educadora. Dissertation (master's in geography)—post-graduate programme in geography, Universidade Estadual Paulista. Rio Claro (SP), p 234

Maia EJP (2014) A Geografia Escolar na Província de Minas Gerais no período de 1854 a 1889. Doctoral thesis, post-graduate programme in education, Universidade Federal de Minas Gerais. Belo Horizonte

Martins MC (2014) Reflexos reformistas: o ensino das humanidades na ditadura militar brasileira e as formas duvidosas de esquecer. Educar em Revista Curitiba 51:37–50. Available at https://www.scielo.br/j/er/a/7yMGWJjk4j7Fr3LLjdjWHDR/?lang=pt&format=pdf. Accessed on 20 June 2021
Menezes ET (2001) Verbete tecnicismo educacional. Dicionário Interativo da Educação Brasileira - EducaBrasil. Midiamix Editora, São Paulo. Available at https://www.educabrasil.com.br/tecnicismo-educacional/. Accessed on 20 June 2021
Monbeig P (2010) O papel e o valor do ensino da geografia e de sua pesquisa. Revista Tamoios 2(2). ISSN 1980-4490. Available at https://www.e-publicacoes.uerj.br/index.php/tamoios/article/view/609/641. Accessed on 16 June 2021. https://doi.org/10.12957/tamoios.2006.609
Oliveira L (1967) Contribuição ao ensino da geografia. Doctoral thesis, Universidade Estadual de Campinas, Campinas. Available at http://www.repositorio.unicamp.br/handle/REPOSIP/281308. Accessed on 20 June 2021
Oliveira AU (Org.) (1989) Para onde vai o ensino de Geografia? Contexto, São Paulo
Pinheiro AC (2003) Trajetória da pesquisa acadêmica sobre o ensino de geografia no Brasil: 1972–2000. Thesis (geosciences Ph.D.), post-graduateprogramme in geosciences, Universidade Estadual de Campinas. Campinas (SP), p 277. Available at http://www.repositorio.unicamp.br/handle/REPOSIP/287394. Accessed on 3 Aug 2018
Pontuschka NN (2009) Paganielli, TomokoYda; Cacete, Núria Hanglei. Para ensinar e aprender geografia. Cortez, São Paulo
Resende MS (1989) A geografia do aluno trabalhador. Loyola, São Paulo
Rocha GOR (1996) A trajetória da disciplina Geografia no currículo escolar brasileiro (1837–1942). Dissertation (master), post-graduateprogramme in education, Pontifícia Universidade Católica de São Paulo. São Paulo
Santos CW dos, Mororó LP (2021) O desenvolvimento das licenciaturas no Brasil: dilemas, perspectivas e política de formação docente. Revista HISTEDBR Online Campinas SP 19:1–19. https://doi.org/10.20396/rho.v19i0.8652339. Available at https://periodicos.sbu.unicamp.br/ojs/index.php/histedbr/article/view/8652339. Accessed on 22 June 2021
Silva JM (2012) A bibliografia didática de Geografia: história e pensamento do ensino geográfico no Brasil (1814–1930). Thesis (doctorate) post-graduate programme in geography, Universidade Federal de Uberlândia. Uberlândia
Vesentini JW (Org.) (1989) Geografia e Ensino: textos críticos. Campinas, Papirus
Vlach VRF (1988) A propósito do Ensino de Geografia: em questão, o nacionalismo patriótico. Dissertation (master) post-graduate programme in geography, Universidade de São Paulo. São Paulo

Chapter 12
The Contribution of Milton Santos to the Theoretical Formation of Brazilian Geography

Mónica Arroyo and Fabio Betioli Contel

Abstract Milton Santos' academic and political career in Brazilian Geography dates back to his first works on the reality of the state of Bahia—in the north-east region of Brazil—during the 1950s. His research and publications on urbanisation in Third World countries in the 1970s earned him international renown and made him a central author for theoretical production in human geography. In this context of great political effervescence and enormous transformations in the human sciences, Santos developed the theory of the two circuits of the urban economy—a critical approach to the current conceptions of urbanisation, which takes the specificity of peripheral countries as its starting point. Another of the author's central contributions during this same period was the concept of socio-spatial formation, which considers space as a fundamental component of the social totality and its movement. More recently, in the 1990s, his conceptual proposals on what he called the technical-scientific-informational milieu—this geographical medium comprising an increasingly intensive presence of informational techniques, one of the main factors which permitted the advancement of the contemporary process of globalisation—were fundamental. This article aims to re-examine some of Milton Santos' contributions, while also attempting to stress the historical contexts in which each of them were created. It is important to note that he is an author with an independent and critical theoretical production that expresses an emancipating vision of the geography of the Global South.

Keywords Milton Santos · Brazilian geography · Metageography

M. Arroyo (✉) · F. B. Contel
Universidade de São Paulo, São Paulo, Brazil
e-mail: mmarroyo@usp.br

F. B. Contel
e-mail: fbcontel@usp.br

R. C. Lois González and M. A. Mitidiero Junior (eds.), *Brazilian Geography*,
Advances in Geographical and Environmental Sciences,
https://doi.org/10.1007/978-981-19-3704-0_12

12.1 Introduction

Milton Santos' academic and political trajectory makes him an author central to theoretical production in Brazilian geography. This year (2021) marks the 20th anniversary of his death, and the relevance of his proposals means that they continue to be critical instruments for interpreting the contemporary world. This relevance derives, in large part, from the fact that Santos was a public intellectual, aware of his historical responsibility and engaged in the problems of his time. As the author himself shows, the work of an intellectual must be a "permanent state of alert in order to, at the same time, obey the imperative of critiquing history as well as critiquing himself, as its interpreter" (Santos 2002, 56, translated).

His research and publications, as well as[1] his classes and his input into the public debates in which he participated, are a rigorous expression of his "intellectual craftsmanship", built throughout his life in Brazil and abroad, such as in the courses, conferences and seminars given in France, Canada, Venezuela, the United States, Tanzania, Peru and Japan, among other countries.

This text intends to retrace part of his personal and intellectual trajectory, highlighting central elements of his biography as well as some of his main ideas for the construction of a geography that is robust and attentive to the main contradictions of the historical present. This is an author with an independent and critical theoretical output, who expresses an emancipatory vision of the geography of the Global South.

12.2 Education in Bahia and His Doctorate in Strasbourg: Regional Geography and Planning

A fundamental aspect of Milton Santos' intellectual trajectory concerns what we could call his "period of education and first geographical reflections" (Grimm 2011), which began with his studies at the Faculty of Law in Bahia (between 1940 and 1944) and lasted until the completion of his doctorate at the University of Strasbourg (France).

After graduating as a lawyer, he sat his first public competitive examination and became a geography teacher at a high school in Ilhéus, in the south of the state of Bahia. The dissertation that he presented for this competition would be published by the Official Press of Bahia, and was entitled *O povoamento da Bahia: suas causas econômicas* [The settling of Bahia: its economic causes] (Santos 1948). This was the first academic book written by Milton Santos, and with it began his projection into Bahia's intellectual and political scene. It is also worth mentioning two other important books from this period: *Zona do Cacau. Introdução ao estudo geográfico*

[1] "The Will to Comprehension", published in the Folha de São Paulo newspaper in 1999. It is important to note that all the texts originally published in this newspaper over two decades were gathered in the book O País Distorcido (Santos, 2002).), organized by Wagner Costa Ribeiro, with an essay as an epilogue by Carlos Walter Porto Gonçalves.

[Cocoa Zone: Introduction To Geographic Study] (Santos 1955) and *Estudos de Geografia da Bahia. Geografia e Planejamento* [Geographical Studies of Bahia: Geography and Planning] (Santos and Tricart 1958)[2].

These three works already showed some important characteristics of Santos' trajectory: (1) A solid background in classical French geography—thc main references in urban geography at the time being Jean Tricart himself, but also Georges Chabot, Max Sorre and Pierre George (Bataillon 2006). In Brazil, Pierre Monbeig and Josué de Castro (two of his main bibliographic sources) were widely read; (2) A strong attraction to working with empirical sources (whether official statistics or surveys carried out in the field) as a way of structuring his geographic explanation; (3) The concern with local or regional issues in the state of Bahia, including what was at the time called "applied geography". In the opening text of the work *Estudos de Geografia da Bahia*, Santos and Tricart define the concept of region in a way very similar to that used in classical French geography:

"The concept of region is a complex one, as it is synthetic in nature. It is mainly based on the landscape, including the essential realities that explain it. The elementary geographic region is a set of the physical environment and its aspects of use by man. To have an objective value, the geographic region must present a certain homogeneity, which implies homogeneity in a physical environment as well as conditions for the use of natural resources by humans of the same essence" (Santos and Tricart 1958, 11/12, translated).

A fundamental detail to be highlighted in Milton Santos' biography is his closeness to Jean Tricart, who he met at the Congress of the International Geographical Union that took place in Rio de Janeiro in 1956. Thereafter, they established bonds of academic cooperation and friendship, which would lead Tricart to Bahia as well as bringing Milton to Strasbourg, in order to pursue his doctorate under Tricart's supervision between 1958 and 1959.

The pursuit of his doctorate at the[3] University of Strasbourg allowed Santos to enjoy a highly dynamic university environment, and to be guided by this geographer who was already one of the main renovators of French geography at the time. Within the context of this shift that the doctorate in France would bring, three characteristics of Tricart's personality stand out, having a direct influence on Milton Santos' education: (1) The extreme rigour that he imprinted onto all his academic activity; (2) The publication, in 1951, of the work *L'HabitatUrbain* [The Urban Habitat], an extremely innovative proposal for urban geography, which was the basis for the writing of Milton Santos' doctorate; (3) Jean Tricart—along with Jean Dresch and Pierre George—also played a fundamental role in organising networks of researchers

[2] During this period, the author also wrote: Estudos sobre Geografia, Salvador: Tipografia Manu, 1953; Regional Studies and the Future of Geography, Salvador: Imprensa Oficial, 1953; and Ubaitaba, Study of Urban Geography. Salvador: Official Press, 1954.

[3] As Maria Auxiliadora da Silva (2005, 71) shows, "Impressed by the intelligence and culture of the young professor, Tricart invites him to take a Doctoral course at the Institute of Geography of the University of Strasbourg, one of the most renowned in Europe. Thus, Milton Santos made his first great crossing of the Atlantic, towards what would later be his second country, when he received him, years later, as an exile."

and professors engaged in the renewal of the discipline—the three being members of the French Communist Party who began a process of introducing Marxism into their work.

His initial[4] training was in the basis of the production of several other books—including his doctoral thesis, translated into Portuguese as *O Centro da Cidade de Salvador*, and also published by the Bahia Official Press in 1959. These works already indicated a theoretical maturity and what the author himself called the "will for comprehensiveness" that he always imprinted onto his academic output. Although a tributary of the central concepts that characterised classical French geography—region, habitat, site and urban function, among others—the critical reading of this same French geography that formed him also stands out. In the study *A Cidade comocentro de região* [The City as a region's centre] (Santos 1959), the author indicates the impossibility of using what he called the "Rochefort method"—alluding to the innovative proposals that Michel Rochefort was making at the time in order to understand urban geography. For Santos (op cit, p. 28, translated), the use of this method would be "impractical" for the case of Brazil, "for what he originally intended—that is, the schematic determination of hierarchies through the use of data relating to categories of the active population".

This intellectual posture, his work as a columnist for the newspaper *A Tarde* (the main vehicle of communication for the Bahia press) between 1954 and 1964, and the prestige he gained from becoming a university professor in Salvador (both at the Catholic University of Salvador—1956–1960—and as a full professor of Human Geography at the Federal University of Bahia—1961–1964) made him an important personality in Bahia (Silva and Antunes 2019). His input into the intricacies of regional policy was also noteworthy: (1) Between 1959[5] and 1961 he held the position of Director of the Bahia Official Press (1959–1961); (2) Between 1962 and 1964, he held the position of President of the Bahia State Economic Planning Foundation Commission; (3) In 1961, he was chosen by the President of the Republic, Janio Quadros, as Deputy Chief of the Civil Office in Bahia—a position he held between 1962 and 1964 (Silva 2005).

This collection of elements linked to his academic biography and his political activity made him one of the main targets of the military regime, right after the coup that took place in 1964. Milton Santos was arrested, spent around 100 days in prison, suffered an onset of a heart attack, and was then placed under house arrest. With an invitation from colleagues at the University of Toulouse to teach at that same university, he managed to loosen control over his house arrest and went into exile in France, where he would live from 1964 to 1971, teaching at

[4] For a better detail of this generation that revolutionized French geography, see Bataillon (2006). These authors were both tributaries of traditional French geography, but also renovators of it; in particular Jean Tricart and Pierre George, who in the early 1950s began to introduce Marxism into the geography they produced.

[5] The articles written by Milton Santos when he worked at Jornal A Tarde, from 1949 to 1963, were gathered in the book Milton Santos: Correspondent of Jornal A Tarde 1950 – 1960, organized by Maria Auxiliadora da Silva and William Antunes and published by Editora da Universidade de São Paulo. Santa Catarina (UFSC) in 2019.

three universities: Toulouse (1964–1967), Bordeaux (1967/1968) and finally Paris I—Sorbonne (1968/1971). During this period, he began the research that would lead to one of his fundamental books: *L'EspacePartagé. Les deux circuits de l'économie urbaine dans les pays sous-devéloppés* [The Shared Space: The two circuits of urban economy in underdeveloped countries] (1975).

12.3 Proposals for Autonomous Thought: The Two Circuits of the Urban Economy and Socio-Spatial Formation

Milton Santos' research concerns, which are masterfully summarised in *L'EspacePartagé*, appeared as early as 1965, with the publication of the book *A Cidade nos países Subdesenvolvidos* [The City in Underdeveloped Countries], in which the author seeks to answer the following question: "What is, after all, a great city in the underdeveloped world?" (Santos 1965, 1, translated). These inquiries are also the basis for the writing of another central book by the author in this period: *Le Métier de Géographeen pays sous-développés* [The Profession of Geographer in underdeveloped countries], published in France (Santos 1971). Therein, Santos proposes new approaches and concepts that would be central to all of his later intellectual trajectory. Based on a severe critique of the classical geography that formed him during his doctoral period, this book begins a fruitful dialogue with other areas of knowledge—above all economics, but also sociology, political science, history, anthropology and philosophy. Thus, it moves further and further away from the canons of classical geography, and gains enormous academic resourcefulness and intellectual autonomy.

The author turns from dialogue with the authors of the political economics of underdevelopment/development—mainly Gunnar Myrdal—to the proposition of debates with two central characteristics: (1) The search for a "metageography" (that is, the use of abstract thinking and systematic theorisation in the discipline—which until then had still been quite empiricist and descriptive); and (2) The critical study of the concrete realities of peripheral countries, so little-considered by the canons of the social sciences of this period. At least three main notions proposed by Santos in *Le Métier de Géographe* would be fundamental for his later theoretical output: the idea that geography would be a "philosophy of techniques", and the concepts of "roughness" and "derived spaces".

This is also the spirit that is synthesised in the work that would become one of the main watersheds in the author's trajectory, the aforementioned *L'ÉspacePartagé: Les deux circuits de l'économieurbaine dans les pays sous-devéloppés* (1975). Originally published in France, the work would be translated in Brazil in 1978, just as in the same year an English version was released (by the English publisher Methuen). The book soon became an obligatory reference for Third World studies, as shown by one of the leading names in urban geography at the time, McGee (1996).

In the wake of his critical reading of the phenomenon of urbanisation, and the search for his intellectual autonomy, chief among the authors strongly opposed by Milton are some sociologists (such as Kingsley Davis, Gideon Sjoberg and Bert Hoselitz), anthropologists (mainly Robert Redfield in his articles on "orthogenetic and heterogenetic cities") and economists. For Santos, these proposals were more adjective than substantive, and departed from theoretical assumptions and empirical examples that had little or nothing to do with the reality of underdeveloped countries. It would therefore be necessary to begin with the concrete conditions of the processes of urbanisation in these countries—and their peripheral geoeconomic structures—in order to build other explanations that are more consistent with these realities.

In the case of his criticism of economists, another fundamental aspect should be highlighted: several of them were concerned with state action and the importance of regional planning (in its broadest sense) in order to overcome the problems of underdevelopment. Opposing this functionalist view of planning was also one of the author's main concerns, and sharp criticisms are made of François Perroux, Jacques Boudeville and John Friedman (three of the main exponents of the theory of polarisation and regional planning at the time).

Last but not least, it is essential to highlight that one of the main tasks that the author proposed was the critiquing of the urban geography that was practised at the time. At least three authors were central to Milton's challenge, when elaborating his theory of the two circuits: (1) Walter Christaller and the "theory of central places"; (2) Mark Jefferson and the "law of urban primacy"; (3) John Alexander and the "theory of the basis of export" (Santos [1975] 1979, translated).

All these proposals that he critiqued, although solid in terms of their internal logical coherence (and formal elegance), had little to offer to the critiquing of issues such as the international division of labour, technological modernisation, poverty, structural unemployment and huge regional disparities, among other characteristics of what Milton himself called "derived spaces". A new theorisation needed to be created, and this is what the author achieved in *L'EspacePartagé*. In this work, he analyses the unequal and selective organisation of space which capitalist modernisation provokes in peripheral countries, mentioning that the proposal was "a theory of economic development in its spatial dimension" (Santos [1975] 1979, 19, translated). As the author explains,

> [...] The existence of a mass of people earning very low wages or living from occasional activities, alongside a minority with very high incomes, creates a division in urban society between those who are able to have permanent access to the goods and services offered and those who, having the same needs, are unable to satisfy them. This creates both quantitative and qualitative differences in consumption. These differences are the cause and effect of the existence - that is, of the creation or maintenance, in these cities, of two circuits for the production, distribution and consumption of goods and services. (Santos 1979, 29, translated).

Being chiefly concerned with the study of the large metropolises of poor countries, Santos shows how these two circuits are formed within cities in peripheral countries: (1) a modern *superior circuit*, composed of technology-intensive firms and activities that are highly profitable and extremely sophisticated in terms of their internal organisation. These firms are much smaller in number (in essence, they are oligopolies or monopolies), and they employ a very small portion of the working-age populations in the peripheral countries where they operate; (2) in parallel with the formation of these oligopolies, there is also the *inferior circuit of the urban economy*, composed of a very large number of small firms and economic activities that are very simple in the techniques they use, are poorly organised—or not very "bureaucratic"—and have very low profitability. It is these firms and activities, in fact, that characterise urbanisation in poor countries (while made invisible by the theorisation proposed by European and North American authors). This inferior circuit constitutes the main "escape valve" for the working-age population of poor countries to develop economic survival strategies in their living spaces. This circuit is comprised of an enormous diversity of companies, such as small repair and maintenance workshops, handicrafts, small businesses, neighbourhood markets, personal service providers, self-employed workers, street vendors, etc. At present, the links that are formed between the circuits of the urban economy are even more complex, as shown by Silveira (2015, 2016) in her research on Latin American cities.

In addition to the international prestige that Milton Santos gained with the publication of *L'EspacePartagé*, it is important to note another central feature of his production at the time: the systematic dialogue with historical materialism, through some of the main Marxist authors of the time. It is possible to identify some inspirations and starting points in his work that are clearly derived from this dialogue, in two main senses: (1) in order to create his metageography, Santos makes use of some propositions from thinkers such as Jean-Paul Sartre (and his proposal of "practico-inert"), Louis Althusser (and his notion of "social instance") and Maurice Godelier (his concept of system as a "set of structures" linked by "rules"). (2) Milton also revisits authors who have already produced a discourse focused on theoretical problems in geography, urbanism or regional economics. This is the case with the reading of Aníbal Quijano, Osvaldo Sunkel, Jose Luis Coraggio, Alejandro Rofman, Alain Lipietz, David Harvey, James Blaut, Richard Peet, David Slater and Edward Soja, among others just as the book *L'EspacePartagé* reflects a seminal contribution[6] by Santos to theorisation in urban geography, another of the author's main

[6] Among the most markedly Marxist articles that Santos wrote at this time, we can highlight the following: 1. Under development and poverty: a geographer's view, The Latin American in Residence Lectures, University of Toronto, Canada, 1972-1973, 1975; 2. Under development in the Third World: I, Socio-Economic Formation and Space, Antipode, v. 9, No. 1, February, 1977, Worcester, Mass. USA; 3. Underdevelopment in the Third World: II; Mode of Production and Third World Urbanization; III, Geography and Planning, Antipode, v. 9, no. 3, Worcester, Mass USA, December, 1977; 4. "Lima, the periphery at the pole". In: H. Ross and G. Gappert (eds.). The Social Economy of Cities, Urban Affairs Annual Review, Sage Publications, Beverly Hills USA, 1975, pp. 335-360; 5. "Space and domination: a Marxist approach", International Social Science Journal, vol. XXVII, no. 2, pp. 346-363 (also in French edition, pp. 368-386), 1975; 6. "Articulation of modes of production and the two circuits of urban economy wholesalers in Lima, Peru", Pacific Viewpoint, nº 3, 1976, pp. 23-36; 7. "The Devil's Totality: How Geographical Forms Diffuse Capital and Change Social Structures", Contexto/Hucitec, São Paulo, November 1977, pp. 31-44. Some of these texts were later published in the book Economia Espacial (Santos, 1979).

contributions at this time was "*Society and Space: social formation as theory and method*", published in parallel in the North American magazine *Antipode*, in the French *Cahiers Internationaux de Sociologie* and in the Brazilian *Boletim Paulista de Geografia* (all in 1977) (Santos 1977a, b, c).

In the article, Santos argues that the role of space in relation to social dynamics was still underestimated, even in geography. This under-dimensioning is related, firstly, with the broad "environmentalist" tradition of the discipline, which demanded that all studies should be carried out starting from the natural framework (the soil, the relief, the urban site, etc.); this environmentalist method made it difficult to systematically insert the social dimension into geographic studies. The author also shows that geography has always been more interested in the *forms*, rather than the *formation*, of space. It was necessary to vigorously incorporate social dynamics and history (as a process) into the geographical explanation; and in order to do this, the Marxist *socio-economic formation* category should be the main theoretical instrument (Mamigonian 1996).

The emphasis that the author places on the necessary distinction between the *mode of production* and *socio-economic formation categories* is also essential: the former would be a more abstract category, which should serve as a theoretical framework for the social sciences as a whole, in order to identify the more general possibilities that the historical period offers for human actions; the socio-economic formation category, in turn, as it refers to concrete historical-geographic realities, would be the main theoretical tool to be used in the renewal of geographic theory. Through socio-economic formations, it is possible to identify the possibilities effectively realised by societies within the territories where they operate. As Santos shows,

> "Modes of production are made concrete on a historically-determined territorial basis. From this point of view, spatial forms would be a language of the modes of production. Hence, they are selective in their geographical determination, thereby reinforcing the specificity of places" (Santos, 1979, 14, translated).

In 1977 Santos returned to Brazil, also due to the distension of the national political situation (under pressure, the military was forced to organise the country's political re-democratisation). He taught at different institutions between 1977 and 1979, until he settled at the Federal University of Rio de Janeiro, where he was a professor from 1979 to 1983. Meanwhile, he also served as a visiting professor at the University of São Paulo's Faculty of Architecture and Urbanism, and in 1978 participated in the important Association of Brazilian Geographers' (ABG's) National Meeting of Geographers in Fortaleza, presenting his recently released *Por uma Geografia Nova* [For a New Geography] (Santos 1978), which became a central reference for the generation of critical geographers who, with him, would produce a profound renewal of Brazilian geography from then on (Moreira 2000). In 1983, via public competitive examination, he became a full professor at the University[7] of São Paulo's Department of Geography, where he would work until the end of his life in 2001.

[7] Regarding this renewal movement, Ruy Moreira (2000, 32) – one of its main protagonists – recalls that "Por Uma Geografia Nova: from the critique of Geography to a Critical Geography, a book launched at the Encontro de Fortaleza, 1978, enters the scene. book with which Milton Santos brings

At the University of São Paulo, he continued his systematic effort to produce a solid geography committed to explaining and transforming the world and Brazil. Embedded in the Department of Geography, Santos found favourable institutional conditions in order to pursue his academic and existential project, training a generation of researchers and proposing important contributions to his conceptual system—foundations of the critical social theory that he built[8].

12.4 The Nature of Space and the Current Conditions of Its Transformation

In his output during the 1980s and 1990s, Santos maintained the critical analysis of the historical present as a central concern, with a strong theoretical and political bias. He sought to create, improve and update concepts, targeting the construction of a renewed social theory that allows a deeper understanding of the contemporary world, thereby contributing to its transformation (Santos 1982, 1985, 1987, 1988, 1993a, 1994, 1996a, b, 2000). He carefully observed the changes that took place in the latter decades of the twentieth century and sought to explain them with a theorisation effort based on geography, while considering the contributions from other areas of knowledge (mainly from philosophy and social sciences in general).

It is interesting to point out that, during the early 1990s in Latin America, neoliberal ideas were on the rise in political debates, in the media, and even in part of the university environment. Within certain intellectual circles, there was also a diagnosis that the State "had failed", and that it was therefore necessary to minimise its actions. Globalisation arose as a panacea to be pursued by Latin American countries, which had to adjust in order to enter a world marked by the end of history, the dissolution of borders, and belonging to a global village, among many other fables disseminated during the period. In the midst of this unique and apparently triumphant discourse, Milton Santos systematically and coherently prepared a critical reading of globalisation, exposing his ideas through articles, books, conferences, lectures, courses and interviews.

Among several initiatives, the organisation (together with Professor Maria Adélia Aparecida de Souza) of a series of scientific meetings to critically discuss globalisation stands out. In[9] 1992, they held the International Meeting "*O Novo Mapa do*

---

to the movement what it lacked: the basis for the systematization of ideas." It is important to point out that Milton Santos was president of the AGB between the years 1963-1964.

[8] We also highlight that the author's personal library, document files and three-dimensional objects make up the Milton Santos Collection, donated to the Instituto de Estudos Brasileiros (University of São Paulo), and available to the public. The Collection houses documents accumulated by the geographer throughout his life, from the years he lived in Bahia, during the 1940s until 1964; in exile, between 1965 and 1977; and after his definitive return to Brazil, in 1977, until his death in 2001 (www.ieb.usp.br).

[9] During this period, Milton Santos and Maria Adelia de Souza were President and Secretary, respectively, of the National Association of Graduate Studies and Research in Urban and Regional

Mundo" [The New World Map] at the University of São Paulo's Department of Geography. In 1993, the debate initiated at this event continued at the ANPUR Seminar "*Território: globalização e fragmentação*" [Territory: globalisation and fragmentation] and, in 1994, at the 1st International Meeting of ANPEGE "*Lugar, Formação Socioespacial, Mundo*" [Place, Socio-spatial Formation, World]. These debates were completed in 1995 at the 1st National Meeting of ANPEGE "*Território Brasileiro e Globalização*" [Brazilian Territory and Globalisation], which took place at the Federal University of Sergipe, in the city of Aracajú. The main intention running through all these events was a deepening of the methodological proposals that geography in particular, and the social sciences as a whole, were confronting in order to understand our times.

In 1996, under the coordination of Professor Maria Adélia de Souza, the University of São Paulo's Department of Geography organised the International Meeting "*O mundo do cidadão, um cidadão do mundo*" [The world of the citizen, a citizen of the world], a tribute to Milton Santos, which was a moment of reflection on the theoretical-methodological contribution of his work. Also in this decade, Santos received important awards and distinctions, in particular the *VautrinLud* (awarded in 1994 at the International Geography Festival in the French city of Saint-Dié-des-Vosges), as well as the title of Doctor *Honoris Causa*, which he received from several Brazilian and foreign universities. Regardless of the occasion, his discipline and rigour stood out in the preparation of[10] each lecture, each class, each award. Precisely searching for ideas, concepts and propositions in order to provide density to the argument: a task that he himself assumed and demanded of his students and collaborators.

The understanding of the social reality, starting with one of its instances, involves an ontological effort that, for this author, implies the definition of geographic space. Based on this perspective, one of the central issues in his writings is the permanent search for the formulation and reformulation of this category. For the author,

> Space has many definitions, depending on who is speaking and what they want to express. Here, the voice is that of a geographer who proposed some ways of focusing on the issue: space as a dialectical meeting of fixed and flux; space as a contradictory collective, formed by a territorial configuration and by relations of production, social relations; and, finally, what will preside over today's reflection, space formed by a system of objects and a system of actions. (Santos, 1994, 110, translated)

Planning (ANPUR) from 1991 to 1993, and of the National Association of Graduate Studies and Research in Geography (ANPEGE) from 1993 to 1995

[10] He received the title of Doctor "Honoris Causa" in foreign universities: Universidad de Buenos Aires, Argentina, 1992; Complutense University of Madrid, Spain, 1994; University of Barcelona, Spain, 1996; National University of Cuyo, Argentina, 1997; Universidad Nacional del Sur, Bahía Blanca, Argentina, 2000; Universidad de la República, Uruguay, 2000. And in Brazilian universities: State University of Southwest Bahia, 1995; Federal University of Sergipe, 1995; Federal University of Rio Grande do Sul, 1996; State University of Ceará, 1996; University of Passo Fundo, 1996; Federal University of Santa Catarina, 1996; Paulista State University UNESP, 1997; State University of Rio de Janeiro, 1998; Federal University of Rio de Janeiro, 1999; University of Brasilia, 1999; Federal University of Pernambuco, 1999; Federal University of Paraná, 2000

Natural, manufactured, computerised objects—all of them submitted to a growing process of artificialisation of nature, in which (live) human work is increasingly superimposed on (dead) human work. Today, natural objects as autonomous entities and with an independent causal logic are increasingly rare, and one can speak of the almost absolute predominance of second nature. In this sphere of social forces and economic, political or cultural actions—deliberate or not, rational or not, hegemonic or not—they are distinguished according to different degrees of intentionality and rationality. Technical objects are a product of the social process and actions are the social process itself—both fully historical and, therefore, in permanent transformation. However, neither can be considered in isolation themselves, as if they were autonomous; on the contrary, according to Santos, they must be understood as a totality: actions carried out through objects and objects being realised based on actions, in the same movement. Far from being linear or mechanical, this movement is complex, full of contradictions, conflicts, contingencies and uncertainties. There are multiple determinations and mediations that can intervene in the permanent process of aggregation in which the movement of the real is involved.

The indissoluble relationship between space and time is another of the author's central concerns, appearing permanently and explored in depth in his writings. In this period in which his metageography was already a project in the process of consolidation, his inquiries are presented with a high degree of complexity and philosophical sophistication. According to Ana Clara Torres Ribeiro, action is a privileged path to accessing the links between past, present and future. In her words,

> In Milton Santos' work, action, in the fullness of its affirmation as praxis, is inscribed in the ontology of space. Action is the carrier of time in the very spatiality of the techniques, insofar as they are manifested, in the same practical and political movement, the historically-inherited conditions and the project of their transformation. (Ribeiro 2003, 31, translated)

The emphasis on technique and science—and the need to question this process from a critical geographic perspective—remained at the centre of his output throughout the 1980s and 1990s. Information, in all its forms, is also recognised as a fundamental driver of the social process and incorporated into the understanding of geographical space. Science, technique and information are at the base of all forms of the use of space, as they likewise participate in the creation of new vital processes and the production of new species (animals and plants). At the International Meeting "*O Novo Mapa do Mundo*" (1992), Santos gave the opening lecture and, by evoking the culminating moments of previous periods, he pointed out that "the contemporary acceleration has imposed new rhythms on the displacement of bodies and the transportation of ideas, but also added new items to the story" (Santos 1993b, 15, translated). New materials emerge as well as new forms of energy, intense urbanisation, exacerbated consumption, the proliferation of symbols, words, images—in short, the empire of speed and vertigo imposes itself in great strides and requires a consideration of the manner in which space translates this movement.

From this perspective, Santos points out that the geographical environment of the contemporary period is updated with a growing content of science and technique, but also of information, which "is both present within things and necessary for action carried out on these things" (Santos 1994, 51, translated). Thus, the concept of *technical-scientific informational milieu* gains centrality in his theoretical and epistemological construction, and through it the author offers a substantive reading of globalisation, in its perverse character, but also in what it carries as potential, as a bearer of other possibilities of existence. The technical-scientific informational milieu is based on the combination of the *technosphere* and the *psychosphere*, which operate in an inseparable way. The technosphere is the result of the growing artificialisation of the environment, formed by technical objects that are increasingly impregnated with intentionality and are precise and sophisticated, present in the city and in the countryside; the psychosphere is comprised of ideas, beliefs, wishes, desires, passions, habits—the scope for the production of meaning. Both are subject to the law of those who impose changes—that is, of the hegemonic agents of the economy, politics and culture. Through them, the geographical environment introduces rationality—as an expression of instrumental action—as well as, at the same time, irrationality and counter-rationality (Santos 1994, 1996a, b).

12.5 The Strength of Place: Towards Another Globalisation

The effort of theorisation in order to conceive the world we live in, from a geographic perspective, gave structure to several research projects coordinated by Santos in the 1980s and 1990s—the results of which he presented for discussion at congresses, seminars and colloquiums. As expressed by Maria Adélia Aparecida de Souza, in the book-homage *O mundo do cidadão. Um cidadão do mundo*, Milton "makes us revisit old concepts of geography – region, place, territory – adjusting them to today's world. In so doing, Milton revitalises geography" (Souza 1996, 32, translated).

When Milton Santos opposes world and place—concepts that he favours in geographic analysis—he shows that they work in a unitary manner: the world as latency, a set of possibilities and place as existence, a set of opportunities. "It is the place that offers the world's movement the possibility for its most effective realisation. To become space, the world depends on the potential of the place" (Santos 1994, 5, translated); these virtualities are expressed in a particular combination of objects and actions, since "each place is defined as much by its corporeal existence as by its relational existence. This is how subspaces exist and are differentiated from each other" (Santos 1994, 6, translated). In this way, the author recovers the actuality of studying what is unique, specific and different in each place as a central task in the tradition of geographic science.

One of the ways to unravel the complexity of reality is precisely to unite the place and the world in the same visible movement. For this, Santos uses pairs of opposing and complementary categories that facilitate the analysis, but also allow the indissoluble, indivisible essence of the unitary reality to be captured. The new and the old, the external and the internal, technosphere and psychosphere, horizontalities and verticalities, allow us to reconstruct a spatial dynamic as an arena of antagonisms and complementarities. Globalisation and fragmentation, metropolisation and de-metropolisation, flexibilisation and rigidity, are processes that coexist under contradictory tension, but respond to the same logic. Global reason and local reason, intelligent spaces and opaque spaces, organic solidarity and organisational solidarity, are phenomena that are qualified beginning with a contrast, which merge with each other and, at the same time, distinguish and distance themselves. This is a matter of a methodical path that translates into a permanent process of deconstruction and reconstruction of the whole. Although these categories refer to extreme, originally opposed situations, each one of them contains aspects of the other. Hence we see the possibility of discovering the variety of intermediate situations, because in all cases there are different combinations, in different densities, with a definite synchronicity. Such situations appear to be superimposed, contrasting with each other, complementing each other, with one having priority over the other, with greater or lesser effectiveness than the other.

In the quest to decipher the multidimensionality of the real, Santos resumes the notion of banal space since, as it is the space of *everyone* and is *all of* the space, it requires all players to be included in the analysis—regardless of their strength or power—as well as all dimensions of the happening, all the determinations of the social totality. With a similar purpose, Santos (1994, 44, translated) invokes the concept of place as an "encounter between latent possibilities and pre-existing or created opportunities" which depends on the combination of different ways of existing and the game of social determinations, and is expressed in a permanent tension. This field of social forces that operates dialectically in each place, oscillating between cooperation and conflict, creates and recreates this dynamic, favouring some forces over others according to the moment. Thus, the possibilities that the world presents are used in different ways, depending on the places.

The place, a meeting point for different temporalities, allows for co-presence, coexistence, contiguity, proximity, the obligatory interdependence generated by face-to-face situations—in short, being together. And so the author comes to everyday life and asserts that "through the place and day-to-day life, time and space, which contain the variety of things and actions, also include the infinite multiplicity of perspectives" (Santos 1994, 39, translated). Plurality and diversity come into play, as tension, negotiation and conflict.

If the speed-based empire of the technical-scientific informational period asserts the power of hegemonic actors, time passing slowly is an ally of the poor, of common men who, immersed in the daily struggle for survival, "end up being faster in discovering the world" (Santos 1994, 85, translated). The experience of scarcity, which is reproduced every day, leads to the perception of each person's situation and becomes a path towards the expansion of consciousness. In everyday praxis, the political action of slow men materialises. This dialectical movement is also imprinted on the concepts of verticality and horizontality, coined by Santos in his writings in the 1990s. For the author,

> Verticalities are vectors of superior rationality and the pragmatic discourse of the hegemonic sectors, creating an obedient and disciplined daily life. Horizontalities are both the place of finality imposed from without, from afar and above, as well as that of counter-finality, locally generated. They are the theatre of a conforming, but not necessarily conformist, everyday

> life and, simultaneously, the place of blindness and discovery, of complacency and revolt. (Santos 1996a, b, 227, translated)

It is precisely in this density of social life, in this thickness of happening, in this contradictory tension, where multiple forms of material and symbolic life coexist, where multiple forms of use and appropriation of territory are practised, that Milton Santos sees the possible limits to perverse globalisation:

> When we accept that we consider technique in conjunction with politics and admit to assigning it another use, we are convinced that it is possible to believe in another globalisation and another world. (…) the interpretation of history cannot be left to the immediate understanding of the technical phenomenon, requiring an understanding of how, in this same situation, technique and politics are related, assigning this the central role in understanding the actions that adapt the present and can make possible another future. (Santos 2000, 125, translated)

These are some of the main proposals elaborated by Milton Santos for understanding the world, which allow an accurate interpretation of contemporary society, starting from a geographic perspective. With his work, the role of geography in social analysis and in the construction of the future is renewed, thanks to the permanent search for signs of the new, of movement, of change—in short, of everything that points to the possible advent of another historical period. A globalisation with a human face, in which all the planet's inhabitants are considered fundamental, in their material and spiritual wholeness. According to the author himself, the definitive end of this perverse globalisation that still envelops us would be the beginning of the *popular period of history*.

References

Bataillon C (2006) Six géographes en quête d'engagement : du communisme à l'aménagement du territoire. Essai sur une generation. Cybergeo 241. https://doi.org/10.4000/cybergeo.1739

Grimm F (2011) Trajetória epistemológica de Milton Santos. Uma leitura a partir da centralidade da técnica, dos diálogos com a economia política e da cidadania como práxis. Faculdade de Filosofia, Letras e Ciências Humanas da Universidade de São Paulo (Doctoralthesis), São Paulo, p 305

Mamigonian A (1996) A geografia e 'a formação social como teoria e como método´. In: Souza MAA (org) O mundo do cidadão. Um cidadão do mundo. Hucitec, São Paulo, pp 198–206

McGee T (1996) Geografia e desenvolvimento: crise e renovação. In: Souza MAA (org) O mundo do cidadão. Um cidadão do mundo. Hucitec, São Paulo, pp 454–464

Moreira R (2000) Assim se passaram dez anos (A Renovação da Geografia no Brasil no Período 1978–1988). In: Geographia (UFF), no 3, 2nd year, pp 27–49

Ribeiro ACT (2003) Pequena reflexão sobre categorias da teoria crítica do espaço: território usado, território praticado. In: Souza MAA (org) Território brasileiro. Usos e abusos. Edições Territorial, Campinas, pp 29–40

Santos M. Sociedade e espaço: a formação social como teoria e como método. In: Santos M (ed) Espaço e Sociedade. Petrópolis, Rio de Janeiro, pp 9–27

Santos M (1948) O Povoamento da Bahia: suas causas econômicas. Imprensa Oficial da Bahia, Salvador

Santos M (1955) Zona do Cacau. Introdução ao estudo geográfico. Imprensa Oficial da Bahia, Artes Gráficas, Salvador

Santos M (1959) A Cidade como centro de região. Livraria Progresso Editora, Salvador
Santos M (1965) A Cidade nos países subdesenvolvidos. Civilização Brasileira, Rio de Janeiro
Santos M (1971) Le métier du geographe en pays sous-développé. Un essay méthodologique. Editions Ophyrs
Santos M (1975) L'Éspace partagé. Les deux circuits de l'économie urbaine dans les pays sous-devéloppé. Éditions M-Th. Génin, Paris (Translation into English: the shared space: the two circuits of the urban economy in underdeveloped countries. London: Methuen, 1979)
Santos M (1977a) Society and space: social formation as theory and method. Antipode 9(1):3–13
Santos M (1977b) Société et espace: la formation economique et social comme théorie et comme méthode. Cahiers Internationaux de Sociologie, vol LX, 1st Sep 1977b, pp 261–276
Santos M (1977c) Sociedade e espaço: a formação social como teoria e como método. Boletim Paulista De Geografia 54:81–100
Santos M (1978) O Espaço dividido. Os dois circuitos da economia urbana nos países subdesenvolvidos. Francisco Alves, Rio de Janeiro
Santos M (1979) Economia Espacial: Críticas e Alternativas. Hucitec, São Paulo
Santos M (1982) Pensando o espaço do homem. Hucitec, São Paulo
Santos M (1985) Espaço e método. Nobel, São Paulo
Santos M (1987) O Espaço do Cidadão. Nobel, São Paulo
Santos M (1988) Metamorfoses do Espaço Habitado. Hucitec, São Paulo
Santos M (1993a) A Urbanização brasileira. Hucitec, São Paulo
Santos M (1993b) A aceleração contemporânea: tempo mundo e espaço mundo. In: Santos M et al (eds) Fim de século e globalização. Hucitec, São Paulo, pp 15–22
Santos M (1994) Técnica, espaço, tempo. Globalização e meio técnico-científico informacional. Hucitec, São Paulo
Santos M (1996a) A Natureza do Espaço. Técnica e tempo, Razão e emoção. Hucitec, São Paulo (Translation into English: The Nature of Space. Durham: Duke University Press, 2021)
Santos M (1996b) De la totalidad al lugar. Oikos-Tau, Barcelona
Santos M (2002) O país distorcido: o Brasil, a globalização e a cidadania. Publifolha, São Paulo
Santos M (2017) Por uma outra globalização. Do pensamento único à consciência universal. Record, Rio de Janeiro (Translation into English: toward an other globalization: from the single thought to universal conscience. Berlin: Springer, 2017)
Santos M, Tricart J (1958) Estudos de Geografia da Bahia. Geografia e Planejamento. Livraria Progresso Editora
Silva MA (2005) No mapa do mundo, um ponto de raiz – a Bahia. In: Silva MA et al (orgs) Encontro com o pensamento de Milton Santos. O lugar fundamentando o período popular da história. Empresa Gráfica da Bahia, Salvador, pp 71–82
Silva MA, Antunes W (orgs) (2019) Milton Santos: Correspondente do Jornal A Tarde 1950–1960. CFH-Universidade Federal de Santa Catarina, Florianópolis
Silveira ML (2015) Modernização contemporânea e nova constituição dos circuitos da economia urbana. Geousp Espaço e Tempo (Online) 19(2):246–262
Silveira ML (2016) Circuitos de la economía urbana. Ensayos sobre Buenos Aires y São Paulo. Café de las Ciudades, Buenos Aires
Souza MAA (1996) Por que ouvir dizer e por querer saber. In: Souza MAA (org) O mundo do cidadão. Um cidadão do mundo. Hucitec, São Paulo, pp 26–34

Chapter 13
Carlos Augusto de Figueiredo Monteiro and the Construction of Brazilian Geographical Climatology

Francisco Mendonça

Abstract Brazilian university geography was founded in the first half of the twentieth century under the philosophical and epistemological frameworks of the French School and the German School. This was one of the events that opened up the institutional modernisation of the country—a way of placing scientific knowledge within the sphere of the country's development. For around 20 years, the nascent geography was dominated by its Franco-German roots, at a time when the first professors and graduates in Brazilian geography were educated. Carlos Augusto de Figueiredo Monteiro was educated within this context, at the end of the 1940s, and subsequently pursued an internship in France with Francis Ruellan, in the Laboratory of Geomorphology/Dinard. Within this scientific context, he made contact with and was influenced by the Norwegian—or frontogenesis—School (in particular the contributions of Serra and Ratisbona), being one of the first to apply (1) the approach of atmospheric circulation and air mass dynamics to the genetic analysis and classification of the climates of Brazil. This contribution inaugurated a new approach (1960s) to climate, based on atmospheric dynamics and their interaction with human activities within Brazil's territory. Echoing the criticisms of Max Sorre regarding the statistical approach to weather, and Pierre Pedelaborde regarding weather types, he developed (2) the rhythm-based analysis of weather types, as required in order to take advantage of the atmosphere as a geographical factor. Concerned about the chaotic urbanisation of Brazil and the subsequent reduction in the quality of life there, he proposed (1970) a (3) theoretical-methodological foundation for the insertion of urban climate into the planning of cities (Urban Climate System; UCS). These three major contributions constituted the formation of a Brazilian Climatology School and a Brazilian Urban Climatology School. However, his contributions also extended to the fields of geographical epistemology, environmental analysis and cultural geography.

Keywords Climatology · Geography · Monteiro

F. Mendonça (✉)
Universidade Federal Do Paraná, Curitiba, Brazil
e-mail: chico@ufpr.br

R. C. Lois González and M. A. Mitidiero Junior (eds.), *Brazilian Geography*, Advances in Geographical and Environmental Sciences,
https://doi.org/10.1007/978-981-19-3704-0_13

13.1 Introduction

Founded in 1934, the USP—University of São Paulo—inaugurated Brazilian university education in geography. A group of young French academics—recent graduates of the Sorbonne University—pioneered university education in the field of humanities in the country, with emphasis on geography, history and anthropology, thereby guaranteeing university status for the institution that was then being founded. The São Paulo economic and intellectual elite took an important step in the country's entry into modernity, even though access to university was already guaranteed almost exclusively to the country's white and middle- to upper-class population.

Pierre Monbeig (a former student of Paul Vidal de La Blache) and Pierre Deffontaines led this endeavour, along with other scientists and intellectuals—notably from the field of law, among others. Geography at Brazilian universities therefore began life with a strong hue of French geography, an aspect that has characterised it since its origin.

Followed by the University of Brazil in the then-federal capital, Rio de Janeiro, the founding of geography in higher education, was recorded in 1935, with the French academic Pierre Deffontaines as its main articulator. The French brand in the genesis of Brazilian university geography thus expanded throughout the national territory, even when it was introduced in 1938 at the then-University of Paraná, a state whose capital had a strong German influence. It was only in the middle of the twentieth century that other influences, such as from the USA, in the field of New Geography, came to have a more significant influence on training in geography within the country.

It was in the context of the first half of the last century, particularly in the 1940s, that Carlos Augusto Figueiredo Monteiro (here in after simply referred to as Monteiro), the illustrious geographer whose contribution to the formation of Brazilian geographical climatology is described in this text, studied at university. He was a student at the then-University of Brazil, an institution that followed—particularly, but not exclusively—French geography according to Lablachian principles. Perspectives from German geography were also present at that university, particularly from the practice of Leo Weibel—which gave the young student's education a somewhat comprehensive perspective.

Monteiro's contribution to the construction of Brazilian geography has been recognised by the country's entire scientific community and several other institutions throughout the world, particularly due to his international exposure via the International Geographic Union (IGU—Environmental Committee) where he worked for a few decades, in addition to spells in several countries as a lecturer, a visiting professor and/or collaborating in research.

In this text we highlight, by invitation, Monteiro's contribution to the formation and consolidation of Brazilian geographic climatology, based on the author's own publications, as well as others emphasising his role in the construction of this scientific field in the country—such as those of Santana Neto (2001), Zavatini (2004) and Mendonça (2012). The drafting of this text also included a recent publication from Monteiro himself (2015), in which he detailed an interesting reflection on his output in climatology, highlighting the paradigmatic character of his constructions; in the same work, three other climatology scholars in Brazil highlighted relevant aspects of his contribution to the understanding and construction of Brazilian geographical climatology. Monteiro ended his research and his scientific output regarding climatology in 1986, the year he retired from his career as university professor at the USP (University of São Paulo).

It must be highlighted, however, that climatology is just one part of his scientific and intellectual construction, since his publications show a vast field of interests covering geographical epistemology, environmental geography, cultural geography, architecture and urbanism, arts, etc. He is, therefore, one of the most important Brazilian scientists and intellectuals of the second half of the twentieth and early twenty-first centuries, Professor Emeritus at the University of São Paulo and Doctor Honoris Causa at several universities in the country.

Retrieving his work and his contribution to Brazilian geography is always a great challenge. As detailed and attentive as one may be, it will be impossible to cover all of the rich legacy that he has provided to it. As such, the present text should be viewed as a partial reading of Monteiro's contribution to the construction and consolidation of Brazilian geography. In the field of climatology, in particular, his contribution is detailed here beginning from three approaches that are highlighted in the work produced by him for approximately 30 years, between the late 1950s and the end of the 1990s.

Some questions encourage a reflection on why Monteiro would have dedicated himself to understanding the country's climatic configuration—and what the originality of his construction was—to the point of supporting the creation of a Brazilian "geographical climatology school" and "urban climatology school" (Mendonça and Danni-Oliveira 2007). The driving context of his scientific creation was marked by aspects that were challenging for a university professor who was concerned with the status of knowledge being offered to students, professionals and society at that time, of which the following were noteworthy:

The weak knowledge about tropical climates until the mid-twentieth century forced a breakthrough in this field of science, given the increasing presence and importance of these areas in the world economy, particularly in relation to agricultural products that are highly dependent on climatic conditions;

- The industrial dispersion through the system of multinationals in the post-war period began to demand more in-depth knowledge of the sites where production units were to be implanted, at the same time as advancing the knowledge of the climatic risks associated with the production and circulation of the international economy;
- The considerable extent of the Brazilian territory, with a significant and unknown detailed climatic configuration, evidenced the need for greater knowledge of this in order to aid the development of all economic and social activities within the country;
- The performance of teacher activity in schools at all levels in Brazil (particularly universities) which were widespread during the middle of the century; in most

of them, geography courses needed teachers with more specific training on the Brazilian territory;

- The deepening of knowledge of the Brazilian territory's characteristics as a basis for the activities of territorial planning and development which, as a result of State policy through the creation and performance of the CNG (National Council of Geography) and IBGE (Brazilian Institute of Geography and Statistics), demanded increasingly sophisticated actions from geographers; and
- The need to update more-advanced scientific and technical approaches that could guide the approach to and understanding of the country's climates—hence the search for new conceptual, methodological and technical approaches, in line with what was done in the more-developed countries of the West.

It was within such a context of challenging demands and questions that Monteiro found himself, very soon after the end of his university education, involved in the analysis of the atmosphere and climatology of Brazil; here, he laid the foundations of a new analytical, conceptual and methodological perspective, as will be seen below.

13.2 A Perspective on Climate in Search of Its True Comprehension

Indigenous people, people of African descent, settlers, mestizos, etc. had, for a long time, already recorded their understanding about the configuration of the air over the different locations of the Brazilian territory. The advance of colonisation based on slavery and the expropriation of indigenous territories had been carried out for centuries, beginning with the exploitation of nature, with agricultural production being one of its most obvious hallmarks; in order to ensure its effectiveness, knowledge of the climate was imperative. Agriculture is not produced without a thorough knowledge of thermal conditions and the availability of rainwater in a given location—thus, for around four hundred years observations were made and knowledge was gained of climates in Brazil, albeit without modern records based on the measurement of its components and without gauging its dynamics.

Since the end of the nineteenth century, scholars linked to the Imperial Observatory produced scientific knowledge about the country's climate, organised under the auspices of modernity, thereby beginning the practice of keeping records in this area within the country. However, although they were modern, but produced in accordance with the prevailing paradigms at the time, the first works of Brazilian climatology show an attachment to climatic determinism and to the static-statistical conception of climate. The setting-up of the first network for the measurement of atmospheric elements in the field of coffee production at the beginning of the century (in the States of São Paulo and Rio de Janeiro), as well as the efforts to describe and classify the country's climates, retained the use of the concepts of Julius Han and Wilhelm Koeppen until the mid-twentieth century. The general conception that underpinned the understanding of climate was that it was the average state of the atmosphere, from a statistical perspective, mainly in relation to aspects of the vegetation.

During the first decades of that century, the search for understanding the dynamism of the atmosphere gained great impetus from the analytical perspectives of the Scandinavian (or frontogenesis) School, and its repercussion was soon visible in Brazil. Considerable advances in the search for gaining an understanding of the climate in its dynamics were recorded in several studies—a fact revealed by the search for the genesis of atmospheric phenomena and the questioning of the understanding of climate as an average state of atmospheric conditions. The genetic classification of climates presented by Arthur Strahler was one of the first to apply atmospheric dynamics to the planet's climatic typology.

But the conception of climate as an average state of the atmosphere was strongly questioned when, with the provocation levelled at geographers by Maximilian Sorre, he urged them to overcome the predominant statistical view of climate. This fact moved the young Monteiro, who worked restlessly on developing the concept introduced by Sorre: that of understanding climate as a normal succession of types of weather over a given place—that is, its manifestation in terms of atmospheric states that are more and less repetitive over places. In this way, an important step was taken towards the understanding of climate in its most palpable manifestation, jumping from the median condition of the atmospheric elements and strongly approaching its concrete and detailed configuration.

Thus, from Monteiro's propositions, notably from the 1960s onwards, the notions of climate began to consider climatology in a synthetic dimension; in this, atmospheric weather came to be conceived beginning from its chronological and repetitive succession, showing what is usual (more and less frequent), and not an average state of the atmosphere. Meteorological weather is characterised by an occasional manifestation (days, hours, etc.) while climate is characterised by the perception of regimes (annual variation); obviously, the use of statistics to represent the magnitude of atmospheric phenomena—both extremes and averages—is taken into account, but these are not what define it in a compartmentalisation of the air associated with biomes or with terrain, as the prevailing perspective that had until then been advocated.

This construction constituted a paradigm shift in the conception of climates in Brazil. It changed not only the dominant perspective of the climate taken as a static factor, by introducing the dynamic conception of the same, but went a long way towards bringing its understanding closer to reality. Atmospheric states represent the climate of a place insofar as the normal dynamics of types of weather result from the interaction between the air (atmosphere) and the surface of the planet, on which human activities take on more and more importance. In this sense, the meso-, topo- and microclimatic scales are increasingly highlighted in the studies of geographic climatology, which reinforces the geographical nature and dimension of the climate.

By betting on the genesis of climatic phenomena, especially on the possibility of associating them with human activities, the approach to the frequency of their occurrence is inserted from the perspective of their usual manifestation, more and less repetitive over places. Within the context of climate dynamics, the focus on their genesis allows a vision of the future to be developed—hence a close approximation with climate forecasting, an aspect of direct use for applied climatology, notably in the field of urban, rural and regional planning for the territory.

Taking on and developing in detail the approach to atmospheric circulation and dynamics from centres of action and air masses (Fig. 13.1), Monteiro took Arthur Strahler's genetic climatic classification as a reference and adapted it, based on circulation and atmospheric dynamics, for the climates of Brazil (Fig. 13.2). This classification, quite innovative in view of Wilhelm Koeppen's hegemonic generic classification of climate types, has been widely applied in studies and territorial planning actions over the last fifty years. But, in addition to this theoretical and conceptual design, it seemed to also be necessary to invest in the construction of a methodological perspective that could account for the scalar refinement of the study of climate; this was how, in the early 1970s, after more than a decade of research, Monteiro published the rhythmic analysis of types of weather aiming to account for the fine temporo-spatial approach to the climate.

Developed by Monteiro in the 1960s, based on the atmospheric circulation and dynamics published by Adalberto Serra and Leandro Ratisbona in accordance with the assumptions of the Scandinavian School. Monteiro refined this perspective by emphasising the geographic approach to the latitudinal and longitudinal movements of oceanic and continental Centres of Action in South America. He outlined, in a very didactic way, the displacements of air masses according to the seasons of the

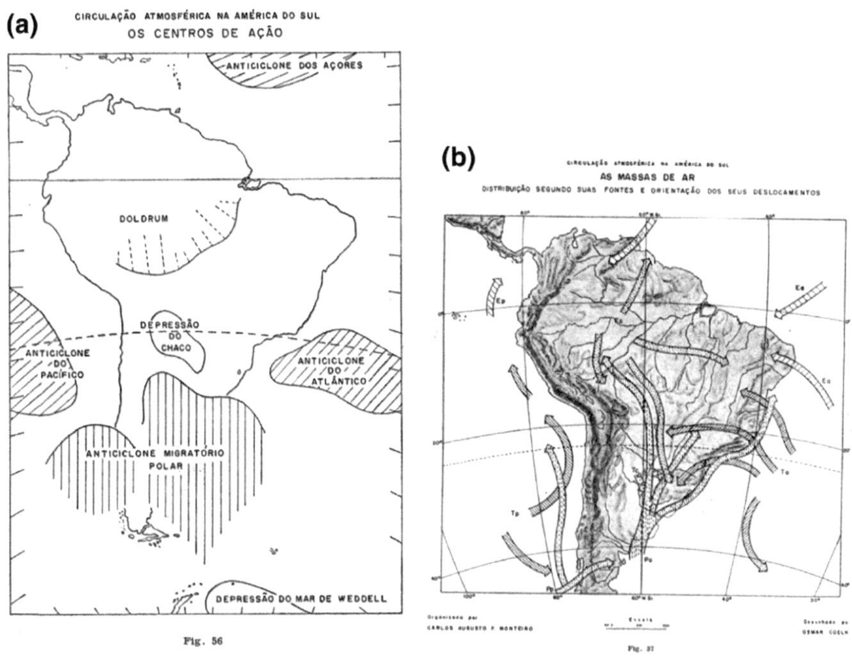

Fig. 13.1 **a** Centres of action and **b** predominant air masses in South America. *Source* Monteiro (1963)

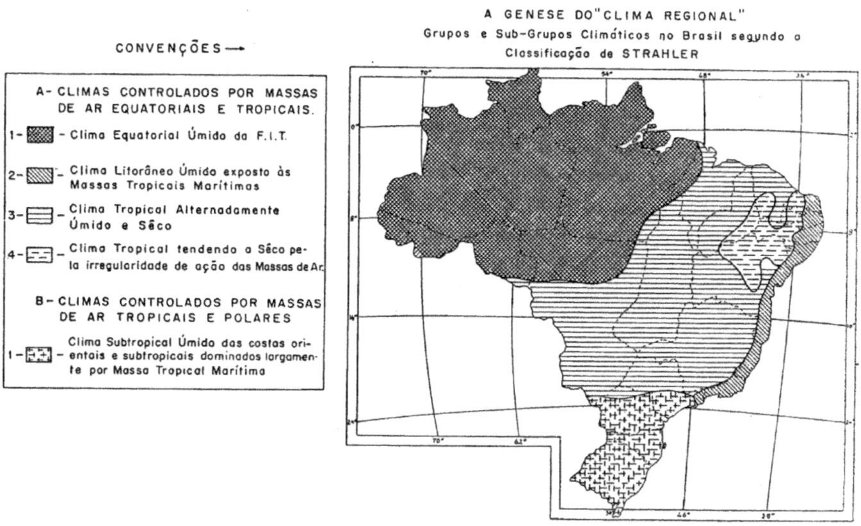

Fig. 13.2 Climatic classification of Brazil and the South Region. *Source* Monteiro (1963)

year and, in particular, explained the importance of frontogenetic processes in the configuration of types of weather and climates in Brazil.

Published in 1968 by the IBGE, this typology of climates in Brazil explains the genetic conception of the country's climates, revealing the paradigmatic conception of climate as the usual state of the atmosphere in interaction with the Earth's surface. By emphasising the role of air masses in the configuration of different climates, and highlighting the interaction between the atmosphere and the surface, in a dynamic process, this classification explains the position in contrast to the then-predominant perspective of climate as an average state of the atmosphere average state of the atmosphere.

13.3 The Rhythms of Time: The Need for an Approach Using a Fine Spatio-Temporal Scale

Dealing with climatic phenomena in their long- and short-term manifestations presented itself as a challenge to Monteiro's curiosity at that moment of intense scientific fruition. Inspired by Pierre Pedelaborde's studies of weather types in the Paris basin, through which the importance of knowing the particular types of atmosphere in that area was made clear, Monteiro inserted the notion of dynamics. In other words, the different types of weather usually follow one another over places as a consequence of the centres of action and air masses that, in interaction with local and regional geographic factors, create an imprint on the climatic configuration of places.

New times instigated the search for an understanding of the air in its movement—thus, analysing the succession of different types of weather appeared as a concrete and real possibility for understanding the climate in its dynamism, as a "moving force". The construction of the rhythmic analysis of types of weather is therefore configured as the necessary step for confirming the outlooks thrown up so far in order to abandon the understanding of climate as an average state, since identifying and analysing the succession of types of weather will henceforth allow the temporal evolution of the atmosphere over places to be monitored.

This approach highlights the chronological decomposition of the states of the atmosphere, taking into account their continuous succession, since they can only be accurately observed in their everyday manifestation. Atmospheric states considered as types of weather manifest themselves daily, and their succession can be observed starting with the variation of climatic elements in interaction with tertiary and regional atmospheric circulation. According to Monteiro (1971: 9, translated), "*the rhythm of the climate can only be understood through the concomitant representation of the climate's fundamental elements in chronological time units, at least daily, compatible with the representation of regional atmospheric circulation that generates the atmospheric states which succeed each other and constitute the foundation of the rhythm*".

Monteiro thus highlights his picture of the rhythmic analysis of types of weather, combined with the conception of the climate as a usual state: "*In our view, a huge advance was made between the numerical (average) valuation of the climatic elements, but the simple identification of the types of weather operating in the region seems to place them in "separate drawers" since, in our understanding, the types of weather should be seen "in progressions". Types of weather should not just be identified and filed; they must be observed in sequences - that is, in 'progressions of types of weather'. Through this, it will be possible to obtain the climatic rhythm, capable of exhibiting the usual trend without neglecting extreme deviations*".

In order to perform the rhythmic analysis of types of weather for a given location, it is necessary to take into account all the climatic elements, the aspect of the sky and the prevailing atmospheric systems (Fig. 13.3). Thus, depending on the day and time, it becomes possible to observe the succession of types of weather and the genesis of interesting phenomena in a particular atmospheric situation. This methodology, then, allows a detailed understanding of atmospheric phenomena from the perspective of their manifestation in a given time and place; it constitutes an advance in the understanding of the rhythm of the climate in a fine timescale. The mapping and analysis of different types of meteorological weather in a given chronology and place allows a secure understanding of the dynamism of the atmosphere and climate for a given location.

This base graph for the development of the rhythmic analysis of types of weather shows the daily recording of temperature, rainfall, wind direction and the atmospheric systems acting on the days on which cases of dengue were recorded. Other climatic elements, such as relative humidity and atmospheric pressure, were recorded in this type of graph when looking at Monteiro's original proposal, published in 1971. The data represented in this type of graph allow us to identify and analyse the types of

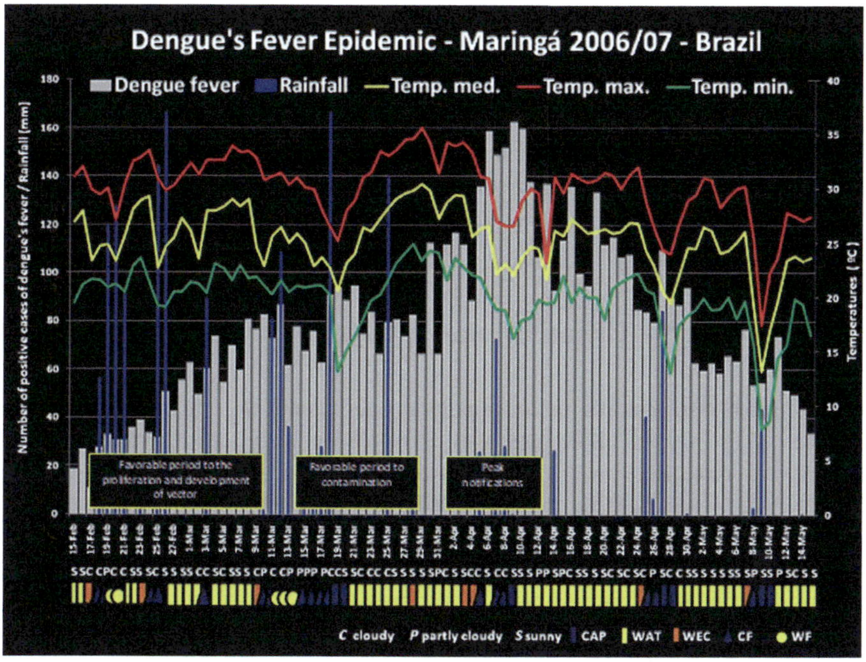

Fig. 13.3 Climatic elements and daily dengue cases in the city of Maringá, Paraná. *Source* Mendonça et al. (2011)

weather and their succession on a daily level, through records using mathematical quantities as well as (beginning from synoptic data) through the atmospheric systems generated by the types of weather. In this example, it is clear that a drop in the daily temperature range below 20 °C (minimum temperature) affects the reduction of dengue cases in the following days; a rise in temperature associated with the incidence of rain favours the proliferation of mosquitoes (dengue vector, *Aedes aegypti*) and the formation of an epidemic of the disease.

The rhythmic analysis of types of weather, proposed by Monteiro, therefore constitutes an approach to climate from its genesis, as it identifies them according to the interaction between local geographic attributes and the regional circulation of the atmosphere. It proved to be very important in studies focused on the environment, agricultural production and urban and regional planning, among others. In a very particular way, it has been successfully applied in studies and interventions related to natural disasters and extreme weather events in certain places.

Once he had advanced in the conception of climate as its usual manifestation over places, and having constructed a method that made it possible to understand climate in its temporal dynamics on a fine scale, another concern and challenge presented itself to Monteiro. It seemed imperative to test these two constructions against the real facts… it was urgent to get out of the laboratory understanding and away from the real climatic configurations; it was necessary to get as close as possible to reality and understand it—as fluid and fickle as it is, it was necessary to try. How about putting these conceptions about the climate and its direct & indirect effects on populations into practice? And where, during those years, was most of the Brazilian population found? The effort to understand and contribute to the solution to urban environmental problems arose as an essential task, and as a possibility for the advancement of geography, having an applied scientific nature in relation to the planning and management of urban environments.

13.4 The UCS—Urban Climate System

In the mid-1960s, Brazil's urban population surpassed its rural population. This process did not happen in the same way across the national territory as it is known. It was very representative of the Southeast, South and Northeast regions of the country—the parts where the country's main economic activities and most-expressive political forces were concentrated.

Completely devoid of economic and social development planning policies and actions, Brazilian urbanisation originated, above all, from a significant rural exodus, later associated with the considerable vegetative growth of the urban population. Forged by profound changes in agricultural production, hordes of migrants suddenly left the countryside and, as if hypnotised by the alluring image of city life, began to thicken the geographic peripheries of medium and large cities in the country.

After the 1960s, Brazilian cities began to register an increasingly accelerated and chaotic growth, resulting in what many understand as urban swelling, since urban growth in the country became a reality in complete dissonance with what might be framed as social development. Favelas have grown and spread frighteningly since then, evidencing a sharp drop in the quality of urban life and the living conditions of city dwellers.

The La Tourette Charter and the Andean Charter, two important documents in modern urbanism drawn up in the mid-twentieth century, already drew attention to the deterioration of urban spaces and the quality of life in cities. They called attention to the worrisome issue of squatter settlement (subnormal occupations, according to the IBGE) in Latin American urban areas, stressing the importance of control by the state and by society—a fact that has not successfully materialised since then.

It was in this context, of exacerbation of urban environmental conditions in the second half of the twentieth century, and bearing in mind their worsening in the future, that environmental movements raised their voices since that decade. The vertiginous and worrisome process of Brazilian urbanisation caught the attention of Monteiro, who brought it to the field of geographic climatology in the early 1970s, opening a new way of approaching urban environmental problems from the perspective of the urban climate.

Even before dealing directly with the climate of cities, his concern turned to the understanding and configuration of the geographic scale of the phenomenon. Until then, Brazilian geography had been using the large areas of the country (region and nation) as its predominant scale of approach, in a generic way, with a more significant focus on the past tense of the landscapes. In his proposal, the local, topoclimatic, climatic and urban microclimatic scales are highlighted, beginning detailed studies in national climatology, while focusing the approach in the present moment with an eye on the future of cities; thus, there was a significant change in the attitude of studies in climatology within the country.

The UCS—Urban Climate System (Fig. 13.4)—was conceived by him as a central element of his professorship thesis, under the title Theory and Urban Climate (*Teoria e Clima Urbano*), presented to the Department of Geography at the University of São Paulo in 1976. This approach reflected important concerns at the national level about the quality of life already popular in cities internationally, but Monteiro's proposal had a very high level of originality, since it highlighted particularities of urban areas in the tropical world.

This figure was presented by Monteiro in 1976; this theoretical and methodological contribution to the study of urban climate encompasses both the perspective of the development of general studies on cities and an application to the planning and management of urban problems. This is an approach grounded in the general theory of systems, applied to the analysis of urban climate, which relates the nature of the urban site (input) to the transformations produced by the city (attributes/transformation), and which is introduced in relation to various problems (output) derived from the interaction between society and nature, such as heat islands, air pollution, floods, landslides. Directly applied to urban planning, this proposal focuses on several parallel effects such as impacts on health, thermal discomfort, quality of life, urban

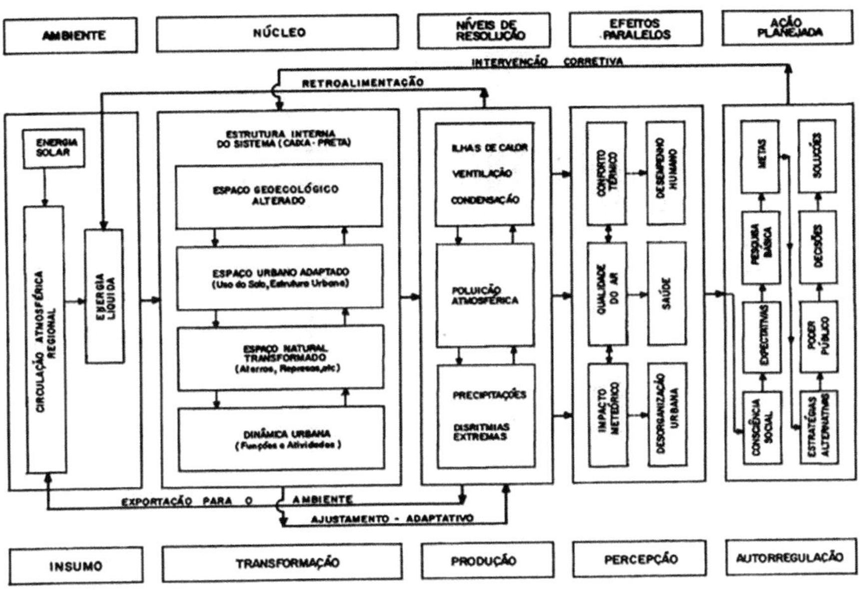

Fig. 13.4 UCS—Urban Climate System

disorganisation. There is a whole flow between the different dimensions of the system, with feedback processes derived from the implementation of planning actions being able to substantially improve problems and life in cities.

Inspired by the Popperian perspective of the production of knowledge, Monteiro (1999, 2015; pp. 93–99, translated) structured his proposal into ten basic statements that ensure the relationship between the normative action of theory and the operative action of research—namely:

1. The urban climate is a system that encompasses the climate of a given terrestrial space and its urbanisation.
2. The urbanised space, which is identified from the site, constitutes the core of the system that maintains intimate relations with the immediate regional environment in which it is embedded.
3. The UCS imports energy through its environment—it hosts a succession of events that articulate differences in states, changes and internal transformations, to the point of generating products that are incorporated into the core and/or exported to the environment, configuring themselves as a whole complex organisation that fits within the category of open systems.
4. Energy inputs into the UCS are thermal in nature (originating from the primary energy source of the entire Earth—the Sun), implying unequivocal dynamic components, determined by atmospheric circulation and decisive for the water component included in this set.
5. The evaluation of this energy input into the UCS must be observed in quantitative terms as well as, especially, in relation to its mode of transmission.
6. The internal structure of the UCS cannot be defined by the simple overlay or addition of its parts (ecological, morphological or functional urban compartmentalisation), but only through the intimate connection between them.
7. The set-product of the UCS presupposes several elements that characterise the urban participation in the system's performance. Since this production is varied and heterogeneous, a qualifying simplification, which must be constituted through channels of human perception, is required. There are three proposed channels of human perception: (a) thermal comfort, (b) air quality and (c) impact meteors.
8. The urban nature of the UCS implies special conditions of internal dynamism, depending on the evolutionary process of urban growth and development, since various structural trends or formal expressions follow each other throughout the urbanisation process.
9. The UCS is recognised as being subject to self-regulation—a function conferred on the urban man who, insofar as he is familiar with it and is able to detect its dysfunctions, can, through his decision-making power, intervene and adapt its functioning, using recycling devices and/or feedback circuits capable of driving its development and growth in line with pre-established goals.
10. Due to the possibility for self-regulating interference, those negative entropy properties are added to the UCS, as an open system, due to their own capacity

for specialisation within growth through adaptive processes, thereby being categorised as a morphogenetic system.

Three basic questions of consistency simultaneously ensure the geographic perspective of the UCS—an aspect evidenced in the interaction between (urban) society and nature in the city, which are explained as follows: 1st question—Order of magnitude and degrees of organisation (scale of analysis inspired by Arthur Koestler's holon); 2nd question—Behaviour patterns and self-regulation (between growth and planning) and; 3rd question—Procedural dynamics and structural patterns (between operator and operand).

Aware of the urban problems arising from the chaotic urbanisation recorded in Brazil, Monteiro highlighted the importance of urban planning as a rational perspective for controlling and solving those related to the climate of cities. Thus, in his proposal, a precautionary perspective is evident, since well-applied planning aims to correct or prevent urban environmental problems. By understanding the urban climate as an open system of interactions, he proposed its analysis in three dimensions—or complementary channels—of human perception: the thermodynamic, the physical–chemical and the hydrodynamic. Once again, the perspective of a unified geography is evident in his stance—one that understands the interaction between society and nature as inherent to the structure of landscapes, as it understands the urban climate as an anthropogenic derivation from human activities over nature.

Problems such as environmental/climatic discomfort, thermal stresses and the formation of heat islands come to prominence within the thermodynamic channel; air pollution, acid rain, thermal inversion, etc. illustrate concerns within the physical–chemical channel; water scarcity, urban floods and overflows, land movements, etc. pertain to the hydrodynamic channel. All these problems, glaring in most medium and large cities in the country, result from the careless interaction between society and nature, and have registered increasingly serious impacts since the middle of the century, when the urbanisation process intensified in a frightening manner throughout the country, especially on its coastal fringe.

During the 1980s, Monteiro also produced several studies on the urban climate—opportunities that gave him the chance to test his theoretical and methodological contribution through the research of some of his direct and indirect advisees and collaborators (e.g. Monteiro and Mendonça 2003). During his work as a visiting professor at the University of Tsukuba (Japan) and the Federal University of Santa Catarina (Brazil), he was able to advance even further in some details of his proposal for the study of the urban climate; this time, he stuck more directly to the scale of the phenomenon, interactions between its component elements and factors (anthropogenic derivations) and the technical/technological resources used to sharpen and bring the studies closer to reality. During this period, however, studies in climatology were already beginning to employ new technologies such as remote sensing, notably satellite images, for climate analysis, opening a new stage in climatology; technology showed an unprecedented advance, but the conceptual and methodological foundations continued to have Monteiro's contribution as their main basis.

In these nearly 50 years of presenting the Urban Climate System (UCS) to the scientific community and city planners, the proposal has been tested and amplified in around three hundred studies on different Brazilian cities (Monteiro 2015). The field of thermodynamics is highlighted in these studies, particularly due to the greater technical ease of its investigation associated with the significant thermal discomfort caused by heat in tropical cities. Then there is the hydrodynamic channel or subsystem, with special emphasis on the process of urban floods and overflows which, as with the previous channel, also present facilities for investigation and appear as one of the problems with the greatest impact on urban populations in the humid and rainy portion of Brazil. The physical–chemical channel recorded very few studies, mainly due to the difficulties related to the high cost of equipment for measuring air quality, even though this is a serious problem documented in a considerable portion of Brazilian cities.

13.5 Conclusions and Final Considerations

Between the early 1950s and 1990s, Monteiro produced one of the most advanced insights into Brazilian climatology. His constructions gave rise to a Brazilian "geographical climatology school" and "urban climatology school", both recognised by the national geographic community as they highlight paradigmatic conditions for studying and gaining knowledge of the atmosphere from its interaction with the surface/human activities.

By contesting the hegemonic conception that took climate as the mean state of the atmosphere over a given place, proposing its understanding as being the usual condition of the atmosphere, with its types of weather and phenomena that are more and less repetitive in places, he introduced a new perspective for the study of climates. In this regard, he detailed his proposal by designing and applying the analysis of types of weather in a sequential and chronological manner, revealing the dynamism of atmospheric states in a geographical vision—that is, as a process involving nature and society… the understanding of the atmosphere as a "moving force".

His theoretical and methodological proposal of the UCS, launched when Brazilian urbanisation was gathering significant momentum, revealing an intense process of socio-spatial segregation, can be considered as far ahead of its time. At the international level, studies on urban environmental problems were just beginning to take shape and, being particularly aimed at cities in developed countries, they did not highlight the importance of the atmosphere as an essential element of urban development. Thus, by proposing an integrative approach to the study of cities' climates, he greatly contributed to the development of research on this theme as well as to the involvement of climate in urban planning and management processes—fundamental strategies for improving the quality of life in cities. This input, along with the two previous ones, formed the core of Monteiro's contribution to the consolidation of Brazilian geographic climatology.

Ending his activities as a university professor in the early 1990s, Monteiro once again focused on climatology and, certainly as a result of his previous concerns, he moved forward with the discussion of extreme climatic events and their impacts on society (Monteiro 1991). The gigantic scale taken by cities across the country and around the world, and especially in non-developed countries, as well as the growing number of populations exposed to natural hazards, drew the attention of climatology scholars. In this sense, his latest contribution emanates from concerns that had already been popular in the fields of science and international political ecology since the late 1980s—namely the intensification of climate risks in the context of global climate change.

As we affirm in part of this text, Monteiro's contribution to science in general, and to geography and climatology in particular, did not merely open perspectives for the understanding of phenomena and processes related to the relationship between society and nature in the elaboration of landscapes. It also allowed for the tracing of strategies and paths for solving problems on a regional and urban scale, since in its elaborations there is always the perspective of theoretical exercise combined with the pragmatism of actions. He is, as the consensus within Brazilian geography agrees, one of the pillars of its construction and consolidation, considered by us as one of the founding thinkers of Brazilian geography—a thinker ahead of his time.

References

Mendonça F (2012) La connaissance du climat au Brésil: entre le vernaculaire et le scientifique. Confins (Paris) 15:5–20

Mendonça F, Danni-Oliveira IM (2007) Climatologia: Noções básicas e climas do Brasil. Oficina de Textos, São Paulo

Mendonça FA, Roseghini WFF, Aquino JR J 2011. Climate change, weather types and dengue: a comparative approach between Giruá (RS) and Maringá (PR)/Brazil. In: UGI—2011: Conferência Geografia Regional—proceedings, vol 1. UGI/FISA, Santiago/Chile, pp 1–12

Monteiro CAF (1963) Geografia do Brasil - Grande Região Sul, vol IV, Book I. IBGE, Rio de Janeiro

Monteiro CAF (1971) Analise rítmica em climatologia: Problemas da atualidade climática em São Paulo e achegas para um programa de trabalho. Climatologia, vol 1. Instituto de Geografia da USP, São Paulo

Monteiro CAF (1991) Clima e excepcionalismo – Conjecturas sobre o desempenho da atmosfera como fenômeno geográfico. UFSC, Florianópolis

Monteiro CAF (1999) O estudo geográfico do clima. Cadernos geográficos, Florianópolis, year I, no 1

Monteiro CAF, Mendonça F (2003) Clima urbano. Contexto, São Paulo

Monteiro CAF (Org) (2015) A construção da climatologia geográfica no Brasil. Alínea, São Paulo

Neto JL (2001) História da Climatologia no Brasil: gênese, paradigmas e a construção de uma Geografia do Clima. Lecturer thesis. FCT/UNESP, Presidente Prudente

Zavattini JA (2004) Estudos do clima no Brasil. Editora Alínea, Campinas, SP

Chapter 14
Aziz Nacib Ab'Saber and the Professionalisation of Research in Geomorphology in Brazilian Geography Courses

Antonio Carlos Vitte and Rafaela Soares

Abstract The history of the discipline of geography in Brazil and its professionalisation is told as one of an exclusive products, inserted into the country's project of modernisation, which would be achieved only with the knowledge and integration of the Brazilian territory. Aziz Nacib Ab'Sáber, along with João José Bigarella, Fernando Flávio Marques de Almeida and Maria Regina Mousinho de Meis, are leaders when it comes to the formation of geomorphology in Brazil as a scientific field of research—in geography and geosciences in their entirety, and with strong impacts on other disciplinary fields. In general, and independently of their technical-scientific options, it can be said that, with greater or lesser influence, their understanding of geomorphology was united by a cross-cutting axis: Lester King's Theory of Pediplanation. Unlike the masters alluded to above, Aziz Ab'Sáber has remained faithful to its paradigmatic principle—that is, that life is the result of a balance between is to siasis database and paleoclimatic variations. This lies within a polycyclic vision, where the glacial phases would lead to the production of pediplains and their correlative deposits, while in tropicality, the carving phase of thalwegs would predominate with the consequent dismantling of the surfaces. In 1956, the 8th International Conference of the IGU Congress was held, and in this congress, Ab'Sáber was introduced to Jean Tricart's thoughts and to Jean Dresch, who during their field visits drew his attention to the role of stone lines as indicators of geomorphological processes. The importance of stone lines for geomorphological studies, associated with the notion of physiology of the Otto Karl Siegfried Passarge landscape, marked not only Ab'Sáber's own reinvention in terms of his conception of geomorphology, but also enabled him to be a leader in the renewal of Brazilian geomorphology from the referential perspective and using the very epistemological structure of Brazilian geographical science.

A. C. Vitte (✉)
UNICAMP Universidade Estadual de Campinas, Campinas, Brazil
e-mail: acarlosvitte@gmail.com

R. Soares
Universidade Federal Do Rio de Janeiro, Rio de Janeiro, Brazil
e-mail: rafaelaniemann@gmail.com

R. C. Lois González and M. A. Mitidiero Junior (eds.), *Brazilian Geography*, Advances in Geographical and Environmental Sciences,
https://doi.org/10.1007/978-981-19-3704-0_14

Keywords Geomorphology · Brazil · Aziz Nacib Ab'Saber

14.1 Introduction

The development and professionalization of a scientific field and its sub-fields of knowledge occur in the middle of an environment marked by conflict between the producers of knowledge, and the same occurs in relation to the process of incorporating theories, approaches and methodologies.

This means that when we look for their incorporation into a given thematic frame by a scientific community, we must take account of more than the just the acceptance itself, the conflicts and strategies that the leadership and the collective of researchers have developed to ensure the success and maintenance of their themes—and even sub-fields of scientific knowledge within a given science—and the context that has influenced their acceptance by practitioners of that science. It is worth noting that, prior to the creation of geography courses in universities, institutional spaces already existed in Brazil that were dedicated to teaching, but mainly to applied research in Brazil. Spaces such as museums, scientific commissions, polytechnic schools and scientific societies emerged in large numbers in the country from the mid-nineteenth century, first on the initiative of the imperial government and then followed by the provincial governments. The main objective of the imperial government was to build national identity through nativism, by encouraging research in the natural sciences and the humanities (Lopes 2009).

The historic milestone for the creation of scientific institutions in Brazil, such as museums and institutes, was the arrival of the Portuguese royal family in 1808, who at that time were fleeing the Napoleonic conquest in the Iberian Peninsula. In order to meet the needs of the royal court that was then established in the colony, the Royal Press, National Theatre, National Library, Real Horto [Royal Garden], Royal Museum, Botanical Garden and Naval Academy were immediately formed in 1808, with the Royal Military Academy—the first technical teaching institution created in Brazil—following in 1810, as well as the colleges of surgeons in Bahia and Rio de Janeiro, which would become the first medical schools in Brazil (Azevedo 1964, p. 526).

Unlike in other Latin American countries[1], universities were only formally created in Brazil after the proclamation of the Republic in 1889—but it was only from the 1920s onward, in fact, that the state effectively assumed control of the creation of universities. From 1889 to 1918, 56 new higher education schools were created—a

[1] The year 1538 creation of the University of Santo Domingo, 1551 San Marcos University in Peru, 1553 University of Mexico, 1662 University of Bogotá, Colombia, 1692 University of Cuzco, Peru, 1728 University of Havana, Cuba, 1738 post independence the University of Santiago , Chile. The first Brazilian university was that of Paraná, created in 1916, the University of Rio de Janeiro began operating in 1920. At that time the United States had 78 7 universities and in Latin America as a whole there were 20 universities. (DANTES, 2005; MAHL, 2012; ENCYCLOPEDIA LATINOAMERICANA, 2021).

number that rose to 150, spread across technical schools, polytechnics, museums, colleges and universities—by the 1930s (Sampaio 1991).

The University of São Paulo (USP) was created in 1934. In 1937, the University of Rio de Janeiro became the University of Brazil and is now called the Federal University of Rio de Janeiro (UFRJ). It was on this occasion that geography courses that followed the guidelines of the French matrix were created. In their curricula, the undergraduate courses in geography included a set of disciplines generically labelled "physical geography"—among them geomorphology, whose teaching and research followed the canons established by Emanuel De Martonne at that time (1873–1955) (Vitte 2011a)[2].

The beginning of the twentieth century was marked by intense debates between intellectuals, scientists and politicians about the paths of scientific research in Brazil. This discussion began in the state of São Paulo, which at that time had the best research structure and organisation in relation to other federated entities. The debate revolved around a pragmatic science at the service of the state, as defended by engineers from the Polytechnic Schools of São Paulo and Rio de Janeiro, led by Francisco Bhering (1967–1924)[3], versus the naturalist conception of Orville Derby (1851–1915) who, although he graduated in geology at Cornell University and made use of the methods of the US Geological Survey, proposed a conception of science based on the propositions of Alexander von Humboldt in which the Ministry of War was associated, where the Rondom Commission, the Engineering Club, the Ministries of Transport and Foreign Affairs, the Telegraph Office and the Rio de Janeiro Society of Geography were found (Duarte 2011).

---

[2] Emmanuel De Martonne published the first version of his work Traité de Géographie Phisique in 1909 and soon after participating in the transcontinental excursion of 1912 led by William Morris Davis (1850-1934), he adopted the concept of relief physiography and as a method of study the conception of the Geographical Cycle of Erosion, prepared by Davis in 1899. De Martonne's work was hegemonic in France until the year 1950. In 1933 De Martonne visited Brazil at the invitation of the Rio de Janeiro Geography Society, gave a conference highlighting the role geography in territorial planning. He then participated in fieldwork in Serra do Mar, in the region of Santos (SP), an activity that was reported in 1933 in the Bulletin de la Association de Géographes Français. Between 1943 and 1944 De Martonne published two articles in the Revista Brasileira de Geografia, where he discussed the geomorphological problem on the coast of Brazil. (CLOUT, 2004; VITTE, 2011)

[3] Francisco Bhering foi o responsável pela produção da primeira carta ao milionésimo do Brasil no período republicano. Em 1909 Bhering foi o representante brasileiro naConvenção Internacional de Londres que discutiu a padronização da produção da cartografia mundial. Ficou acordado entre os países que a escala seria a de 1:1.000.000 e a projeção policônica. A carta foi publicada em 1922 como parte das comemorações do centenário da independência do Brasil. (DUARTE, 2011). Até então, apenas a Carta Geral do Império, coordenada por Charles Frederick Hartt (1840-1878), apresentada na Exposição Universal de Viena em 1873 e publicada no Brasil em 1875, na escala 1:3.370,20, tendo como meridiano central o Rio de Janeiro. (BIAGGI, 2015). Em 1875 foi criada a Comissão Geológica do Império, coordenada por Hartt, mas foi abruptamente encerrada em 1879 pelo imperador Pedro II. O objetivo da Comissão era realizar o levantamento topográfico e geológico do paísvoltados a exploração dos recursos naturais do país. Neste curto período coletou mais de 500.000 amostras de rochas, plantas e animais, enriquecendo sobremaneira o conhecimento da natureza brasileira. (BIAGGI, 2015, p. 13-14).

The more technical view of science emerged between 1900 and 1902, when Francisco Bhering harshly criticised the quality of the work of the Geographical and Geological Commission of São Paulo, coordinated by Orville Derby (Silva 2006, pp. 28–30). With the advance of coffee growing towards the interior of the state of São Paulo—at the time called the "*sertãodesconhecido*" [unknown outback]—the São Paulo government entrusted the commission with carrying out an extensive survey of the state's natural resources for the purpose of expanding coffee production. Between 1887 and 1888, the commission carried out the triangulation and topographic survey of approximately 17,000 km^2, in addition to detailed descriptions of hydrography, climate, vegetation, agricultural aptitude, population and indigenous ethnography (Silva 2006, pp. 25–27).

In addition to the delay in finalising and publishing the letter, Bhering questioned the technical quality used by Derby, pointing out the imperfection in the contour lines and in the location of geographic features (Silva 2006, pp. 24–25). For Francisco Bhering, who was supported by the Engineering Club and the Rio de Janeiro Society of Geography, surveys of natural resources and mapping of the Brazilian territory needed to be supported by modern techniques and technologies, which would only be possible via the work of the professional engineer-geographer—at that time the most skilled in the domain of innovations in the fields of cartography, topography, astronomy and surveying. Furthermore, only the polytechnic schools would be able to train professional engineer-geographers, which for Bhering were fundamental for the territorial development of Brazil (Biaggi 2015, pp. 10–12).

This same political position was defended by the engineers of the São Paulo Polytechnic School, at the time of the creation of the history and geography courses at the University of São Paulo's Faculty of Philosophy, Arts and Sciences in 1934. In the process of creating the University of São Paulo (USP), two views collided—one that defended the production of "detached" scientific knowledge as a function of higher education, while another view postulated that the university should be geared towards techniques for solving national problems (Ferreira et al. 2008, p. 52), a proposal that was defended by the engineers of the São Paulo Polytechnic School.

The solution found by Fernando Azevedo, USP's first dean, was a mix between the two proposals—teaching for the training of professionals focused on school teaching, as well as detached research. Despite this conciliatory solution, the critics continued in a clear misogynist position—in addition to leading campaigns whose objective was to make the geography course unfeasible. The strategy found was the creation of professorships and the incorporation of elite and middle-class women and men who sought social prestige in their university degree, in line with their social roles in the established order (Biaggi 2015, pp. 10–221). As the demand for the geography course was low among men, the solution found was to invite women as well as teachers from the state education network to join the Faculty of Philosophy, Arts and Sciences at USP, particularly in the geography course (Limongi 1989).

Conflicts and questions about research in geomorphology carried out by teachers of geography courses intensified, since USP was created in order to carry out detached research and train professional staff for work in primary education. For the professionals of the polytechnic school, however, research needed to be pragmatic and interested in the needs of the state, while these professionals also strongly criticised the prospects of women who were also carrying out research. Critiques were made, and interrogations carried out mainly by scientists linked to or coming from the scientific committees and mainly from the polytechnic schools that until then held the thematic authority over geological and geomorphological knowledge.

This situation became even more tense, and until the present day, it still remains in conflict, because from the mid-1940s—but mainly in the 1950s with the creation of Petrobras and the geology courses—the conflicts turned into a struggle for professionalisation, in the case of geology, with consequent absolute control of the field of geomorphological knowledge. It is into this context that we must insert the construction by Aziz Nacib Ab'Saber (1924–2012) of geomorphological knowledge within the field of geographic science, and the consolidation of which was achieved starting from a struggle between geologists, geographers and engineers for control of geomorphology within the context of professionalisation and social recognition of their respective sciences. This dispute is inserted into the control of a scientific field, geomorphology, in parallel with the process of the professionalisation of scientific research in Brazil.

14.2 Institutional Spaces Prior to the Creation of Universities in Brazil

In general, until the 1980s in Brazil, the historiography of sciences was marked by a positivist conception linked to a conception of narrative and a linear view of the facts, and the focus of which was the concern of the travellers-explorers who passed through here and recorded facts and events—and who were often considered the founders of science in the country. Even though European metropolitan centres were the producers of knowledge from the seventeenth century onwards, this was only possible because they gathered information collected in the New World. This means that science and knowledge were also being produced in the periphery since the start of the world's process of westernisation (Burke 2003).

The first major attempt to present an overview of the history of science in Brazil was carried out by Fernando de Azevedo (1894–1974), who was concerned with associating scientific development with the country's economic, cultural and political development in the 1940s and 1950s. Also according to Azevedo (1994), it was precisely during the periods 1914–1920 and 1939–1945 (both marked by the two world wars) that, given the difficulties in exporting and mainly in importing products from Europe and the United States, governments created policies to incentivise the industrialisation of Brazil (Brito 2015).

But, it was from the 1970s onwards that there was effectively an increase in the production of works on the history of science in Brazil, which included new methodological, thematic and approach-based contributions, as well as sources and research materials. Such research sought to extract, from the historiographical horizon, the hitherto-dominant view of a pre-scientific phase and another later phase, in which the formation of a scientific field with clearly delimited lines of research within the postgraduate programmes in history would, in fact, have initially taken place (Ferri and Motoyama 1979–1981).

Following independence in 1822, in addition to inheriting the institutions created from 1808 onwards, the Astronomical Observatory (1827), the Pedro II Academy (1837), the Historical and Geographical Institute of Brazil (1838) and the Central School (1855) were created (Dantes 2015).

But, it was during the Second Empire (1840–1889), whose economic conditions were favourable due to the growth of coffee production that the state began to encourage scientific development in the country, with a nativist bias whose objective was the construction of Brazilian national identity (Lopes 2009). In this context, institutions and museums were created, such as the Emilio Goeldi Museum (1866), the former Botanical Garden of Grão-Pará (created in 1798), the Geological Commission of the Empire (coordinated by the naturalist Frederick Hartt (1840–1878) and which operated between 1871 and 1875), the Imperial Agricultural Institutes such as the Imperial Fluminense Institute of Agriculture (1860) and the Imperial Agronomic Station of Campinas (1887). The Rio de Janeiro Polytechnic School (1874), the Ouro Preto Escola de Minas (1875) and several scientific societies were also created (Lopes 2009).

The creation of these institutional spaces by the imperial government also encouraged provincial governments to formalise the creation of their "scientific systems" with a view to promoting better knowledge of their territories, with the objective of the economic and social development of the respective provinces (Figuerôa 1987; Lopes 2009). With regard to the state of São Paulo, specifically, in the nineteenth century, the São Paulo government had been investing in the creation of institutes, isolated schools and scientific commissions for the purpose of recognising and surveying nature and its potential and improving the quality of agriculture in the state, in addition to creating a technical-scientific workforce specialised in both agriculture and industry.

The Geographical and Geological Commission (1886)[4] was a result of the development and internalisation of coffee farming, which gave rise to the need to really understand the soil, climate, geomorphology, geology and hydrography, in addition to producing cartographic mappings about the territory of São Paulo.

Entrepreneurs from São Paulo associated with the provincial government, and using the German technical schools as a model, created the polytechnic school in 1893 which, with courses in civil, industrial and agricultural engineering and mechanical arts, created a material strength laboratory in 1899 from which the Institute of Technological Research (IPT) originated. In addition to the IPT, what would become the current institutes of the University of São Paulo—such as the Institute of Astrophysics and Geophysics and the Faculties of Architecture and Urbanism, Fine Arts,

---

[4] In 1975, state decree 6,822 transformed the former Geographic and Geological Commission of São Paulo into a Geological Institute. (GEOLOGICAL INSTITUTE OF SÃO PAULO, 2019)

Physics, Chemistry and Animal Science—were created within it. With the foundation of the Republic (1899) shaped by positivist and science-oriented ideology, the idea that only science would be able to modernise, and indeed rebuild, the country and the nation took effect among intellectuals (Sevcenko 2003).

In order for these objectives to be achieved, the strategy followed two intertwined paths: one was the reformulation of the objectives of the old institutes and academies formed under the empire (1822–1889) and the creation of new universities with new undergraduate courses in order to generate new scientific fields from the dismemberment of the activities that were carried out within the old scientific committees, research institutes and polytechnic schools.

In 1901, the Luiz de Queiroz School of Agriculture (ESALQ) was inaugurated and later incorporated into the University of São Paulo in 1934. Aiming to improve the quality and productivity of bovine herds, in 1905, the São Paulo government created the Zootechnics Institute, and in 1907, the HortoFlorestal (Forestry Garden) was created, which from 1970 became the Forestry Institute, and is now the Forestry Foundation. In 1927, the Biological Institute was created—its objective being agricultural research and animal and plant health (Molina 2011).

This was a historical period in which the Nation State effectively began to consider the "geographical question" as a strategy for the development of the country and its insertion into the modern world. This is how, under the first Vargas government (1930–1945), the National Council of Geography was created in 1933, followed by the National Council of Statistics in 1934 (Vitte 2008, 2011).

In 1936, the two councils were merged, giving rise to the Brazilian Institute of Geography and Statistics (IBGE), which began to absorb young geographers who had graduated or were still studying geography at the then-University of Brazil. The relationship between the IBGE and the University of Brazil was very close, not only because the IBGE's technical staff came from the university, and on many occasions foreign researchers who came to the IBGE were connected to the University, offering undergraduate courses and developing research activities with the teachers (Machado 2000; Adas 2006).

The second Vargas government (1950–1953) deepened the country's strategic need to dominate the exploitation of natural resources such as oil, which was at that time extremely highly valued throughout the world—a consequence of the greed of the large North American and British oil companies.

As the Brazilian subsoil had already been nationalised in the 1930s with the creation of the Vale do Rio Doce Company, the second Vargas government created Petrobras, a national and strategic company for the exploration and refining of oil and natural gas. The undergraduate course in geology at USP was only created within the Faculty of Philosophy, Arts and Sciences in 1957, during the government of President Juscelino Kubscheck (1956–1960). With the 1969 university reform carried out during the military dictatorship (1964–1985), the Institute of Geosciences was created, to which the geology course was transferred (https://igc.usp.br/institucional/historia).

At the University of Brazil, the undergraduate course in geography maintained a close relationship with the IBGE—a situation that facilitated the incorporation of scientific research into its daily life. At the University of São Paulo, the situation of the undergraduate course in geography was already different, due to the strong criticism from researchers, mainly from the polytechnic school, and their attempts to make it unfeasible. The researchers' arguments were considered competent for research in geomorphology due to their having the necessary technical and scientific capacity in this area, and the fact that the undergraduate course in geography received a strong contingent of female students meant that it would not, for this reason, be in a position to develop a scientific reasoning suited to the demands of São Paulo society.

This is a fact that unfortunately allows us to deduce that, for the scientists of the polytechnic school, women were not suitable for scientific research, leaving them only with the task of teaching—since one of the goals of the creation of the USP, and in particular the Faculty of Philosophy, was detached research, in addition to training professionals for school teaching. In the opposite sense, since the mid-nineteenth century, the positivists had argued that research should be pragmatic—an about-turn from the previous position (Sampaio 1991).

Therefore, we can consider that during the historical period (the end of the 1940s, and mainly during the 1950s) when Ab'Saber began his research activities (focused on geomorphology, but bearing the mark of the French School of Geography and its close relationship with the Annales School)[5], the Brazilian academic and university environment, especially in São Paulo, was wrestling with a great dispute over professionalisation, which affected the domain of thematic areas—such as in the case of geomorphology.

14.3 Aziz Nacib Ab'Saber's Contribution to the Formation of Geographic Geomorphology in Brazil

According to Vitte (2008), geomorphology in Brazil was developed from the influence of Emanuel de Martonne (1873–1955) and Pierre Monbeig (1908–1987). It ended up benefiting the development of a firm methodological perspective for geography. For Monbeig, geographic analysis needed to produce regional monographs, in which regional delimitation was based on the relationship between the natural and the social (Abreu 1994).

Historically, this period coincided with the expansion of coffee production in Southeastern Brazil, particularly in São Paulo, as well as the process of São Paulo's industrialisation and urbanisation and the change in the regional orbit, particularly

[5] Alves (2010, p. 185-189) reports that since the late 1930s, French historians and geographers influenced by the Annales School method, such as Fernand Braudel and Pierre Monbeig, who at the time were in Brazil as professors of the French Mission in creation of the History and Geography courses at USP, began to publish themes about Brazil in their publications, especially Monbeig, who dedicated a large part of his stay at USP to understanding the pioneering areas of the State of São Paulo. Also according to Alves (op.cit) the first publication by a Brazilian geographer in the Annales was Aroldo de Azevedo, who in 1948 published an article on oil.

between the northeast and the southeast (Oliveira 1981; Cano 1994). In other words, geomorphology at USP and at the former University of Brazil was developed from a secondary reading of the Davisian cycle, particularly at USP, with a strong influence of the Annales method by Pierre Monbeig, where the notions of history and occupation were important for delimiting a region/compartment (Vitte 2011b).

In 1958, Ab'Saber drew attention to the enormous production of Brazilian geomorphology—the result of the expansion of geography courses in Brazil and the internalisation of the country's economic development. From the 1950s onwards, Brazilian geomorphology underwent a major paradigmatic rupture, with the emergence of Pediplanation Theory and associated major transformations within geology, particularly with regard to sedimentology and stratigraphy. In addition to the emergence of new techniques for the representation and acquisition of information, there would be a paradigmatic rupture in Brazilian geomorphology (Vitte 2008). From a political and economic point of view, the 1950s were marked on the global stage by the intensification of the Cold War and the Chinese Revolution. In Brazil, this was the era of Juscelino Kubitscheck de Oliveira (JK) and, through the implementation of national-developmentalist ideas, saw the construction of Brasília, the establishment of the automobile industry and the opening of highways.

For the Earth Sciences, the 1950s were declared the decade of the oceans, in which Earth Sciences researchers sought to unravel continental processes through the study of ocean floor sediments. This was the moment when the knowledge of sedimentology and stratigraphy began to help geomorphological studies. Added to this was the discovery of the earth's climatic variations and the possibility of associating the evidence of these variations with continental sediments and, from there, arriving at an age for the landforms. Even from the 1950s, we see the use, albeit timid, of aerial photographs for geographic and geomorphological research, enabling a three-dimensional view of shapes and their associations in scales—which, associated with fieldwork, would allow the construction of more consistent hypotheses to explain the geomorphological phenomena in an intertropical environment (Vitte 2008). It was in this cultural context that the Brazilian community of geomorphologists encountered the Pediplanation Theory elaborated by the South African geologist Lester King (1907–1989, 1956)—which, according to Abreu (1982), arose from the influence of the Chicago congress of 1936, which was dedicated to the work of Walter Penck (1888–1923).

The beginning of the 1950s, until approximately 1957, was marked by a process of transformation in geomorphological research—not exactly a rupture, but a transitional phase due to epistemological obstacles (Bachelard 1996), such as the works in geology that were more advanced in the empirical knowledge of the Brazilian reality than those in geomorphology, which, guided by an anachronistic model and incompatible with the Brazilian tropical reality, ended up providing no significant advances in knowledge of the genesis of the Brazilian terrain.

During the first seven years of the 1950s, intense regional studies with genetic concerns were carried out by Fernando Flávio Marques de Almeida (1916–2013) and Aziz Ab'Saber. These works were fostered through significant advances in geology, the dissemination in Brazil of the works carried out by the French in Africa and mainly through the influence of the reflections of Lester King and von Englen, which were carried out from 1940 onwards, shortly after the Chicago Congress, which discussed the work of Walter Penck (Abreu 1982). An interesting example from this period in Brazilian geomorphology is Aziz Ab'Saber's doctoral thesis, "*Geomorfologia do SítioUrbano de São Paulo*" [Geomorphology of the Urban Site of São Paulo] defended in 1957 (Ab'Saber 2007)—a thesis supervised by Aroldo de Azevedo (1910–1974), with Fernando Flávio Marques de Almeida as one of the examining members. This work marked a profound transition and at the same time a reconstruction of the interpretative model of relief and its genesis (Vitte 2008).

There was no paradigmatic rupture, but the interpretative change brought about by new bibliographical sources, as in the case of Oscar Dietrich von Englen (1881–1965) and mainly by the obstacles that geology—particularly the works of RuyOsório de Freitas that drew attention and began to demand analytical work with deep correlation between the elements of nature, such as the role of tectonics and lithologies in structuring drainage and defining the geomorphological enclosure (in this case the São Paulo basin). Another striking influence on Aziz's work was the text "*O PlanaltoPaulistano*" [The São Paulo Plateau] by Fernando Flávio Marques de Almeida, published in 1954 by AGB in the book "*A Cidade de São Paulo*" [The City of São Paulo] (Vitte 2008). Aziz's doctoral thesis is paradigmatic because in it, in addition to the change of conception on the genesis and evolution of the terrain, a timid methodological essay can be clearly perceived that would move towards Aziz's work of 1969 (geomorphology at the service of quaternary research).

It is a work of geomorphology, but of an essentially geographical nature, given the author's concerns in constructing a spatiality of the terrain—the São Paulo basin—and of its highly complex genesis, but it is also concerned with the historical questions related to the construction of space—in this case the urban site, as the terrain decisively influenced the options for occupying the land and its real estate valuation. In 1956, the UGI Congress was held in Rio de Janeiro, where the internal discussions were intensified along with those developed in the "post-congress" fieldwork, led by Jean Tricart (1920–2003), Jean Dresch (1905–1994) and Ab'Saber. The central focus of the discussions was the issue of materials on the hillsides, mainly for detrital paleo pavements and their paleoenvironmental and geomorphological significance (Vitte 2008).

It is a work that definitively marks the birth of a master of Brazilian geography, clearly demonstrating the influence of Deffontaines (1894–1978) and Monbeig, with the regional and historical concern and, from the point of view of geomorphology, the influences of Francis Ruellan (1894–1974), von Englen, Jean Dresch and Tricart. In 1964, João José Bigarella (1923–2016), a chemical engineer by training and a disciple of Reinhard Maack (1892–1969) by choice and passion, published a work "*VariaçõesClimáticas no Quaternário e suasImplicações no RevestimentoFlorístico do Paraná*" [Climatic Variations in the Quaternary and their Implications in the Floristic Coatings of Paraná] (*Bol. Paranaense de Geografia*, n. 10–15, 1964), which became a very important landmark for chrono-geomorphological studies, and for the first time in Brazil, the validity of the theory of biostasis and resistasis was demonstrated in order to explain the evolution of Brazilian terrain (Vitte 2008).

At that time, Aziz worked at USP as an assistant to Professor Aroldo de Azevedo in the Brazilian geography chair, while Edgar Kuhlmann (1882–1958) taught biogeography to geography students and hypothetically detailed the mechanisms that could explain phytogeographic differences in the Brazilian territory. This was the time when the Forest Refuges and Temporary Fortifications Theory was structured from a geomorphological point of view by Aziz, in 1979, at the same time creating the morphoclimatic domains of Brazil (Ab'Saber 1967), starting from an association between the formulations of Tricart and Cholley with their notion of erosion systems, in addition to the reflections of Kullmann, Monbeig and Aroldo de Azevedo (Vitte 2008).

The Forest Refuge Theory represented an immense revolution in Brazilian geomorphology in a global context, since Aziz imprinted in its elaboration the need to consider geomorphological compartmentalisation as a *sine qua non* condition in order to understand, on the one hand, the complexity of the biogeographic fabric of Brazil, and on the other the very specificity of the aforementioned refuges. Starting from the Forest Refuges Theory, the climatic geomorphology was rendered dynamic. It then became possible to specify the relationships between the Würm-Wisconsin variations, for example, with the distribution of forest cover, and the existence and persistence of landforms and correlative deposits in morphoclimatic environments that differ from or even contrast with the current conditions (Vitte 2008).

Thus, one of the greatest revolutions in global climatic geomorphology was constituted, with input from Aziz Ab'Saber, João José Bigarella and Maria Regina Mousinho (Vitte 2008), who would work together on many occasions to form the political and scientific structure that guaranteed the maintenance of the climatic paradigm in the interpretation of the Brazilian terrain (Bigarella Marques Filho; Ab'Sáber (1961).

So, based on the specialisations of geology and the new techniques and theoretical-methodological foundation of the Pediplanation Theory and the Bio-Resistance Theory, geographers-geomorphologists were awakened to the study of surface materials and chiefly to the possible role of "stone lines" and gravel pits as a record of climate change in Brazil (Ab'Saber 1962).

In the late 1960s, Brazilian geomorphology witnessed two major revolutions. Firstly, with Ab'Saber—the result of a long reflection and a lot of experience in the field, which had already begun for him during his doctoral thesis in 1957, "*O sítiourbano da cidade de São Paulo*", and which gained an epistemological structure in the classic 1969 work "*Um Conceito de Geomorfologia a Serviço das Pesquisassobre o Quaternário*", a methodological work that continues to influence geomorphological research to the present day (Vitte 2008).

This conception of geomorphology was a theoretical and methodological landmark among works of geomorphology, and at the same time, it places Ab'Saber as the person who incorporated and developed the propositions of the Germanic epistemological lineage (Abreu 1982). In Ab'Saber's (1969a, b) conception, geomorphological analysis should be centred on the quaternary and should involve three stages, with terrain being the product of a complex interaction, woven by endogenetic and exogenetic forces.

Thus, in a work of geomorphology, we must consider the "topographic subdivision" as the first level of analysis, which involves not only the analysis of the topography, but mainly the influence of geology and structure in this subdivision, which is regionally defined by the remnants of the process of flattening.

At the second level of analysis, the geomorphologist must consider the "surface structure of the landscape", which corresponds to the soils, but mainly to the colluvium, the colluvial slopes and in this case the possibility of gravel pits and "stone lines"—not only in rock-to-colluvium contact, but also with lines embedded in the colluvial package. The physical, chemical and micromorphological analyses allow us to deduce the processes (and the quality thereof) that acted in the destruction or even in the reshaping of the past forms (Vitte 2008).

The correlation of the first two levels already allows a compartmentalisation of genetically homogeneous forms to be established, with great usefulness in environmental planning. The third level of analysis by Ab'Saber (1969a, b) is the "landscape physiology", understood by the author as the expression of the current functioning of the geosphere. In this case, it corresponds to the current processes that act in the modelling of forms. With this methodological proposition, Ab'Saber (1969a, b) detached himself from the problems arising from the adoption of the taxonomy of landforms, such as those proposed by Tricart (1982). Now, the forms are the product of past and current processes, in a framework in which geology as well as climatic and paleoclimatic forces participate (Vitte 2008).

Despite few questionings of the Aziz and Bigarella model, the 1960s were fundamental for constructing a true paradigm in Brazilian geomorphology. Therefore, a theoretical, methodological and interpretative structure of the terrain and its processes was set up, together building a true geographic geomorphology. The model's hallmark is Aziz's 1969 article, "*A geomorfologia a serviço das pesquisas do quaternário*" [geomorphology at the service of quaternary research], which it still continues to exert a strong influence in geomorphological research in Brazil, while nothing else has been built in theoretical and methodological terms to seek more precise and profound analyses on the genesis of the Brazilian terrain (Vitte 2008). Perhaps, this is one of the biggest problems of Brazilian geographic geomorphology, because this model and this developed method exerted such power over time for generations of geographers-geomorphologists that scientific creativity itself, on the part of geography, has been affected to such an extent that today we are having great difficulty in maintaining geomorphology within geography and developing models that are more suited to the current stage of scientific development in Brazil.

14.4 Final Reflections

In recent years, concerns about the history of geomorphology in the world (and in Brazil in particular) have increased—a fact that can be observed in the thematic sections of congresses and meetings on physical geography and geomorphology. The driver of this process is related to the transformations that geographic science and geology itself are enduring, starting with the crisis of modern science and intensified by the processes of globalisation and environmental degradation. These facts are catalysing the process of re-discussion on the foundations of modern rationality—a context into which geography and geology are structured.

It is in this context that the construction of lines of research in the history and epistemology of geomorphology was considered opportune, as well as the expansion of discussions on this topic in congresses, seminars and symposia, as well as postgraduate courses in geography in Brazil.

The quantitative and qualitative increase in discussions will help in the construction of parameters for the new geomorphology, its cognition and representation, since a new surface of the earth and a new level of complexity and organisation of natural processes are under construction. The earth and the world are mutating, and terrain is an aggregating element of the discussion about a new geosphere that is under construction—now no longer earthy as in the nineteenth century, but based on the rarity of nature and new patterns of spatio-temporal interrelationships of the geomorphological processes that are triggered by human intervention in the construction of the world.

The brief historical and epistemological overview of Professor Aziz Ab'Saber's reflections demonstrates the struggle for the professionalisation of scientific fields in Brazil and how rich the history of geomorphological thinking is in the context of Brazilian geographic science.

References

Ab'Sáber AN (1958) Meditações em torno da notícia e da crítica na geomorfologia brasileira. Notícia Geomorfológica 1:1–6

Ab'Saber AN (1962). Compartimentação topográfica e domínios de sedimentação Pós-Cretáceos do Brasil, Tese (Concurso para a cadeira de Geografia Física) - Depto. Geografia,UFPR,1962

Abreu MA (1994) Estudo geográfico da cidade no brasil: evolução e avaliação (contribuição à história do pensamento geográfico brasileiro. Revista Brasileira de Geografia 56(1/4):21–123

Ab'Sáber AN, Menezes C (2007) O que é ser geógrafo. Bertrand Brasil, RJ

Ab'Sáber AN (1967) Domínios morfolclimáticos e províncias fitogeográficas do Brasil. Orientação 3

Ab'Sáber AN (1969a) A Depressão Periférica Paulista: um setor das áreas de circundesnudaçãopóscretácica da Bacia do Paraná. Geomorfologia 15:26

Ab'Sáber AN (1969b) Províncias geológicas e domínios morfoclimáticos no Brasil. Geomorfologia 15:15

Ab'Sáber AN (1970) Províncias geológicas e domínios morfoclimáticos no Brasil. Geomorfologia 20

Ab'Sáber AN (1961a) Revisão dos conhecimentos sobre o horizonte subsuperficial de cascalhos inhumanos no Brasil Oriental. Boletim Universidade do Paraná Geografia Física 2:32

Ab'Sáber AN (1961b) Superfícies aplainadas do primeiro planalto do Paraná. Boletim Paranaense de Geografia 4/5:116–125

Ab'Sáber AN (1972) Participação das depressões periféricas aplainadas na compartimentação do Planalto Brasileiro. Institute of Geographic USP, Geomorfologia, no 26

Adas S (2006) O campo do geógrafo: colonização e agricultura na obra de Orlando Valverde (1917–1964). Doctoral thesis (Ph.D. in geography), SP, FFLCH-USP
Alves FG (2010) Folheando páginas, descobrindo Histórias. A Revista de Históreia e a difusãod-ifusão da historiografia dos Annales no Brasil (1950–1960). Master's dissertation (master's in history), João Pessoa
Azevedo F (1994) As ciências no Brasil, vol 1, IBGE, Rio de Janeiro
Bachelard G (1996) A formação do espírito científico. Contraponto, RJ
Becker B, Eagler C (1993) Brasil—Uma Nova Potência Regional na Economia-Mundo. Rio de Janeiro, Bertrand Brasil
de Biaggi E (2015) Tradições cartográficas e fixação de fronteiras na independência brasileira. Terra Brasilis (Nova Série) RJ 4:1–20
de Brito SHA, Cardpso MA (2014) Fernando de Azevedo (1894–1974): uma mesma obra e suas várias leituras. Revista HISTEDBR Online Campinas SP 14(60):134–157
BigarellaJoão J, Marques Filho P, Ab'Sáber AN (1961) Ocorrências de pedimentos remanescentes nas fraldas da serra do Iqueririm (Garuva-SC). Boletim Paranaense de Geografia 4, 5:71–85
Burke P (2003) Uma História Social do Conhecimento 1: De Gutenberg a Diderot, São Paulo: Editora da Unesp
Cano W (1994) Raízes da concentração industrial em São Paulo. Instituto de Economia, Campinas
Clout H (2004) Lessons for experience: French geographers and the transcontinental excursion in 1912. Progr Human Geogr 12(5):597–618
Dantes MAM (2015) As ciências na história brasileira, SP:SBPC. Ciência e Cultura 57(1):26–29
de Abreu AA (1982) Análise Geomorfológica: reflexão e aplicação. São Paulo, FFLCH, USP, professorship thesis
de Almeida FFM (1964) Fundamentos geológicos do relevo paulista. Geologia do Estado de São Paulo (IGG Bol 41). Instituto Geológico E Geográfico, IGG, São Paulo, pp 167–262
de Martonne E (1943) 1943–1944. Problemas morfológicos do Brasil tropical atlântico I. Rio de Janeiro. Revista Brasileira de Geografia 4:3–26
de Martonne E (1944) Problemas morfológicos do Brasil tropical atlântico II. Revista Brasileira de Geografia 5(4):523–550
de Oliveira F (1981) Elegia para uma re(li)gião. Paz e Terra, RJ
da Silva DP (2006) Orville Adelbert Derby e o mapeamento de São Paulo (1886–1905). Master's dissertation (master's in history), PUC, SP
Duarte RB (2011) Incógnitas Geográficas. Francisco Bhering e as questões territoriais brasileiras no início do século XX. Master's dissertation (master's in geography), SP, FFLCH-USP
Enciclopedia Latino-Americana (2021) Universidades http://latinoamericana.wiki.br/verbetes/u/universidades, acessado por Antonio Carlos Vitte, 05/11/2021
Ferreira LO, Azevedo N, Guedes M, Cortes B (2008) Institucionalização das ciências, sistema de gênero e produção científica no Brasil (1939–1969). História Ciêncais e Saúde Manguinhos suplemento RJ 15:43–71
Ferri MG, Motoyama S (Coord) (1979–1981) História das Ciências no Brasil, vols 1–3. São Paulo, Edusp/Edu/Cnpq
Figuerôa SFM (1987) Modernos bandeirantes: a Comissão Geográfica e Geológica de São Paulo e a exploração científica do território paulista (1886–1931). Master's dissertation (master's in history), SP, FFLCH-USP
Gregory KJ, Gurnell AM, Petts GE (2002) Restructuring physical geography. Transc Inst Br Geogr NS 27:136–154
Gregory KJ (1992) A natureza da geografia física. Rio de Janeiro, Bertrand Brasil
Instituto Geológico De São Paulo (2021) Instituto Geológico comemora 133 anos de criação da Comissão Geográfica e Geológica. https://www.saopaulo.sp.gov.br/sala-de-imprensa/release/instituto-geologico-comemora-133-anos-de-criacao-da-comissao-geografica-e-geologica. Accessed by Antonio Carlos Vitte, 10/06/2021
Instituto De Geociências, USP—História Institucional. https://igc.usp.br/institucional/historia. Accessed by Antonio Carlos Vitte, 10/06/2021

Lester K (1956) A geomorfologia do Brasil oriental. Rev Bras Geogr 18(2):147–265

Limongi F (1989) Mentores e clientela da Universidade de São Paulo. In: Miceli S (org) História das ciências sociais no Brasil, vol 1, Vortice, SP

Lopes MM (2009) O Brasil descobre a pesquisa científica, 2nd edn. Hucitec, SP

Machado LO (2000) As idéias no lugar. O desenvolvimento do pensamento geográfico no Brasil no início do século XX. Terra Brasilis (Nova Série), RJ, vol 2, pp 1–14

Mahl ML (2012) Orville Adelbert Derby: notes for the study of his intellectual scientific performance in the state of São Paulo (1886–1905). Revista de História SP 167:295–300

Molina RS (2011) Escola Agrícola Prática "Luiz de Queiroz" (ESALQ/USP). sua gênese, projetos e primeiras experiências—1881–1903. Faculdade de Educação, Unicamp, Master's dissertation (master's in education)

Oliveira F de (1981) de Elegia para uma re(li)gião. Paz e Terra, Rio de Janeiro

Sampaio H (1991) Evolução do Ensino Superior Brasileiro: 1808–1990. SP: Núcleo de Pesquisas sobre Ensino Superior da Universidade de São Paulo, Working Document, no 8. https://nupps.usp.br/downloads/docs/dt9108.pdf

Schwartzman S (1979) Formação da comunidade científica no Brasil. Cia. Ed. Nacional, Finep, São Paulo, Rio de Janeiro

Sevcenko N (2003) Literatura como missão. Cia das Letras, SP

Tricart J (1982) Ecogeografia. IBGE, RJ

Vitte AC (2008) A geografiafísica no Brasil: um panorama quantitativo a partir de periódicos nacionais (1928–2006). Revista Da Anpege 4(4):39–49

Vitte AC (2011a) A construção da geomorfologia no Brasil. Revista Brasileira De Geomorfologia 12(3):91–108

Vitte AC (2011b) Breves considerações sobre o papel de Pierre Monbeig na formação do pensamento geomorfológico uspiano. Confins, 11, p 1–15. https://doi.org/10.4000/confins.6954

Vitte AC (1999) O texto no contexto. A tese de doutoramento de João Dias da Silveira "Estudo Geomorfológico dos Contrafortes Ocidentais da Mantiqueira". I Seminário Nacionalde História do Pensamento Geográfico, Rio Claro: Unesp, Anais

Part II
Brazilian Geography, a Geography of the Street

Chapter 15
The Right to the City and the Housing in Brazilian Cities

Arlete Moysés Rodrigues

Abstract This article will analyse the general conditions of housing in Brazilian cities and housing policies, with an emphasis on the current situation. The right to adequate housing forms part of the right to the city, but the right to the city is broader; it consists of a different way of producing the city. The lack of adequate housing for workers is inherent to the capitalist production method. Housing includes land ownership—one of the foundations of the capitalist production method—and produces income for owners, given the collective production of the city. Buildings require a construction process and produce profits for the construction sectors. In Brazil, since its beginnings, land as a form of wealth has not been permitted for all. In 1850, the Land Law characterised land as a commodity in the capitalist production method. As such, access to this commodity can only be gained through purchasing and/or leasing—something which workers must guarantee with their own (always insufficient) salaries, as confirmed by the rate of the minimum wage. State housing policies for workers in the 1930s made employment conditional on the ability to rent and/or purchase. In 1966, the income brackets were established, and working would thereafter allow the acquisition of housing through the BNH-FGTS. From then on, the world of work began to subsidise urban development. This policy came to an end in 1986 with the dissolution of the BNH. There was financing, but no housing policy. The housing policy was reintroduced in 2009, with the Minha casa, minhavida [my house, my life] housing programme, set to expire in 2020. In any event, there is no guarantee of the right to housing as expressed in the Federal Constitution—something which evidences the socio-spatial inequality.

Keywords Right to the city · Right to housing · Housing insecurity · Public housing policy · Brazil cities

---

A. M. Rodrigues (✉)
Universidade Estadual de Campinas, Campinas, Brazil
e-mail: moysesarlete@gmail.com

R. C. Lois González and M. A. Mitidiero Junior (eds.), *Brazilian Geography*, Advances in Geographical and Environmental Sciences,
https://doi.org/10.1007/978-981-19-3704-0_15

15.1 Introduction

Housing is a commodity of the capitalist production method—bought, sold and/or rented and, in general, paid for in cash. It is a fundamental requirement for living, as no one can live without occupying space, and it is there that family life, rest, food, hygiene practices, etc. take place. It is the place for the reproduction of life. How can this commodity, a basic need, be paid for? For the majority, which includes workers in general, payment for this commodity is made through wages received—since, under capitalism, the means of production (capital/money, raw materials, driving force, land) are in the hands of the capitalists, separated from the workers who have only their labour power to rely on for survival. Workers, when selling their labour power, receive cash wages, the value of which should be sufficient to meet the basic requirements for their reproduction, including housing.

In Brazil, since the beginning of the 1930s, housing has been one of the items that must be considered in the calculation of the minimum wage, in order to allow the reproduction of the workforce. When salaries do not allow the cost of living to be covered, it is necessary to institute public housing policies. As wages and public housing policies are insufficient, a large portion of city dwellers in Brazilian cities live in conditions of insecurity—in favelas, tenements and in areas lacking infrastructure and public services equipment.

The right to housing is included in the 1996 Habitat II Agenda. Furthermore, in Brazil, it is expressed as a social right in Article 6 of the 1988 Constitution, along with other social rights, although there is no mention of how such rights can be served. The right to the city counts only as an ideation of government programmes and has nothing effective in terms of its concreteness.

How can a commodity be a right? In what ways can the state provision this vital need, this right? Are public policies sufficient for those who cannot afford housing? Unravelling the form and content of public housing policies is fundamental to understanding housing as a need and as a right. How may this right to housing materialise into the utopia of the right to the city? Discovering whether or not the right to housing has materialised through public policies is an important task for geographers who believe that it is necessary to combine a theoretical framework, analytical methodology and empirical data, allowing the production and reproduction of urban space to be analysed.

In order to contribute to the understanding of the right to housing and the city, we present some characteristics of housing insecurity in Brazilian cities and highlight elements of urban land and its income as the basis on which the production of urban space is founded, with work-wage relations establishing parameters for the supply of housing and the public policies on affordable housing that try to meet this need.

15.2 Overview of Housing Insecurity in Brazil

Housing is a fundamental requirement for living and a right that integrates another—the right to the city—which is broader and constitutes another way of producing and consuming the city. Having a right to housing is recognising that the production of—and within—the city needs to acknowledge that there is no possibility of living without a roof. Embedded within the right to the city, housing as a right symbolises that the place of dwelling must be classified as a value for life—the value of using and enjoying urban space.

The focus of this text is the presentation of some causes of housing insecurity that engender struggles for the right to housing and the city. The lack of adequate housing needs to be understood in its connection with capitalist labour relations, with the production and reproduction of space in its complexity and with the contradictions and conflicts inherent in capitalist urbanisation. We begin with some data that indicate the lack of housing and its insecurity for a large portion of those who live in Brazilian urban centres. We then focus on urban land ownership and the production of housing by the state. Finally, we deal with social struggles around the right to housing, as well as the process of its institutionalisation.

Since the 1960s, the urban population has been predominant in Brazil, reaching 85% of the total population in 2015, as shown in Table 15.1. The urban population is mainly concentrated in the southeast region, where the largest number of metropolises and large cities is also found and which is, in general, a gauge of progress and development. Within the country, we can say that this progress appears in islands of wealth, surrounded by oceans of poverty and insecurity.

Merely appearing in surveys under urban population does not imply having urbanity—that is, it does not mean having an adequate standard of urban living or participating in democratic life. A large part of the urban population lives in insecure houses, in places without infrastructure, without public transport, without basic sanitation, without collective equipment and without a guarantee of continuity when occupying areas and building their shacks/houses, running the risk of being displaced when these areas/neighbourhoods become of interest to the real estate sector. We can

Table 15.1 Urban and rural population of Brazil in greater regions, 2015

	Urban population		Rural population	
Brazil and greater regions	Total	%	Total	%
Brazil	173,566,000	84.72	31,294,000	15.28
North	13,145,000	6.42	4,379,000	2.14
Northeast	41,414,000	20.22	15,227,000	7.43
Southeast	80,020,000	39.06	5,897,000	2.88
South	25,076,000	12.24	4,214,000	2.06
Central-West	13,911,000	6.79	1,578,000	0.77

Source IBGE and PNAD Contínua—Elaboration: Flávio Lima

Table 15.2 Estimate of occupied households in subnormal agglomerations per greater region of Brazil in 2019

States and federal district	Number of houses occupied in subnormal clusters[1]	Houses occupied in subnormal clusters in relation to the total number of occupied houses (%)
North	918,498	17.91
Northeast	1,459,486	28.46
Central-West	127,175	2.49
Southeast	2,321,963	45.28
South	300,625	5.86
Brazil/total	5,127,747	

Source IBGE—Elaboration: Flávio Lima

only artificially separate the living conditions of a place of dwelling from the working conditions, but we focus our analysis on the housing conditions.

One of the data items that demonstrates the absence of urbanity is the fact that a significant portion of the urban population lives in favelas, termed as subnormal agglomerations by the Brazilian Institute of Geography and Statistics—IBGE, concentrated in percentage terms in the southeast region, as shown[2] in Table 15.2.

Estimates for 2019 indicated the existence of 5,127,747 favela shacks in 13,151 clusters, spread over 734 municipalities 3[3]. In 2010, the number was 3,224,529 shacks in 6329 favelas and 323 municipalities, which reveals the increase in poverty between these two periods. To reflect on the causes of impoverishment in the country, we can mention several actions such as the approval of Constitutional Amendment no. 95, December 2016 (Brazil 2016), which froze investments in health, education and social assistance for 20 years; the 2017 labour reform (Brazil 2017b), which made labour relations insecure; the freezing, since 2016, of the real value of the minimum wage—MW; other laws and rules related to the use of urban land that will be highlighted in the text.

When comparing the total number of houses with the number of shacks, the dimension of poverty becomes even more evident. As can be seen in Fig. 15.1, in Belém and Manaus—metropolises in the northern region of the country—more than 50% of the housing units are shacks in favelas. In national metropolises with great economic significance, such as Rio de Janeiro and São Paulo, around 20% of homes are shacks in favelas. Considering the population of the two metropolises, the number of favela dwellings is extremely high. In Rio de Janeiro, there are 453,571 shacks, and

[1] Subnormal is a term used to designate precarious houses (shacks), while subnormal cluster refers to a set of these precarious houses

[2] Favela: a group of precarious units–shacks–that occupy, in a disorderly manner, dense public or private areas, without title deeds and without access to essential public services. We use, in the text, as synonyms of favela: domicile, habitation, dwelling, shack, depending on the source used, with the meaning of house, place of residence, roof.

[3] The data are underestimated, as the survey is carried out in clusters of more than 50 shacks and there are a large number of favelas with less than 50 households in the vast majority of cities.

in São Paulo, 529,921 (IBGE 2020a, b). It is estimated that there are approximately 25,000,000 people in small shacks, usually located in areas without infrastructure and without legal guarantee of continuity of residence.

Another piece of data which gives us visibility of the socio-spatial inequality is the calculation of the housing deficit, which means the need to produce new homes. In 2019, this deficit corresponded to almost 6 million units, as shown in Table 15.3. Its calculation is made up of insecure homes—improvised and rustic; cohabitation—over-occupied units and single-room houses; and the excessive burden of rent when this is greater than 30% of the family income for insecure income units of up to 3 times minimum wage (MW) (Fundação João Pinheiro 2020). Those who live on the street, without a roof over their heads, are not included in the calculations.

The research carried out by the Fundação João Pinheiro (2020) points out that the deficit is on the increase, even with the production of low-income housing units 5[4]. Housing is not a commodity like any other that sees its value decrease with scaled-up production. On the contrary, as urban expansion increases more and more, the price of land and buildings becomes higher.

In order to show that the right to housing is not being served, it is common to compare the deficit to the number of vacant houses that are of the same scale. This comparison shows how real estate production serves the purposes of obtaining income, interest and profits and places highest priority on exchange value, with little consideration from the capital operating in the sector for the needs of those who live in poverty and whose goal is to obtain a roof over their heads.

The largest quantitative and percentage deficits are found in the southeast region, considered the most 'developed' in the country, showing the contradictions between economic development and living conditions, and between the advance in real estate production in general and the insecure conditions of living in cities. Families receiving up to 3 times MW6 live in very insecure conditions and represent 42% of the total deficit. When the level[5] of insecurity of a house is looked at, it turns out that 72% of families earn up to 1 minimum wage, which shows the ranges of income by type of insecurity.

In contrast to the insecure housing conditions of low-wage workers, in São Paulo only 1% of owners (22,400 people) account for 25% of all registered properties in the city, representing 45% of the municipal real estate value, with calculated wealth at around R$749 billion. When only these figures are considered, in urban properties such owners possess an average of R$34 million per person, which is around 600 times the national average of equity distribution (Oxfam Brazil 2017).

While a small number own urban properties, more than 14,000 families (from March 2020 to May 2021) were evicted, and another 84,000 are threatened with eviction (Campanha Despejo Zero 2020). Entire families are thrown on the streets

[4] We use with the same meaning: affordable housing/housing and social housing, depending on the source researched.

[5] The 2021 Minimum Wage is R$1,100.00 per month, equivalent to US$192 dollars, as of June 6, 2021.

Table 15.3 Housing deficit in Brazil and greater regions in 2019

Brazil and greater regions	Components						Overall total
	Insecure housing		Cohabitation		Excessive rent burden		
	Total	% of overall total (%)	Total	% of overall total (%)	Total	% of overall total (%)	
Brazil	1,482,585	25.23	1,358,374	23.11	3,035,739	51.66	**5,876,699**
North	310,312	43.12	258,206	35.88	151,120	21	**719,638**
Northeast	634,070	35.64	449,418	25.27	695,477	39.09	**1,778,960**
Southeast	280,267	12.25	461,144	20.16	1,545,710	67.59	**2,287,121**
South	160,423	25.92	95,150	15.38	363,299	58.70	**618,873**
Central-West	97,514	20.66	94,456	20.01	280,132	59.33	**472,102**

Source Fundação João Pinheiro and IBGE—Elaboration: Flávio Lima

and their shacks destroyed, infringing on the[6] right to housing as set out in the 1988 Brazilian Constitution (Brazil 1988) and in the City Statute (Brazil 2001).

In the middle of the COVID-19 pandemic, when houses, in addition to providing shelter, have the purpose of individual and social protection, there have been evictions on the one hand and, on the other, growth in property sales, particularly high-end properties. It should be noted that the urban building construction sector has not ceased its activities, even during periods of greater social isolation. The real estate industry claims that, with the pandemic, low interest rates and the expensive dollar, real estate sales in February 2021 saw a 49% increase compared to the same period in 2020 (Secovi 2021). During the pandemic, the sale of luxury properties increased significantly, boosting the real estate sector. This occurred due to the increase in the importance of the role of housing and the "availability of low-interest credit" (Estadão 2021, p. 4, translated). The class struggle is evidenced in the dispute over urban land, where exchange value is given highest priority at the expense of usage of and the right to housing. In order to understand the contradiction between insecurity for the majority and wealth for a minority, we highlight land ownership as an explanatory element of this socio-spatial inequality.

15.3 Urban Land Ownership: A Factor in Socio-Spatial Inequality

Land ownership as a factor in inequality dates back to the colonisation process and is directly related to the power of social and territorial control. In colonial Brazil, the 'distribution' of land had the aim of occupying the territory and cultivating virgin lands, however what happened was the concentration of huge tracts of land, giving rise to large estates. In Imperial Brazil in 1850, with the Land Law, land had already become a commodity and was given a price to be bought and sold at in the market. The fictitious starting price contained in the law defined the monopoly of access and the absolute rent.

The private ownership of land seems to be natural and has been naturalised: some are born with the right to be owners, while others can buy—if, of course, they have the resources to do so. Where might the resources to buy or rent decent housing come from? In Brazil, ownership—of even a small piece of land to cultivate and/or live on—is a privilege. The Brazilian philosopher MarilenaChaui distinguishes privileges, needs and rights. For the author, privileges are for the few—those who, for their maintenance, massacre those who do not have privileges, such as via evictions in the middle of a pandemic. Need refers to the many, and the state should attend to it, as the right is universal—it represents the entire society governed by the rights (Chaui 2012). Following from this, we understand that the rights to housing, mobility,

[6] The Zero Eviction Campaign, articulated by housing movements and national and international entities, has acted to stop evictions during the pandemic.

health, education, social assistance, employment, land to cultivate[7]—as made explicit in Article 6 of the 1988 Brazilian Constitution (Brazil 1988)—are rights for all Brazilians and not just for those who are able to pay for housing. With an approach towards universality, social movements fight for the right to housing and the city.

In every version of the Constitution of Brazil, property has always been an absolute and inalienable privilege. In the 1988 Constitution (Brazil 1988), the social function concerns the use of property in the urbanisation process—that is, if land or property is left vacant or under-occupied, it may be subject to taxation and, ultimately, be expropriated with payments of public debt bonds, which has never happened. Land ownership is one of the pillars of the development of capitalism. For large landowners, in terms of real estate development, the land and the buildings on it are most important in terms of exchange value, while for those who own land for their personal use, the usage value of the land prevails. Land[8] ownership allows the owners to obtain monopoly rent, absolute rent and differential rent from it.

Monopoly rent is directly related to the title of ownership and can also stipulate a 'quality' which interests the market. Absolute income is related to the production of, and in, the city—with each new building, subdivision of land, extension of urban areas, implementation of infrastructure and collective equipment increasing the price that results from the general urbanisation process. Differential income is a product of the unequal production of urban space, the location within the space of infrastructure, equipment and collective means of consumption, the extensions of lots, the quality of construction—in short, the characteristics resulting from the production and reproduction of urban space. To the rent from the land, we must add the profit obtained from production as well [9] as the interest from the financial capital, which marks it out as an expensive commodity that is not accessible to the majority.

Land is a 'gift' from nature, a non-reproducible treasure that is assigned a price by its owner which is altered by the production and reproduction of urban space. It is the basis for building in urban areas. Since no one can live without occupying space, conflicts over its use/possession/ownership are constant and heated. Due to its characteristics, it is of fundamental importance for capital accumulation and, as David Harvey points out, since the second half of the twentieth century, the city has been the place where capital surpluses are largely applied (Harvey 1982, 2014). In São Paulo, a survey by FIPE-ZAP (FIPE—the Institute for Economic Research)[10] 11 indicated that, from 2008 to 2018, the price of urban properties for sale increased by 208.94%, while the price of properties for lease increased by 114.98%—that is, the production from places where capital surpluses are applied has been extremely advantageous for the capital invested in them (FIPE-ZAP 2021).

[7] See Rodrigues (2019)

[8] See Jaramillo (2009)

[9] See Bastos (2020)

[10] The FipeZap index is calculated by ZIPE (Fundação Instituto de Pesquisas Econômicas) based on real estate advertisements (ready apartments, rooms and commercial conjuntos up to 200 square meters) published on ZAP Imoveis and other internet portals.

Under capitalist laws and regulations, control over the types and uses of property is exercised by the capitalist state, which constitutes an arena for conflict in which class contradictions are expressed12[11]. In Brazil, in addition to the constitutions that define property as an absolute value, the rules on subdivision and construction limit and prevent access to land/housing for a portion of city dwellers who receive low or no wages. As a result of the lack of access to housing, and since a place of dwelling is a basic necessity for living, there are collective occupations of land, everyday and individual occupations in favelas, and the super-congestion of residents in rooming houses and tenements. These are ways for the working class to have a shelter in order to survive.

Those who individually or collectively occupy land/houses without legal legitimacy and those who fight for the right to housing are viewed by landowners as enemies. In reality, these supposed enemies do not put property and power at risk—they simply seek to obtain the right to have rights—the right to housing, with greatest emphasis on its usage value. They do not question property in general—they question the predominance of its exchange value, they fight to stay on the occupied lands, they enter into conflicts to enforce the principles of the social function of property, of the right to housing, with the goal of ensuring that the city can be enjoyed by everyone and not just by a minority—that is, they seek right to the city.

The naturalisation of property as a class privilege means that any response to claims for the right to dwell is opposed by the dominant sectors. When small achievements are gained, the legislation is changed—as is currently the case in Brazil in relation to the regularisation of social-interest land tenure, with programmes for the production of affordable housing extinguished and evictions authorised, even if the properties are not fulfilling their social function. Those who fight for the right to have rights seem to threaten the ideal of land ownership as wealth and the resulting power—that is, the privileges. It has been mistakenly asserted that squatters impede urban development—thus, favelas, collective occupations and insecure housing must be 'moved' to distant areas.

The opposition between the class of owners and that of non-owners can be exemplified by the adoption of terms such as 'the landless' and 'the homeless' to designate those who struggle for land/housing. They have no land to cultivate, and they have no land to build their homes on, as they only have the sale of their labour power to rely on in order to meet their needs.

15.4 The World of Work and the Production of Affordable Housing

Capitalism, despite advancements in the forces of production, has never fulfilled the idea that, by selling their labour power, workers can have a decent existence. The separation of the means of production between the owners and controllers, depending

[11] See Mascaro (3013), Gramsci (1988); Poulantzas (19974); and Hirrsch (2010).

exclusively on the sale of their labour power in order to survive, is the cause of the need to fight for the right to housing. Access to housing depends on wages that, historically, do not allow houses to be rented or bought, as highlighted by Engels in 1872 when he analysed the situation of workers and the housing issue (Engels 1995).

The production of housing and of the city is linked to the interests of landowners who obtain rents; the real estate sector that makes a profit; and the financial sector that earns interest. The workers, in turn, who build houses and cities brick by brick, have only their wages to rely on for their reproduction. While landowners, the real estate sector and the financial sector, together with the state, intensify the rapid transformation and expansion of cities, increasing in price with each production, wages do not keep up with the appreciation of real estate that results from the urbanisation process. This capital appreciation is not related to the needs of the working class. Agnes Heller, when analysing Marx's theory of need, points out that need is interwoven with the production of value (Heller 1978) that occurs with each new public and/or private action. The author also points out that the need for workers is relegated to the reproduction of life, arising from wages and not from the production of value.

As already stated, dwelling is one of the basic needs—however, it needs to be paid for. How can it be paid for by selling labour power? In Brazil, in 1943, the Consolidation of Labour Laws (CLT) established that "Minimum wage is the minimum consideration due and paid directly by the employer to every worker [...] capable of satisfying, at a given time and in a given region of the Country, their normal need for food, housing, clothing, hygiene and transport" (Brazil 1936, translated, our emphasis). The definition of what the MW must provide to workers is maintained over time. In 1988, the Brazilian Constitution decided that the minimum wage should meet the basic vital needs of workers and their families "for housing, food, education, health, leisure, clothing, hygiene, transport and social security, with periodic adjustments that preserve purchasing power, and its being bound for any purpose shall be prohibited" (Brazil 1988, translated, our emphasis).

In relation to housing, the terms of the MW reflect the rent which would be paid monthly to the owner of the property after receipt of salaries. Before that, however, a place of dwelling is necessary in order to even have a job. In addition, one does not live only on the day/month when one has a job: when unemployed, people have no place and means to live (Rodrigues 2017). Even being employed and receiving up to three times MW, workers are unable to pay the rent for an adequate house, as the price of land/housing is defined by the general urban production, which interferes with the price of rents. Thus, the rental price is defined by the price of the property and is not based on its usage.

Under the logic of the capitalist production of space, no relationship exists between prices resulting from production in general and the reproduction of labour power. Each sector, each capitalist, seeks to extract the greatest possible profit from their activities. In the case of housing, as mentioned above, the rent from land with the continuous appreciation of the city in the process of urbanisation is added to the profits from production. In 2021, the rate of rent increase was calculated at 23.4% (Fundação Getúlio Vargas 2020), while the MW had a readjustment of 5.26%. In addition, we highlight that the "payment of tax-free profit rose by 129% from 2007 to 2019, while taxable income, including salaries, increased by 41% in the same period" (Poder 360—Newsletter 2021, p. 1, translated). Shareholders' profits and dividends are exempt from Income Tax in the country (Brazil 1995). DIEESE—the Inter-Union Department of Statistics and Socio-Economic Studies—calculates that the 2021 MW should be R$ 5330.69 (DIEESE 2021). Therefore, workers who earn less than 5 times MW are unable to adequately reproduce life. In Table 15.3, mentioned above, we can see that 84% of those who pay excessive rent receive up to 2 times minimum wage, showing the direct relationship between the world of work and housing.

As a result, we reaffirm that workers seek to 'remedy' living through occupying empty areas, forming favelas, collective land occupations and occupying empty and abandoned buildings that do not fulfil their social function and struggle to avail of housing as a right. When possible, they buy land, which is almost always irregular, and overwork—building insecure housing on the outskirts of cities, constituting the self-construction and binomial irregular subdivisions that a large portion of Brazilian cities have produced and continue to produce (Oliveira 2003). Thus, overexploited workers produce the urban space and their housing with usage value given highest priority—however, when the area where they live is consolidated, they tend to be absorbed by real estate development (Holston 2013). Another way of trying to solve the housing problem is the fight to convince the state to produce affordable housing, with subsidised prices, giving workers access to purchasing units based on their usage value.

15.5 The Institutional Process of Recognising the Right to Housing

The debate over the right to housing emerged as a guiding principle in the Universal Declaration of Human Rights in 1948 (UN 1948). In 1966, the International Covenant on Economic, Social and Cultural Rights clarified that housing is a fundamental need (UN 1966). In 1976, Habitat I defined that urban land, due to its importance, could not be in the hands of the market and speculators and that states should intervene to give land a social purpose (Habitat I Agenda 1976). These premises were incorporated in Brazil through the Movement for Urban Reform, which proposed them to the 1988 Constituent Congress. Through this, the principles of the social function of property and of the city were included in the Brazilian Constitution of 1988 (Brazil 1988).

In 2001, the City Statute (Brazil 2001) regulated the constitutional articles that deal with this subject. At that time, it was determined that urban property does not fulfil its social function if it is unoccupied or underutilised. Therefore, in order to apply the principles, properties that do not fulfil their social function must be identified in Municipal Master Plans—as such, this principle became an instrument of municipal planning (Rodrigues 2004). Thus, there is a shift from social struggles towards urban development tied to municipal planning. In order to guarantee areas for affordable housing, urban popular movements have had to participate in debates on municipal master plans, which at the same time elicited greater knowledge about how the city's production shifts them from the concrete struggle towards planning. After 20 years of the City Statute, in a few cities, it has been possible, specifically, to apply some instruments in order to enforce the social function of property and the city.

The right to housing, as a human right in the universal perspective, was only included in the Habitat II Agenda in 1996 (Habitat II Agenda 1996). The explanation of the right to housing in this document is due to the participation and mobilisation of popular movements around the world that were present in Istanbul—among them, the Urban Reform Movement. In addition to establishing the right to housing as a human right, the text defined principles in relation to forced evictions (Rodrigues 2000). In Chapter III, item 39(a), the Habitat II Agenda (1996) states that governments should be "Protecting all people from and providing legal protection and redress for forced evictions that are contrary to the law, taking human rights into consideration; when evictions are unavoidable, ensuring, as appropriate, that alternative suitable solutions are provided". In reality, in addition to evictions carried out by public authorities with the aim of promoting modernisation, most evictions are the result of the action of private owners and aimed at expanding urban designs and denying the achievement of a right. These removals/evictions are called repossessions of ownership when, in reality, they are repossessions of property, because the ownership is with the occupants. The purpose of the evictions is, for the most part, targeted towards real estate valuation—land, with exchange value given highest priority. They are carried out without being backed by any housing policies—even during the pandemic, as mentioned above.

Although the right to housing, as a human right, was approved in Habitat II and included in the Brazilian Constitution, it is still an abstraction. It does not materialise insofar as the right to land ownership, in its absolute sense, continues to prevail. Although it is a basic need—a right—contradictorily, housing is a commodity of the capitalist production method, with its exchange value prevailing over its usage: it is a need, and it is a nationally and internationally expressed right, but it is a commodity that needs to be paid for. How can we correlate human rights with a commodity?

In any case, the principle of the right to housing and of the social function of property and the city allows for the coordination of movements that fight for an end to forced evictions, for the production of affordable housing and for the regularisation of land with a social interest. The ways in which popular urban movements make it possible to access the right to housing include the production of affordable housing by the state in its various instances, in addition to the regularisation of land with a social interest and the urbanisation of favelas. Owning a home means being able to have a place of dwelling, even when unemployed and/or with low wages. If housing is a commodity, there is no reason not to hold the property title.

15.6 The Production of 'Social-Interest' Housing by the State

In order to understand the struggle directed at the state for the right to housing, it is necessary to remember that one of the state's tasks, in almost every country in the world, is to meet social needs through public policies, in order to maintain balance between the social classes that comprise it. The capitalist state is a contradictory distillation of the social classes that constitute it. It is able to meet the demands of the working classes and thus minimise insecure living conditions, as long as they do not alter the structures of power. The affordable housing movements address the constituted state as a way of trying to obtain housing as a right, despite paying for it as will be presented.

By acting in the production of housing, the state also promotes the appreciation of value and the increase in the price of land and buildings. In this way, contradictorily, it provides housing for one portion of workers and makes it more expensive and inaccessible for other portions. This is a real contradiction, considering what has already been said about land rent, about the commodity of land/housing and about wages. It is up to the state to act in order to foster capital accumulation, as seen with real estate production, and at the same time to meet a basic need for capital and for work. In any case, the state promotes public housing policies that, although essential, are always insufficient. They do not reach the essence of the issue of land ownership, as rules on land ownership and real estate development which favour the exchange value continue to prevail. Such public policies do not impose limits on real estate speculation—one of the objectives of the principles of the social function of property—and do not scale the value of the MW in a way that allows for an adequate reproduction of life. In addition, the workers pay a substantial share of the resources to finance the housing, directly in instalments and indirectly in the production of housing, through funds from the world of work.

In Brazil, the first major housing production programme for workers was instituted in 1936, with the creation of a property portfolio for the production of housing projects for workers, together with the creation of the Social Security System (Brazil 1936). The resources came from employer and employee contributions to social security and to the property portfolio, which was authorised to allocate 50% of its reserves to the financing of housing construction. The sale and/or lease of units took place according to a worker's professional category, as established by the legislation during that period. It is the world of work which finances the production of units via compulsory contributions, in addition to the fact that payment for the financing (in the case of purchase) and the rent (in the case of leases) came from wages.

With the increase in pensions, the resources in the portfolio were exhausted and, as a result, it ceased production. In 1964, after the military coup, the BNH—the National Housing Bank—was created. It was supposed to manage and finance a policy for construction and home ownership, particularly for lower-income social classes, as well as to expand employment opportunities and boost the civil construction sector. The cornerstone of the BNH, the Financial System for Housing (SFH), was intended to provide resources that would guarantee the execution of the National Housing Plan, in addition to investing in urban areas. Until 1966, the BNH had scarce resources, and in order to obtain them, labour legislation was changed, creating the Term of Service Guarantee Fund (FGTS) in December 1966.

This change meant the end of job stability13[12]. Responsibility for the payment of severance benefits, which had been handled by the employer, was transferred to the BNH—that is, to the state—with substantial changes in labour legislation. The BNH was, for a period, the second-largest Brazilian bank, with money coming from the world of work (Rodrigues 2015). In this way, urban space and housing are produced for the benefit of workers who, in turn, also pay for the purchase over long periods of time. The large housing complexes produced during the BNH period were located on the outskirts of urban areas, without infrastructure for services and equipment for collective consumption. This promoted the urbanisation of distant areas, with large empty spaces between the urban continuum and housing complexes, thus contributing to land price speculation and the expansion of urban areas.

With the increase in unemployment, withdrawal from the FGTS by workers, and the decrease in the amount of resources deposited, in addition to investments in major works and several other internal issues, the BNH was dissolved and the resources are now managed by the Caixa Economia Federal [Federal Savings Bank] (Brazil 1986). The closing of the BNH signified the end of a national housing policy. From 1986 to 2008, there were no national housing programmes—a fact that had a direct impact on the increase in the housing deficit, the increase of favelas and shacks and the expansion of the insecure periphery with the self-construction of units in irregular subdivisions. With the increase in insecure housing, popular urban movements intensified their private and collective struggles—participating, as we have seen, in the 1988 Constituent Congress for an Urban Reform, which contained the principles of the social function of property and the city. The world of work is not dissociated from the production and consumption of housing for the working class, although wages are completely dissociated.

The resumption of a national housing programme took place in 2009 with the PMCMV—the Minha Casa Minha Vida [my house, my life] Programme (Brazil 2009), with the aim of producing housing for various income groups and boosting the economy after the 2008 global crisis. The resources originated from the FAR (the Residential Lease Fund), the FDS (the Social Development Fund—funds for low-income families, arising from a law based on a popular initiative for the creation of the National Housing Fund) and the FGTS (the Term of Service Guarantee Fund). The smallest-scale production was directed towards families earning up to 3 times MW, although it is within this range that the biggest deficit exists. Despite some

[12] Until the edition of the FGTS, when employees were dismissed, compensation was made on the basis that for each year worked, the dismissed employer would receive one month of the last salary. With 9 and a half years of work, he would receive 20 months of the last salary, that is, double. Since the creation of the FGTS, employees deposit 8% of the employee's monthly salary so that, when he is fired, he can withdraw from his account with amounts corrected by bank fees. The layoffs became more frequent and interned by the state

resources coming from the Federal Government, most of the funds came from the world of work.

Production was high until 2017, when resources began to run out, and continued to decrease until the programme ended. By 2019, 6.1 million units had been contracted, with 5.1 million already delivered (Brazil 2020). Although production was high, it did not reduce the deficit, nor the number of residents in insecure areas, favelas, tenements—which shows that production will always be insufficient to meet needs. Conversely, it caused an increase in the price of land and buildings, particularly in large cities—in a direct relationship with absolute income. This increase was caused by the expansion of urbanisation in general, as well as by the demand for land for the implementation of cooperatives which were undertaken by the private sector (Bastos 2020).

Many studies and pieces of academic research highlight the fragility of national public housing policies, identifying the following as problems: the location of low-income housing complexes in areas without infrastructure, leaving large empty spaces for real estate speculation; its insufficiency for those earning low wages, despite high production, because the resources for the lower-income groups were small compared to the middle- and high-income groups; the increase in the price of land and real estate when the state assumes a share of the production of housing; and the non-resolution of the housing issue. We understand that these critical analyses are correct in relation to their own terms, insofar as they focus on the government's agenda for serving the principles of the right to housing and the social function of land in the city. In addition, they highlight the insecure nature of how a vital need is served.

The PMCMV was dismantled in 2020, but it had been practically paralysed since 2018. It was replaced by the Casa Verde Amarela [Green-Yellow House] Programme (Brazil 2021), which has no resources for the range in which the deficit is greatest. Despite the complexity, insufficiency and problems—particularly the location in peripheral areas without infrastructure—the production of affordable housing has been demanded by movements that fight for housing. Such movements put pressure on the state so that, in addition to the production of new units, it promotes the regularisation of the tenure of land with a social interest and the urbanisation of favelas in order to prevent the forced removal of populations.

15.7 The Relentless Struggles for the Right to Housing and the City

The struggles for the right to housing are historic, as already presented, with demands for continuity of residence in occupied areas, the implementation of infrastructure, collective equipment and the production of new housing. In the late 1980s, popular movements that had previously been dispersed according to their place of residence and type of demand joined together as a collective in order to fight for an Urban

Reform that would meet the needs of the majority. In doing so, these movements proposed the popular amendment for Urban Reform, guided the approval of the City Statute and developed the Popular Initiative Law in order to create a popular housing fund. In addition, they took part in the United Nations Conference on Environment and Development in 1992, were actively present at the United Nations Habitat II Conference on Human Settlement in 1996 and were part of the Council of Cities, from its beginnings in 2003 until its dissolution in 2019. They have also been involved in the Cities Conferences since 2003 and take part in the preparation of master plans. Furthermore, they organise and are present in various struggles to remain in the occupied areas: they were even responsible for the creation of the Campanha Despejo Zero, as mentioned above.

They fight for continuity of residence in the occupied areas with the principle of the regularisation of land with a social interest. Regularising occupied areas implies recognition of the state's inability to produce adequate housing for workers and highlights housing as a right by recognising that the production of the house and the city is carried out by everyone—and not just by typically capitalist agents. The confrontations happen directly with the state, which intermediates the regularisation, either on public lands or on private lands that do not fulfil the social function. The regularisation of land tenure recognises the right to decent housing, as it promotes the urbanistic and legal regularisation of tenure. The Council of Cities, created in 2004 (Rodrigues 2010), emphasised land regularisation in order to serve the right to housing and prevent forced evictions. Chapter III of the 2009 PMCMV (Brazil 2009) explains the parameters—for the social function of the property and, mainly, for the regularisation of land with a social interest. As already stated, the struggle for housing as a right with emphasis on the regularisation of land does not put land ownership in dispute—it only points towards giving highest priority to usage value.

Despite the small areas that are usable for land regularisation, the dominant sectors fear that this process will change the property ideology that gives primacy to exchange value, and they therefore changed the rules for its implementation soon after the 2016 coup. Law 13 465/2017 (Brazil 2017) defines new parameters for the regularisation of land tenure, having the delivery of public lands to private interests as one of its preconditions. It removes the potential for land regularisation tenure with highest priority given to usage and moves towards having legal regularisation as its objective—that is, the title of ownership, which expands the perimeter within which the real estate sector is able to act. What prevails under the law is the regularisation of land with a specific interest, which aims to legalise irregular areas of walled subdivisions and condominiums (Rodrigues 2014). Current legislation is aimed towards the title of ownership and the granting of highest priority to exchange value. In a society like Brazil's, where private property is one of the preconditions for inequality, the re-urbanisation of specific interests legalises the formerly illegal occupation carried out by real estate development. On the other hand, the occupation of land for housing—which is a constitutional right—has been violently resisted, and occupants have been removed using a strong police apparatus, even when occupying public areas.

There have been many years of relentless struggle in order to have the right to housing and the city. Several institutional achievements, that were not always facilitated, were attained during the period that extends from the late 1980s until 2016. Among them, as already mentioned, are the definition of property and of the city's social function, the recognition of the right to housing as a human right and the approval, after 12 years, of a bill based on the popular initiative for the creation of the National Housing Fund (Brazil 2005), with resources from the Union. With the National Housing Fund Law, the National System of Popular Housing (SNHIS) was created, managed by a Council linked to the Council of Cities until its dissolution in 2019. The Council of Cities, dissolved in 2019 (Brazil 2019), had wide participation from civil society, particularly from popular movements whose representatives were elected at the Conferences of Cities. It is important to emphasise that it was via proposals from the Conferences of Cities and the Council that the National Urban Development Policy (Brazil 2004a) and the National Housing Policy (Brazil 2004b) were drawn up, highlighting priorities for serving families with income of up to three times MW (Rodrigues 2010). The historical struggles for the right to housing and the city that became institutionally recognised 'disappeared', but the struggle for housing as a human right remains and has no respite, because it is about the struggle for life.

15.8 Final Considerations

The right to housing is contained in the right to the city, with the principle that it is another way of producing and consuming the city. This is a utopia to be built daily which, in order to be realised, depends on one of the fundamental rights—that of housing—being served. In the current historical setting, up to the completion of this text in June 2021, around half a million deaths caused by COVID-19 were recorded in Brazil. In addition, around 19 million Brazilians are starving, and around 55.2% of households, which corresponds to 116.8 million people, are experiencing some food shortage (Agência Brasil 2021). The struggle for the right to housing and the city becomes more pressing when considering the deaths, the 15 million unemployed and the 76.5 million outside the workforce (Agência Brasil 2021). Even without data, we must consider those who are working as 'self-employed'—many of them in the digital platform services sector. For them, the city is the factory floor: they use their own work tools, have no formalised relationship and have come to prominence during the pandemic due to the importance acquired by their work. For workers in an irregular situation, the housing issue is even more problematic, since it is not even possible to observe the relationship between minimum wage and housing, as is done with those who are working under a formal contract.

As shown throughout the text, life is insecure for a huge portion of the urban population. The COVID-19 pandemic lifted the veil that prevented some of the contradictions from being seen, such as the dismantling of the historic achievements of the working class and the principles of the right to housing. However, there is still a lot to be done in order to actually achieve housing as a right. Housing is a necessity and at the same time a commodity that needs to be paid for: a commodity that increases in price with production from and in the city—a production achieved through work in general, but under a system in which, with their wages, workers have no way of accessing decent housing.

Is the commoditisation of land/housing contradictory to the premise that it is a human right? In the 1970s, Florestan Fernandes stated that, in relation to the perpetuity of oppression from the beginnings of capitalism in Brazil to the detriment of the reproduction of social life and noting its permanence: "it seems incredible that such a systematic kind of oppression can exist today; and even more, that it and the terrible mechanisms of repression, which it needs to resort to, may be reconciled with the egalitarian ideals—of respect for the human person, for the fundamental rights of man" (Fernandes 2020, p. 372, translated). Fifty years later, we can see that the disrespect for the fundamental rights of humans has only increased. Among the fundamental rights, housing plays a key role in the reproduction of the workforce, but also in the reproduction of capital. It is essential that housing is recognised as a right, because it emphasises that the house and the city must be considered with primacy given to their usage. It is a matter of fundamental rights that, although theoretical, reinforce the guarantee of life and reveal that workers must guarantee their own reproduction—and that, in order to do this, they have only insufficient wages.

Is it contradictory for a commodity to be a right? Everything in capitalism is a commodity, so it is necessary to clarify that, even though it is a commodity, it is a vital need that must be served. It is a contradiction arising from the fact that, under capitalism, labour power is a commodity and the wages earned from work allow for the purchase of other commodities necessary for survival and reproduction.

The principle of housing as a right allows for the coordination of movements that fight for an end to forced evictions, for the production of affordable housing, for continuity of residence in occupied areas and for less unfair jobs and wages. The right to housing, as an integral part of the world of work and the production and reproduction of urban space, makes it possible to commit to the right to the city in its fullest potential. The fight for the right to housing and the city is a way to promote the empowerment of workers in relation to property and to challenge the entrenchment of the divine right in relation to land/housing ownership. Having a roof to serve as a shelter is an inalienable right in the urban space.

References

Agência Brasil (2020) Pesquisa revela que 19 milhões passaram fome no Brasil no fim de 2020. Rio de Janeiro, 06 Apr 2021. Geral. Retrieved on October 31, 2021 https://agenciabrasil.ebc.com.br/geral/noticia/2021-04/pesquisa-revela-que-19-milhoes-passaram-fome-no-brasil-no-fim-de-2020

Bastos RD (2020) Superespeculação imobiliária: a elevação excepcional e generalizada dos preços dos imóveis urbanos. In: Suave Â (ed) O urbano em questão: Reflexões Críticas. Liber Ars Press, São Paulo, pp 1–25

Brazil (1936) Law 185 of 14 January 1936. Establishes commissions on the minimum wage. Official Gazette of the Union: Section 1, Brasília, DF, 21 Jan 1936. Retrieved on October 31,

2021 https://www2.camara.leg.br/legin/fed/lei/1930-1939/lei-185-14-janeiro-1936-398024-publicacaooriginal-1-pl.html
Brazil (1986) Decree-law 2.291 of December 1986. Dissolves the national housing bank (BNH) and takes other measures. Official Gazette of the Union: Brasília. DF, 24 Nov 1986
Brazil [1988 Constitution] (1988) Constitution of the federative Republic of Brazil. Official Gazette of the Union: Brasília, DF, 05 Oct 1988
Brazil (1995) Law nº 9.249/95. Changes the legislation on corporate income tax, as well as the social contribution on net income, and other measures. Official Gazette of the Union: Brasília, DF, 27 Dec 1995
Brazil (2001) Law nº 10.257/2001. City Statute. Official Gazette of the Union: Brasília, DF, 10 July 2001
Brazil (2004) Ministry of cities. National urban development policy. Booklets. Urban Development. MCidades. Brasília, DF
Brazil (2005) Law nº. 11.124 of 16 June 2005. Provides for the national social interest housing system (SNHIS), creates the national social interest housing fund (FNHIS) and institutes the FNHIS management council. Official Gazette of the Union: Brasília, DF, 17 June 2005
Brazil (2009) Law 11.977 of 7 July 2009. Provides for the Minha Casa, Minha Vida Programme (PMCMV) and the regularisation of land titles for settlements located in urban areas, and other measures. Official Gazette of the Union: Brasília, DF, 08 July 2009
Brazil (2016) Constitutional amendment no 95 of 14 December 2016. Amends the transitory constitutional provisions act, in order to institute the new fiscal regime, and other measures. Official Gazette of the Union: Brasília, DF, 16 Dec 2016
Brazil (2017) Law 13.465 of 11 July 2017. Provides for the regularisation of rural and urban land, and other measures. Official Gazette of the Union: Brasília, DF, 08 Sep 2017
Brazil (2019) Decree Nº 9.759 of 11 April 2019. Dissolves and establishes guidelines, rules and limitations for collegiate bodies of the federal public administration, 70ª edn. Official Gazette of the Union: Section 1—Extra, Brasília, DF, p 5, 11 Apr 2019
Brazil (2020) Ministry of the economy. Special secretariat for finance. Secretariat for assessment, planning, energy and lottery. Minha Casa Minha Vida Programme. Evaluation report. Brasília, DF, Dec 2020. Retrieved on October 31, 2021 https://www.gov.br/cgu/pt-br/assuntos/noticias/2021/04/cgu-divulga-prestacao-de-contas-do-presidente-da-republica-de-2020/relatorio-de-avaliacao-pmcmv.pdf
Brazil (2021) Law Nº 14.118 of 12 January 2021. Establishes the Casa Verde e Amarela [green-and-yellow house] Programme. Official Gazette of the Union: Brasília, DF, 13 Jan 2021
Campanha Despejo Zero (2020) Despejo zero: em defesa da vida no campo e na cidade. Retrieved on October 31, 2021 https://www.campanhadespejozero.org/
Chaui M (2012) Democracia e sociedade autoritária. Comunicação Informação 15(2):149–161. Retrieved on January 21, 2021 from https://revistas.ufg.br/ci/article/view/24574/14151. Âncora Publications, Coimbra
de Oliveira F (2003) Crítica à Razão Dualista: o ornitorrinco. Boitempo Press, São Paulo
DIEESE (2021) Departamento Intersindical de Estatísticas e Estudos Socioeconômicos. Salário Mínimo nominal e necessário. Retrieved on October 31, 2021 https://www.dieese.org.br/analisecestabasica/salarioMinimo.html#2021
Engels F (1995) The housing question. Pathfinder Pr, Atlanta (original work published 1872)
Estadão (2021) Venda de imóveis de luxo dispara, com pandemia, juro baixo e dólar alto. Retrieved on October 31, 2021 https://economia.estadao.com.br/noticias/geral,venda-de-imoveis-de-luxo-dispara-com-pandemia-juro-baixo-e-dolar-alto,70003681835
Fernandes F (2020) A revolução Burguesa no Brasil. Ensaio de Interpretação Sociológica. Contracorrente Press, São Paulo (original work published 1974)
FIPE-ZAP—Fundação Instituto de Pesquisas Econômicas (2021) Fipezap index. Fipe. Available at https://www.fipe.org.br/pt-br/indices/fipezap/. Accessed on 25 May 2021

Fundação Getúlio Vargas (2020) IGP-M acelera para 2.09% em dezembro de 2019, aponta FGV IBRE. Retrieved on October 31, 2021 https://portal.fgv.br/noticias/igp-m-acelera-209-dezembro-2019-aponta-fgv-ibre
Fundação João Pinheiro (2020) Déficit Habitacional no Brasil—2016–2019. FJP Press, Belo Horizonte
Gramsci A (1988) Maquiavel, a Política e o Estado Moderno. Civilização Brasileira Press, Rio de Janeiro
Habitat I Agenda (1976) Report of the United Nations conference on human settlements. The Vancouver declaration on human settlements. Report. Vancouver, Canada
Habitat II Agenda (1996) Report of the United Nations conference on human settlements (Habitat II). Istanbul, 3–14 June 1996. Retrieved on October 31, 2021 https://www.un.org/ruleoflaw/wp
Harvey D (1982) Thelimits to capital. Basil Blackwell Press, London
Harvey D (2014) Seventeen contradictions and the end of capitalism. Oxford University Press, Oxford
Heller A (1978) Teoria de las necessidadesem Marx. Península Press, Barcelona
Hirsch J (2010) Teoria Materialista do Estado. Revan Press, Rio de Janeiro
Holston J (2007) Insurgent citizenship: disjunctions of democracy and modernity in Brazil. Princeton University Press, Princeton
IBGE—Instituto Brasileiro de Geografia e Estatística (2012) Quase dois terços das favelas estão a menos de dois quilômetros de hospitais. IBGE—2021 census. Retrieved on October 31, 2021 https://censo2021.ibge.gov.br/2012-agencia-de-noticias/noticias/27728-quase-dois-tercos-das-favelas-estao-a-menos-de-dois-quilometros-de-hospitais.html
IBGE—Instituto Brasileiro de Geografia e Estatística (2019) Estimativa de domicílios ocupados para operação do Censo Demográfico. In: IBGE. Aglomerados Subnormais 2019: classificação preliminar e informações de saúde para o enfrentamento à COVID-19. Retrieved on October 31, 2021 https://biblioteca.ibge.gov.br/visualizacao/livros/liv101717_apresentacao.pdf
Jaramillo GS (2009) Hacia umateoria de la renda del suelo urbano. EdicionesUniandes, Bogotá
Mascaro AL (2013) Estado e Forma Política. Boitempo Press, São Paulo
OXFAM Brazil (2017) Nós e as Desigualdades 2017. Research. Oxfam Brazil, São Paulo, Retrieved on October 31, 2021 https://www.oxfam.org.br/um-retrato-das-desigualdades-brasileiras/pesquisa-nos-e-as-desigualdades/pesquisa-nos-e-as-desigualdades-2017/
PNAD. Pesquisa Nacional por Amostra de Domicílios Contínua (PNAD) (2020). Retrieved on October 31, 2021 https://www.ibge.gov.br/estatisticas/multidominio/condicoes-de-vida-desigualdade-e-pobreza/17270-pnad-continua.html?=&t=o-que-e
Poder 360—Newsletter (2021) Distribuição de lucro de empresas aumenta mais que salário. Retrieved on October 31, 2021 https://www.poder360.com.br/economia/distribuicao-de-lucro-de-empresas-aumenta-mais-que-salario/
Poulantzas N (1986) Poder Político e Classes Sociais. Martins Fontes Publications, São Paulo
Rodrigues AM (2000) A utopia da Reforma Urbana e a Habitat II. In: de Souza ÁJ (ed) Cidadania e Globalização. Saraiva Press, São Paulo, pp 183–190
Rodrigues AM (2004) Estatuto da Cidade: função social da cidade e da propriedade. Alguns aspectos sobre população urbana e espaço. Cadernos Metrópoles 12(1):9–25. Retrieved on October 31, 2021 https://revistas.pucsp.br/index.php/metropole/article/view/8807
Rodrigues AM (2010) Conselho das Cidades: uma avaliação. Revista Terra Livre (AGB Nacional) 32(1):223–230. Retrieved on October 31, 2021. https://publicacoes.agb.org.br/index.php/terralivre/issue/view/32
Rodrigues AM (2014) Propriedade fundiária urbana e Controle socioespacial. Retrieved on October 31, 2021 http://www.ub.edu/geocrit/coloquio2014/Arlete%20Moyses%20Rodrigues.pdf
Rodrigues AM (2015) Política Habitacional e a Intervenção no Espaço Urbano. In: Ferreira Á, Rua J, Mattos RC (eds) Desafios da Metropolização do Espaço. Editions Consequência, Rio de Janeiro, pp 253–272
Rodrigues AM (2017) A moradia nas cidades brasileiras. Contexto Press, São Paulo (original workpublished 1988)

Rodrigues AM (2019) A propriedade da terra como um elemento da desigualdade territorial nas cidades brasileiras. In: Jacinto R (ed) Novas Fronteiras, Outros Diálogos: Cooperação e Desenvolvimento Territorial, pp 195–206
Secovi SP (2021) Departamento de Economia e Estatística. Report. Retrieved on October 31, 2021 https://www.secovi.com.br/pesquisas-e-indices/pesquisa-mensal-do-mercado-imobiliario. UN. Universal Declaration of Human Rights (General Assembly resolution 217 A). Retrieved on October 31, 2021 https://www.un.org/en/about-us/universal-declaration-of-human-rights
UN (1966) International covenant on economic, social and cultural rights. Retrieved on October 31, 2021 https://www.ohchr.org/en/professionalinterest/pages/cescr.aspx

Chapter 16
The Long March of the Brazilian Peasantry: Socioterritorial Movements, Conflicts and Agrarian Reform

Ariovaldo Umbelino de Oliveira

The Red flag waved
The people are taking a stand
Leave the fear of everything for later
Pull the knife, disarm your hand
Keep calm to fight
Unleash the invasion line
The reform is coming slowly
Landing in the river of reason
Shot from cows and oxen
The people are taking a stand
The people are taking direction.
Zé Ramalho (Song "Sem-Terra", Zé Ramalho, CD *Eu sou todo nós*, Rio de Janeiro, EMI, 1998.).[a]

Abstract The history which marks the long march of Brazilian peasants is written in the almost always bloody struggles of this social class. On covering it, I make clear that my understanding of the logic of capitalist development is based on the understanding that said development is performed in an unequal and contradictory manner. That is to say, I assume that the development of capitalism—and its subsequent expansion to the countryside—is produced unevenly, complexly and, therefore, plurally. This theoretical reference framework, therefore, is the opposite of that which views the uniform, total and absolute expansion of salaried work in the countryside with the founding characteristic of capitalism. Therefore, I believe that capital works with the contradictory movement of inequality in the process of its development. In the case of Brazil, capitalism is acting by simultaneously developing—in the sense of implementing salaried work—the production of several crops in different areas of

Ariovaldo Umbelino de Oliveira é professor titular do Departamento de Geografia da Faculdade de Filosofia, Letras e Ciências Humanas da Universidade de São Paulo (FFLCH-USP).

[a]Música "Sem-Terra", Zé Ramalho, CD *Eu sou todo nós*, Rio de Janeiro, EMI, 1998.

A. U. de Oliveira (✉)
Universidade de São Paulo, São Paulo, Brazil
e-mail: ariovaldo.umbelino@gmail.com

R. C. Lois González and M. A. Mitidiero Junior (eds.), *Brazilian Geography*, Advances in Geographical and Environmental Sciences,
https://doi.org/10.1007/978-981-19-3704-0_16

the country. This is also the case, for example, with sugar cane crops, reforestation and oranges, partially in Sao Paulo, soy in Mato Grosso, etc. On the other hand, this same capital develops peasant production in an articulated and contradictory manner. This means that I also assume that peasants are not social subjects outside of the scope of capitalism, but rather social subjects within it and, therefore, a social class of it. As such, I have broken capitalism down into four social classes: bourgeoise, proletariat, landowners and peasants, as well as, of course, indigenous people and the descendants of Quilombos.

Keywords Brazilian peasantry · Class struggles · Landowner · Quilombolas · Indigenous peoples

16.1 Principles and Assumptions

The history that marks the *Long March* of the Brazilian peasantry is written in the often-bloody struggles of this social class. In addressing it, I make it clear that my understanding of the logic of capitalist development is based on the understanding that it is done in an unequal and contradictory way. In other words, I assume that the development of capitalism—and its consequent expansion in the field—is made heterogeneous, complex and therefore plural. This theoretical reference framework, therefore, is in contrast to the one that sees the homogeneous, total and absolute expansion of wage labour in the field with a founding characteristic of capitalism.

In this way, I believe that capital works with the contradictory movement of inequality in the process of its development. In the Brazilian case, capitalism is simultaneously developing, in the direction of the implantation of wage labour in the field in several cultures, in different areas of the country. This also occurs, for example, with sugar cane, orange and soybean crops, as well as others. On the other hand, this same capital develops peasant production in an articulated and contradictory way. This means that I also assume that the peasant is not a social subject from outside capitalism, but a social subject from within it. In other words, in Brazil there are at least four social classes: bourgeoisie, proletariat, landowners and peasants.

Another important theoretical presupposition to be emphasised refers to the profiteering nature of capitalism in Brazil. This is to say that, in Brazil, the development of the capitalist mode of production is mainly achieved by fusing together the capitalist and the landowner into the same person.[1] This process, which originated in slavery, has been increasingly consolidated, since the transition from slave labour to free labour, particularly with the Law of the Earth and the end of slavery. However, it was in the second half of the Twentieth Century that this fusion expanded significantly.

Thus, the development of agriculture will not act in the direction of the transformation of landowners into capitalist entrepreneurs, but instead it has transformed

[1] J.S. Martins, in his works *O cativeiro da terra e o Poder do atraso*, both edited in São Paulo, by Editora Hucitec, develops this conception that I take as a reference.

the industrial and urban capitalists—especially the Central-South of the country—into landowners (*latifundiários*). The policy of fiscal incentives of the Development Superintendency of the Northeast and the Superintendency for the Development of Amazonia (Sudam) were the instruments of economic policy that made this fusion possible. Thus, urban capitalists became the largest landowners in Brazil, having areas with dimensions never recorded in the history of humanity. The most classic example is the famous Jari Project. The project was implemented by the multimillionaire Daniel K. Ludwig, and was "nationalised" at the end of the Figueiredo government, when it moved to a group of about 25 companies, led by the Azevedo Antunes group. After the creation and operation of the Executive Group for the Lower Amazon Region (GEBAM), the occupied area had a surface area of more than four million hectares.[2] This process revealed two contradictory aspects of these capitalists: The same automobile industry that practised the most advanced working relations of capitalism in the Central-South, in the Amazon, on the contrary, practised "peonage" in its agricultural properties, a working relationship also called "white slavery". In other words, the same company operated differently in different regions of this country.

In Brazil, this alliance has meant that, instead of the bourgeoisie acting in the sense of removing the obstacle (irrationality) that the private property of the earth brings to the development of capitalism, it acted in order to solidify, even more, the private property of the earth. It was as a result of this same alliance that in the Constituent Assembly of 1988, the only chapter of the Constitution to receive practically the unanimity of the votes of the representatives of these elites was the vote on Agrarian Reform. It should be noted that this did not occur in relation to other chapters of the Brazilian Constitution. Thus, the concentration of private land ownership in Brazil cannot be understood as an excrescence based on the logic of capitalist development. On the contrary, it is a constitutive part of the capitalism that is developed here. It is a capitalism that reveals its contradictory duplicity: One advanced in its backwardness and one delayed in moving backward. This is why I insist on the thesis that land concentration in Brazil has *sui generis* characteristics in world history. At no time in the history of humanity were there private properties the size of those found in Brazil. The sum of the area occupied by the 27 largest private properties in the country is equal to the total area occupied by the state of Sao Paulo. If that total is added to the area occupied by the 300 largest private properties in the country, it is equivalent to twice the total surface area of the same state.

The data for 1992, disclosed by Incra, showed that there were 3,114,898 rural buildings in Brazil and, among them, 43,956 (2.4%) with an area over one thousand hectares, occupying 165,756,665 ha.[3] Meanwhile, another 2,628,819 properties (84.4%), with an area of less than 100 ha, occupied only 59,283,651 ha (17.9%). Studies carried out revealed that if the Incra applied in full the precepts of Law 8.624, which defines what is productive and unproductive land in the country, there would

[2] Body directly linked to the National Security Council.

[3] *Atlas Fundiário Brasileiro*, Brasília, Incra, 1996.

be something around 115,054,000 ha (20% of the total area) as unproductive properties. The *Atlas FundiárioBrasileiro (Brazilian Land Atlas)* indicated that 62.4% of the registered real estate area was classified as non-productive and only 28.3% as productive. This information revealed, therefore, the contradiction represented by the private property of the land in Brazil, retained for non-productive purposes. Even in practice, the only social commitment that landowners had was to pay the rural territorial tax (RTI), but that is not the case.

These large land extensions are concentrated in the hands of numerous economic groups because in Brazil, they now function as a reserve of value, or as an equity reserve. That is, they function as collateral instruments for access to bank financing, or as a set of government incentive policies. Thus, we are facing a violently concentrated land structure as well as a capitalist development that has generated an enormous set of miserable people. The contradictory logic is one single logic: the capitalist development that concentrated the land, at the same time, has pushed an increasing share of the population into urban areas, generated in them an increasing mass of poor and miserable people. At the same time, however, this exclusion also reached the countryside itself. Most children of the peasants, whose properties are less than 10 ha, will certainly never be able to become peasants in the land of their parents. They will only have one option: the road—that is to say, the road that will take them to the city, or the road that will lead them to the struggle to reconquer the land.

Migration has thus been one of the main characteristics of the Brazilian population. The overall picture has shown, contrary to many interpretations, that the rural population grew in absolute and total terms by 1970, when it reached just over 41 million people. Recorded in the later demographic censuses, there was a decrease that caused it to reach 38.5 million in 1980; 35.8 million in 1991; 33.9 million, in 1996; 31.8 million in 2000 and in 2010, a total of 29.8 people lived in rural areas. However, now, in 2017, IBGE launched a new proposal that indicated that 76% of the Brazilian population resided in urban areas in 2010, while, according to the classification adopted in 2010, the total was 84.4%. Thus, we now have in 2021, approximately 51.3 million people living in the rural areas of the country. This was not only true in the border regions, but also in the core area of Brazilian capitalism, as is the case in the state of Sao Paulo. Many times, the new rural settlements derived from Agrarian Reform are at the root of this process. Moving around the country in a true migrant adventure, as Mr. Pedro Casaldáliga said, the Brazilian peasants, in their own way, were inserted into the countryside.

The data available from the IBGE agricultural census reveal its general situation and importance today. The agricultural establishments grew from 1950 (2,064,642) to 2017 (5,073,324), that is, the establishments increased by 3,008,682, an increase of 145.7%. In 1970, the total of 4,924,019 was reached in number, after which, there was a shift from 5,801,809 in 1985 to 4,859,865 in 1995–96.

The occupied area had also increased significantly from 232,211,106 ha in 1950 to a total of 351,289,816 ha in 2017, which means that they had an increase of 119,078,710 ha, or 51.3%. In these nearly 70 years, there has been an increase to 374,924,929 ha in 1985, and 333,680,037 ha in 2006. Thus, the data show that there is no relationship between the decrease in the number or occupied area of the establishments as the data have been updated in the last 70 years. On the contrary, there was an increase in the number of cases in Brazil and in the occupied area. This is the case for small establishments, i.e. with an area of less than 100 ha, which increased in number (1,763,491 establishments in 1950, and 4,524,385 establishments in 2017). In other words, the number of establishments in the country increased by 156.6% (Figs. 16.1 and 16.2).

The country had only one change in the total number of establishments that shifted downward in the last 50 years, from 5,225,162 in 1985, down to 4,318,861 in 1995–96. In the past 70 years, medium establishments have also increased from 268,159 establishments in 1950 to 420,719 establishments in 2017. Thus indicating a total increase of 56.9%. Large establishments also increased from 32,628 establishments

Fig. 16.1 Brazil: number of establishments

Fig. 16.2 Brazil: occupied area

in 1950 to 51,203 establishments in 2017, which is to say, they had an increase of 57% (Fig. 16.3).

As for the occupied area, there was a total of 38,588,119 ha in 1950 for those establishments (fewer than 100 ha), while in 2017, they reached an area of 71,804,615 ha, which is an increase of 86.1%. This growth was most notable between 1950 and 1970, when it rose from 38,588,119 to 69,153,199 ha. Then there was a change from 79,551,798 ha in 1985 to 70,575,779 ha in 1995–96 (Fig. 16.4).

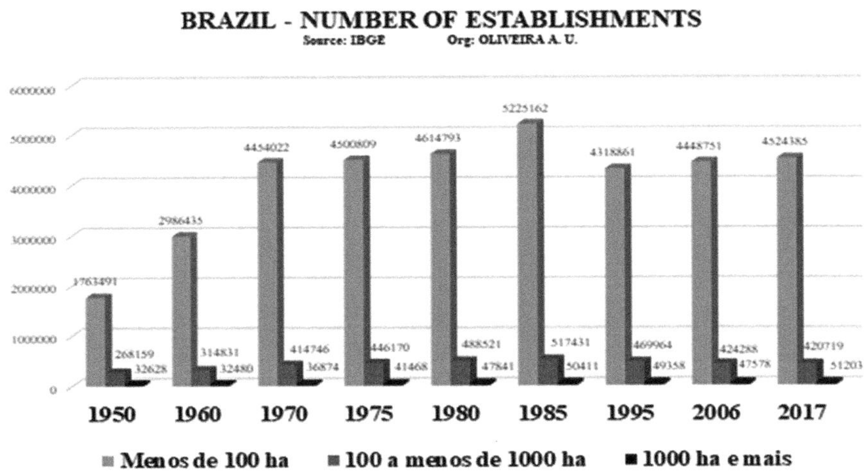

Fig. 16.3 Brazil. Number of establishments

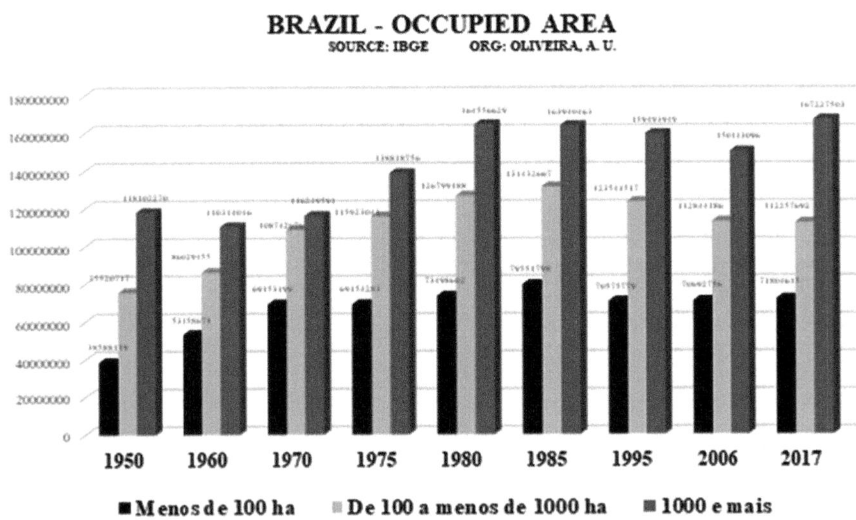

Fig. 16.4 Total Occupied Area, according to the surface of the exploitation

Meanwhile, medium establishments (from 100 to 1000 ha) rose from 75,520,717 ha in 1950 to 112,257,692 ha in 2017, which is an increase of 48.6%. They had a change from 69,153,199 ha in 1970 to a total of 131,432,667 ha in 1985. Finally, the same was true for large establishments (1000 ha and more), rising from 118,102,270 ha in 1950 to 167,227,503 ha in 2017, which is an increase of 41.6%.

In contradiction, this process did not occur only because of the crisis experienced by Brazilian agriculture, but above all because of the processes derived from the growth of the struggle for land, which increased significantly in the country. Moreover, they increased greatly from the year 1998, when they reached 152 conflicts, or 0.42 conflicts/day, to 4.32 conflicts/day in 2020, that is, more than four conflicts per day in the country.

The data of land conflicts in Brazil clearly showed that if we compare the period from 1985 to 2000, we will find a significant drop, represented by the data of 636 conflicts in 1985 and 168 in 2000. Then there was an increase in the number of conflicts that reached 777 conflicts in 2005 and culminated in a fall in 2008 when it reached 459 conflicts. Since then, there has been a new period of increasing conflict, which reached a total of 1576 conflicts in the year 2020 (Fig. 16.5).

It is important to remember that if 2008 is taken as a reference, we will find coinciding behaviour of the curves of the number of land conflicts, occupations/take-backs and new camps. However, after 2008, there was a 12-year sequence, in which there was a lag of data on land conflicts, occupations/take-backs and new camps. This is because the first has an upward curve of 459 conflicts per land in 2008, to 1576 conflicts in 2020. Whereas, the two curves of occupations/take-backs and new camps plummeted downward, reaching the miniscule numbers of 29 occupations/take-backs and five new camps in 2020.

Fig. 16.5 Brazil: number of land conflicts 1985–2020

This oscillation in data was a reflection of the process that occurred with the movement of Landless Rural workers (*TrabalhadoresRurais Sem Terra*—MST), which had a victory at the 5th Congress held in Brasilia, for the faction represented by those who wanted to see the movement marching towards socialism, as a first battle. At that point, the MST's flag of struggle was no longer the struggle for agrarian reform, and became the struggle for socialism, and everything "hidden". That is why the MST began to look for militants among the children of the enlisted Landless Rural workers, to base part of the struggle in the youth, as it mounted the Rise Up movement (Movimento Levante), a faction to dispute policies in the student movement. Moreover, it went on to hold the University Day in Defence of Agrarian Reform (JURA) to commemorate 17 April, the date of the massacre of Eldorado dos Carajá, as a kind of day of struggle for agrarian reform, which was now longer a reform, but a struggle for socialist agrarian reform. This is how the MST became a movement that fought for agrarian reform, and stopped participating in Brazilian peasants' struggle to conquer the land. Today, Brazilian peasants have followed their fate of not having an organised movement that understands their struggle, the struggle for the land, to continue to stand with a social subject in the Brazilian countryside. However, does that mean the MST is a movement that does not support the peasants' struggle? Indeed, it does support it. However, the peasants understand that they are not a social class, but rather a social category. They are part of the working class as it appears in the many texts of the current MST leaders. So that's where the knot is. This is where the difference between the MST and the peasants continues in their historic rage, fighting to be able to put themselves as a social subject, as a social class placed in agriculture among the others, both bourgeois and proletariat classes. That is why, I repeat, Brazil has at least four social classes: Bourgeoisie, proletariat, landowners and peasants.

Another fact that we can use in agricultural studies is the condition of the farmer. There was a numerical drop in the establishments run by lessees,[4] partners and squatters, who accounted for less than 30.6% of this fall between 1995–96 and 2006, and 32.1% between 2006 and 2017. This means that the number of lessees, partners or squatters decreased to 1995–96 from 1,497,752 to 1,040,032 in 2006, and to 706,376 in 2017. Thus, the total number of owners rose from 3,603,624 in 1995–96, to 4,135,604 in 2006, and in 2017 increased to 4,366,948. This means an increase of 14.8% between 1995–96 and 2006, and 5.6% between 2006 and 2017. These data represent an increase of 70.7% in 1995–96, 79.9% in 2006 and 86.0% in 2017, among landowners. This fact reveals that the social pressure exerted by the social movements for the struggle for Agrarian Reform has led the owners to no longer cede their land to the lessees, partners or squatters. In the 4.3 million establishments with an area of up to 100 ha in 1995–96, about 88% of the staff was of family origin, that is to say, paid employment accounted for only the remaining 12%. This is an opposite and

[4] Lessees: individuals who take something on lease.Partners: the contract by which one of the parties (owner) transfers land to another, so that it can cultivate it, dividing the profits in the proportion stipulated, the average, a third, a quarter, a percentage, etc.Squatters: who occupies an abandoned or returned land to cultivate it.

contrasting reality with that of establishments of more than one thousand hectares, where employed labour accounted for 81%.

Among the 4.4 million establishments of family origin, in 2006, there was a total of 74.38% of employed persons of family origin, that is to say, paid employment accounted for only the remaining 25.62%. The 2017 Census of Agriculture presented an increase in paid labour in relation to family labour when it reached 67%, leaving employed labour at 33%. Thus, there has been a relative increase in wage labour compared to family work. In 2006, there were 16,568,265 employed staff, and in 2017, the number reached 15,105,125 people; family staff reached 12,323,110 people in 2006, while employed staff accounted for 4,245,095 people in 2006. In the meantime, the family staff amounted to 10,115,559 people in 2017, and employed staff reached 4,989,566 people. Thus, there was a decrease in the number of family employment of 2,207,551 people, that is, 21.8%. The total number of employed persons decreased by 1,463,080 people, or by 9.7%, while the salaried individuals increased by 744,456 people, or 17.5%.

In Brazil of the Twentieth Century, this structural combination defined the Brazilian countryside: In peasant units, the predominance of family employment; in capitalist units, the dominant presence of salaried work. In 1995–96, peasant agriculture in Brazil occupied an area of 70.5 million hectares (18%), and agriculture has been building its place in Brazilian society. In 2006, it reached 80.1 million hectares (24%), and in 2017, it reached 80.9 million hectares (23%). Thus, although the data indicated a drop in percentage, it was in fact an absolute increase of 9.6 million hectares between 1995–96 and 2006, and between 2006 and 2017 an increase of 0.8 million hectares. These gains from peasant agriculture in the country have been obtained through access to rural credit, although it was difficult, because in 1995–96, only 5% had access to it, leaving only 30% of the total. In 2006, the result was 15.1% had access to rural credit, accounting for 29.93% of total credit. In 2017, the result was 11.9% that had access to rural credit, with 19% of the total.

As for technology, in 1995–96, the picture was no different, since only 10% had a tractor, 38% used fertilisation and 1% had combines. In 2006, the situation was as follows: 8.6% had tractor, 31.9% used fertilisation and 1.6% had combines. Finally, in 2017, 8.8% had a tractor, 38.5% used fertilisation and 1.2% had combines. However, the use of pesticides reached more than 33.7% of the establishments in 1995–96. In 2006, the use of pesticides reached 23.98% of the establishments, and in 2017, the use of pesticides reached 25.5% of the establishments.

Even so, in 1995–96, these stubborn peasants were responsible for more than 50% of the production of English potatoes, beans, tobacco, cassava, tomatoes, agave, tree seed cotton, banana, cocoa, coffee, cashew, coconut, guarana, black pepper, grape and the absolute majority of horticultural products. They also produced more than 50% of pig farming, poultry, eggs and milk. In 2006, peasants were responsible for producing more than 50% of the pineapple, pumpkin, garlic, peanuts, onions, fava, beans, tobacco, sesame, jute, malva, castor bean, cassava, watermelon, ramie, sorghum broom, tomato, avocado, açai, acerola, agave, tree seed cotton, plum, blackberry, banana, cocoa, coffee, cashew, kaki, star fruit, coconut, cloves, mate-grass, fig, custard apple, guava, soursop, guarana, jabuticaba, jaca, mountain apples, kiwi, lychee, lime, lemon, laurel, mango, passion fruit, nectarine, quenching, pear, peach, pepper, Surinam cherry, pomegranate, tangerine, urucum, grape, peach palm and cupuaçu. In 2017, peasants were responsible for producing more than 50% of the pineapple, pumpkin, garlic, peanuts, onion, fava, tobacco, mauve, castor bean, cassava, watermelon, ramie, sorghum broom, avocado, açaí, acerola, agave, tree seed cotton, plum, blackberry, atemoya, olive, banana, cocoa, coffee, cashew, kaki, star fruit, clove, mate-grass, fig, custard apple, guava, soursop, guarana, jabuticaba, jaca, mountain apples, kiwi, lychee, lime, laurel, mango, passion fruit, nectarine, quenching, pear, peach, pepper, dragon fruit, Surinam cherry, pomegranate, tangerine, urucum, grape, peach palm and cupuaçu.

It is also worth remembering that in 1995–96, the medium establishments (100 to 1000 ha) and the large ones (more than 1000 ha), occupying 283 million hectares (82% of the total), accounted for more than 50% of the production volume of tree seed cotton, rice, sugar cane, corn, soybean, wheat, Indian tea, orange, apple and papaya. In 2006, the medium and large establishments, occupying 263 million hectares (78.8% of the total), accounted for more than 50% of the production volume of tree seed cotton, rice, white oats, English potatoes, sugar cane, rye, barley, rape, pea, sunflower, melon, maize, soy, sorghum, wheat, Triticale, rubber, Indian tea, palm oil, orange, apple and papaya. In 2017, the medium and large establishments, occupying 279 million hectares (79.6% of the total), accounted for more than 50% of the production volume of cotton in herbaceous stone, garlic, rice, white oats, English potatoes, sugar cane, rye, barley, rape, pea, beans, sesame, sunflower, flax, melon, maize, soy, sorghum, tomato, wheat, triticale, rubber, Indian tea, coconut, palm oil, orange, apple and papaya.

The same reality appeared in the data on the value of agricultural production in 1995–96, since the units with an area of up to 100 ha produced 46.5% of the total, i.e. 18% of the agricultural area generated almost half of the wealth coming from the field. Meanwhile, establishments with more than 1000 ha produced only 21.2% of the production value, although they occupied 45.1% of the total area. In 2006, the data were as follows, since the units with an area of up to 100 ha produced 43.35% of the total, that is, 21.9% of the agricultural area generated almost half of the wealth coming from the field. Meanwhile, establishments with more than 1000 ha produced only 28.3% of the production value, although they occupied 45% of the total area. In 2017, the data were as follows, as the units with an area of up to 100 ha produced 31.1% of the total, i.e. 20.4% of the agricultural area generated almost half of the wealth coming from the field. Meanwhile, establishments with more than 1000 ha produced 41% of the production value, although they occupied 47.6% of the total area.

It is because of this set of reasons that the peasants stubbornly struggle in Brazil on two fronts: first, to enter the land and become landowners; and second, the struggle to remain on earth as farmers of foods that are fundamental to Brazilian society. They are therefore a class in constant struggle, because the various governments have not considered them in their public policies. For this reason, the struggle for land developed by peasants in Brazil is a specific struggle, a particular characteristic of the Twenty First Century. I understand that the last century was, par excellence, a time of formation and consolidation of the Brazilian peasantry as a social class.

Thus, these peasants are not obstacles to the development of productive forces, preventing the development of capitalism in the field; on the contrary, they have virtually never had access to the land, and are therefore unearthed, "landless". They struggle to gain access to the land. It is within these contradictions that the social movements of struggle for the earth have emerged, bringing conflicts and violence along with it.

16.2 A History of Fighting and Violence

The social conflicts in the Brazilian countryside and its unique brand—violence—are not only exclusive to the Twentieth Century. They are constant marks of the development and occupation process of the country. Indigenous peoples were the first to know this process. For more than 500 years, they have been subjected to a truly historical ethno/genocide. The capitalist territory in Brazil has been the product of the conquest and destruction of indigenous territories. As already mentioned in other articles, this struggle between the indigenous nations and European capitalist society previously,[5] and with national *versus* international characteristics, for a time at the end of the Twentieth Century, and still today, and in the midst of this world fusion that is beginning to be consolidated, it never ceased in the history of Brazil. The indigenous people have been cornered. They fought, fled, died, and today they are resurrecting in their struggle. In their escape, they left a migration route, clashes between peoples and new adaptations. The Amazon is certainly their last stronghold. However, the globalised capitalist Brazilian society insists on their capitulation. The indigenous "reserves"—fractions of capitalist territory to imprison the indigenous free territory—are demarcated, and often disregarded. At the same time as the indigenous peoples' struggle, the black slaves' struggles against the profiteering farmers were born. From these struggles and the escape of slaves, the Quilombos were born—the real land of freedom and the work of all within the colonial capitalist territory. Many Quilombolas died as a result of the real war promoted by the slave lords. Today, belatedly, Brazilian society has begun to recognise the rights of the remaining Quilombola lands.

The squatters are another part of the landless peasants, who have historically been fighting on one end against their expropriation, and on the other, against the gunmen, "plantation gendarmes" of speculator landowners and land invaders. They have had many movements: Canudos, Contestado, Trombas e Formoso are part of these many stories of the struggles for land and freedom in the Brazilian countryside. They are also memories of the capacity for resistance and social construction of these expropriated people in the search for a portion of the territory and memories of the destructive capacity of capital, capitalists and their repressive governments.

[5] A.U. Oliveira, A geografia das lutas no campo, 10 ed., São Paulo, Contexto, 2001.

In the 1950s and 1960s, Peasant Leagues shook the north-eastern camp and gained national attention, but many of their leaders were murdered. The Confederation of Agricultural Workers (Contag) was created and the government of John Goulart began a process of agrarian reform, creating SUPRA. However, the violence of the 1964 military coup suffocated the yearning for freedom by the residents on the lands of the armed landowners of the Brazilian North-east and many the landless peasants that the coffee crisis and the beginning of industrialisation were generating. The military extinguished SUPRA and created the Brazilian Institute of Agrarian Reform (IBRA), but they never carried out agrarian reform, even after enacting the Land Statute in November 1964.

Thus, violence has been the main characteristic of the struggle for land in Brazil. The data in the following chart on the number of killings in conflicts in the countryside show that this violence seems to have no end. In the period between 1964 and 1971, most of the deaths occurred in the northeast region, in the landowners' intention to slow down, by violence, the ideals sown by the Peasant Leagues. Since 1972, the largest number of murders in the countryside were in the north region, although they were also present in the north-east, central-west, south-east and south. The 1970s was marked above all by the struggle of the peasant farmers in the northern region. The military government with its territorial policy focused on tax incentives for entrepreneurs on the one hand and on the other, also in the north region, colonisation as an alternative to agrarian reform in the regions of former occupation (north-east, south-east and south), which created the scenario for violence. To gain access to tax incentives, businessmen had to implement their agricultural projects in the region, which was occupied by the Indians, Quilombolas and peasants. Indigenous peoples were subjected to genocide or ethnocide. The squatters were no better off: Either they were pushed into new areas on the expanded border, or they were expelled from their possessions and migrated to cities that were established in the region.

The states that received public colonisation projects were Pará, Rondônia, Acre, Roraima and, in part, Mato Grosso, which was characterised by the presence of private colonisation. In this way, Indians, squatters, colonists and land invaders became characters of the conflicts. The Missionary Indigenist Council (CIMI) was born in defence of the Indians, and the Pastoral Commission of the Land (CPT) was established for the squatters and settlers. The violence, which turned indistinctly against the squatters, settlers and Indians, also extended to their defenders: priests, pastoral workers, lawyers and trade union and non-trade union leaders (Fig. 16.6).

Special attention must be paid to the growth of violence in the 1980s, resulting from the increase in the social pressure made by the peasants in their struggle for land. The so-called modernisation of agriculture was having the opposite effect. To counteract the "conservative modernisation", the peasants increased their struggle for land. Civil society moved towards political openness. Anistia (Amnesty), DiretasJá (Rights Now), the formation of the Central Única dos Trabalhadores (CUT—Single Workers' Centre), Partido dos Trabalhadores (PT—Workers' Party) and other left-wing parties opened up support fronts for the struggle waged by the peasants. The National Conference of the Brazilian Bishops (CNBB) placed the question of land at the centre of the 1980 Fraternity Campaign: "*Land of God, land of brothers*". A

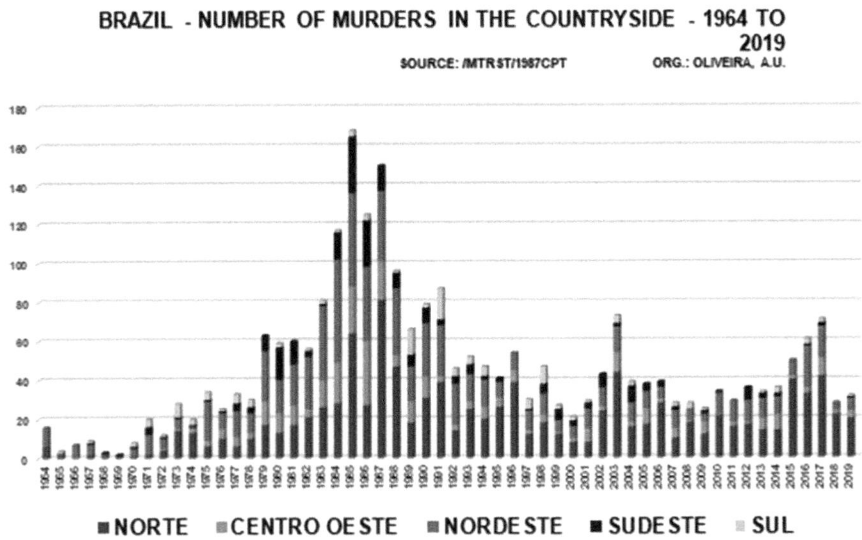

Fig. 16.6 Brazil: number of murders in the countryside 1964–2019

document on the land was produced to support discussion in the Ecclesiastic Base Communities (CEBs). The discussion about the poverty situation that the majority of the population was living in was discussed in the poor outskirts of Brazilian cities. In the CEBs and CPT, a set of community leaders began to discuss their future and their utopias. The conquest of the land was one of them. Thus, with the increase in social pressure, the violence of the landowners also grew, which at that time was practised as an extreme resource to retain the capitalist private property of the land.

The Land Statute, made into law by the military regime, was a dead law. Colonisation in the Amazon appeared as an authentic counter-reform. As Octavio Ianni wrote, after 20 years, the military did not even allow the Statute to come out of a national plan for agrarian reform. It was the "New Republic" that was entrusted with this historical mission, without, however, obtaining the support of its allied base in the political parties of the PMDB and PFL, or the landowners. Finally, it was forgotten that for the Statute to be made into Law, the gap controlled by the profiteering speculators had to be overcome. More than that, the gap was controlled by the alliance between the national sectors of the globalised and now territorialised capital.

The statistics of the dead in the land battles were growing, doubling, tripling, quadrupling. The UDR (Ruralist Democratic Union) was born under the leadership of Doctor Ronaldo Caiado, an entity that joined the landowners in defending their property and in forming a fund to elect constituent congressmen to defend their interests in the Constitution. They won and made the agrarian reform chapter a more limited legal text than the Land Statute itself.

However, if violence generates death, it also generates forms of fighting against death. The movement of the Landless Rural Workers (MST) was born as a product of this contradiction. As I stated in my book *A geografia das lutas no campo* (The

Geography of Rural Battles), the denial of expropriation was no longer exclusive to distant squatter migrants. It came to be designed, articulated and executed in the city, with the presence of the migrants whom the city/society insisted on denying the right to citizenship. Direito agora (Rights now) was built and conquered in the struggle to recapture the territory, lost in the historical trajectory of expropriation/re-appropriation.

Occupation, camps and settlements became the new forms of struggle for those who were already fighting or who had decided to fight for the right to free land and free work.[6] The land that will allow workers and peasants—owners of the time that capital stole and builders of the community and/or collective territory that capital failed to hold back or due to pressure—to reimpose/reproduce within the territory of general capitalist reproduction. In the occupations, camps, peasants, peons and *boias-frias* (rural workers) found the political forging of a historical alliance out of their needs and struggles. More than that, the transformation of the organised action of the new leaders opened up new prospects for the workers and peasants. Rural strikes in the city to seek social conquests in the countryside are still components located in the Brazilian countryside, an unequivocal sign that these workers and peasants, in spite of everything, still struggle.

This moment lived by capitalist development in Brazil is fundamental to the countryside, because the bases for its industrialisation are being made. What everyone observes is capital acting as a compressor roll, crushing everything in the direction of accumulation and expanded reproduction. It is in the contradictory logic of this direction that social conflicts and the struggle for land in Brazil should be understood. The recent occupation of the Amazon is therefore a synthesis and antithesis of this violent process. If the opening of ownership by the squatter peasant derives from the conscious denial of proletarianisation, then colonisation has been the relief valve of the pressures that the concentration and the rewrapping of the earth bring with it, but the reality of the Amazon forest and the lack of public policies to tie men to the earth make a profit. The returned colonists were studied by José Vicente Tavares dos Santos. The pressure that capital exerts in one place is not the same in other places, partially releasing portions of the territory of these shares. From this pressure and counter pressure, the movement of landless rural workers was born, a peasant movement that crossed into the third millennium.

[6] Occupation: taking or being in possession of land without authorization from the landowner;Camps: land where there are tents or tents for temporary accommodation of people; Settlement: is a set of agricultural units, installed by Incra in a rural property. Each of these units, called parcels or lots, is intended for a family of a farmer or rural worker without the economic conditions to acquire a rural property.

16.3 "He Who Knows the Time, Does Not Wait for it to Happen"[7]

This was the main reason for the urgent need to understand the peasant struggle for the land, within which it was undeniable that the movement of landless rural workers should take a prominent place. The MST, as the most organised rural social movement at the end of the Twentieth Century and the beginning of the Twenty First Century, represented, in the whole of the recent history of this country, another step in the *Long March* of Brazilian peasants in their daily struggle for land. This peasant struggle revealed a new side to anyone interested in agrarian matters. It was not a struggle to leave the land; rather, it was a struggle to enter the land. Land that had been kept unproductive and privately appropriated to serve as a value reserve and/or equity reserve for the dominant classes. It was, therefore, a struggle by the expropriated people, which, most of the time, experienced urban or rural proletarianisation, but decided to build the future based on the denial of the present. It was not, therefore, a struggle that only revealed a new option of life for this poor part of Brazilian society; more than that, it revealed a strategy of struggle believing that it is possible today to build a new society. A new society with justice, dignity and citizenship.

Thus, this contradictory struggle did not exclude even the interior of the state of Sao Paulo, where the development of capitalism ended its most spectacular expansion in recent decades. For this very reason, that part of the urban and rural proletarian peasant workers began to deny this condition. As a product of this denial, they organised themselves to fight for a piece of land, so that they could regain the autonomy to work that they had lost and now reconquered in the collective experiences in the countrysides conquered in the struggle.

The profound transformations through which Brazilian agriculture passed at the end of the Twentieth Century and the beginning of the Twenty First Century revealed the contradictions present inside the agrarian structure, and its contemporary component: The struggle for agrarian reform. More than that, it revealed the organic relationship between the struggle for land and the conquest of democracy by those excluded. The conquest of democracy that was consumed in the conquest of the land, in the conquest of its peasant identity, in the end, in the conquest of citizenship.

As I mentioned in my article *MST: terra, sobrevivência e inclusão social* (*MST: Land, survival and social inclusion*), it was within these processes of the struggle for land that the MST was born.[8] However, it is important to emphasise that the MST was not the only social movement in the history of Brazil, and it was not currently the only one in the Brazilian countryside, emphasising that there was a large number of fighting movements in the Brazilian countryside. It is sufficient to remember the struggle of indigenous peoples by the demarcation of their territories; the struggle for the land of work carried out by the squatters taking place in various parts of the country; the fight of peons against peonage ("white slavery"). This process took

[7] Verse by Geraldo Vandré in the song *Caminhando ou Para não dizer que não falei de flores.*

[8] MST: terra, sobrevivência e inclusão social, em Milton Santos, *Cidadania e globalização*, São Paulo, Saraiva, 2000.

place in the Amazon, especially in large farms, but also in the reforestation areas of the central-west, south and south-east of Brazil. There was also the struggle of the peasants against land expropriations for the execution of the great works of the State. It was noted that the Movement of the People Affected by Dams (MAB) was born exactly from this process of struggle of the peasants against these expropriations and, in particular, against the derisory value for their remuneration. There was the movement of the peasants against the subordination practised by the industry in the poultry sector; the movement of the tobacco farmers, who joined and rebelled against this subordination, which held them hostage to these industries that acquired their raw materials; The Brasiguayo and the Brasiliano movements, emphasising that more than 250,000 Brazilians were in Paraguay and more than 40,000 in Bolivia. Some of these Brazilians—especially the Braziguayos—were aligned and interwoven with the MST in Mato Grosso do Sul, and there were also occupations, camps and settlements that were products of this articulation. The Boias-Frias movement was also notable, in which strikes were held in the interior of the state of Sao Paulo as well as struggles for better working conditions. The CUT's affiliate Contag, the Grito da Terra Brasil (Cry for Brazil's Land) Movement, as well as the rubber tappers in the Amazon joined their struggles and the demarcations of the extractive reserves.

The MST was part of this struggle of the Brazilian peasantry and was certainly the most major of these movements, as it had a more solid organisation and was nationally recognised. It was the one forging the possibility of victory of the struggle of these different sectors that formed the heterogeneous Brazilian peasantry. The MST, therefore, was a young social movement, which was born in the early 1980s, whose action strategy was based on the occupation-camp-settlement logic. Anyone who wants to know and understand the MST must understand this process of struggle in the occupations and camps, and therefore, the struggles in the settlements. Thus, the MST was a movement that simultaneously articulated the spread of the struggle, combining it in contradiction with the territorialisation of this very movement in the settlements. It possessed and gave importance to its democratic, basic, effectively mass organisational structure, the organisational structure that respected the differences of these movements in various parts of the country, and that had a national collective representing the different regions where the movement was active. It was a distinct movement, because it respected the decisions made collectively. It was one of the few places in this country where disagreements of a particular concept or decision-making process were debated. However, once a proposal was won, it was embraced by all and put into practice by everyone. This practice, unfortunately, did not occur in political parties, nor did it occur in sectors of the trade union movement.

The MST, with this new component in its organisation, was born as a mass movement, fighting against the State not complying with the Agrarian Reform Law. One of the ways to understand it is to analyse its watchwords. When the MST was formed in the 1980s, the motto was "*Land for those who work on it*" (1979–83). When they began to face resistance to access the land, a new slogan emerged: "*Land is not won, land is conquered*" (1984). By strengthening and advancing, especially during the Sarney Government, and seeing that the first National Agrarian Reform Plan was not being implemented, the mottoes became: "*Without Agrarian Reform, there is no democracy*" (1985), and "*Agrarian reform now*" (1985–86). With the increase in violence, which not only affected workers and peasants, but leaders, lawyers, politicians, religious leaders, etc., the MST changed its watchwords to: "*Occupation is the only solution*" (1986), "*While the landowner wants war, we want land*" (1986–87) and, on the occasion of the Constituent Assembly, "*Agrarian Reform: In law or in the ditch*" (1988), and "*Occupy, Resist, Produce*" (1989), after the settlements began to be conquered. This process showed that politically the movement was not only consolidated, not only articulated at the national level, but also changed qualitatively from a political point of view.

In the 1990s, during the Collor Government, the MST changed its political fighting strategies and the watchwords became: "*Agrarian Reform: This struggle is ours*" (1990–91) and "*MST, now it is worthy*" (1992–93). With the election of Fernando Henrique Cardoso, the following motto emerged: "*Agrarian Reform: A struggle for all!*" (1995). The change in the watchwords represented the change in the political strategy of the movement: "*Agrarian Reform: A struggle for all!*" had a particularly important political significance. It showed the awareness that it was necessary to connect the movement with society as a whole. This was a period, as will be seen in this text, of growth and acceptance of the movement in Brazilian society as a whole. In the year 2000, the motto became "*Agrarian Reform: For a Brazil without large landowners*", in a clear allusion to the historical need for the end of unproductive lands and the fulfilment of the constitutional legacy that the land had to fulfil its social function.

16.4 Conflicts and Land Occupations

The analysis of the Brazilian agrarian reality at the end of the Twentieth Century and the beginning of the Twenty First Century showed, in a complete way, the presence of land conflicts. If, on the one hand, the "conservative modernisation" expanded its areas of action, the social movements equally increased the social pressure on the state in the struggle for land. The figure regarding the evolution of the number of conflicts in the Brazilian countryside between 1985 and 2020 shows five distinct periods. The first is the second half of the 1980s, which peaked in 1988 when the conflicts were generalised throughout all Brazilian regions. The end of this five-year period indicated a reduction in the actions of the movements, perhaps motivated by the unrealised historical possibility of the victory of Lula and PT in the first free presidential elections after military governments in 1989. The second period coincides with the first half of the 1990s, when the number of conflicts was reduced to half the previous period, revealing a change in the strategies of struggles and the need to re-accumulate forces. The number of conflicts in the north-east remained, and conflicts in other regions began to grow in relative terms: north, central-west, south-east and south (Fig. 16.7).

The third period corresponds to the second half of the 1990s, coinciding with the government of Fernando Henrique Cardoso, which presented new growth in conflicts,

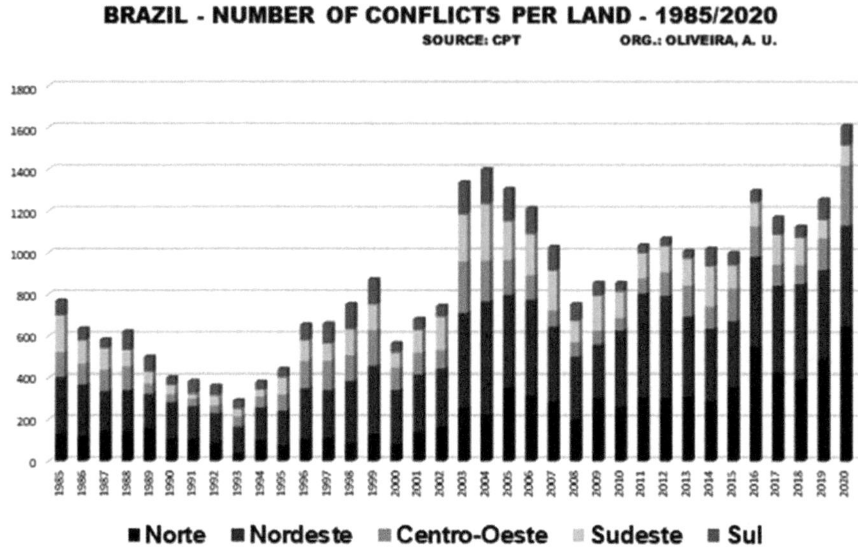

Fig. 16.7 Brazil: number of conflicts per land 1985–2020

reaching a level higher than the entire decade of the 1980s. There were 870 conflicts throughout the country in 1999. There was also an increase in the occurrence of conflicts in the oldest occupation regions: north-east, south-east, central-west and south. Some states appeared as concentrators of these conflicts, such as the case of Paraná in the south region; Minas Gerais and Sao Paulo, in the south-east; Goiás and Mato Grosso do Sul, in the central-west; Pernambuco, Alagoas and Bahia, in the north-east; and, Pará in the north. The fourth period, from 2003 to 2006, coincided with the first term of Lula's PT administration, which created the 2nd National Agrarian Reform Plan (NARP 2). After much struggling by the socioterritorial movements, NARP 2 was implemented, but it did much less agrarian reform than the Plan required. In his first term of office, Lula settled 150,000 families, and in the second, 110,000 families. In the Dilma government, agrarian reform was completely forgotten, and she only settled 19,000 families. The fifth and final period was marked by an increase in the number of conflicts, which after falling to 459 land conflicts in 2008, jumped to 1576 conflicts in 2020. However, the decline in the occupations/take-backs and the new camps continued, falling from 180 occupations/take-backs to 29 in 2020. Meanwhile, new camps went from 15 in 2012, to just three in 2020. These gaps between the curves were very clearly explained by the change of the MST, which in the 2007 Congress, decided to abandon its position in the struggle for agrarian reform, and began to adopt the radical position of fighting for a socialist revolution. Thus, the land occupations became a typical practice of the social movements in the struggle for the land until 2007, which then decreased due to the transformations of the MST (Fig. 16.8).

Fig. 16.8 Brazil: number of occupations/take-backs 1985–2020

However, returning to the Government of Fernando Henrique Cardoso (FHC), his response to the increase in conflicts was to increase police repression. This government made history for the violence it used, which hitherto was unprecedented in Brazil: State police forces were used to kill the peasants in the struggle for the land. The massacres of Corumbiara and Eldorado dos Carajás were examples of the FHC government. Both massacres represented the position of the Brazilian landowner elites to not give an inch, even on the issue of land and agrarian reform. The ruralists' support for the political support of the FHC government was the counterpart of two government practices: The first, its repressive position towards the social movements; the second, in the economic sphere, an extension—it is still not known until when—of the debts of these landowners, who did not settle them.

16.5 The Brazilian State and Agrarian Reform

As already mentioned in this text, in Brazilian society, the land is a special commodity. Much more than a value reserve, it is an asset reserve. The land is not retained for the purpose of putting it into production. This is why most of the land in this country remains unproductive. More than that, this unproductive land is retained for the purpose of being an instrument from which elites will have access to state policies. Thus, elites have not allowed the state to implement any agrarian reform policy in Brazil. Analysing the policies of the Brazilian State and the possibilities and/or attempts of agrarian reform paints a very interesting picture. Taking the decades of the 1950s and 1960s as a historical cut-off, the countryside was being shaken by the Peasant Leagues. The government of João Goulart, under political pressure, created the Superintendency of Agrarian Reform (SUPRA), and initiated a process of agrarian reform, one of the reasons for which he was overthrown in April 1964. The government of Castelo Branco—and his Minister of Planning, Roberto Campos—approved the Land Statute, the legal instrument of agrarian reform in Brazil. Roberto Campos informed members that they would approve the Land Statute. It was approved, but it was not implemented. After the period of the military government, it can be seen that, in fact, it has not been implemented. It was up to José Sarney to draw up the 1st National Agrarian Reform Plan, provided for in the Land Statute approved by the military.

Since social pressure increased since the end of the 1970s and the beginning of the 1980s, the New Republic government accepted the preparation of the NARP 1. During Tancredo Neves' visit to the Vatican, the only thing the Pope asked for was agrarian reform. For this reason, Nelson Ribeiro, linked with the Church, was the prime minister of agrarian reform.

NARP 1, announced at a conference of rural workers, initiated the opposite movement of the ruralist sectors that were part of the government of the New Republic in order to prevent its implementation. Violence in the countryside grew brutally, due to the landowners' reaction, led by Ronaldo Caiado. In order to proceed with cattle auctions, UDR was created, which practically "militarised" the landowners in the attempt to stop the implementation of the plan. As a result, there was a succession of ministers at the Ministry of Agrarian Reform, which culminated in the death of Marcos Freire, and the entire senior management of Incra, in the episode at the Carajás Airport in the south-east of Pará, when their plane crashed. JaderBarballho, then-Governor of Pará—a state that had the highest number of murders in the countryside in this country, the absolute majority of which went unpunished—took over the Ministry of Agrarian Reform and Development (MIRAD) and, on a continuing basis, extinguished the National Institute of Colonisation and Agrarian Reform (Incra). He then established the Legal Institute of Rural Land (INTER), extinguishing MIRAD and INTER, recreating the National Institute of Colonisation and Agrarian Reform (Incra). In other words, it dismantled the minimum organisation that had been set up for the implementation of the Plan. At the end of the Sarney Government, the results of the NARP 1 were as follows: Only 8% of the land foreseen was expropriated, and 10% of the families were settled. Thus, the dream of settling 1.4 million families, which had been announced in 1985, was reduced to just over 140,000 families.

In the Collor government, the UDR practically took control of agrarian reform in Brazil. Minister Antonio Cabrera Mano took over the Ministry of Agriculture and promoted the complete abandonment of agrarian reform. The fall of Collor and the rise of Itamar Franco changed practically nothing, because he had been Collor's right-hand man. Until 1994, the result of the State's actions regarding rural settlements was: by 1963, 140 families were officially settled in colonisation projects in Brazil; from 1964 to 1984, 162 families were settled between colonisation and settlements; from 1985 to 1994, 53 families were settled. These data allowed us to confirm that, from the policies of the Brazilian State, a policy of access to land was never implemented for the peasants.

16.6 The Fernando Henrique Cardoso Government and Agrarian Reform

Comparing the government of Fernando Henrique Cardoso with the previous ones (Sarney, and Collor/Itamar), it can be verified by the data provided by Incra that in six years, 373,210 families were settled in 3505 rural settlements. These settlements included land regularisations (the take-overs), Quilombola remnants, extractive settlements, Casulo and Rural ID projects, and agrarian reform projects. The pressure from the social movements with the expansion of occupations pressured the FHC government to expand settlements. This fact showed that in addition to being a pro-positive government policy, agrarian reform also reflected the need to respond to social pressure. However, the participation of the North Region in all settlements was as follows: 223,368 families or almost 60% of the total. If we look at the number of land occupations in that region, it represented just over 10% of the total. Meanwhile, most of the camps in the different regions were still waiting for agrarian reform to arrive. It was estimated that there were about 100,000 camps.

Analysing the settlements year by year between 1994 and 2017, it can be verified that there was growth in the number of families settled by 1998, when it reached just over 101,000, with a significant reduction in 2003 (with the settlement of 36,000 families. Therefore, according to Incra data, there was a declared policy of reducing settlements by the FHC government. Later, in the Lula administration, there was an increase in the number of families settled, when the total reached 136,000 families settled. From there, it was a steep descent down, with the Temer government having only 1205 families settled (Fig. 16.9).

In this way, the policy of agrarian reform has been going through historical times and contrasting strategies. While the MST policy was to lay bare the unproductive land and the land-grabbing by landowners, the response was police violence or the criminalisation of the leaders. These were the cases of the Pontal do Paranapanema, in the state of Sao Paulo, the Corumbiara massacre in Rondônia and Eldorado do Carajás, in Pará.

In Pontal, since 1957, the state knew that farmers illegally occupied those lands. In other words, more than one million hectares of land should return to state control, and this has only occurred slowly. In this way, a peculiar situation occurred, because the movement of landless rural workers was in the case of Pontal (SP), Corumbiara (RO) and Eldorado do Carajás (PA), causing a new political component to emerge in the struggle for land, which was the denunciation of the land-grabbing by the landowners. This was the case in Corumbiara: the lands of the farmer who claimed to be the owner should have already been taken back by the State because the farmers did not comply with the Law that allowed access to that land. In Eldorado do Carajás, the same occurred: The farmer who was said to be owner of the Macaxeira Farm, in fact had an authorisation to exploit chestnuts but not the title of land ownership. Thus, the MST brought to light this new discussion, and it was clear that at the time, the elite pact on the land was at its root and, in particular, its legal basis. As

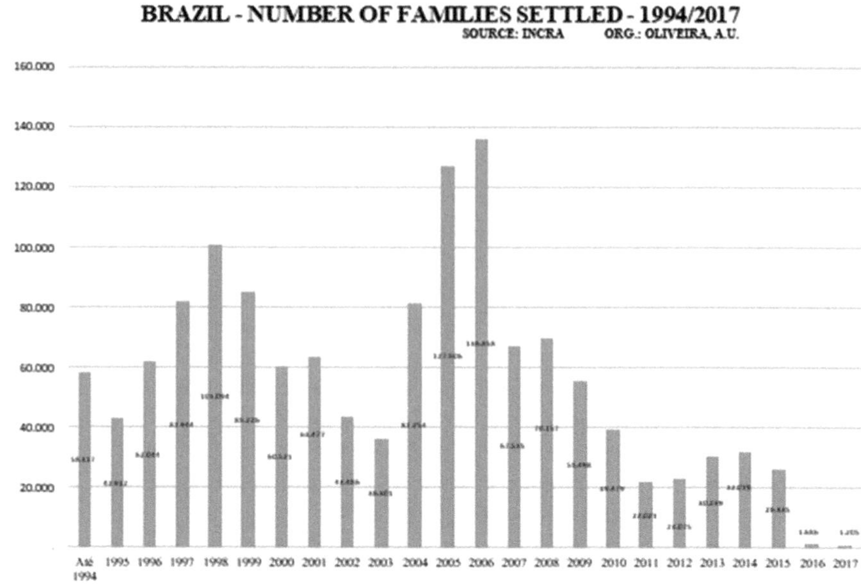

Fig. 16.9 Brazil: number of families settled 1994–2017

a counterpoint, the State sought to criminalise the leadership of the MST. This was therefore a first political strategy to confront social movements.

The second strategy was the legal change that was being implemented by the Ministry of Agrarian Development. First, the debts of the ruralists were guaranteed, then the progressive Rural Territorial Tax (ITR) was created, but it never implemented. Later, the Land ID and Land Bank Projects were created in order to implement an authentic agrarian reform of the market, as the leaders of the social movements liked to affirm. Finally, two more coercive measures: MP 2109, which prohibited the survey for two years of occupied buildings and MDA Ordinance No 62 of 27/03/2001, which excluded the settlers from the agrarian reform due to "*invasion or robbing of rural buildings*". The last measure was the registration for agrarian reform settlements by mail, broadcast with television advertising and print stating that the "*door is open for agrarian reform, it's just a matter of going in and registering*".

Another political strategy to confront social pressure by settlements was to stimulate the creation of new social movements that did not adopt the occupation tactic as a strategy of struggle. These new movements adopted the exclusive tactic of the so-called negotiation. Several sympathetic trade union centres to the government were involved in these actions to create new social movements, aiming to weaken the social base of the MST.

A fourth strategy that was put into practice by the Ministry of Agrarian Development (MDA) was the holding of meetings and seminars with intellectuals who studied the agrarian question, to assist in the elaboration of policies and actions of government and, mainly, to form a kind of intellectual action front of criticism of the movements and their organic intellectuals. MDA has also created the Centre for Agricultural Studies and Rural Development (NEAD), to conduct studies and actions aimed at so-called family farming.

However, the fifth action was one that was marked by spectacular bids, according to the so-called society of the spectacle. Action in the media was mobilising the government, movements and public opinion. There were reports seeking to depict the MST leadership as satanic, and counter-propaganda was organised by large press agencies that featured unproven claims, coaching of journalist teams, conducting public opinion surveys on the MST, the production of virtual material via the Internet, etc. These actions generated in the media created a significant amount of news that was mainly aimed at removing the image of support that the population had formed of the MST and agrarian reform after the March to Brasilia. Certainly, this action was due to the fact that the media had not reported anything about the significant decline in the number of families settled since 1998. However, at the beginning of August 2001, the National Confederation of the Associations of the Incra Servers (CNASI) disclosed data proving the decline.[9]

16.7 The Future ... in the New Twenty First Century

The history of the agrarian question in Brazil today revealed that the MST was the new face of Brazil, the part of this country that was in struggle. As strange and extemporaneous as many might find it, the movement from the city to the countryside contradicted the general movement from the countryside to the city, but it was also a movement that strived for the construction of a new society. The settlements sought to establish collective and/or community production, or even individual production. The problems were numerous and ranged from barriers on access to credit, to bureaucratic bossy-ness, to Stalinist impositions and non-understanding of the peasant ideology of farming their own land and the freedom of labour. However, it was these landless people, now in the midst of the Landless movement, who marched through the streets and cities of this country, occupying public places and buildings. The MST was practically the only social opposition force to the Fernando Henrique Cardoso government. For that reason, there was a campaign to try to destroy it.

Even so, history has been relentless with those who tried to ignore it. In Brazil, it was almost a consensus that any alternative to removing social exclusion in the country was through agrarian reform. It therefore had a social objective. That is to say, it was the way to remove at least a part of the poor from social marginalisation. However, agrarian reform was also economic, because it would lead to an increase in the supply of agricultural products from these small units to the market. Agricultural reform, however, had to be political as well. It had to be an instrument through which this portion of the population would gain its citizenship.

---

[9] *Jornal dos Trabalhadores Rurais Sem Terra*, ano XIX, n. 213, p. 14, ago. 2001.

In the Landless camps, I always heard the peasants who camped there say things like "*I prefer to die fighting for a piece of land, to die worthily, than die as an indigent on the outskirts of the city*." Therefore, the path of many of these poor people to citizenship was through agrarian reform. However, it was also a proposal for agrarian reform that had to be accepted as a proposal to transform this society in the search for justice, dignity and solidarity.

For this reason, landless peasants are re-teaching the ideals of nation, homeland and patriotism in this early Twenty First Century, full of visions of a world in which citizenship is conquered by a few. I conclude this text with the verses of a landless peasant:

It is for love of this country—Brazil,
that we follow each other in rows.

Order and Progress, Zé Pinto.[10]

[10] Music "Order and Progress", Zé Pinto, CD *Arte em movimento* – MST, São Paulo.

Chapter 17
Land and Food: The New Struggles of the Landless Workers Movement (MST)

Bernardo Mançano Fernandes

Abstract This article aims to analyse this new struggle by the MST and contribute towards an understanding of the emergence of the struggle for land and food. The MST, recognised as one of the most prominent peasant movements in the struggle for land and agrarian reform in Brazil, has also set up a food movement through participation in the creation of popular and institutional markets, where it sells part of the production from its territory in the construction of a peasant food system. The struggle for land in the third decade of the twenty-first century is very different from the struggle for land in the 1970s/80s, when the MST was founded. The characteristics of the struggle against large estates and agribusiness, including multinational corporations, are analysed. There is a new struggle for land, which is approaching the cities and has appropriated food sovereignty and agroecology. The struggle for land has its roots in other struggles. The struggle for healthy food is inseparable from the struggle for land and agrarian reform. A new direction has been taken. This theoretical essay, based on various studies, including monographs, dissertations, theses, books and research reports, will analyse the ecological production of rice in MST agrarian reform settlements in the state of Rio Grande do Sul, the production of coffee in settlements in Minas Gerais, as well as other foods in settlements in other regions of Brazil. With the struggle for land and foods that have provided peasant farmers with an income, we will show the creation of new peasant food system which is resulting in the sustainable modernisation of agriculture. We will argue that this is an experience not limited to the MST, but rather which forms part of the globalisation of the agrarian question.

Keywords Struggle for land · Agrarian reform · MST · Food

B. M. Fernandes (✉)
Universidade Estadual Paulista, São Paulo, Brazil
e-mail: mancano.fernandes@unesp.br

R. C. Lois González and M. A. Mitidiero Junior (eds.), *Brazilian Geography*,
Advances in Geographical and Environmental Sciences,
https://doi.org/10.1007/978-981-19-3704-0_17

17.1 Current Situation of the Struggle for Land and Agrarian Reform

Under capitalism, the peasant farmers' struggle is, above all, a territorial struggle, because the existence of peasant farmers is inseparable from peasant farming territory. In the history of peasant farming communities, it is their struggle that guarantees their existence as communities dedicated, in particular, to food production. The multidimensionality of the struggle may be understood by peasant farmer territorialities in different countries. These struggles explain the persistence demonstrated in the face of theories produced since the nineteenth century regarding the end of peasant farming. In the current context of the global agrarian question, the persistence of peasant farming communities takes two forms: resistance and subordination. Both make possible the spatialisation, territorialisation and de-territorialisation of peasant farming communities around the world. Subordination is the condition created by capitalism to control the peasant farming communities' territories, their space and time, their will and dreams, or to destroy them. Resistance is the condition under which the peasant farming communities seek to take control over their members' lives, their territories and construct them in their own way.

The large plantations, which began with the colonisation of continents such as America, Africa, Asia and Oceania, are one of the territorial expressions of capitalism in the formation of agribusiness, which was consolidated in the mid-twentieth century, when capitalism organised a complex system formed by the agricultural, livestock, industrial, mercantile, technological and financial sectors, expanding its hegemony from the agrarian world. Perhaps, in all of human history, peasant farming communities have never faced such difficult times in their resistance. However, the hegemony of agribusiness produced by deforestation, the production of large-scale monocultures, the intensive use of pesticides and the standardised production of ultra-processed foods, reached far beyond its own limits, reached the limits of the planet, producing an era which Wolford (2021) called the Plantanioncene.

It is against this backdrop that we can understand the current struggle for land and agrarian reform (Fernandes et al. 2018). The concentration, standardisation and globalisation of agribusiness are at the same time characteristics of hegemony and the limits of agribusiness. Theories of the end of peasant farming communities based solely on economics do not allow us to understand the struggles of peasant farmers at the beginning of the twenty-first century, with the emergence of other territorial struggles, such as those for education, health, housing and, as we analyse in this article, for food. It is not possible to separate the struggle for land from the struggle for food, as they are part of the same process. Brazil is an excellent laboratory for this study, as the Landless Workers Movement (MST) was formed in this country, one of the most active socio-territorial movements (Fernandes 2005; Halvorsen et al. 2019) that contributed to the creation of the Via Campesina, a worldwide organisation of peasant movements.

To guarantee its existence, the MST needs to create new spaces and transform them and new territories, territorialising its activities. At the same time, it is threatened with de-territorialisation by agribusiness. The genesis of the MST occurs with the creation of a space for political socialisation, in a house or in a church or in a trade union, for example. The space for political socialisation is a gathering of people with the goal of sharing their life experiences in order to transform them. The first stage of the space for political socialisation is when the gathered people introduce themselves and get to know each other. It is a communicative space where they state the reason for being at a meeting in which they intend to occupy the land. Land occupation is an act of confrontation with agribusiness and large estates that concentrate most of the land. It is also a dispute for territory. The second stage of the political socialisation space is the exchange of experiences and knowledge. It is an interactive space in which families who have already participated in occupations and taken over land report their experiences to families who are going to participate in an occupation for the first time. These are difficult moments of decision-making that will change people's lives. During the third stage of the space for socialisation, decisions have already been taken and the families present themselves in an organised manner to society with the occupation of a property, demanding agrarian reform, creating a space for struggle and resistance (Fernandes 1996).

The creation of communicative, interactive spaces of struggle and resistance comprise the space for political socialisation that forms part of the process of spatialising the struggle for land. The occupation turns into a camp with dozens, hundreds or even thousands of families (Photo 1). The occupation creates a space for dialogue between the landless, the government and the landowners to find a solution to the problem of concentration of land ownership. The spatialisation process can take months or years, depending on the political situation and the actions that the parties involved take. Landowners may use private militias, or they may file applications for repossession to the Judicial Authorities. The landless pressure the government for the expropriation of land for the settlement of families. The government, under pressure, seeks to negotiate a way forward in relation to the agrarian question.

The result depends on the correlation of political forces between the classes and the government. During the negotiations, the families hold various demonstrations with the aim of obtaining political support from various institutions in society. When families win a territorial dispute, they are settled, and the scene of struggle and resistance is transformed into a territory. Thus, the spatialisation process becomes part of the territorialisation process of the struggle for land. However, families may lose the territorial dispute and will be forced to vacate the land and transfer the encampment to the sides of a road, where they will remain encamped until they either win or give up the struggle (Fernandes 1996). This whole process of creating spaces and transforming them into territories makes the MST a socio-territorial movement.

The MST is the peasant socio-territorial movement responsible for most of the actions geared towards the struggle for land and the struggle for agrarian reform (DATALUTA Brasil Report 2020), and has promoted new territorialities in all dimensions of territorial struggles, especially in the struggle for food. As we have stated, the struggle for food is not separate from the struggle for land and forms part of its territorialisation. In Fernandes (2000) we analyse this process across Brazil. The MST is one of more than a hundred peasant movements in Brazil involved in the struggle for land. In Table 17.1, we observe that the Northeast region of Brazil is a "peasant" region, in the sense that it is where most of the struggle for land takes place, and the state of Pernambuco is where the largest number of occupations have taken place, involving the largest number of families.

The State of São Paulo, the territory of which is mainly dedicated to the production of sugarcane, is the state with the second highest number of occupations and families involved in the struggle for land. In the last three decades, more than one million one hundred thousand families have occupied land in Brazil. This is certainly the most significant struggle for land and agrarian reform in the world. In chart 1, we can analyse the different political backdrops to the struggle for land, with the Fernando Henrique Cardoso administrations, 1995–2002, and the Luiz Inácio Lula da Silva administration, 2003–2011, which experienced the most intense forms of popular pressure in the struggle for land and were also the governments that settled the most families under agrarian reform settlement projects (Fig. 17.1).

Figure 17.2 contains the spatial profile of this struggle, where we can observe concentrations of occupations and repossessions in all Brazilian regions, with greater or lesser intensity. The repossessions are actions of indigenous populations who claim their territories were invaded by landowners and/or agribusiness corporations.

Table 17.2 contains a set of data that demonstrate the territorialisation of the struggle for land with the settlements secured in the struggle for agrarian reform. Settlements are territories formed by peasant farming communities, most of which are the result of land occupations. Almost one million one hundred thousand families settled in an area covering more than eighty-two million hectares. These numbers are close to 10% of national territory, which has allowed some scholars to claim that, even though Brazil has had two agrarian reform projects, the first during the Sarney government (1985/1990) and the second during the Lula government (2003/2010), they were never implemented, and has allowed other researchers to claim that Brazil is carrying out an agrarian reform, albeit conservatively (Fernandes 2013).

Figure 17.3 helps us understand how the advance of occupations put pressure on governments to carry out agrarian reform; however, they need to be governments that dialogue with socio-territorial movements that have projects and that negotiate the territorial dispute within the context of the correlation of forces between agribusiness and peasant farming communities.

Figure 17.4 is a contrast to Fig. 17.2. It observes that, while the struggle for land is concentrated in the north-east, south-east and south regions, most families were settled in the Amazon. There, there was the settlement of squatters, families that had possession, but not ownership, of the property, and were included under the agrarian reform policies.

In this part, our objective was to show a panoramic view of the struggle for land and reflect its current situation. The charts show waves that are read by the economic cycles and do not indicate the end of this process. The meaning of the present situation lies in the fact that, however hegemonic it may be, agribusiness does not account for all agricultural activity, despite agribusiness ideologues attempting to include all food produced in Brazil as if it formed part of agribusiness. The fact is that peasant

Table 17.1 Brazil—number of land occupations and repossessions and families by state and macroregion 1988–2018

Region/state	No. occupations	%	No. families	%
North	**851**	**10.72**	**108,513**	**9.85**
AC	56	0.71	5347	0.49
AM	10	0.13	2536	0.23
AP	3	0.04	140	0.01
PA	505	6.36	74,782	6.79
RO	132	1.66	14,613	1.33
RR	12	0.15	1253	0.11
TO	133	1.67	9842	0.89
North-east	**3048**	**38.38**	**406,247**	**36.87**
AL	397	5.00	47,238	4.29
BA	721	9.08	103,930	9.43
CE	114	1.44	13,941	1.27
MA	123	1.55	19,234	1.75
PB	191	2.40	21,917	1.99
PE	1205	15.17	155,395	14.10
PI	76	0.96	9479	0.86
RN	122	1.54	17,829	1.62
SE	99	1.25	17,284	1.57
Midwest	**1187**	**14.95**	**191,895**	**17.41**
DF	47	0.59	8570	0.78
GO	391	4.92	61,636	5.59
MS	571	7.19	83,188	7.55
MT	178	2.24	38,501	3.49
South-east	**1813**	**22.83**	**224,310**	**20.36**
ES	94	1.18	12,497	1.13
MG	566	7.13	60,154	5.46
RJ	76	0.96	11,780	1.07
SP	1077	13.56	139,879	12.69
South	**1043**	**13.13**	**170,980**	**15.52**
PR	608	7.66	81,670	7.41
RS	247	3.11	60,972	5.53
SC	188	2.37	28,338	2.57
Brazil	**7942**	**100**	**1,101,945**	**100**

Source CEDOC DOM Tomás Balduíno—CPT.**ORG**. Dataluta Network

Photo 1 Land occupation in the municipality of Mirante do Paranapanema – SP in 1996 (Bernardo Mançano Fernandes)

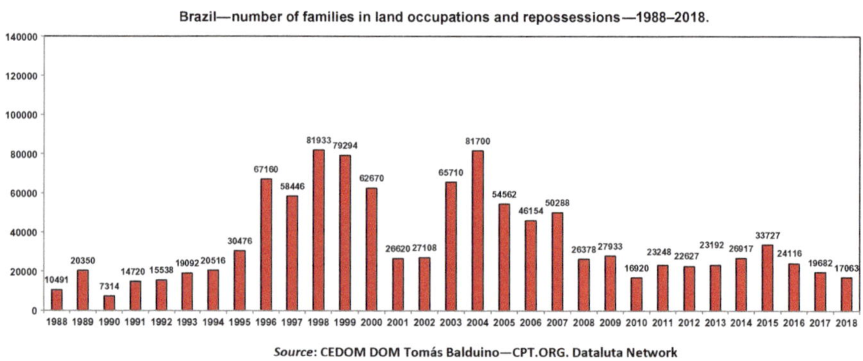

Fig. 17.1 Brazil—number of families in land occupations and repossessions—1988–2018. *Source* CEDOM DOM Tomás Balduino—CPT.ORG. Dataluta Network

agriculture is a model of development that, in addition to fighting for land, is now also fighting for food.

Table 17.2 Brazil—number of rural settlements—1979–2018

Region/state	Settlements	%	Families	%	Area	%
North	**2175**	**22.8**	**503,147**	**45.9**	**61,214,830**	**74.2**
AC	159	1.7	29,953	2.7	5,227,145	6.3
AM	143	1.5	72,707	6.6	27,322,548	33.1
AP	58	0.6	19,073	1.7	2,260,868	2.7
PA	1148	12.0	291,352	26.6	20,110,098	24.4
RO	217	2.3	42,250	3.9	3,601,031	4.4
RR	67	0.7	22,215	2.0	1,445,927	1.8
TO	383	4.0	25,597	2.3	1,247,214	1.5
North-east	**4373**	**45.9**	**329,910**	**30.1**	**10,522,458**	**12.8**
AL	179	1.9	14,451	1.3	114,385	0.1
BA	728	7.6	60,059	5.5	2,093,205	2.5
CE	457	4.8	25,788	2.4	917,107	1.1
MA	1033	10.8	140,943	12.9	4,411,080	5.3
PB	314	3.3	15,218	1.4	292,204	0.4
PE	612	6.4	3577	0.3	560,725	0.7
PI	501	5.3	34,521	3.2	1,389,363	1.7
RN	305	3.2	21,531	2.0	523,848	0.6
SE	244	2.6	13,822	1.3	220,541	0.3
Midwest	**1273**	**13.4**	**164,905**	**15.1**	**8,327,263**	**10.1**
DF	20	0.2	1052	0.1	7727	0.0
GO	462	4.8	27,361	2.5	1,363,746	1.7
MS	206	2.2	32,280	2.9	718,147	0.9
MT	585	6.1	104,212	9.5	6,237,643	7.6
South-east	**869**	**9.1**	**57,196**	**5.2**	**1,602,035**	**1.9**
ES	95	1.0	4667	0.4	52,337	0.1
MG	416	4.4	25,150	2.3	1,056,083	1.3
RJ	75	0.8	7125	0.7	127,984	0.2
SP	283	3.0	20,254	1.8	365,631	0.4
South	**837**	**8.8**	**40,192**	**3.7**	**827,923**	**1.0**
PR	329	3.5	20,360	1.9	429,771	0.5
RS	345	3.6	13,840	1.3	293,655	0.4
SC	163	1.7	5992	0.5	104,497	0.1
Brazil	**9527**	**100**	**1,095,350**	**100**	**82,494,509**	**100**

Source INCRA—National Institute of Colonisation and Agrarian Reform/**DATALUTA**—Struggle for Land Database, 2018. www.fct.unesp.br/nera

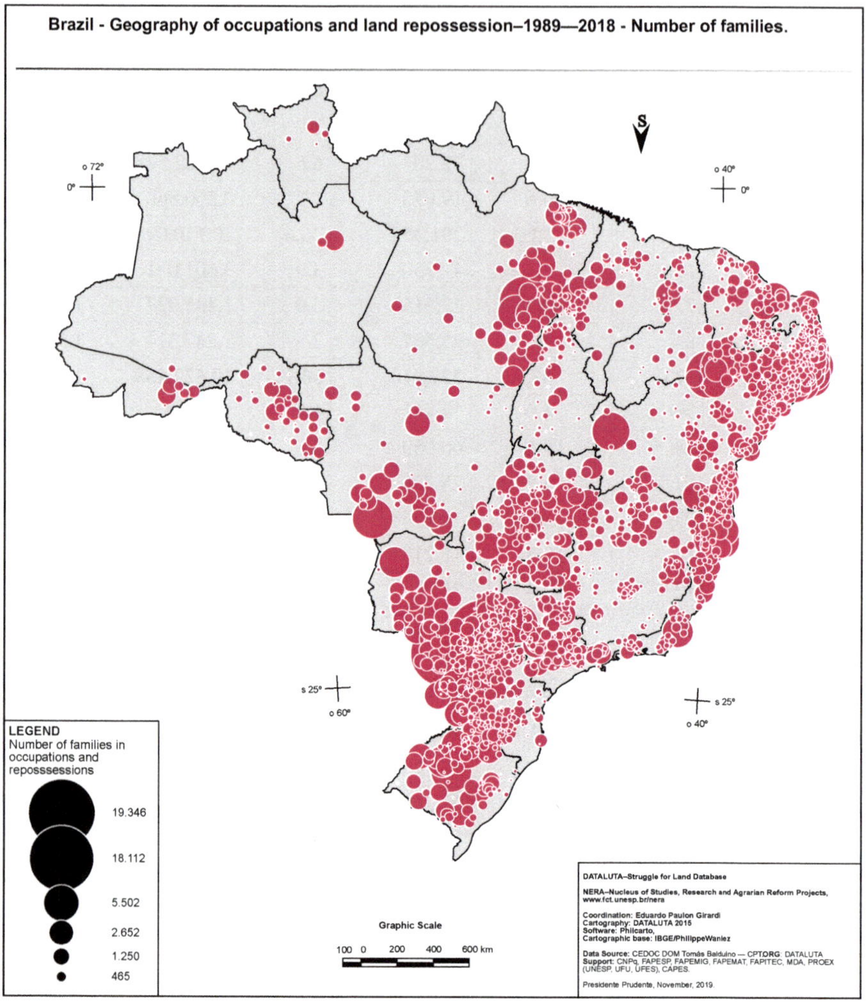

Fig. 17.2 Brazil—geography of occupations and land repossession—1989–2018 number of families. *Legend* Number of families in occupations and repossessions. **DATALUTA—Struggle for Land Database**, **NERA**—Nucleus of Studies, Research and Agrarian Reform Projects, www.fct.unesp.br/nera, **Coordination**: Eduardo Paulon Girardi, **Cartography**: DATALUTA 2015, **Thematic Mapping Software**: Philcarto, **Cartographic base**: IBGE/Philippe Waniez, **Data source** CEDOC DOM Tomás Balduino—CPT.ORG: DATALUTA Network, Support: CNPq, FAPESP, FAEMIG, FAPEMAT, FAPITEC, MDA, PROEX (UNESP, UFU, UFES), CAPES, Presidente Prudente, November 2019

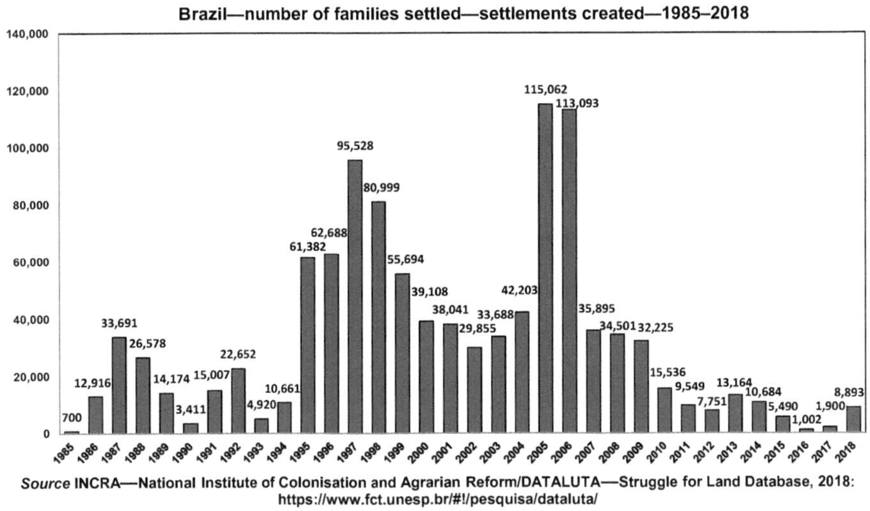

Fig. 17.3 Brazil—number of families settled—settlements created—1985–2018. *Source* INCRA—National Institute of Colonisation and Agrarian Reform/DATALUTA—Struggle for Land Database, 2018. www.fct.unesp.br/nera

17.2 Land and Food: The MST's Struggle for the Price of Ecological Rice

The MST, recognised as one of the main peasant movements in the struggle for land and agrarian reform in Brazil, has also become known as a food movement in the construction of popular and institutional markets, where it sells part of the production of rural settlements. In 2020, faced with an increase in the price of rice (which, together with beans, forms the basis of the Brazilian diet), the MST decided to maintain the price of ecological rice in accordance with its commitments to cooperative families and consumers. In the first part, we stated that the struggle for land is not separate from the struggle for food. In this part, we show the emergence of the struggle for land and food. This theoretical essay is based on several studies: monographs, dissertations, theses and books on ecological rice production in MST agrarian reform settlements in the state of Rio Grande do Sul. We seek to reflect on price and peasant income and the peasant food system in the sustainable modernisation of agriculture.

On 12 September 2020, the news item "Amid rising prices, MST organic rice remains fairly priced" published on the Landless Workers Movement (MST)[1] website drew attention. Having studied the MST for forty years in the struggle for land, for agrarian reform and for various other public policies, this news suggests that something new is happening. A few days later, the Folha de São Paulo newspaper also reported the fact: "MST says it will maintain a fair price for settlement rice", leading

[1] https://mst.org.br/2020/09/12/em-meio-a-alta-de-precos-arroz-organico-do-mst-segue-a-preco-justo/

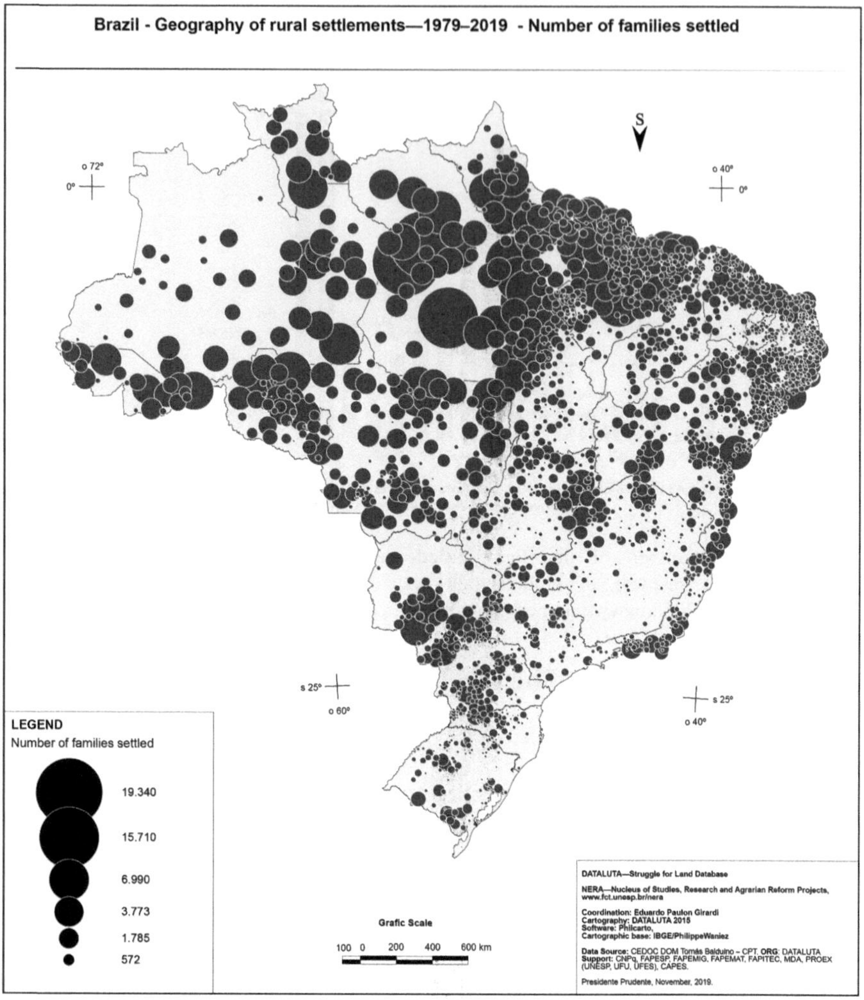

Fig. 17.4 Brazil—geography of rural settlements—1979–2019 number of families settled. *Legend* Number of families settled. **DATALUTA—Struggle for Land Database, NERA**—Nucleus of Studies, Research and Agrarian Reform Projects, www.fct.unesp.br/nera, **Coordination**: Eduardo Paulon Girardi, **Cartography**: DATALUTA 2015, **Thematic Mapping Software**: Philcarto, **Cartographic base**: IBGE/Philippe Waniez, **Data source**: CEDOC DOM Tomás Balduino—CPT.ORG: DATALUTA Network, **Support**: CNPq, FAPESP, FAEMIG, FAPEMAT, FAPITEC, MDA, PROEX (UNESP, UFU, UFES), CAPES, Presidente Prudente, November 2019

to the question: does this mark the beginning of a new MST struggle?.[2] In these four decades of studying the MST, we have researched different types of struggles: for land, for agricultural and livestock policies, for education, health, housing,

[2] https://www1.folha.uol.com.br/mercado/2020/09/mst-diz-que-mantera-preco-justo-por-arroz-de-assentamento.shtml

infrastructure and services, for development models, etc., but never a struggle for a lower fair price. The struggles of peasant farming communities for a fair price tend to be upwards, to receive a price that allows them to continue producing food and reproducing as a social class in their territories. This has been the struggle against capitalised income from the land that lies at the heart of the agrarian question. In order to explain the MST's stance in not accepting the increase in the price of ecological rice[3] and keeping the price in line with its moral, economic and political values, we analyse peasant income and the peasant food system, taking their territorialities as a reference point.

Our analyses of these struggles and disputes start from a territorial perspective, because the existence of the MST, as a peasant socio-territorial movement, is based on land, a fraction of the territory, on being and belonging to a class that is creating its own food system based on food sovereignty and agroecology using sustainable modernisation process as a starting point. Since its genesis, the MST has been territorialised throughout Brazil, producing an enormous range of territorialities (Fernandes 2000). These territorialities are of active resistance (Martins 2017) in which the MST rejects the hegemonic food system of agribusiness, which consists of and derives from a development model of capitalist agriculture.

This article is the result of a history of joint work between the Movement and the University. Our analyses are shared through cooperation agreements that forge closer ties between the São Paulo State University (UNESP) and the MST. Both institutions are committed to the sustainable development goals of the UN 2030 Agenda (United Nations 2020), the first two of which are to end poverty and hunger. Some references of this sharing are the undergraduate courses in geography and the postgraduate course in geography as part of the cooperation between UNESP, the MST, Via Campesina, the Florestan Fernandes National School, the Latin American Council of Social Sciences (CLACSO) and the UNESCO Chair in Territorial Development and Education for the Countryside.

The special undergraduate course in geography was only taught once, from 2007 to 2011, and trained 46 male and female geographers who work in the schools and cooperatives in the settlements.[4] This course was supported by the Federal Government's National Programme for Education in Agrarian Reform (PRONERA). The first postgraduate (master's) course in geography with focus on Territorial Development in Latin America and the Caribbean was offered in 2011 and has since become a regular course, now in its tenth year. Three of the MST's works on rice production emerged from these two courses (Almeida 2011; Oliveira 2016, Oteiro e Silva 2020).

Almeida (2011) studied the territorial dispute between agribusiness and peasant farming communities in the Santa Rita de Cássia settlement, in the municipality of Nova Santa Rita, Rio Grande do Sul. In addition to a history and description of the settlement, Almeida analysed the agribusiness strategies to produce rice through the leasing of peasant families' land. The territorial dispute became even more heated when the families decided to release the tenants from their situation

[3] The term ecological rice used by MST cooperatives is equivalent to the term organic rice.

[4] https://www.unesp.br/aci_ses/jornalunesp/acervo/273/ensino

of dependence by suspending leases, resuming planting and starting to produce ecological rice. The publication of a collection on the impacts of rice production in the Metropolitan Region of Porto Alegre (Guasselli and Medeiros 2015) offered environmental analyses and the contradictions generated by territorial disputes.

Oliveira (2016) studied labour and income relations in several settlements, including ecological rice producing settlements in the metropolitan region of Porto Alegre. An analysis was conducted of monetary income, self-consumption income, initiating discussions on the formation of peasant income. The researcher demonstrated that control of the production and commercialisation process increases the income of the settled families from the perspective of the creation of food sovereignty policies. Oteiro e Silva (2020) focused on studying the types of territorial rents, deepening the discussion on peasant income, MST cooperation systems and markets as disputed territories.

The relationship between the University and the Movement, i.e. between UNESP and the MST, is one of many experiences of research projects and academic training of the MST with various other Brazilian universities, especially with postgraduate programmes in geography. Coca et al. (2020) demonstrates the participation of Brazilian agricultural geography in studies on food sovereignty. For example, in his doctoral thesis, produced under the Postgraduate Programme in Geography of the Institute of Geosciences of the Federal University of Rio Grande do Sul, Martins (2017) studied the ecological production of rice in the settlements of the metropolitan region of Porto Alegre, starting with the characteristics of the landscape with occupation of the floodplains. In his study, Martins also emphasised the territories of resistance and dispute against agribusiness. He deepened the studies on peasant autonomy based on participatory management, through ecological rice management groups. He studied the various production collectives from seed production, production, processing, certification and marketing, which make up what we call the peasant food system, producing a territoriality of inclusion of most members of cooperative families.

Origuéla (2019), also in a doctoral thesis in Geography at the UNESP Presidente Prudente campus, analysed the production of territorialities in disputes in the territories of rural settlements in Rio Grande do Sul, studying forms of subordination, emancipation and autonomy, taking how the capitalist food system uses and expropriates peasant territories as a starting point. Expropriation has been resisted through the creation of a new food system, based on food sovereignty and agroecology, with the experience of rice production in Rio Grande do Sul being a reference point.

In addition to rice production, Origuéla also studied the production of vegetables (Photo 6) to deepen his research into the peasant use of the territory and contribute towards the discussions concerning peasant income. Cooperation was highlighted as a structural condition for bolstering the peasant food system. The series of activities and collectives based on cooperation and knowledge sharing created the conditions for the establishment of the peasant food system, controlling productive processes and including people of different ages. All these research projects were carried out through what we call militant research, which involves researchers and producers,

Photo 6 Production of vegetables Settlement Santa Rita de Cássia II, municipality of Nova Santa Rita – RS (Photo Alex Garcia)

i.e. various thought collectives, with the goal of building methodologies, methods, theories, techniques and open technologies to understand the realities studied.

This series of works shows that the MST contributes towards the understanding of the creation of a food system and has a reference in the production of ecological rice, which began at the beginning of the twenty-first century with the securing of the settlements in the metropolitan region of Porto Alegre. Struggling for land and agrarian reform against large estates since its founding, from the 1990s onwards, it also began to face off against agribusiness as a hegemonic capitalist model of agricultural development. On a global scale, the capitalist food system is on the verge of collapse with the increase in the production of commodities, hunger, obesity and global warming (Preiss and Schneider 2020). Agribusiness has always presented itself as the only development model, and most studies on agricultural development define family or peasant agriculture as part of agribusiness.

This text breaks with this definition by analysing the territorialities of the peasant food system as a series of productive processes in the construction of food sovereignty policies for agroecological production, and the territorialities of agribusiness, or capitalist food system, producing global commodities, with intensive use of pesticides and ultra-processed foods, which are international public and environmental health problems. We are faced with different territories and social relations that produce different territorialities.

The increase in agricultural production from agribusiness has been accompanied by an increase in people going hungry (Silva 2020). In 2018, Brazil returned to the UN hunger map. In 2020, more than 19 million people were going hungry (AgênciaBrasil 2020). Why is hunger greater in rural areas, where food is produced? Silva, 2020, claims that hunger does not result from a lack of food, but from a lack of access to food. The question is: what are the forms of access to food? Obviously, it is not just shopping at the supermarket. Hunger is also related to a lack of opportunity to produce food. The human right to food is also a territorial right. People who live in the countryside and who are going hungry are unable to produce food because land and agricultural policies are concentrated in the hands of agribusiness corporations. These people are excluded from the processes of knowledge building, primarily from knowing how to produce and having the opportunity to do so.

In response to the concentration of land, production and agricultural policies, Brazil has created different experiments with institutional and popular markets. An example of a successful institutional market is the Food Acquisition Programme (PAA), which bought more than 367 different types of food, reflecting the enormous potential of the almost 5 million peasant families in Brazil, who occupy only 25% of the agricultural territory and produce most of the food consumed by Brazilians, ensuring food sovereignty in various municipalities and regions. Some examples of popular markets are the agroecological fairs and baskets and the rural warehouses created by the MST. These are initiatives by governments and movements in the creation of policies that strengthen the food system of peasant farming communities. The Bolsonaro government, committed to landowners and agribusiness, reduced PAA resources by 95% (Brasil de Fato 2020).

17.3 Ecological Rice Production and Price Struggles

The MST is the largest producer of organic rice in Latin America, according to the Rio Grande do Sul Rice Institute (IRGA). In 2020, the Movement estimates a harvest of 312 thousand 60-kilo sacks over an area of 3215 ha, through various cooperatives: Charqueadas Agricultural Production Cooperative (Copac); Sete de

Julho Agricultural Production Cooperative (Coopal); Viamão Organic Producers' Cooperative for Agrarian Reform (Coperav); Terra Livre Agrarian Reform Workers Cooperative (Terra Livre); Central Cooperative of Agrarian Reform Settlements of Rio Grande do Sul (Coceargs), Settled Workers Cooperative of the Porto Alegre Region (Cootap); Agricultural Production Cooperative of Tapes Settlers (Coopat); and the Nova Santa Rita Agricultural Production Cooperative of (Coopan).

The creation of this peasant food system is based on a principle of food sovereignty, which is to guarantee production in as many places as possible. In the case of rice in general, 80% of national production is concentrated in the state of Rio Grande do Sul and settlements form 88.5% of the ecological management areas in the state. The MST is strengthening rice production in Piauí, Maranhão and Bahia as a way of territorialising ecological rice production to other states and regions. The MST's struggle against the rise in the price of rice in general, and ecological rice in particular, starts from an analysis of the following factors: capitalist speculation, based on the concentration and monopoly of production; the abandonment of public policies for peasant agriculture by the Bolsonaro government, which has contributed not only to an increase in the price of rice, but also of beans and milk; the decrease in public stocks and the increase in exports, causing shortages (Porto 2020).

In the 2020 harvest, the MST's fight for the price of ecological rice started with a price of R$ 64.00 for a 50-kilo sack of paddy rice, as defined by the Rio Grande do Sul Rice Institute—IRGA. While MST cooperatives took the decision to pay their members R$ 64.00, a monopoly of ten companies paid producers around R$ 50.00 per sack (Stedile 2020). The average price per kilo of ecological rice in MST stores is around R$ 6.00 and for organic rice in supermarket chains, around R$ 10.00. The difference between the price of a kilo of ecological paddy rice paid to settled families and the price charged to consumers in MST stores is approximately R$ 4.72. The difference between the price per kilo of organic rice paid to the producer and the price charged to the consumer in supermarket chains is around R$ 9.00.

The difference between the prices on the popular market and the price on the capitalist market evidences the MST's struggle for a fair price, i.e. the price determined by moral, economic and political values that allows its reproduction, under dignified conditions, as a social class in its territories. This fair price is not only composed of monetary quantities, but also of other quantities and qualities. It contains the work of the families, the use of the land, water and appropriate technologies, the struggle for land, the struggle for agrarian reform, the settlement won where the families produce healthy food, public policies with food sovereignty, agroecological production for a model of sustainable development of agriculture. This is the fair price the peasant farming communities charge for the rice that arrives at the table.

The price of the peasant farming communities takes on the capitalist price of capitalised income from the land, which promotes speculation, the concentration of land, wealth and power, market monopolies and development policies, which causes discontinuity in public policies for the peasantry. These are reasons for the MST not to keep up with the rise in the price of rice. What explains its reaction is its history in the creation of a model of sustainable agricultural development. The ecological rice experiment in Rio Grande do Sul is one of those that are contributing towards the creation of a peasant food system.

Evidently, the MST could not react to the price increase if it were subordinated to the agribusiness food system, i.e. if it were producing rice solely for the companies that monopolise the purchase of rice in Rio Grande do Sul. This is a first element to consider when referring to the peasant food system. There are various struggles that form this system. The struggle for land is the first step in its constitution. The struggles for territorial development: infrastructure and services, agricultural policies, housing, education and health are fundamental to its development. For this reason, it is not enough just to have land, territorial sovereignty is inseparable from food sovereignty. It is necessary to have appropriate technology, agroecological knowledge, popular and institutional markets. These are conditions that contribute towards the formation of peasant income, made up of family, community or cooperative or collective work, monetary income, self-consumption income and other income.

Classes produce territories and are produced by them. Classes is the last chapter volume 5 of Das Kapital. Marx did not complete it. He stated that "The owners merely of labour-power, owners of capital, and land-owners constitute then three big classes of modern society" (Marx 1988, p. 297). Peasants can own labour-power and land, or they can lease the land, or they can occupy the land. They can be de-territorialised through the expropriation of their lands, or they can be territorialised through the struggle for land and agrarian reform. After all, classes are not only constituted of social, economic and political relations, but also of their territories, spaces where all the actions that produce and are produced by their respective territories take place.

Territorial control by social classes happens in different ways. Oliveira (1991) analyses the territorialisation of monopoly capital, which means the control of the territory by capital and monopolisation of the territory by capital, when capital has the monopoly of peasant territory, for example. This format of territorial control, Fernandes (2008) called territoriality of the capitalist mode of production in peasant territory. We may observe in the studies on the production of ecological rice that the MST, when building its food system, defined the peasant territoriality in its territories, driving out the territoriality of agribusiness. The construction of the peasant system is not dissociated from several other struggles, with land, technology and markets, among many others. This condition allowed greater control over the formation of peasant income and strengthened a model of sustainable development, which even enabled the fight for a fair price.

17.4 Conclusion

This study contributes towards the updating of rural or agrarian geography research on peasant movements. In the twenty-first century, the agrarian question is also global and the struggle for land cannot be analysed separately from the struggle for food. All policies created by the agribusiness model to end hunger have failed. The role of peasants in the world takes on new meaning when the importance of their participation in sustainable development is understood.

This is our analysis of how peasants struggle to gain autonomy in their territories in the sense of trying to control production, processing and marketing in the creation of an appropriate food system, in their history of struggle for land-territory and development policies. This makes the MST a food movement that promotes the sustainable modernisation of agriculture, without the use of pesticides, without the exploitation of work, guaranteed by the production of food and the struggle for land, the territorialisation of the Movements.

This essay raises questions into which further research is suggested to continue studies on peasant income, the peasant food system, popular markets, modernisation and sustainable development models. It also reinforces the idea that the so-called modernisation of capitalist agriculture, through the complex model of agribusiness systems, is an accumulation of environmental problems on a global scale and an increase in social inequalities.

It also presented an original analysis based on agrarian geography for the studies of peasant movements from their actions in the creation of spaces and transformation into territories. This reading of the different uses of spaces and territories enables the study of the processes of spatialisation, territorialisation and de-territorialisation of the peasantry.

References

Agência Brasil. Pesquisa revela que 19 milhões passaram fome no Brasil no fim de 2020. https://agenciabrasil.ebc.com.br/geral/noticia/2021-04/pesquisa-revela-que-19-milhoes-passaram-fome-no-brasil-no-fim-de-2020

Almeida JC (2011) A disputa territorial entre agronegócio e campesinato no assentamento Santa Rita de Cássia II em Nova Santa Rita—RS. Monograph (Geography) Course in Geography. Faculdade de Ciência e Tecnologia—Universidade Estadual Paulista (UNESP), Presidente Prudente

Brasil De Fato (2020) Com queda de 95% em 8 anos, aquisição de alimentos chega a menor nível com Bolsonaro. https://www.brasildefato.com.br/2020/10/29/com-queda-de-93-em-8-anos-aquisicao-de-alimentos-chega-a-menor-nivel-com-bolsonaro

Coca EFL, Santos LL, Souza RD, Salvaterra JR (2020) A soberania alimentar na Geografia Agrária brasileira. Revista Terra Livre

Fernandes BM (1996) MST: formação e territorialização. Hucitec, São Paulo

Fernandes BM (2000) A Formação do MST no Brasil. Editora Vozes, Petrópolis

Fernandes BM (2005) Movimentos Socioterritoriais e Movimentos Socioespaciais: contribuição Teórica para Uma Leitura Geográfica Dos Movimentos Sociais. Revista NERA, issue 8, p 14–34

Fernandes BM (2008) Entrando nos territórios do Território. In: PAULINO, Eliane Tomiasi; FABRINI, João Edmilson. Campesinato e territórios em disputa. Expressão Popular, São Paulo

Fernandes BM (2013) A reforma agrária que Lula fez e a que pode ser feita In: Sader E (org) 10 anos de governos pó-neoliberais no Brasil. Boitempo Editorial, Flacso Brasil, São Paulo, p 191–206

Fernandes BM, Rincon F, Kretschmer R (Org) (2018) La actualidad de la reforma agraria en América Latina y el Caribe. Clacso, Buenos Aires

Guasselli LA, Medeiros RM (Org) (2015) Impactos da produção de arroz naRegião Metropolitana de Porto Alegre/RS: Análise territorial e ambiental. Impressa Livre, Porto Alegre

Halvorsen S, Fernandes BM, Torres FV (2019) Mobilizing territory: socioterritorial movements in comparative perspective. Ann Am Assoc Geographers 109(5):1454–1470
Martins AFG (2017) A produção ecológica de arroz nos assentamentos da região metropolitana de Porto Alegre: territórios de resistência ativa e emancipação. Doctorate (Geography) Postgraduate Programme in Geography at the Instituto de Geociências da Universidade Federal do Rio Grande do Sul
Marx K (1988) O Capital, vol 5. Nova Cultural, São Paulo
Oliveira AU (1991) A agricultura camponesa no Brasil. Contexto, São Paulo
Oliveira VS (2016) Análise das relações de trabalho e renda em assentamentos de reforma agrária nas regiões norte, sul e metropolitana de Porto Alegre—RS. Dissertation (Geography) Postgraduate Programme in Territorial Development in Latin America and the Caribbean—Territorial of the Instituto de Políticas e Relações Internacionais—IPPRI da Universidade Estadual Paulista—UNESP, São Paulo
Origuéla CF (2019) Território e territorialidades em disputa: subordinação, autonomia e emancipação do campesinato em assentamentos rurais no Rio Grande do Sul. Doctorate (Geography) Postgraduate Programme in Geography at the Faculdade de Ciência e Tecnologia—Universidade Estadual Paulista (UNESP), Presidente Prudente
Oteiro e Silva M (2020) Questão agrária, cooperação e mercados: novas perspectivas para o MST. Master's (Geography) Postgraduate Programme in Territorial Development in Latin America and the Caribbean—Territoral of the Instituto de Políticas e Relações Internacionais—IPPRI da Universidade Estadual Paulista—UNESP, São Paulo
Porto S (2020) Arroz: uma crise anunciada. Le Monde Diplomatique, 25 Sept 2020
Preiss PV, Schneider S (2020) Sistemasalimentares no século 21: debates contemporâneos. Editora da UFRGS, Porto Alegre
Relatório Dataluta Brasil/Coordenação Sobreiro Filho, J. Girardi, E. P. no 21 (2020). NERA, Presidente Prudente
Silva JG (2020) Por que o Brasil voltou ao mapa da fome? Valor Econômico, p 15. https://diplomatique.org.br/arroz-uma-crise-anunciada/
Stedile JP (2020) Alta do preço do arroz, especulação capitalista e soberania alimentar. https://www.poder360.com.br/opiniao/brasil/alta-do-preco-do-arroz-especulacao-capitalista-e-soberania-alimentar-por-stedile/
United Nations (2020) Transforming our world: The 2030 agenda for sustainable development. https://sustainabledevelopment.un.org/post2015/transformingourworld.
Wolford W (2021) The plantationocene: a lusotropical contribution to the theory. Ann Am Assoc Geogr. https://doi.org/10.1080/24694452.2020.1850231

Chapter 18
Geography and Indigenous Peoples: Struggles of Resistance

Márcia Yukari Mizusaki and José Gilberto de Souza

Abstract This study aims to analyse the processes associated with indigenous struggles, particularly in relation to the Guarani and Kaiowá in Mato Grosso do Sul state, Brazil. Such processes embody the advancement and intensification of the contradictory movement of class and non-class antagonism, expressed by capitalists and land owners against indigenous peoples. In America, the processes of commodification and incorporation of labour around the axis of capital emerged from racial division and other structuring elements, such as patriarchalism and religion, which became dominating powers, closely linked to the socio-spatial relations that have (re)produced them. In Brazil, the process culminated in the constitution of privately owned land, expropriation and violence, consolidating windfall profits as a result of turning nature into a monopoly and commodities—a specific class power unfolding in the production of value, either by expropriation or by the violent proletarianisation to which indigenous peoples have been subjected. This process fosters capital territorialisation and impacts the life of the indigenous peoples, whose individuals contradictorily re-frame their (re)existence. Indigenous labour has been incorporated into the axis of capital, from overexploitation to the capital-centric focus, while the indigenous workers contradictorily refer to their ancestry and cosmology, taking a stand as indigenous people. In these struggles, the indigenous identity becomes an identity of resistance, and the socio-spatial practices aim towards the preservation of knowledge. Such knowledge and practices gain relevance and reaffirm the state of being and living in the world, opposing the metabolic rift between man and nature, even if the commodification of life and common assets leads to the self-destruction of humankind.

Keywords Indigenous peoples · Guarani-Kaiowá · Resistance · Territory · Time and space

---

M. Y. Mizusaki
Universidade Federal da Grande Dourados, Dourados, Brazil

J. G. de Souza (✉)
Universidade Estadual Paulista/Rio Claro, São Paulo, Brazil
e-mail: jg.souza@unesp.br

R. C. Lois González and M. A. Mitidiero Junior (eds.), *Brazilian Geography*,
Advances in Geographical and Environmental Sciences,
https://doi.org/10.1007/978-981-19-3704-0_18

18.1 Introduction

This article presents reflections on the impacts of capitalist society on the lives of indigenous peoples, outlining issues related to their struggles of resistance. This work is embedded in the context of the expansion of the global market for commodities and the demand for land, which has repercussions (as has been seen historically) on the forms of occupation and social formation in Brazil at the cost of the commodification of land, dispossession, and genocide. A period marked by continuous expulsions, which increased exponentially as a result of neo-developmentalist and neo-extractivist policies throughout Latin America, and which greatly affected the native peoples, the peasants and the quilombolas (inhabitants of quilombos—hinterland settlements founded by people of African origin) (Svampa 2013, 2019; Souza and Cabero Diegues 2012; Souza and Borges 2017; Sassen 2015). Despite the difficulties and limitations imposed in understanding the historical processes of social formation of native peoples and the lack of research on their geographies, the objective is to analyse their contemporary socio-spatial dynamics. The process of geographical expansion of capital reached their forms of production and social reproduction at a dizzying rate, colliding directly with their shamanic conception of the world, especially when capital itself leads societies towards a one-dimensional logic of existence, linked to commodities, determining them as the new reason for the world—the reason[1] of value (Dardot and Laval 2016) that obliterates the multidimensionality of relationships so that they become increasingly contradictory. This is a trajectory fundamentally driven by the logic of the expanded reproduction of capital and commanded by the economic power of large international financial corporations, made complex by the development of means of communication and information technology, which impose new rhythms, dynamics, processes, objectivities and subjectivities.

A reflection is therefore proposed, in order to consider the contradictory particularities of the meeting of worlds (Krenak 2019) and their geography cities which, in the dialectical relationship between different ways of being, conflict and unfold in a given series of transformations starting from a multiplicity of interactions and determinations[2]. We carry out such a reflection from a point of view of not dichotomising or assuming all-encompassing procedures. It is undertaken starting with the singular,

---

[1] According to Meliá (1990, p. 41) the ideal of the shaman plays an important role in the Guarani way of being. "Guarani shamanism is, in its essence, we believe, the consciousness possessed by the divine and made word of a good way of being that is fully lived in the religious festival and in the community invitation". Mura (2006, p. 303) also highlights that "... the shaman plays an important role in the interpretation of the characteristics of the cosmos and its changes, as well as in the formation of a specific morality and ethics (teko porã), and also in the definition and emotional-affective stability of the kaiowa person (for means of mongarai, healing processes and in diagnosing and combating witchcraft)".

[2] According to Moreira (2004, p. 34) "Geographicality is, thus, the spatial being-being of the entity. It is the ontological state of being in space-time. A valid principle attributable to any entity – it can be man, a natural object or space itself (...) – whatever the character of its quality. His reflection is presented in the question of the being of the entity. In the case of man, geographicity is the way in which hominization as the essence of metabolism expresses its existence in the form of space. The

understanding this as the bearer of movement for the individual, and the universal, the ontological primacy of the real—which means recognising it as existence itself (Lukács 1967), which is the existence of the indigenous Guarani and Kaiowá peoples in the state of Mato Grosso do Sul, Brazil. An analysis is made of territorial conflicts, and their struggles and forms of (re)existence in the face of the advance of expansion and incorporation, and the appropriation of their traditional territories by capitalist society. This state has the second-largest self-declared indigenous population in the country (73,295 inhabitants), behind only the state of Amazonas (168,680 inhabitants), according to data from the demographic census carried out by the Brazilian Institute of Geography and Statistics (IBGE 2010). The territorial conflicts occurring in the state between native peoples and agribusiness sectors, as well as the resistance struggles undertaken by the indigenous people—in this case, the Guarani-Kaiowá are indicative examples of the deeper territorial contradictions within this process of capitalist socio-spatial formation, particularly in Brazil.

Territory, as a suitable space resulting from power relations, becomes an expression of the socio-spatial practices of subjects in deep contradiction, as an expression of the diverse human social dimensions in class confrontation, historically produced beginning from the different mechanisms for the production of inequalities (gender, race, class, …) with a view to feeding and maintaining the classes[3] that are in power (Federici 2017). And precisely because of their relational and historical dimensions entangled in capitalism, even though they are not class structures, it is imperative to recognise that indigenous peoples are involved in class confrontation, and for this very reason, they cannot be analysed as essentialisms, but as movements against forms of domination and exploitation, given the power relations.

As stated, the territory has its expression, and its epistemic-categorical centrality in power relations, since this dimension is present in all societies and cannot be disregarded in the internal formations of indigenous societies and in their relations with class-based society[4]. Territory and power are categories that have, therefore, a relational and dynamic expression—product and process of the contradictory reproduction mechanism of life, in which human existence is realised through the mediation of work, in its broad, ontological sense, in the relationship that societies establish among themselves and with nature[5].

geographicity of man is then the way in which freedom from necessity emerges and takes shape through the concrete form of spatial existence in society".

[3] What is considered, therefore, is that there are human dimensions that are in contradiction to the hegemonic forms of production and reproduction relations, but whose centrality is not in the class unit, as a social category, it is where the original forms of production and of socio-spatial reproduction relationships, the quilombolas, the LGBTQIA+ communities, women, among other human dimensions. These are primeval forms of ontological existences and realizations, which object, which are opposed to the unicist forms of social representation and power that gain prominence in the face of development in capitalism and show racism, machismo, xenophobia in direct forms of domination, patriarchy, among others.

[4] The expression or epistemic-categorial centrality refers to its explanatory capacity of the real, as a movement of thought in abstracting the essentiality of the real. (Souza and Fulino, 2020).

[5] The work here does not refer exclusively to the production of goods, but to the primary and secondary teleological metabolic processes, in which thought and action are unified, thought is

For the methodological steering of the research, a bibliographic survey was carried out on the subject, with data collection at agencies such as FUNAI (the National Indigenous Foundation) and analysis of documentaries and news sites of agencies and entities that traditionally act in defence of the rights of movements struggling for land/territory, such as the Federal Public Ministry (MPF/MS)[6], the ISA (Socio-environmental Institute), the CIMI (Indigenous Missionary Council), the CPT (Pastoral Land Commission) and the APIB (Articulation of the Indigenous Peoples of Brazil). Photographic records were taken, and informal conversations also took place with indigenous leaders in the indigenous lands visited from April to August 2018 and January 2020, namely Ñu Vera I and II, Boqueirão, ÑuPorã, Apika'y, Pakurity, PassoPiraju; Panambizinho (municipality of Dourados); Kurusu Amba I, II and III (municipality of Coronel Sapucaia); Guapoy and Guyraroká (municipality of Caarapó); GuyraKambi'y (municipality of Douradina); LaranjeiraNhanderu (municipality of Rio Brilhante); and Guaiviry (municipality of Aral Moreira).

We seek to highlight some historical elements, to situate ourselves in a spatiotemporal context, noting the difficulties and limits that such studies present. We highlight some aspects of the Guarani social organisation in the pre-contact period, and the transformations that took place; we approach the advance of the capitalist expansion fronts on the traditional territories of the original peoples and on the forms of land appropriation that were configured in the process of constituting the current agrarian structure in the state, and finally, we analyse the consequences of this process for the indigenous peoples present in the current state of Mato Grosso do Sul and their strategies of r-existence.

18.2 The Expansion Fronts in Guaraní Lands

The state of Mato Grosso do Sul has an area of 357,147.9 km^2 and is located in the western portion of Brazil, with an estimated population of 2.8 million people (CIDADES/IBGE 2020), about 85% of whom live in urban areas. The state also stands out as a result of its large indigenous population, as highlighted above. More recent data indicate an indigenous population of 83,000 people, according to the Special Indigenous Health District (DSEI 2017).

From a total rural area in the state of 30,549,179 ha, 27,011,325 ha are concentrated in 11,617 out of a total of 71,164 settlements (IBGE, Agricultural Census 2017). In other words, 88.4% of the land is concentrated within just 16.3% of the state's agricultural production units. The Guarani-Kaiowá, in turn, occupy an area of around

work and metamorphosing natura-naturans/naturata is work and are intertwined, as processes of formation of the be social.

[6] MOCULTURE of faith (2017). Production by Joana Moncau, Grabriela Moncau, Izaque João and Spency Pimentel. Brazil: Futura, 2017. Documentary. (23 min). Available at: https://canaisglobo.globo.com/assistir/futura/curtas/v/6769937/. Accessed on: June 17, 2021.

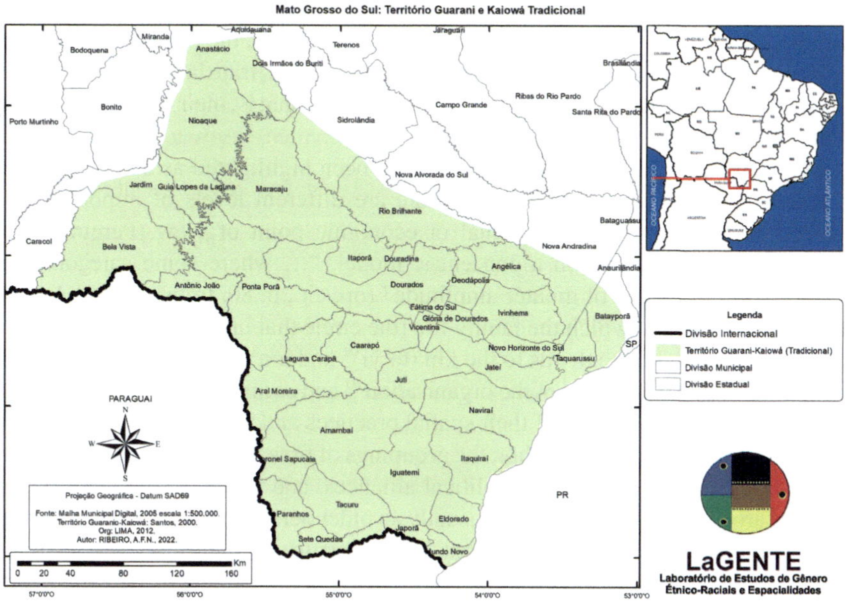

Fig. 18.1 Traditional Kaiowá and Guarani Territory in Mato Grosso do Sul. *Source* Lima (2012)

30,415 ha (Cavalcante 2013), or 0.1%[7] of the state's rural area—a fact that allows us to understand the territorial configuration within the state of Mato Grosso do Sul and the class power within this unit of the federation (Oliveira 2008).

Around eleven indigenous ethnic groups currently live in the state, among which we highlight, for this work, the Ñandéva, who in Mato Grosso do Sul call themselves Guarani (Guarani language, Tupí-Guarani linguistic family, Tupi stem) and the Kaiowá (Kaiowá language, Tupi-Guarani language family, Tupi stem), who constitute the largest population.

In the past, the Guarani would have occupied the best lands in the foothills of the Cordillera and in the basins of the Paraná[8], Paraguay and Uruguay rivers (Meliá 1990). There is a theory that the Kaiowá are descendants of the Guarani from the former mission of Itatin, and that the Ñandéva, in turn, are descendants of the Guarani from the former province of Paraná and Guairá, and came to present-day Mato Grosso do Sul starting in the seventeenth century (Mura 2006). In Fig. 18.1, we can see what has been indicated as the location of their traditional territories in the current state.

[7] It is worth noting that the number of establishments does not refer to the number of owners, since a person or business group may own more than one establishment, which shows a more acute situation of land concentration in the state of Mato Grosso do Sul.

[8] The spelling for indigenous groups, as in the case of the Kaiowá and Guarani, will be guided by the Convention on spelling of names for indigenous populations, which are capitalized. When they are found as an adjective, they will be written in lower case (Pereira, 1999; Lima, 2012).

According to Antonio Brand, some traits are common to the Guarani way of being, such as "… the tekoha (village), the system of leadership, the extended family, the reciprocity-based economy, and others" (Brand 1993, p. 2, translated). For the author, this set of cultural categories defines their ñandereko—that is, their own way of living, their collective way of being.

In this system of relationships, kinship has been highlighted as a relevant factor in Kaiowá social organisation, where there are different levels of importance for subjects from a political, ceremonial or economic point of view (Pereira 1999). An organisation is based on a concentric model, "… where some categories are encompassed by others of greater amplitude" (op. cit., p. 80, translated). A level of social organisation in which the family fire (the circle that includes parents, children and in-laws) is found 9 extends to the kin (te'yi)[9] and, on a larger scale, the tekoha.

Pereira's (1999) analyses of the organisation[10] and composition of the te'yi make us think about the dynamics of their spatial practices and their relationship with the territory[11]. According to the author, the groupings of fires occur not only due to blood ties, but also due to affinity and political alliances, and reveal a configuration that revolves around a head of kin (hi'u), usually an older person, considered the mainstay of the home, where it is highlighted that "… the territory is identified with the figure of its controller" (op. cit., p. 86, translated). The economic and social decisions taken by the te'yi in their daily lives are independent from those of other te'yi.

Based on Pereira's observation (op. cit), it is clear that the scope of a te'yi, which is under the leadership of a head of kin, is clearly identified by its co-residents, where economic relations and exchanges are established from the principle of reciprocity, also constituting a basis for political representation. Such representation, however, is given based on "… their discursive capacity, their charisma, kindness and generosity" (p. 88, translated), while their authority can be dissolved where their practices are no longer consistent with socially accepted conduct. "… a leader deprived of power, relying only on their discursive capacity, their charisma, kindness and generosity. The use of violence is vehemently repudiated by members of the te'yi. By their very nature, the hi'u will barely think of exerting violence against someone in their group, otherwise their authority will suddenly disappear" (Pereira 1999, p. 88, translated).

Souza (2009, p. 108, translated) highlights that power practices cannot be understood only "from the perspective of repression and denial. On the contrary, power is also exercised due to its positivity—it generates transformative relationships, and

[9] According to Pereira (1999, p. 81) "... groups of close relatives, gathered around a family fire, where the meals consumed by the members of the fire are prepared". Family fire is very dynamic in its form and composition and is accompanied by the principles of mutual cooperation between close relatives, political arrangements and marital dynamics. Belonging to a fire is fundamental to Kaiowá existence.

[10] "The te'yi are residential units that bring together a varied number of family dwellings" (Pereira, 1999, p. 85).

[11] Tekoha (teko + ha) means a place where they live according to their customs. The realization of teko takes place in tekoha: "It is where land, territory, subsistence, social relations and religious festivals are confused and merged. It means and produces an interrelation between the social, economic, political and religious, essential for the Guarani" (Brand, 1993, p. 95).

subjects recognise and emancipate themselves". The conservation of a kin will depend on the ability of the hi'u to maintain it, and its composition is very dynamic, resulting from marriages, separations, alliances and changes. Thus, the organisation of the kin is based on the belief that blood ties deepen relationships and reduce conflicts—something that does not always happen, as highlighted by the author. The Kaiowá seek cohesion and solidarity through closer ties of kinship, "... co-residence, political alliances, economic cooperation and the formation of religious units" (Pereira 1999. p. 98, translated).

The te'yi are the stage where collective efforts are made and the economy of reciprocity takes place, with land being highlighted as the centre of the socio-economic system and, in this sense,

> Guarani economic production does not seek to accumulate a surplus, but to redistribute it free of charge, being certain of reciprocity through hospitality and celebrations. (...) As they do not seek to accumulate more than what is necessary for their family's self-sufficiency and for redistribution, the time devoted to productive work is limited by the need for consumption. As a result, there was time for hunting and fishing and, above all, for celebrations and rituals. Therefore, the work for the Guarani was also not characterised by exploitation, as is the case in our society. It was a work that reached its peak through cooperation (Brand 1993, p. 113, translated).

Cooperation is expressed in the mutirão (potyrô)—the work party, the basic unit of the Guarani economy. It has two dimensions: that of the extended family, where economic activities are carried out (such as work in common farmsteads, construction of houses), and the other, the tekoha, which has a political-religious character. This economy of reciprocity (oreva) happens on a more restricted level, within the extended family, generally involving two or three family homes. In this Guarani form of social organisation, reciprocity is valued and constitutes one of the elements that allow them to maintain their teko, which is linked to the ideal of behaviour, attitudes, laws and norms. The positive ethical and moral values (tekoporã) are what make it possible to maintain the behaviour considered sacred (tekomarangatu). And in this sacred conduct, human actions cannot allow the Land (Yvy) to suffer evil (Pereira 1999; Mura 2006). For Mura (2006), this conception of Land (Yvy) gives it a meaning quite different from the idea of private property. In his words[12]:

Yvy must be understood as the part of the Cosmos created by ÑaneRamõi (Our Grandfather) and allocated by his son, Ñande Ru (Our Father), to the care of the Indians.

> Thus, land takes on a special meaning for the Indians and, unlike in a Western conception, it cannot be considered as a parcel or property, whose possession would be in the hands of an individual or groups of individuals. On the contrary, the Guarani insistently indicate that they belong to the land, and their own actions are central to its conservation. Therefore, the

[12] For the Guarani, yvy is simultaneously earth (inorganic matter), world and soil. The distinction between one or another characteristic is made through the linguistic contextualization of the word yvy or, in the case of soil, through the adjective that allows to differentiate them; for example, yvy morotí (white earth), yvy pytã (red earth), yvy hú (black earth) and yvy sayju (yellow earth), each with specific properties for agriculture, an activity that allows and gives meaning to this classification" (Mura, 2006, p. 104).

shamanic and ritualistic activities somehow synthesise the conditions that each community lives with, and the difficulties in maintaining such a cosmic balance (Mura 2006, p. 104–105, translated).

In this sense, it is observed that, for the Guarani, land is not understood as a spatiality of production, but of social reproduction, as a way of producing culture—that is, land for the Guarani is not an economic resource, as Meliá (1990) points out. The concept merges existence and production in oikos, "for the enjoyment of all" (Brand 1993, p. 158, translated). Each family fire has autonomy to produce. Pereira (1999) observes that it is common for associations between fires to occur,

.... forming a unit of production and consumption. These nuclei are coordinated according to ideological principles, which govern the economy of reciprocity and are usually composed of father-in-law and son-in-law, father and son, a group of siblings, etc.. They tend to cultivate a continuous area together, dividing the crop into plots at harvest time, when each one can use their produce as they like. Many fires prefer to cultivate their fields alone, but the option implies restrictions from the point of view of the number of relationships established with other fires (Pereira 1999, p. 89–90, translated).

The author clarifies that the garden is divided into portions of varying sizes, according to the needs and the reputation of those involved. It is observed, then, that within the kin, cooperation during production occurs with a small number of families (fire). As this expands, this cooperation is lost and starts to acquire a political character, while this cooperation on a broader scale would take place more on the religious plane.

As the kins have relative autonomy from a political and religious point of view, associations between them are common. Pereira calls this association tekoha. Thus, "... the formation of a tekoha implies the meeting and cooperation of various kins—te'yi, allies or relatives" (Pereira 1999, p. 99, translated). This scalar movement of social space, political space and space of representation acts in an opposite spectrum to that of capitalist societies, since equality on the capitalist plane takes place via representation, in politics, and does not touch production. This change in direction is expressed precisely in the structure of kin that is achieved via the dimension of equality.

Regardless of the form of social organisation, or its way of being, Mura (2006) aptly observes that it cannot be understood as something static, immutable, but rather as a historical—and we would add, geographical—fact. It cannot be ignored that the model of accumulation through dispossession (Harvey 2004) had a great impact on the lives of native peoples, including the elimination of their presence in large spaces.

In favourable situations—that is, until the first decades of the twentieth century—the te'yi preferred to establish themselves at the sources of rivers and streams, with nuclear families being distributed along and around these river courses or wellheads. The intervening space between the domains of one extended family and another tended to follow the characteristics of the river network, and it could therefore happen that the groups were located very far away from each other. What physically united these family groups was the aforementioned network of trails (tape po'i), via which indigenous people communicated and maintained a high level of foot traffic, following the logic of guata (walking) (Mura 2006, p. 111, translated).

Information on the current Guarani-Kaiowá occupation in Mato Grosso do Sul is more accurate from the second half of the nineteenth century, and even so, not very clear—from the point of view of the territorial domain of the peoples, what is accentuated in this process is their nomadic trajectory, in which existence is processed in a spatial dimension of a territoriality that is not fixed on being: rather, as its central focus is existence, territory is always an expression of being in movement, of becoming.

The Guarani in the state of Mato Grosso do Sul were experiencing the destructive dynamics and the incorporation of common goods and indigenous work into the circuits of capital accumulation. The appropriation and subsequent constitution of large private property has historically conferred class power on landowners, implying the destitution of this territory-movement, establishing ontological ruptures in the original peoples, whether through spatial mobility or the dimensions of production and social reproduction (oikos), politics and representation of the world.

The colonisation process in the south of the former province of Mato Grosso, the current state of Mato Grosso do Sul, began in 1829, carried out by newcomers who came from Minas Gerais and São Paulo. In Brazil, between the end of the Sesmarias Law in 1822 and the Land Law of 1850[13], the land tenure system was the means of territorial occupation and, through this mechanism, expeditions took over large tracts of land in the region (Moreno, 2007).

Although the Land Law stipulated that landowners should legalise land until 1854, in the former province of Mato Grosso the applicability of that Law was extended until 1889, when the first Land Law in the former province—Law no. 20 of 1892—was drafted, allowing the continuation of the appropriation process and its subsequent legalisation, benefiting the holders of large tracts of land. The transfer of responsibility for the demarcation of land from the Union to the States, after the Proclamation of the Republic in 1889, meant the victory of the rural oligarchy, which was able to legislate its interests more directly. The formation of a regional elite based on large estates legitimised expropriation and violence against native peoples (Moreno 2007).

Enslavement, the dynamics of occupation, the constitution of private land ownership and the War with Paraguay (1864–1870) were markers of territorial determinations that impacted the social organisation of the Kaiowá and Guarani, and took on greater significance when, in 1882, Tomaz Laranjeira[14] gained a concession monopoly for the exploration of the yerba mate herb present in the region, which was found in the forests occupied by the Guarani and Kaiowá. With the concession, the Mate Laranjeira Company was created in order to explore for the herb,

[13] Law no. 6012 of 09/18/1850 became known as the Land Law. "It sought to characterize what are vacant lands and aimed to be a legal instrument to discriminate between public and private lands, in addition to preventing access to vacant land, except through purchase". The Land Law is considered "... a milestone in the process of transition from slave to free labor" (Motta, 2005, p. 279).

[14] Thomaz Larangeira, a merchant from Rio Grande do Sul, supplied food to the commission in charge of demarcating the boundaries between Brazil and Paraguay, after the War of the Triple Alliance (1864-1870). Noting the large presence of herbs in the region, he managed to obtain permission to lease herbs, thanks to his political relationships (Jesus, 2004).

using a workforce made up of indigenous and Paraguayan people, as highlighted by Brand et al. (2008). Records indicate the levels of overexploitation of work present in the activity during the operations of the Company—which, with the support of the police, possessed the necessary apparatus to prevent the escape of workers. From 1915 onwards, the Company lost political strength, and lease contracts were reduced increasingly until, in 1943, Getúlio Vargas created the Federal Territory of Ponta Porã and annulled the Company's contracts.

Between 1915 and 1928, 8 indigenous reserves were created by the defunct Indigenous Protection Service (SPI), with the purpose of releasing land[15] for private appropriation. The indigenous people were compulsorily taken to these reserves, aggravating the process of de-territorialisation of the Guarani and Kaiowá. Chamorro (2018) reports that the driving factor for the creation of these reserves resulted from the demand of non-indigenous people to occupy indigenous lands located on the expansion fronts. She observes that there are abundant records from the nineteenth century about the Kaiowá occupation in the localities close to the Ivinhema River, although reserves were not created in the region because, at the time, doing so was not in the interest of economic expansion.

In the 1950s, the occupation that began with subdivisions by private colonisation companies propelled the appropriation of land and the presence of logging companies. This policy of colonisation, which continued even after President Getúlio Vargas was deposed in 1945, was focussed on their traditional territories. Lands were sold to colonising companies at minimal prices and resold at much higher prices. In return, companies had to deliver infrastructure—a commitment that most of the time was not met. Thus, through false powers of attorney, the circumvention of legislation, and irregularities in land divestiture, large properties were constituted and the political power of this class was strengthened (Moreno 2007).

The expansion fronts and private appropriation of land happened alongside overcrowding in the reserves, particularly in the municipality of Dourados, and the aggravation of conflicts and impacts of all kinds, making the traditional way of life unfeasible, increasing dependence on the market, the state, and welfare policies. They have no forest to hunt in, rivers to fish from, firewood to keep them warm in winter, or thatch to build their houses, consolidating imbalances within their Kaiowá social organisation.

> The reduced earth corresponds to a weakened state of being (teko). Consequently, a weakened teko cannot adequately contribute to the maintenance of cosmic balance. In this sense, the loss of access to land by the Indians, due to the conditions of colonial rule to which the Kaiowá were constrained, implies a growing risk of catastrophe. The struggle to recover lands traditionally occupied by these Indians brings with it the need to continue a process of constant relationship with Yvy, so that it does not become ill, seeking to re-establish, as far as possible, the conditions of the indigenous social morphology that allow for the manifestation of a proper Kaiowá way of life (tekoporã) (Mura 2006, p. 108, translated).

[15] According to Brand (1993).

The loss of territory meant fragmentation at various levels of their social organisation, such as kinship, religion, the material production of living conditions, celebrations, and everything that shapes their community relations. Currently, living in small spaces, the Kaiowá can no longer organise their tekoha according to their needs, living with other kins. This has generated tensions, conflicts and disputes, since such fragmentation and disputes mean that not all kins are allies. One of these difficulties was created with the institution of the figure of the captain in the reserves, whose authority, imposed by the state, generates tensions in his relationship with his relatives and rival groups (Pereira 1999). Even in this context of difficulties, the author emphasises that the structure and function of the kin remain, regardless of the area they occupy:

> ... the Kaiowá insist on gathering the houses around the place of residence of the head of the kin - hi'u - occupying a continuous space. (p. 95, translated)

18.3 Tehoka Renewals and Guarani R-existences, An Ontological Turn

From the end of the 1970s, with the intensification of the expropriation process, the Guarani and Kaiowá began the movement to recover their territories. Chamorro (2018, p. 307, translated) pointed out that, in the case of the Guarani and Kaiowá, even in "(...) 1978, the date of the first recovery, it can be said that the Kaiowá and the Guarani lived in the time of ñemosarambipa or spreading". Later, they began to react, supported by achievements such as the Federal Constitution of 1988, among others, initiating what the indigenous people would call the "time of law".

Large assemblies, called aty-guasu, began to be taken up again among the Kaiowá, as an indication of political organisation. The first regional aty-guasu took place in 1978, according to Barbosa and Mura (2011)—the year which saw the first revival for the recovery of their territories. The movement to recover the territory of the Guarani and Kaiowá took place especially beginning with the repossessions and encampments, which they called tehoka, generally occupying properties where their tekoha were located. When carrying out occupations and repossessions, they came into direct confrontation with landowners and capitalists—that is, with the owners of the means of production, receiving class-focussed reactions and offensives, such as violence and crimes against indigenous peoples; dialogue from the rural caucus attempting to curb the demarcation process; auctioning to raise funds in order to hire armed security; and class organisation.

It is observed that although tekoha is highlighted by some authors as a concept that refers to the notion of place, which incorporates the spatial dimension from a symbolic-cultural perspective, authors such as Cavalcante (2013, p. 80, translated) have highlighted that,

> The tekoha category[16] gained emphasis through the indigenous discourse as an exclusive territorial space for a group of large families (village) from the point when, due to the historical situation they experienced, the Guarani and Kaiowá were forced into a political battle for the demarcation of indigenous lands.

This perspective of struggle and resistance draws attention to the dimension and characteristic of this struggle. In the tekoha category, the territorial dimension gains a different connotation, from the point of view that its appropriation begins to incorporate more precise limits, with its realisation, mediated through conflicting relations with landowners, implying the demarcation of the territory. This need for contextual insertion of the concept of tekoha is also addressed by Thomaz de Almeida e Mura (2003). The authors assert that, before the arrival of the non-Indians, the question of precise limits and boundaries—which became urgent after the arrival of the colonisers—was not posed to the Guarani. The authors draw attention to the consideration of the native category not as something inherent, but in its historical context, which does not dispense with spatio-temporal and territorial relations or, therefore, with power relations.

The territorial appropriation model given from the colonial period and consolidated with the constitution of the National State, imposing private forms of appropriation mediated by merchandise, does not represent the only possible territoriality—something which is also shared by Ladeira (2008). The author highlights, however, that this model of territorial appropriation beginning from the state—which imposed a fixed territorial base on indigenous peoples as a strategy of control and domination—created a structural dependence for them in relation to the state, which we consider to be reverberated in the form of becoming, in an ontological shift that poses a deep reflection on the dimension of place and kinship for native peoples. Considering the achievements of the repossessions, tehoka, and the realisation of their coming-to-be, tekoha, from a confederation of places for everyone, or from a sedentism and form of particular spatial appropriation, we understand this process as a condition of r-existence.

The current situation of the Guarani and Kaiowá peoples in Mato Grosso do Sul—be they those who live in areas reserved and recognised by the state for their use; those who remained in the background at the ranches that formed around their old tekoha; those living on the outskirts of cities; those from the camps/repossessed areas—reveal conditions of existence given starting from historical–geographical relational determinations in the process of the surrounding capitalist society's consolidation[17]. This spatial and lifestyle-based distribution was incorporated by the indigenous people into a becoming that had a more unified space-territory-space dimension, becoming fragmented through the imposition of an ontological rupture which, procedurally, was being reframed. Within the reserves, there is a "social reorganisation" resulting

[16] In making this observation, the author agrees with Mura (2004 and 2006 apud Cavalcante, 2013), who originally would have made such an observation.

[17] The land situation of the repossessions is quite diverse, as there are lands that have been homologated, regularized, interdicted, declared, delimited and under study, as already highlighted by Crespe (2015).

from compulsory confinement and the introduction of other agents such as the head of the Indigenous Post, the figure of the captain, and evangelical pastors, in addition to the actions of external agents resulting from proximity to the urban area, such as schools and Non-Governmental Organisations.

Among the Guarani Kaiowá, the repossessed areas are called tehoka[18]—future tekoha (Pereira 2012; Chamorro 2015; Crespe 2015). This category was used for the first time in 2011, at the 1st Meeting of Indigenous Encampments, to make it clear to the "whites" that these are indigenous lands. Even though its use is not consensual, we understand that tehoka is quite enlightening in terms of revealing the construction of a category in the process of struggle and r-existence, as well as the clarity and understanding that the Kaiowá and Guarani have of it. Another important element of organisation in the indigenous movement is the aty-guasu, whose origin can be better explained by Tonico Benites, Guarani-Kaiowá leader and researcher:

As a reaction to these truculent acts by farmers and their thugs, a political movement, little known in the rest of the country, emerged in the [19]80s—the great Guarani and Kaiowá assembly, AtyGuasu. The aim was to face up to the systematic process of ethnocide, the expulsion and forced dispersion of large indigenous families from their traditional territory (Benites 2012, p. 2, translated).

Cavalcante (2013) highlights that, after 1980, 12 Indigenous Lands totalling 24,383 hectares were conquered, recognised and fully owned by the Kaiowá and Guarani, with another ten areas totalling 6032 ha that are partially owned by the indigenous people. The repossessed indigenous/tehoka areas are spaces of struggle and resistance, where they seek to re-territorialise their way of life.

On 7 December 2013, the landowners held an auction, which became known as the "Auction of Resistance"[19], to raise funds and set up security on farms, in order to cover legal and other expenses in the confrontation with the indigenous people. This auction had the support of several parliamentarians representing the sector and, despite indigenous communities and social movements having filed an action against it, Famasul (the Federation of Agriculture and Livestock of Mato Grosso do Sul) and

[18] Crespe (2015, p. 163) observes that tehoka speaks of time and space, of past and future, simultaneously, where "... the suffix rã is an indication for the future. Tehoka is the tekohaque that needs to be again". The author emphasizes, however, that the other terms were not completely abandoned, with no consensus even among the Kaiowá and Guarani.

[19] Despite referring to the Landless Rural Workers Movement, we believe it is possible to apply the same concept used by Fernandes (1996, p. 238-239) to the indigenous movement: "The space of struggle and resistance is therefore the result of a project of struggle [of landless rural workers], of their survival as a historical subject. (...) The encampment is, in its concreteness, the space for struggle and resistance, it is when [the workers] go into direct confrontation with the state and the landowners. (...) In this space, the struggle and therefore the confrontation with the state, through political negotiation, and with the landowners, through direct conflict, is put into question. Depending on the form of the encampment and the power relationship between the political forces, different situations of struggle will arise. With regard to the state, eviction occurs frequently, through the use of violence by the police force. Regarding the landowners, the violent confrontation against the jagunços hired to do the work".

Acrisul (the State Association of Livestock Farmers) appealed and managed to hold the auction[20].

One company that became known for acting as "security" on the properties was GASPEM Segurança Ltda. In 2013, the Federal Public Ministry in Mato Grosso do Sul (MPF/MS) filed a lawsuit for dissolution and cancellation of the company's registration, accusing it of operating in properties that were involved in land conflict, and of committing crimes against indigenous communities (MPF/MS, 08/28/2013). In Mato Grosso do Sul, several indigenous and non-indigenous leaders have been receiving death threats, and in many cases, these have actually turned into a reality. Thus, rentier capitalism is territorialised in the countryside of Mato Grosso do Sul, with violence and interactions in all instances of power.

The territorialisation of sugar cane capital and soy bean and corn monocultures more directly exposes monopolistic industrial capital and landowners in the conflict over land. In May 2010, Raízen, a company resulting from the merger of the Shell/Cosan group, was accused by the Federal Public Ministry of buying sugar cane illegally cultivated in the Guyraroká Indigenous Land, in the municipality of Caarapó. In June 2012, it signed an agreement with FUNAI to no longer buy sugar cane grown in Indigenous Lands. Additionally, Bunge's Monte Verde plant leases the Santa Luzia, Guarida and Três Marias ranches, which fall on the Jatayvary indigenous area (MPF/MS, 11/05/2010).

18.4 Guarani and Kaiowá R-existences: Impasses and Challenges

The occupations and repossessions (tehoka) have been accompanied by conflicts and violence, transforming (particularly in the south of the state) into a battleground with genocide of the original peoples. The increasing restriction of territories for their exclusive use (Barbosa and Mura 2011) led to a reduction in hunting, fishing, foraging and agricultural practices, making them increasingly dependent on money and the market. The process that culminated in the transfer of part of the Kaiowá into the reserves began to represent an impasse for what is their "ñandereko". According to Pereira (1999, p. 95, translated):

> ... the Kaiowá insist on gathering the houses around the place of residence of the head of the kin - hi'u - occupying a continuous space. Compulsory cohabitation with other kin, not always allied from a marital, political or religious point of view, generates constant tension and an endless series of conflicts and disputes over resources, territory and power.

This search by the Kaiowá to maintain the social organisation of the kin, their own way of life in the face of concrete material conditions of existence, reveals the contradictory relationship of the way of life—that is, the way of living permanently

[20] G1 MATO GROSSO DO SUL. "Resistance Auction" raises BRL 640,500 in Mato Grosso do Sul. In: https://www.g1.globo.com/mato-grosso-do-sul/noticia/2013/12/leilão-da-resistencia-arrecada-r-6405-mil-em-mato-grosso-do-sul,html. 12/08/2013.

(ser) is also a way of living temporarily (estar). It is the existence of being in their spatio-temporality. In the case of existence inside the reservations, where they do not have enough space to carry out their way of life, such contradictions become acute in the extreme, unfolding into territorial conflicts in their multidimensionality—that is, political, ethnic, religious, economic, social conflicts, and conflicts between and within extended families, etc. They also reveal the struggles and conflicts experienced by the Guarani and Kaiowá in their process of r-existence.

Data collected by Brand (1993) indicate that until the beginning of the 1970s, the indigenous people who lived on reservations managed to live off the traditional plots of land. However, the agro-forestry management system collapsed; they could no longer practise hunting; agriculture became reduced, since the need for earning a wage displaced the indigenous people to other properties (not only in the region, but in other states)—as in the case of apple harvesting, during the period in which they should have been working to develop their plots. Furthermore, salaried employment occurs without any system of legal protection, where contractors pay what they want.

Through testimonies collected from indigenous teachers in the Dourados Indigenous Reserve, in the villages of Jaguapiru and Bororo, Lima (2012) demonstrates important aspects of the Guarani spatiality in the context in which the indigenous people lived during the expansion of fronts into the space that came to constitute what is now the state of Mato Grosso do Sul. Using memory-based resources, she describes their spatial, individual and community trajectories, in what the author calls memory-based cartography. In this, she demonstrates a spatial and historical trajectory present in previous generations, where links with ancestral territories that go beyond the institutional limits of Brazil are evident, also reaching Paraguay. She also demonstrates a cartographic example (Fig. 18.2) of the Guarani spatiality that remains today. The author shows that teachers from the villages of Jaguapiru and Bororo maintain matrimonial alliances, relationships and ties of kinship within the extended family, even if the families are located in villages in other municipalities—that is, even in the face of this context of fragmentation of their traditional territories: "The village/tekoha networks are basically articulated by ties of kinship maintained through matrimonial and political alliances that shape intra-ethnic social relations" (Lima 2012, p. 172, translated).

The aforementioned map is also quite illustrative of the r-existence of Guarani territoriality in the face of the fragmentation of its spaces in relation to private property and the linear political-administrative divisions of the Nation-State, which often unfold in struggles for territorial reconfiguration, with the repossessions/tehoka. Pereira (1999) highlights that kinship continues, nowadays, to be a strong element in the realisation of their teko way of life, which unfolds into a continuous factor of interaction, cooperation and sharing.

Another factor present among the Kaiowá that interferes with their form of organisation concerns the presence of Pentecostal churches. In the case of the Dourados Indigenous Reserve alone—an area of 3,475 hectares and around 17,000 indigenous people—77 churches were identified in 2013 (Pereira and Chamorro 2018). In this work, they highlight the role of missionary activity in helping "submission,

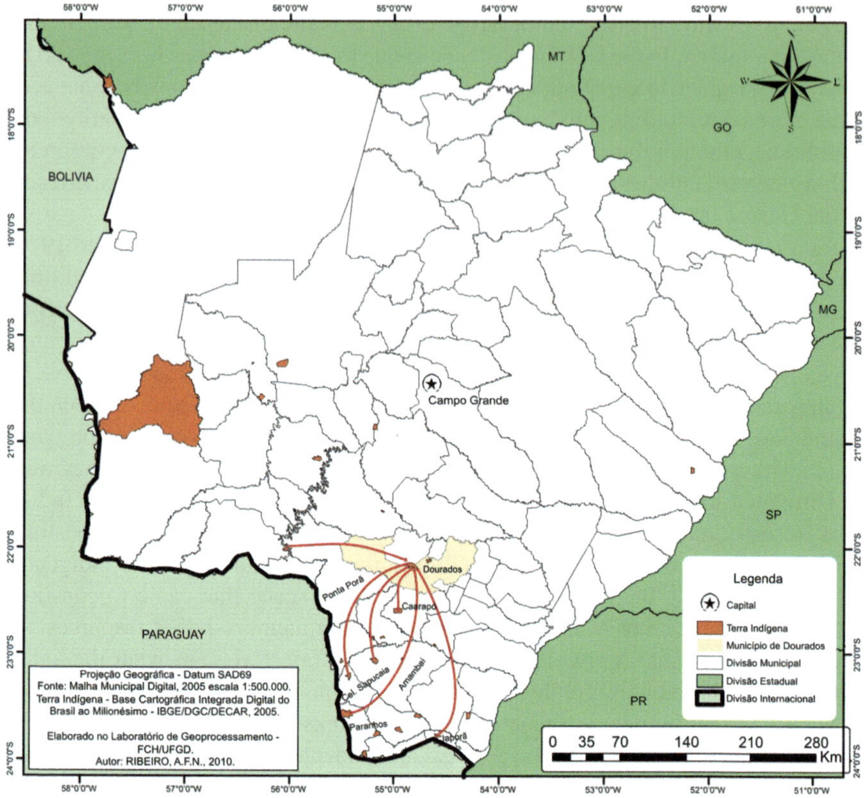

Fig. 18.2 Network of villages and life trajectories. *Source* Lima (2012)

and convincing the indigenous people to occupy the place reserved for them in the project of nation…" (op. cit. p. 633, translated).

One can get an idea of the presence and influence of Pentecostalism within the villages considering that, in the Dourados reserve alone, the figures updated for 2017 show that there were records of 89 churches in the reserve and only two prayer houses, one of which was destroyed in 2019[21]. In the documentary[22] "Monocultura da fé" directed by Joana AranhaMoncau and Gabriela Moncau, based on the works of Kaiowá historian and researcher Izaque João and anthropologist Spensy Pimentel, there are reports[23] of religious intolerance, aggression and dilemmas experienced by

[21] According to data contained in the documentary Monocultura da fé (Monocultura da fé, 2017).

[22] In 2020 the house of prayer was rebuilt, with the support of the Ecumenical Service Coordination (CESE), a Christian philanthropic civil society, which brings together a group of Christian churches (Igreja Evangélica de Confessão Luterana no Brasil; Igreja Presbiteriana Independente do Brasil; Igreja United Presbyterian Church of Brazil; Anglican Episcopal Church of Brazil; Roman Catholic Church – CNBB Available at: https://www.cese.org.br/quem-e-a-cese/a-cese/Aliança de Batistas do Brasilfile:///C:/Users/DESKTOP/Downloads/ESTATUTO_CESE_2019.pdf

[23] Monocultura da fé, 2017.

the Guarani-Kaiowá people. The research was carried out in 16 villages in the south of the state, where Izaque interviewed worshippers, pastors, and indigenous leaders.

Addressing the causes of the persecutions and attacks on worshippers and prayer houses, in the aforementioned documentary Izaque João highlights three reasons: (a) economic interest in raising money among the indigenous population; (b) disqualification of the worshippers' knowledge, so that children leave their cultures and come closer to Pentecostalism; and (c) disorganisation of the Kaiowá social structure, as a way to dominate the Guarani-Kaiowá community, which has been seeking recognition for its traditional territories, in a context where they are surrounded by transgenic corn and soy bean plantations, making these offensives confluent with interests of the landowners themselves.

As can be seen, it is in the process of capitalist appropriation of land, and its subsequent commercial use, that we are going to verify processes of expulsion of the Guarani and Kaiowá, the destruction of their tekoha, confrontation with private security guards, police, and agricultural aviation companies, exposure to pesticides, and all forms of violence against them. In these conflicts, we verify that the indigenous people themselves identify the subjects in conflict, starting from their own cultural systems, but reinterpreted through contact with capitalist society, such as the identification of the "karaí ranchers". It is interesting to note that in this terminology attributed by them, the nature of the conflict is identified. First, the non-indigenous karaí are highlighted, and among the karaí, they qualify the karaí subjects involved in territorial conflicts: the figure of the rancher. In Brazil, according to the Dicionário da terra (Dictionary of the land), the term fazenda (ranch) was related, during the colonial period, to the type of economic activity carried out on the property (raising livestock). Over time, it became associated "with the idea of a large extension of territory" (Motta 2005, p. 221, translated). "Karaí rancher" refers to the non-indigenous owner of this large extension of territory, who we conceptually call the landowner.

By claiming their traditional territories, their tekoha, conflicts over land are territorialised in specific places—that is, they are located chiefly in repossessed areas (tehoka). In a capitalist, class-based society, human existence is not fully realised, as it is rendered impossible, regulated, blocked and controlled by the holders of the means of production, where the common good, transformed into private property, becomes the usage value, exchange value and capital, imprinting a dynamic of existence under the economic determinants of capital accumulation (Moreira 2004). The mediation of merchandise and private property, as well as the territorialisation of this process, increasingly imposed new relationships on indigenous peoples—but not in a linear way, since it was also accompanied by its opposite, in a becoming where r-existences were signified and re-signified.

More recently, at the end of 2018, some camps/repossessions in the municipality of Dourados were the target of violent attacks against the Kaiowá and Guarani, involving private security guards hired by landowners in the surroundings of the Dourados Indigenous Reserve. The attacks, which targeted children, young people and the elderly, included rubber bullets, torture, a grenade wound, and injury from the destruction of their shacks, including injuries to a 75-year-old elderly woman[24].

In this sense, Mura (2006) highlights that the ethnic element has become a new component in the indigenous spatial configuration. Using the memory of the past and narratives with the group, they seek to reconstruct the territory, culturally elaborating what they call the spatio-temporal map, which underlies their current demands. The tekohaguasu, which could be similar to the guára[25], a territorial unit of the Guarani identified in the past, is a category that was defined as a function of present demands which are currently represented in expressions, from the perspective of the Kaiowá, of their contemporary territorial configuration. According to the author:

> the fact that the Kaiowa's descriptions take account of clear physical boundaries of the community space, and that this is religiously attributed to them by the creator god, highlights more than anything the current effort to spatially conceptualise their own social relations - an event made possible, in good measure, via recourse to the memory of the past. (Mura 2006, p. 116, translated)

Despite being the original peoples of America, their rights have been usurped—they are not free to territorialise their way of life. By claiming tekoha, they seek to territorialise other relations—non-capitalist and outside of capitalism. They want to be able to fully realise their way of life in the world, which is not based on the capitalist, class logic—and this plays a fundamental role in the Guarani resistance. Pereira (1999) argues that the profound mysticism that is present in the Guarani culture, and its emotional and sentimental attitudes, does not completely mischaracterise the capitalist economic rationality—and also does not introduce it in a mechanical and homogeneous way. As a creator of value, indigenous labour is appropriated and incorporated into the circuits of capital. However, the Guarani-Kaiowá continue to resist and fight to maintain their way of life, which clashes with the movement for reproduction of capital. The understanding of this way of life is outside the sphere of the political economy, but when they r-exist, they confront the interests of the owners of the means of production and are targeted by offensives from that class. It is therefore not possible, as Quijano (2006) pointed out, to analyse the struggles and resistance of indigenous peoples while bypassing power relations. While seeking to transform the indigenous people into a reserve army within these reserved spaces, the Guarani and Kaiowá r-exist, for the recovery of their traditional territories, re-signifying the meaning of the struggles.

18.5 Final Considerations

The indigenous people denote their resistance by seeking ways of being, existing and resisting, forging a new territoriality and thereby enhancing a new territorialisation

[24] Conf. CIMI, 16/11/2018; 24/09/2019; 07/11/2019;18/09/2019; 07/01/2020).

[25] Guára: ""coming from", in the colonial period, territories based on the local hydrography from which certain Guarani groups come" (Mura, 2006, p. 500).

by giving fluidity to the border imposed by the private ownership of land and the power relations which permeate the realisation of this way of life. In other words, the non-Indians were not able to imprison the Guarani (and along with them their way of life) inside the Reserve—but private property is a siege that goes beyond the limits of this spatiality, imposing limits on their existence and implicating it in forms of land use. Within this logic, indigenous people, workers and peasants are victimised by the violence from such a legal institution.

In the state of Mato Grosso do Sul, in view of this complexity and diversity of tensions and conflicts, we highlight three categories of social subjects who are directly affected in this process—the indigenous people, the peasants and the quilombolas—while only a few questions referring to the Guarani and Kaiowá reality were presented by us. The movements to recover their territories have revealed not only the Guarani and Kaiowá prominence, but also the perverse face of the agrarian debate in Mato Grosso do Sul. The land conflict highlights the issue of land monopoly that is at the centre of territorial disputes in the state, reconfiguring and expanding the debate on the meaning of class, land and territorial struggle. The struggles cannot be detached from their historical, social, economic, political and cultural content. These plans interrelate and complement each other.

The Kaiowá form of social organisation highlights important questions to us when reflecting on daily practices in the struggle for rights, regarding their ability to become human rights. Likewise, among the Kaiowá and Guarani, the recognition and maintenance of their agro-ecological practices, their knowledge, and their social organisation based on elements of solidarity and reciprocity, teach us a lot about (re)existing.

It is noteworthy that in this trajectory of the native peoples' confrontations with social class structures, the ruptures caused by a space-territory-space way of life are constantly being imposed. For the Guarani and Kaiowá, whose existences are demarcated by kinship, such forms of production and social reproduction are gradually being confronted and introducing demands and r-existences in the maintenance of an ethos. The question is to what extent the impositions of a class-based society can produce a territorial ontological turn, obliterating the concepts of shared living, of a tekoha that is unique—even if spatially fragmented by agribusiness and the structures of a class-based society.

References

Barbosa PA, Mura F (2011) Construindo e reconstruindo territórios Guarani: dinâmica territorial na fronteira entre Brasil e Paraguai (séc. XIX–XX). Journal de la société des américanistes [En ligne], 97-2

Brand A (1993) O confinamento e seu impacto sobre os Pai/Kaiowá. PUCRS-IFCH, Pontifícia Universidade Católica do Rio Grande do Sul, Porto Alegre. Dissertation

Brand AJ, Ferreira EML, de Azambuja F (2008) Os kaiowá e guarani e os processos de ocupação de seu território em Mato Grosso do Sul. In: Almeida RA (ed) A questão agrária em Mato Grosso do Sul. UFMS, Campo Grande

Cavalcante TLV (2013) Colonialismo, território e territorialidade: a luta pela terra dos Guarani e Kaiowa em Mato Grosso do Sul. Assis-SP:UNESP, 470 f. Doctoral Thesis
Chamorro G (2015) História Kaiowá—das origens aos desafios contemporâneos. Nhanduti Editora, São Bernardo do Campo
Chamorro G (2018) Povos indígenas falantes no atual estado de Mato Grosso do Sul (séculos XVI–XXI). In: Chamorro G, Combés I (orgs) Povos indígenas em Mato Grosso do Sul. Editora UFGD, Dourados/MS, pp 293–322
Crespe AC (2015) Mobilidade e temporalidade Kaiowá. Do tekoha à reserva, do tekoharã ao tekoha. UFGD, Dourados-MS. Doctoral Thesis
Dardot P, Laval C (2016) A nova razão do mundo. Boitempo, São Paulo
Federici S (2017) Calibã e a bruxa. Mulheres, corpo e acumulação primitiva. Elefante, São Paulo
Fernandes BM (1996) MST—Formação e territorialização. Hucitec, São Paulo
Harvey D (2004) O novo imperialismo. E. Loyola, São Paulo
Jesus LC (2004) Erva-mate: o outro lado. A presença dos produtores independentes no antigo sul de Mato Grosso. 1870–1970. Dissertation (Master's in History). UFMS, CPDO, Dourados-MS
Krenak A (2019) Ideias para adiar o fim do mundo. Cia das Letras, São Paulo
Ladeira MI (2008) Espaço geográfico Guarani-Mbya. Edusp, São Paulo
Lima SCA (2012) Concepções de natureza e território na visão dos professores Guarani da escola indígena de Dourados/MS. UFG, Goiânia/GO. Doctoral Thesis
Lukács G (1967) Estética. Garijalbo, Barcelona/México
Melià B (1990) A terra sem mal dos Guarani—economia e profecia. In: Revista de Antropologia, n 3
Meliá B (2004) La novedad Guarani (viejas cuestiones y nuevas preguntas) Revisita bibliográfica (1987–2002). Revista de Indias V. LXIV(230):175–226
Moreira R (2004) Marxismo e geografia (a geograficidade e o diálogo das ontologias). GEOgraphia. Ano 6, n 11. Available at: https://periodicos.uff.br/geographia/article/view/13466
Moreno G (2007) Terra e poder em Mato Grosso. Contextualizando cenários e personagens. EDUFMT, Cuiabá
Motta M (org) (2005) Dicionário da terra. Civilização Brasileira, Rio de Janeiro
Mura F (2006) À procura do "Bom viver"—território, tradição de conhecimento e ecologia doméstica entre os Kaiowá. Museu Nacional/UFRJ, Rio de Janeiro. Doctoral Thesis
Oliveira AU (2008) O nó da questão agrária em Mato Grosso do Sul. In: Almeida RA (org) A questão agrária em Mato Grosso do Sul. Ed. UFMS, Campo Grande-MS, pp 05–14
Pereira,LM (2012) Expropriação dos territórios Kaiowá e Guarani: Implicações nos processos de reprodução social e sentidos atribuídos às ações para reaver territórios—tekoharã. Revista de Antropologia da UFSCar 4(2), São Paulo
Pereira LM (1999) Parentesco e organização social Kaiowá. UNICAMP, Campinas, Dissertation
Pereira LM, Chamorro G (2018) Missões pentecostais na Reserva Indígena de Dourados—RID: origens, expansão e sentidos da conversão. In: Chamorro G, Combés I (orgs).Povos indígenas em Mato Grosso do Sul. Editora UFGD, Dourados/MS, pp 729–744
Quijano A (2006) El "movimiento indígena" y las cuestiones pendientes en América Latina. www.sisbib.unmsm.edu.pe/bibvirtualdata/publicaciones/san_marcos/n24_2006/a01.pdf
Sassen S (2015) Expulsiones: brutalidad y complejidad en la economía global. Katz Editores, Buenos Aires
Souza JG (2009) Limites do território. Agrária (São Paulo. Online), (10–11), 99–130. https://doi.org/10.11606/issn.1808-1150.v0i10-11p99-130.
Souza JG, Borges ACG (2017) As determinações territoriais da lógica do valor e do autovalor análise da produção de soja no Mato Grosso. Brasil. In: Jacinto R (org) Iberografias: Outras Fronteiras, Novas Geografias: Intercâmbios e Diálogos Territoriais, 32edn, vol 32. Ancora Ed./Centro de Estudos Ibéricos (CEI), Coimbra, pp 247–270
Souza JG, Cabero Diegues V (2012) Por uma desglobalização da produção alimentar: commoditização da agricultura e diversidade produtiva—uma análise de Espanha. Geographia (UFF) 14:63–81

Souza JG, Fulino R (2020) A questão do método. Caminhos da Geografia (UFU. Online), vol 1, pp 1–15
Svampa M (2013) Consenso de los commodities y linguajes de valoración en América Latina. Revista Nueva Sociedad, n 244, marzo-abril
Svampa M (2019) Las fronteras del neoextractivismo en América Latina: conflictos socioambientales, giro ecoterritorial y nuevas dependencias. FLACSO Ecuador, Quito.
Thomaz de Almeida, Rubem Ferreira, Mura, Fábio (Oct 2003) Povos indígenas no Mato Grosso do Sul. In: INSTITUTO SOCIOAMBIENTAL. *Povos Indígenas no Brasil*. Modified Aug 2018. Available at: https://pib.socioambiental.org/pt/Povo:Guarani_Kaiowa. Accessed 25 Jun 2021

Documentary

Monocultura da fé (2017) Produção de Joana Moncau, GrabrielaMoncau, Izaque João e Spency Pimentel. Futura, Brasil. Documentário.(23 min). Available at: https://canaisglobo.globo.com/assistir/futura/curtas/v/6769937/. Accessed on 17 June 2021

News Site

Acrítica (2011) General Heleno lança, na Famasul, ONG Recovê. 18 Mar 2011. https://www.acritica.net/editorias/geral/assista-general-heleno-lanca-na-famasul-ong-recove/35076/. Accessed on 1 July 2021
Conselho Indigenista Missionário (2018) Kaiowás sofrem ataques em menos de um mês na Reserva de Dourados-MS, último ocorreu durante visita da CIDH. In: https://cimi.org.br/2018/11/kaiowa-sofrem-quatro-ataques-em-menos-de-um-mes-na-reserva-de-dourados-ms-ultimo-ocorreu-durante-visita-da-cidh/. 16 Nov 2018. Accessed on 10 Jan 2020
Conselho Indigenista Missionário (2019a) Video: Guarani e Kaiowá relatam caso de tortura durante ataques a retomadas em Dourados. In: https://cimi.org.br/2019a/11/guarani-e-kaiowa-relatam-caso-de-tortura-durante-ataques-a-retomadas-em-dourados/. 7 Nov 2019a. Accessed on 10 Jan 2020
Conselho Indigenista Missionário (2019b) Retomadas Guarani Kaiowá acumulam ataques recentes com destruição, feridos e morte. In: https://cimi.org.br/2019b/09/no-coracao-das-trevas-retomadas-guarani-kaiowa-acumulam-ataques-com-destruicao-feridos-e-morte/. 18 Sept 2019b. Accessed on 7 Jan 2020
Conselho Indigenista Missionário (2019c) A maior violência contra os povos indígenas é a destruição de seus territórios, aponta CIMI. In: https://cimi.org.br/2019c/09/a-maior-violencia-contra-os-povos-indigenas-e-a-apropriacao-e-destruicao-de-seus-territorios-aponta-relatorio-do-cimi/. 24 Sept 2019c. Accessed on 10 Feb 2020
Conselho Indigenista Missionário (2019d) Incêndio destrói casa de reza em Dourados. Available at: https://cimi.org.br/2019d/07/incendio-destroi-casa-de-reza-guarani-kaiowa-na-reserva-de-dourados/. Accessed on 10 July 2019d
Conselho Indigenista Missionário. Notícias (2020a) Laranjeira Nhanderu é atacada, tem casa de reza incendiada e indígenas Guarani Kaiowá vivem terror na virada do ano. In: https://cimi.org.br/2020a/01/laranjeira-nhanderu-e-atacada-tem-casa-de-reza-incendiada-e-indigenas-guarani-kaiowa-vivem-terror-na-virada-do-ano/. 2 Jan 2020a. Accessed on 7 Jan 2020a
Conselho Indigenista Missionário (2020b) Caveirão, tiros e feridos: o ataque de seguranças privados e policiais às retomadas Guarani e Kaiowá.In: https://cimi.org.br/2020b/01/caveirao-tiros-feridos-segurancas-policiais-retomadas-guarani-kaiowa/. 7 Jan 2020b. Accessed on 10 Jan 2020b

Conselho Indigenista Missionário (2020c) Laranjeira Nhanderu é atacada, tem Casa de Reza incendiada e indígenas Guarani Kaiowá vivem terror na virada do ano. 2 Jan 2020c. Available at: https://cimi.org.br/2020c/01/laranjeira-nhanderu-e-atacada-tem-casa-de-reza-incendiada-e-indigenas-guarani-kaiowa-vivem-terror-na-virada-do-ano/. Accessed on 30 June 2021

Conselho Indigenista Missionário (2020d) Laranjeira Nhanderu é atacada, tem Casa de Reza incendiada e indígenas Guarani Kaiowá vivem terror na virada do ano. Available at: https://cimi.org.br/2020d/01/laranjeira-nhanderu-e-atacada-tem-casa-de-reza-incendiada-e-indigenas-guarani-kaiowa-vivem-terror-na-virada-do-ano/. 2 Jan 2020d. Accessed on 30 June 2021

Emily D (2018) Sob cerco evangélico, guarani-kaiowá sofrem com intolerância em Mato Grosso do Sul. In: Brasil de Fato. 25 maio 2018. Available at: https://www.brasildefato.com.br/2018/05/25/sob-cerco-evangelico-guarani-kaiowas-sofrem-com-intolerancia-no-mato-grosso-do-sul.

Maisonnave F (2013) Fazendeira vira antropóloga e faz laudos contra índios. In: Jornal Folha de São Paulo. Poder. 9 June 2013. Available at: https://m.folha.uol.com.br/poder/2013/06/1292183-fazendeira-vira-antropologa-e-faz-laudos-contra-indios.shtml?cmpid=menutopo. Accessed on 05 July 2021

Midiamax (2012) Após ameaça de fazendeiros contra índios, Polícia Federal vai investigar fazendeiros de MS. In: http://www.midiamax.com.br/noticias/812493-apos-ameaca-de-guerra-contra-indios-policia-federal-vai-investigar-fazendeiros-de-ms.html. 21 Aug 2012

Ministério Público Federal (2010) MPF/MS questiona BNDES por financiar usinas que compram cana cultivada em áreas indígenas. 11 maio 2010. Available at: http://www.mpf.mp.br/ms/sala-de-imprensa/noticias-ms/mpf-ms-bndes-financia-usinas-que-compram-cana-cultivada-em-areas-indigenas-em-ms-1. Accessed on 20 June 2021

Ministério Público Federal (2013) Mato Grosso do Sul. Notícias. MPF/MS requer fechamento de empresa de segurança envolvida em morte de lideranças indígenas. 29 ago 2013. Available at: http://www.mpf.mp.br/ms/sala-de-imprensa/noticias-ms/mpf-ms-requer-fechamento-de-empresa-de-seguranca-envolvida-em-morte-de-liderancas-indigenas-2. Accessed on 05 July 2021

Ministério Público Federal (2015) Mato Grosso do Sul. Notícias. MPF/MS acompanha Comissão de Direitos Humanos da Câmara em visita a aldeias guarani-kaiowá. 05 Jun 2015. Available at: http://www.mpf.mp.br/ms/sala-de-imprensa/noticias-ms/mpf-acompanha-comissao-de-direitos-humanos-da-camara-em-visita-a-aldeias-guarani-kaiowa. Accessed on 05 July 2021

Santiago B (2020) Após incêndio criminoso, casa de reza Guarani/Kaiowá é reconstruída em Dourados (MS). Available at: https://www.cese.org.br/apos-incendio-criminoso-casa-de-reza-do-povo-guarani-kaiowa-e-reconstruida-em-dourados-ms-2/.08out 2020. Accessed on 17 June 2021

Chapter 19
The Geography of Labour Under Construction: Theoretical Challenges and Research Praxis

Antonio Thomaz Junior

> *work for us,*
> *you work with us.*
> *You'll be your own boss.*
> *(Ricky Turner. In: Sorry we missed you—Ken Loach,* 2019*).*

Abstract An already bad situation, with the added effects of COVID-19, has unveiled an explosive and cruel reality for workers, society and the environment in general. What has now become a reality, after being reshaped for the past 70 years, especially in Brazil, has revealed the simultaneous coexistence of regressive and ghostly forms (based on technological advances), associated with more rigid practices of labour control. As such, based on Taylorist-Fordist-Toyotist foundations, and under the auspices of technological development and information technology (IT), and now in the second decade of the twenty-first century of digital platforms, the brand of what we call systemic labour degradation is being imposed. The proliferation of outsourcing and platforms are imposing the dismantling of protection rights orchestrated by the State and other bodies linked to the interests of capital and the bourgeoisie. In the course of our research, the forms of (de)realisation, exploitation, subordination, etc., have allowed us to capture the contradictions and territorial dynamics of class conflicts, revealed through the existing plasticity, whether they are peasants, workers, employees in general, Quilombolas, men, women, formal employees or informal, temporary, self-employed, subordinate freelancers, etc. It gives us no pleasure to envisage that within the context of the COVID-19 pandemic,

This text is the product of several Research Projects: 1) Territorial Class Movement, Plasticity of Labour and the Impacts on Workers' Health (territorial conflicts and struggle for access to land and water). Validity: March 2016 to February 2020. Funding Agency: PQ-1/CNPq (Process: 304514/2015-2) 2) Thematic Project/FAPESP "Mapping and Analysis of the Sugarcane Agrohydrobusiness Territory in Pontal do Paranapanema-São Paulo-Brazil: Labour relations, conflicts and forms of land and water use, and environmental health" (Process: 2012/23959-9).

A. Thomaz Junior (✉)
Universidade Estadual Paulista/Presidente Prudente, São Paulo, Brazil
e-mail: thomazjrgeo@terra.com.br

R. C. Lois González and M. A. Mitidiero Junior (eds.), *Brazilian Geography*,
Advances in Geographical and Environmental Sciences,
https://doi.org/10.1007/978-981-19-3704-0_19

the number of unemployed workers could rise by 30 or 40% by the end of 2020, totalling 40 million unemployed and disillusioned, and the situation could be even more explosive in the post-pandemic era. The Geography of Labour produced under the CEGeT has been applied to studies and critical analyses in order to understand the objectified mechanisms of the complex process guided by the extraction/appropriation of the surplus manpower by capital on a global scale, in times of structural and systemic crisis. Nevertheless, only research will allow for an understanding and argumentation based on a dialectical praxis of the denial/affirmation of labour and its transcendence, based on experiences and learning which arise from the territorial movement of labour, and from the historical practices based on precepts of solidarity, registered around the world.

Keywords Geography of labour · Plasticity of work · Systemic degradation of labour · Working class

19.1 Introduction

In our research trajectory, the forms of exploitation, subordination, (de)realisation of labour, etc., have allowed us to gather the contradictions and territorial dynamics of class conflicts and the current stage of civilisation in the barbarism of capital, in a frank debate with Menegat (2006). More than that, through a "geographic reading" of the last 25 years, we have been able to add to the analyses and the theoretical and public debate understandings that bring us closer to the contradictions and redefinitions of the world of labour, through the concepts of our collective work, within the scope of the CEGeT Network of Researchers (Centre for the Study of Geography of Labour), that is, the plasticity of labour, the systemic degradation of labour, and the territorial movement of labour.

Explanatory elements of the objective reality of the geographic dynamics of labour and its plasticity, which is continually remade, have had a great impact on labour, as a consequence, the changes in/of social and political-ideological roles, which contextualise the territorial movement of the working class. Thus, we have been involved in critical studies and analyses, seeking to understand the reified mechanisms of the complex process guided by the extraction/appropriation of surplus labour and surplus value by capital (Marx 1982), on a global scale, in times of structural and systemic crisis.

However, only research has enabled us to understand and argue based on the dialectical praxis of the negation/affirmation of labour and its transcendence, based on the experiences and learning that emerge from the territorial conflicts of the class struggle. We have covered peasants, men, women, workers, wage earners in general, formal employees, whether they are informal, intermittent, independent or subordinate self-employed, etc., from intrinsic manipulations to the neoliberal precarisation of labour in the twenty-first century.

Paying attention to the living totality of labour, given the intensity of changes and contradictions carried out by the productive restructuring of capital, on a world scale, in response to the phenomenon of globalisation, requires theoretical-methodological measures in relation to the rearrangements of the metabolic order that adopts forms and procedures derived/combined from Taylorism-Fordism/Toyotism and, more recently, from the total flexibilisation and the consummation of the systemic degradation of labour, with Revolution 4.0. Added to this are the disastrous consequences that the COVID-19 pandemic imposes on society in general, on workers in particular.

Although the official figures on unemployment reveal an upward movement, it is disheartening that there are more than 31 million Brazilian workers who could be inserted into the job market (the number of unmotivated, underemployed/unemployed due to insufficient working hours, reached approximately 6.8 million). Unemployment hit 15 million Brazilians, especially young people between 18 and 24 years old, who accounted for 4.2 million. One reflection of this nefarious course is the increase in self-employed workers (also called micro-entrepreneurs) who take risks in search of income and survival, which increased by 4.0% from August–October 2020 to February–April 2021, amounting to 23.8 million (IBGE 2021).

In recent years, millions of men and women, or inhabitants of the periphery of the system, have been marginalised in the world of labour, with the informal workforce increasing, which in Brazil will reach 34.1 million workers in 2020, or 40% of the employed population, not counting those who experience turnover, low remuneration—even with formal contracts—and those who suffer prolonged unemployment, for more than six months (IBGE 2021).

The repercussions of this tragedy impacted the increases noted from 2017–2018 to 2020—under the effect of the COVID-19 pandemic—with regard to food and nutritional insecurity, which spiked from 10 to 19 million people, respectively. If that was not enough, according to the Family Budget Survey (POF)/PNAD/IBGE, in March 2021, 55% of the population lived in conditions of food and nutritional insecurity. Moreover, in August 2020, 47% of people of working age were employed, and in March 2021, only 41%, that is, more than half of workers were out of the workforce. However, let us not forget that according to the IBGE/PNAD-C (first quarter of 2021), of the total Economically Active Population (EAP) (out of a total of 100 million people), only 40 million have a permanent employment relationship. In other words, 12 million in the different dimensions of the public service, and as for the distribution by sectors of activities, just over 20% are in the primary sector, 21% in the secondary sector and 59% in the tertiary sector. Of these, 5 million people worked via digital platforms, especially as delivery and Uber drivers (the Uberised), and were totally devoid of rights.

Our objective in this text is to offer, in these few pages, founding ideas/positions/understandings of the Geography of Labour gathered here, considering the completed and ongoing research within the scope of CEGeT, including the systematisations already available. In addition, this publication also stems from the position of an essayist, such that it serves us to critically reflect on the different realities that are being built/destroyed by the impositions of capital and that are directly reflected in the incentivised/forced and constant mobility in occupational/labour redefinitions. Our research has persuaded us to defend the transcendence of labour, that is, to put the living totality of labour on stage and, through the praxis of research, apprehend the specific contents of the capital vs labour conflicts, the geographical meanings of this process and the consequences for the studies in Geography of Labour.

Even more, it has made us think about the effects of outsourcing, flexibilisation, demonstrations against the super-exploitation of labour, class oppression, occupational reinsertion, the manipulative impositions of sophisticated 4.0 technologies—associated with ICTs and algorithms, which are verbalised through Uberisation (on demand/delivery/face-to-face), as well as crowdworking (home office/digital) (Filgueiras and Antunes 2020)—as well as the resistance actions on lands, for life against agrochemicals, occupations of land, and the struggle for urban and rural land, for agrarian reform, etc.

In this sequence, we have reserved a place for the reality experienced by workers, by the working class, the misfortunes caused by the COVID-19 pandemic, even more so in the face of the genocide announced by Bolsonaro's (anti-)government. Thus, through renewed knowledge as an empirical-theoretical expression of the political centrality of labour, or the engine of transformative praxis, we have established the main pillar of the reflections that support the Geography of the Labour we practise, within the scope of CEGeT. The impacts that directly affect the reduction of the traditional industrial working class, the expropriation of millions of peasants, the growing number of unemployed people and the profound redefinition of the labour market demand that we explain these living realities, with the aim of understanding the universe of labour, increasingly fragmented, flexible, heterogeneous, without losing its centrality.

19.2 The Geography of Labour: A Theoretical Construction in Motion

More than 30 years of research, conducted in a training environment, classrooms,[1] political militancy, from a very young age, made it possible for us, by prioritising the prevalence of objective conditions for the existence of labour, to advance towards the understanding of the being who works under the bias of territorial dynamics.

Hence, the incessant search to strengthen the assertion that reality must be transformed and not simply understood, manipulated and managed, which contributed to

[1] Lessons learned as part of academic research, which also produce significant developments in the scope of university outreach and our actions with social movements. It is enough to emphasise that the personal research projects, the collective projects, the orientations both at the undergraduate level (Monographs, Scientific Initiation), which represent a significant amount, as well as at the master and doctorate levels, which total 62 completed works and 13 in progress, and post-doctoral internships. These are only justified when there is a multiplication of individual/collective knowledge and learning, so that the result is socialisation through perennial actions, preferably.

recognising the centrality of labour, as a political centrality, and the praxis of research, without ever giving up on the methods of kneading clay.[2] This metaphor can be read by the poetic message of the Brazilian musician/composer Milton Nascimento: "[…] Every artist has to go where the people are".[3]

It is opportune to recognise the importance of the Lukacsian interpretation of the interrelationships between distinct complexes that are formed and modified (Lukács 2012), considering that, under the productive restructuring process of capital, they guide new lines of expression of the territorial labour conflict, at the beginning of the third decade of the twenty-first century, not restricted to the classic capital vs labour format, but involving other forms of configuring class domination.

In this perspective, at the same time and in a connected way, we base the geographical "reading" of labour on the dimension of regulating society and space. However, we subject our thinking to two connected and simultaneous movements: the construction and destruction of the Geography of Labour. If, on the one hand, we intend to make labour a constant theme of Geography, on the other hand, it is not another new discipline, current or branch, but a field of research that prioritises the universe of real labour and not the academic limits and the understanding of the State as labour categories.

Here it is worth emphasising that, if the mode of constructing society is the same as its space, in capitalist society, divided into social classes, it is necessary to understand that there is a cohabitation of opposite subjects/classes, based on a permanent conflict expressed between hegemony/counter-hegemony, so that there is simultaneously a domination/subordination and a resistance. That is, it is a societal structure whose spatial content contains class inequality and also counterspace, that is, the subjects of resistance, emancipatory (counterspace) struggles, who are, in the sphere of labour, under different forms of control and domination of capital, either through the exploitation of surplus labour, or through the subordination of profiting the land for capital, or through other forms and relationships. Now, it is essential that we know how to understand the conflicts that indicate, from a simple questioning to the forms of resistance of workers to proletarianisation, exploitation, subordination, etc., the defence of general guarantees, such as democracy, freedom of expression, political ties, organisational entities, ideological affiliations and class purposes.

It is precisely in this contradictory coexistence, not always visible/apprehensible, of class conflicts that Moreira (2016) defines the cohabitation between space and counterspace, concluding that there is no space without counterspace and vice versa. Heck (2017, p. 70) interprets this dialectical cohabitation, arguing: "Understanding workers as creative social subjects, who can interfere in space through counterspace struggles cannot do without the revolutionary centrality of labour".

We reinforce this position by explaining the class alliances of the bourgeoisie that support the entire process of capital accumulation and reproduction. Chesnais (1996), in A Mundialização do Capital, was able to capture the essential processes

[2] This expression designates the researcher's need to dedicate himself to empirical investigations, arising from direct research activity on the subjects in question.

[3] We refer to the lyrics of the song "Nos bailes da vida"!

that took place in the orbit of capital, in the first half of the 1990s, which are valid for the time. In the midst of this process, capital—as a globalised social relationship—or the bourgeois fractions that compose it, was able to dispose of the ideological discourse capable of validating it in the scientific-academic environment as well as the political environment, through the ideological concept of globalisation, which quickly added to the capital's manipulative system.

These are the founding pillars that have allowed us to reiterate the indispensability of not distancing ourselves from the history and corresponding territorial movements of workers, around the changes and rearrangements that capital builds worldwide, exactly to make us feel more secure, in order to understand the repercussions at play, whether objectively or subjectively, as well as the implications of this complex of complexes.

The perverse effects of the proliferation of formal, yet extremely exploited jobs, such as in call centres, telemarketing operators (also called infoproletarians), must be observed before the generalisation of informality, outsourcing and job flexibility. This is associated with an increase in the number of women in the labour market, that is, on the basis of intense precariousness, which has repercussions on the growing mass of workers excluded from the formal labour market.

In any case, the precarious proletarian, or "[...] the lowest paid and exploited fraction of the urban proletariat and agricultural workers" (Alves 2014), or those millions that move towards increasing precariousness, with losses of purchasing power/income, which fluctuate with varying degrees of misery—earning a maximum of two minimum wages—which will add to what is beyond the reserve workforce, as it is used as pressure on those who already have a job (in those already precarious conditions), become a surplus or discarded workforce or one that is at the limit, which Marx called the "relative super population" (floating), which guarantees the precariat the most degrading levels of the twenty-first-century proletariat. Some attribute it to composing the ignorant and apolitical mass that is so desired by the bourgeoisie, as if this was something peculiar to a niche and not systemic of class society.

The perverse synchrony between the implosion of the productive forces and their imminent disposal transforms contingents of workers into scrap—who are simply discarded—and, as a result and at the same time as a cause, we have the fact that labour, as a measure of wealth, already in its abstract form, undergoes profound changes, but it has not ceased to be a reference for value, for the law of value, for commodification and for the accumulation of wealth.

We think that it is not theoretically convincing to defend the framing of labour in the category of semi-proletarian, sub-proletarian (as we will see below) or lumpen, as a demarcation to continue maintaining the distance of these contingents from the composition of the working class, as well as the defence of their decentring or non-transcendence. That is, maintaining this blockade will make us miss the understanding of the rich and contradictory process of redefining the struggles and the content of the current geographical dynamics of labour, of the renewal of the composition of the working class in the twenty-first century, exponents of marginalisation carried out by total flexibilisation.

For that, it is essential that we associate the real meanings of the uncontrollability of capital and the fact that society can only be transformed through class struggle, which demands from us attention to the current situation of the social fabric of labour, which is intensely fragmented and the favourite object of the capital manipulation system. Obviously, it is imperative to consider the ruptures and tensions around the search for alternatives in life and labour, beyond the pre-defined formats such as proletarianism (or the condition of proletarianism).

Added to this Harvey's (1992) formulation, which, in my view, contributes to the understanding of the moment we are living, when he suggests the expression of space–time compression, in which he purports that there are two opposite and simultaneous phenomena happening with regard to labour. On the one hand, fragmentation and spatial and temporal dispersion and, on the other hand, under the effects of information technologies, the compression of space, that is, everything happens here, without distances, differences or borders, and the compression of time: everything happens now, with no past and no future.

Therefore, based on the definition of proletarianism, the existential condition of proletarianism (job insecurity at the limit) is separated from the identity of the "social class of the proletariat", because it is necessary to "[…] break with the positivist incrustations that impregnate the sociological (and Marxist) elaborations of the concept of social class (Alves 2011, p. 59).

Perhaps there is still time to think about possible metamorphoses or opportunities to get to know more closely the processes and conditions of life and labour, in which workers are subjected and subordinated to the effects of the productive restructuring of capital, and to the path of globalisation, as stated by Alves (2011, p. 73), "The masses and the people constitute themselves as a class in the full sense of the concept when they become indignant, resist individually or collectively, or even organise and fight, by themselves and for themselves, as a historical class subject capable of social change against the condition of proletarianism".

It offers us little scope to explain the diversity and complexity with which these attacks on labour and the consequent tensions put back on the scene forms of resistance and struggles that require, from researchers, union bodies, social movements, etc., the eyes of a lynx and willingness to face the political debate, getting rid of models or cast archetypes of interpretation, as if we were giving a blow to dialectics and, worst of all, speaking on behalf of it. The breadth of this subject allows us to point out the entire theoretical heritage of Brazilian authors, including myself, as we have seen, formed in these ranks and who agreed, disagreed and, in their own way, added more elements to the debate, offering rich intellectual material.

However, given the scenario of the twenty-first century, the questions that the metabolism of capital presents, the debate and the references inherited from that past are not responding to the challenges of the reality of labour and the working class. Here we recount the considerations elaborated in the thesis submitted to the teaching career promotion contest,[4] based on the understanding that in this synthesis—the

[4] Cf. THOMAZ JUNIOR, 2009.

Geography of Labour in Brazil—offers thoughts linked/potentiated by the praxis of research, over the last 25 years.

The praxis that we believe to be the most correct procedure for apprehending this movement and the mediations that constantly redefine it is also the resource for understanding the mutations internal to the universe of labour. More than that, it is even worth capturing the plasticity that exists between the different geographical forms, which express nothing more than the externalisation of labour in places, their topological, geographical meanings, etc.

Therefore, when it theoretically and politically defended the proletariat, because it is the only class in bourgeois society that lives from the wealth it produces, because it is united by multiple relationships and opposes the interests of other classes, it is therefore the only one capable of bringing together the other segments of society for the radical critique of capital, towards the construction of socialism (Lessa 1997). It is a matter of are we able of convincing ourselves. Thompson's (1998) considerations indicate that "[…] a class cannot exist without any kind of self-awareness" (Thompson 1998, p. 105), considering that class formation and consciousness of class are faces of the same process. Failure to understand this condition may endorse a staged class development model, in which consciousness would be a sort of derivation from the higher stage of the class.

The understanding of class as a historical category supports the critique of the understanding that makes theory prevail over the historical phenomenon, thus making class a static category. It is present in Thompson's (1998) position, on the static component of class, the criticism of the existence of a vanguard that would know, more than the class itself, what would be the convenient or correct class consciousness.

The whole polemic characteristic of this subject is not the focus at this time; rather, it is the fact that these two dimensions (relationship and organisation/class interests) refer to the Marxian concept of social class, so we think it is more appropriate to consider them in connection. In this sense, it is not necessary to give in to the appeals of Bourdieu (2008), when he proposes ruptures with Marxist theory, in view of the restriction of the scope of social analysis and the insufficiency to explain the social world, basically because it gives privilege to the economic sphere. We do not agree with this stance, not least because Marx did not set out to build a theory on classes and also because the immobilisation of this category of analysis makes it an obstacle to the exercise of scientific research and transformative political action. This does not mean that we fail to recognise that there is a gap in the intellectual-academic-political production of studies on social classes, in the face of social differentiation, currently, heterogenisation, cleavages of great depth in labour, etc.

Following this line of argumentation, we are concerned with responding to the results of the territorial mobility of labour, linked with labour change and the respective impacts on the worker's subjectivity. The concept of deterritorialisation opens the possibility of making visible the pre-established links between territorialisation-deterritorialisation-de territorialisation (TDR), which tell us little about the social plots and the respective territorial identities, at each stage.

Our effort has been to qualify the movement of labour, or the aspects and contents of the negation/affirmation of the forms of (de)realisation of the mobilities that each stage of the TDR expresses and the means of control that capital exercises an expression of the plasticity of labour in each of them. It is in this environment of reflections, in which we deal with the understanding of the movement of (de)realisation and the new identities arising from the redefinitions based on labour/occupational reinsertion and their respective territorial dynamics that we arrive at the concept of labour plasticity. Nevertheless, a critical and transformative epistemological reference is necessary, capable of capturing these signs that break the spatial order given by the delimitations of professions/qualifications, due to the impacts of the new demands of capital accumulation and also to situate the new emancipatory calls and pulsations within this movement of (de)realisation of labour.

This research path, theoretically and methodologically referenced in the critique of the sociometabolism of capital, allows us to apprehend the actuality of the centrality of labour or the territorial contents and dynamics experienced by workers, which can lead the transformation/emancipation of society and print another historical course, in addition to capital Mészáros (2007). In other words, these choices have helped us to discuss the issue of the transcendence of the social subject inscribed in the complex plots of the world of labour, which, from our point of view, are not limited to men and women who live from the sale of the workforce or classic salaried workers, based in the stronghold of productive workers, despite being of fundamental importance, as they are the dynamic centre of the generation of more value (surplus value).

However, our view does not stop at this vertex, but proposes to make visible the arc of confrontation between labour and capital expressed, for example, through territorial conflicts under the protagonism of workers who are salaried, unproductive, outsourced, in the service sector, self-employed, subcontracted, PJotised,[5] platformised,[6] intermittent (or fake salaried), sporadic, peasants, considering the

[5] This expression originates from the acronym CNPJ (National Registry of Legal Entities), meaning whether or not an employee registers for tax purposes. Without this corresponding number, it is not possible to open a business

[6] Word from the noun platform, which in the scope of the nascent concept is used to understand the organisation of service providers related to information and communication technologies (ICT) etc, which control workers and labour processes via centralised systems, via the internet, artificial intelligence, etc.

complex spectrum of specific forms, starting with the tenants, squatters,[7] sharecroppers,[8] ranchers,[9] among others, Quilombolas, indigenous people, artisanal fishermen, coconut harvesters, etc.

Nevertheless, the search for the identity of men and women who are part of the working class does not make us fall into the mistake of a priori definitions, which were already discussed by us throughout the 2010s and in other texts that followed. It is imperative to continue dealing with this issue,[10] since the concept of the working class (industrial workers) no longer represents, on a massive scale, the workers (men and women), who, despite still having a prominent place in the dynamics of the production of more value and in the struggles, they are not exclusive protagonists of resistance and emancipatory actions. This is because the workers involved in the activities of providing services "[...] have been increasingly participating in the process of capital appreciation, increasingly integrated in production value chains" (Antunes 2019, p. 18).

In this way, we must not remain insensitive to the changes that have taken place within the class and that have caused changes of expressive depth and breadth in its composition, in the same way as in the degrading conditions of life of the workers, especially due to the disrespect and non-compliance with the labour legislation and the constitution itself, in the Brazilian case, which have repercussions on occupational and environmental health. By way of example, the peasantry, the proletarian, the platform of delivery men and drivers, as well as the other forms of materialisation of labour, are not a priori, but living expressions of the contradictions of the class struggle. Therefore, they have to be understood from the point of view of the actions they carry out in time and space, daily breaking the borders and pre-defined fragmentations of the Taylorist/Fordist proletarian of the twentieth century. Furthermore, these territorial dynamics of labour in the twenty-first century are the result of different processes that are interconnected, and therefore, in tune with the struggle for survival, the struggle against capital, against the stratospheric power of banks and financialisation; against proletarianisation, in favour of autonomy, to the detriment of the denial of heteronomy; against unemployment, hunger, etc.

Hence, the questions and hypotheses that we have been defending and, at the same time, trying to answer, through our research in the scope of Geography of Labour that we are helping to build, that is, the theory cannot do without the reality of class conflict (praxis), since the actions of resistance and opposition to capital are not limited to

[7] Employment relationship that identifies the worker's non-legal possession of the land, or the absence of a property title.

[8] Employment relationship that indicates the possession of half of the fruits of agricultural production, but invariably defined by previously defined tasks and costs, such as, what is the responsibility of the farmer and the landowner with regard to the equivalent of the transfer of housing, animal care, harvesting etc.

[9] Regional denomination that originally designates local conditions of links with the land and its exploitation, in the North of Minas Gerais, which identifies the workers and their families, who are dedicated to the extraction of plant roots and secondary processing, in order to produce medicines, ointments, teas etc, which may be associated with treatments based on healing arts, blessings, etc.

[10] We refer to the Tese de Livre Docência (Free Teaching Thesis): Cf. THOMAZ JUNIOR, 2009.

odd forms of labour (Thompson 1981). From this we extract that the territorial (and geographical) dynamics of labour, expressed in the process of construction/formation and resistance, require us to position ourselves with the objective of identifying the protagonists of the struggles of confrontation with capital, with the State and with the dominant sectors of the bourgeoisie. There are constant redefinitions, which impose the need to assume that the meaning of class—from the initial notions based on seriality to the advanced stages of self-consciousness—is not defined a priori based on the experiences collected by workers, in line with class struggles and clashes, raw material for the transformation of class consciousness, which, in unison, raise revolutionary maturity as a product of the contradictions that reside in production relations (Luxembourg 1999).

In summary, we have provided a debate on the need to rethink the theoretical-methodological references that we have accepted/adopted, in order to understand, explain and offer critical knowledge about the complexities that inhabit the world of labour. It is not a matter of putting ourselves ahead of unattainable pretensions, but of participating, based on the results of research that disturb us and challenge us to understand the convulsed world of labour, in full movement of denial/affirmation, dialectically extrusive/intrusive and full of professional, managerial, protective, controlist and disidentity/desubjectivity redefinitions.

Finally, our point of view is that the intense changes that affect the fraying of the borders of typical employment—the foundation of the pure worker/proletarian—are a theoretical-empirical parameter to define what fits in the working class. That is, it is not a matter of denying the theoretical, political and social protagonism of the proletarian, nor of defending their resignation and, as a result, trade unionism/trade union/worker movement as a whole, much less distorting or deforming Marxism as a theoretical-political-scientific reference; rather, it is to make visible the fact that labour and the working class have been redefined, at the expense of intense derealisation and job insecurity.

The redefinition or expansion of the concept of the working class calls into question the dynamic content of the Marxist dialectic itself, since the exclusivity of its composition, limited to manual workers, salaried workers, producers of surplus value, increasingly disregards expressive contingents of men and women, and the group of wage earners (productive, unproductive) and non-wage earners, as well as the material/immaterial binomial that increasingly inhabits the world of labour. The broad notion of the working class requires much more studies. However, henceforth, we shall understand that the configuration mentioned has its ontological foundation in the social function and in the process of reproduction of social life.

It is imperative to examine the processes of struggle for survival, struggles against capital, against proletarianisation, in favour of autonomy, against private ownership of land, water, means of production, seeds, against the stratospheric power of banks, financialisation, structural unemployment, the lack of jobs for young people, as well as the struggles of the working class. Contrary to those who defended/continued to defend the idea that labour has become superfluous, that it came to an end, the proletariat was (is) no longer a revolutionary subject or loses its energy and transcendent force, what is posited is that the theory of value not only is not eliminated, but it expands and overflows into activities, occupations, places and spaces that meet the demands of capital and its dominating, destructive and revitalised vocation in the twenty-first century. According to Antunes (2019), the opposite of the complete elimination of labour by the informational-digital machinery is the monumental expansion of the new proletariat of the digital age in the service sector.

It would be the same as admitting that the geographic expansion of outsourced/flexible companies, largely represented by service providers, in response to the phenomenon of globalisation, as a strategy for the reproduction of capital, is increasingly materialised in new spatial configurations of "industrial services", such is the case of Foxconn (based in Taiwan), a Chinese company, outsourced and specialised in several countries, with more than 1.5 million workers, which manufactures and assembles electronic products for Apple, Nokia and other transnational companies in the sector.

These companies, exemplified as such, in addition to start-ups and transnational conglomerates affiliated with platformised 4.0 companies, founded on information and communication technologies (ICT), make great use of disrespecting labour laws, by betting on informality and total flexibilisation. They are "[…] increasingly integrated into the value production chains, and are decisive participants in the value generation process of the capitalism of our time." (Antunes 2019, p. 18). These companies make use of the overflow of the marketplace standard, of intermediating itself, that is, "[…] they go beyond the confines of commerce and start to dedicate themselves to the offer, regulation and management of the provision of services," (Carelli et al. 2020, p. 67) bluntly denying that the platformised worker is autonomous.

Digital or sharing platforms themselves go beyond the limits of intermediation between claimants and workers and standardise prices. These platforms impose goals and working hours. They punish, control and use other expedients that assimilate principles of the overexploitation of labour. However, they refuse to comply with the legal procedures for hiring workers and all the charges, imposing even more abysmal instances in the context of social, territorial and class inequalities. After all, this subject holds us back in the elaboration of a specific article, which is worth exploring further in depth, as well as unprecedented approaches arising from the research in progress.

In this perspective, in view of the profound changes experienced by capitalism in the digital-informational-financial era, and the relevant role in "[…] material production and the realisation of more value", Antunes (2019, p. 18), expressly argues that "[…] it is crucial to offer up-to-date insights into the role of labour in services to create more value. […] We have already indicated earlier that we are witnessing the birth of new ways of extracting more value, especially in the service and non-material production sectors that are expanding at the same time […]".

The cybertariat, according to Huws (2017), or the infoproletariat (Braga and Antunes 2009) corresponds to the significance on a planetary scale of service activities for workers, who increasingly participate in the process of capital appreciation. As shown by Antunes (2019, 2020), they transform into digital slaves, victims of algorithm complacency. They are represented by employees of call centres, fast-food restaurants, offices, service industry sectors, telemarketing, increasingly interspersed with self-employed platform workers in home offices, in delivery and Uber services. These social subjects recompose themselves within the scope of resistance struggles and confrontation with capital and redesign the composition of the working class, in contemporary times. In other words, the perverse mark of the destructive development of production forces, in the context of the structural crisis, brings together both their technical-scientific formation and their disposal and is at the base of the labour value law.

What is called into question is the fact that the labour has not ended, it has not lost prominence, nor political centrality or transcendent potential. We are concerned with stating that it is essential to see and understand that what is at stake is the reduction of formal employment, with registration in the portfolio, protected and subject to regulations, as well as legal and constitutional protections, that is, the aftermath of actions that destroy forms of labour with rights.

However, the political-theoretical-ideological disputes remain, and the controversies are gaining signs of vitality, in the academic and political environment in general, which cannot be discussed in depth here, but the indication for some intellectuals is that labour has moved from the centre of society that is built through the extraction of labour, labour value and surplus value to a normative law. That is to say, as modernity ended, we are faced with a society of workers without work. Therefore, labour loses its centrality and is no longer a revolutionary subject (Arendt 2005). With the end of modernity, social struggles revolve around rights and no longer private property, the production of goods and the sphere of consensus between classes prevails, and revolution is no longer necessary (Habermas 1990).

For Gorz (1982), the structural changes are of such a scope and depth that they have produced people who do everything and who come together, but they are not a class, or even a "non-class of non-workers", in which workers do not have an identity through work, and therefore, they are not a transcendent subject. Capital, collapsing in the face of the gigantism of self-valorisation, via financialisation, eliminated the subjects (Kurz 2004). Workers no longer join together in the proletarian identity, because there is no longer a bourgeoisie and a proletariat (Bell 1977).

It is in this myopic perspective that the decrepit intellectuals decreed the end of labour and, as a result, partially or entirely, the theory of value, without paying attention to the fact that the corporate order continues to be built on the accumulation of wealth arising from the exploitation/subordination of labour. However, the transcendence of labour, as a class subject and potential emancipator, requires us to assume its centrality, as an ineliminable component of dialectics and history.

19.3 Contemporaneity of the Total Lack of Protection of the Subjects of Labour and the Working Class

The reciprocal and interdependent relationships that are established in the era of global capital, or financialised capitalism, or the hegemony of the aristocracy of fictitious capital, based on neoliberalism and Toyotist forms of flexible management, adopt logics in tune with the practices that are disseminated, and at the same time, contrary to what was assumed to be characteristic of the system's periphery. They are spatialised in the four corners of the planet.

This new geographical and strategic organisation of capital, anchored in the generalisation of the overexploitation of labour and also in specific forms, became commonplace on a global scale, in the twenty-first century, once reserved for Marini's (1977) reflections for Latin America and Africa. However, it is important to assert that, despite the fact that this wound has become planetary, it is on the periphery of the system that the most abyssal inequities are reproduced and amplified. Thus, super-exploitation, contaminated by the modernity of microelectronics, recovers the creative capacity of the worker, the savoir fair, in the moulds of the initial stages of manufacturing, still superimposed on the initiatives of artisans, or the treasures of the workforce, but under new references.

The vilification of this Toyotist strategy is the insight that allows capital, at the same time, to guarantee levels of productivity under ideological assumptions of workers' participation in process decisions. However, it is based on the intensification of the journey, on multifunctionality, on the flexibility of contracts or hiring precarities, along with the elimination of Taylorist-Fordist career plans, etc. There are also bonus-wage mechanisms that form part of salaries, such as profit sharing (PLR), associated with sophisticated procedures for centralised/decentralised workers' control.

That is, lean production, freeze drying, total quality, Just in Time (JIT), in the form of intermittent wages, coerce workers into forced fulfilment of goals, under strict self-control, as in the Quality Control Circles (QCC), instead of Factory Commissions. In short, capital destroys labour collectives and imposes collaboration collectives. In fact, they pave the way for the consented imposition of worker versatility and the worker's life to be reduced to the exclusive significance of odd working hours, through its human-generic de-effectiveness and objectification to the parameters of transfiguration into merchandise, in the workforce.

Despite the technologies, in certain cases, they make routine tasks less heavy. However, as a rule, they intensify the rhythms, the operations, in view of the mechanisms of downsizing workstations, as well as the times (duration, including with extra hours) and workload. The consequences are, as a rule, illnesses, injuries, illnesses, absences, early retirements and even suicides. Finally, the Toyotist mechanisms are imposed, accepted and/or agreed/negotiated, and they obscure the rawness of workers' exposure, the constancy of endemic unemployment, marginalisation, etc.

In addition, they constitute systems specific to each company, conglomerates, etc. However, they are encouraged as strategies of capital in general, that is, actions that undermine the resistance of workers, through forced and calculated generational change, hastened by voluntary lay-off programmes (VLP).

In practice, they teleguided the removal of older workers from the scene who had experienced processes of resistance struggles and confrontation with capital, members of the extinct Factory Commissions (CF), trade unionists, activists, etc., not to mention, as a result, the impacts on subjectivities, on desolidarisation or contingent class consciousness, and the manipulation of symbolic representations of workers in their organisations, unions, union centrals, the union movement, and generally speaking,[11] in the class struggle, and in conducting organisational actions of the workers, etc.

After all, the main objective of capital—that is, the internal competition for workers—is interposed by the mechanisms of self-control and fulfilment of goals, de-collectivisation of labour struggles and salary campaigns, restricted by company, incentivise behaviours in tune with the prevalence of individualism, to the detriment of the public memory of workers, of the labour movement, which jeopardises the links of labour collectives under strict control of the flexible management package of the labour process, etc. Based on our studies, we arrived at the marks of the systemic degradation of labour (Thomaz Junior 2018a, 2019), which, therefore, are mechanisms that start to be ruled by territorial pillaging (Perpétua and Thomaz Junior 2016), the territories of degradation (Heck 2017), that is, the control processes that indiscriminately subordinate workers, inside and outside of the workplace.

In view of the above, the achievement of the productive restructuring of capital challenges us to significant changes for workers, since it empowers intense and rapid territorial rearrangements, in the production processes themselves (immediate production, distribution, circulation and consumption) and of management and control of labour, in line with the preference to make flexible contracts, remuneration, amidst the amplification of neoliberalism (unlimited), of the privatised or increasingly minimal State, for public policies, for the benefit of workers.

The perennial dismantling of the organisational structure of companies and the repercussions of extreme latitude that affect the labour process and, as a result, the destruction/elimination of formal jobs (with a formal contract and social security and protective rights, etc.) overwhelming the proportion and magnitude, the organisation, structure and routine of the unions, especially in the elimination of these institutions, as well as result in the political weakening of those that remain associated, in different scales according to their political-ideological options.

As an example, we are intensifying studies with attention to the impacts of capitalist interventions, incorporating "technical progress" into the central foundations of monopolising the best lands (fertile, flat, aided by favourable logistics), with water availability, summarising what we call sugarcane agrohydrobusiness. However, they gain energy from new research that considers more details and depth, regarding the controls exercised by the production chains and the intensification of investments by

[11] See ALVES, 2011, 2014.

foreign groups in agriculture—especially in commodities—or the internationalisation commanded by new interests, depending on the raw materials and historical and political conditions, to the detriment of food production.

Thus, immersed in this rapidly and intensively remodelled environment, also called restructured environments, in allusion to the structural, spatial, logistical changes and reorganisation that directly affect companies, in response to the leading role in the productive restructuring of capital, the research projects were and are elaborated and executed, in order to provide knowledge intended in response to advances towards the demarcation of a priority field of investigation. That is, it aimed at approaching labour, not only within the narrow framework of social, economic and strategic references, but also, and necessarily, of labour processes and relationships, of the conflicts that guide and threaten workers, the labour environment and the health/disease relationship, in short, the elements that make up the complex network of working class identity relationships, as we have already seen.

If the social tragedy intrinsic to the scene that is reserved for workers in these restructured environments was not enough, the consummation of the 2016 coup and what happened to the 2018 elections add to the importance of what we have already called the demarcation of the times of lack of protection (Thomaz Junior 2019).[12]

Without being conclusive, we only reiterate how much research is essential for the continuity of the construction of the Geography of Labour to which we are trying to contribute, in order to build references capable of capturing the territorial dynamics of labour, and even more, the contradictions through which we prioritise understanding the conflict of capital vs labour, capital vs capital, labour vs labour, that is, the clue that can illuminate the aridity that imposes itself that is currently so threatening for the entire human species, in the face of the inequities exacerbated by the cruelty of the COVID-19 pandemic.

The geographical "reading" of the dynamics of labour reveals that the main protagonists of the struggles of resistance and confrontation with capital, the State and dominant sectors of the bourgeoisie are no longer restricted to the formal proletariat, with the guarantees of the Social Welfare State, the Fordist factories. However, factories (outsourced, flexible, Fordist-Toyotists, startups) extend to services in general, to the experiences of informalisation that multiply uncontrollably (working from home, street vendors, self-employed, entrepreneurs, etc.), to digital platforms and applications, etc.

The subjects who carry out anti-capital actions are diluted in the various popular demonstrations, on the streets, such as the recurring cases of phone application-based workers for deliveries, in peripheral regions, in medium and large urban centres, in rural land occupations, in virtual environments, in Internet social networks, in new expressions of social movements, as well as in the scope of already known organisations, etc. Furthermore, these new forms of struggle and rebellion go beyond

[12] I am referring to the process that consummated a coup d'état that deposed President Dilma Roussef, legitimately re-elected in 2014. A coup carried out by sectors of the right and extreme right political spectrum in Brazil sponsored by the US government, associated with the mainstream press, the National Congress, as well as the surrendered sectors of the Brazilian bourgeoisie, who are unhappy with the fourth defeat at the polls for the Workers' Party (PT).

the well-known landmarks of the trade union sector, despite still having political and historical importance.

The actions are spatialised, through the alternative praxis of the unions by categories that resist the dismantling resulting from the 2017 Labour Reform, to the supracategory platforms of the Trade Union Centrals, in addition to the alternative initiatives of the workers, such as factory organisations with or without ties to unions, associations, worker cooperatives linked with the provision of services, autonomous and dispersed organisations, as is the case of phone application workers. In addition, there are those linked with social movements, such as Via Campesina, Landless Rural Workers Movement (MST), Homeless Workers Movement (MTST), Unemployed Workers Movement (MTD), Workers Movement for Popular Sovereignty in Mining (MAM), Movement of Workers Affected by Dams (MAB), Popular Peasant Movement (MCP), Web of the Peoples of Maranhão, among others.

It is worth reiterating that their central and unifying element is questioning. This is despite different elements composed of different degrees of identity and class consciousness, but which claim ties to the working class, for example, disrespect for protective, labour rights, private ownership of the means of production, social wealth appropriated by capital, etc., which go beyond known territorial limits.

These negative facets of the society of capital are being expressed on the streets, in digital environments, in corners where conflicts are resolved by shootings, with the prevalence of the extermination of entire families of peasants, the fear that prowls through threats, etc. However, communities, workers and social movements do not remain silent, while several thinkers/researchers remain entrenched in university offices, passionate readers of the mainstream press, which offer them distorted, unreal and forged information.

Quite possibly, they are not participating in the popular demonstrations in Brazil that began to gain dimensions with concerning scales in 2013; therefore, they are not following the movement of labour in the real world. That is, the coup of 2016, the reforms on the agenda in the National Congress, which are sweeping away decades of struggles and achievements of workers, such as the Labour Reform, are not being prioritised in the studies. Even less, despite health impediments, demonstrations in the street and in virtual environments, by workers against the atrocities of Bolsonaro's (anti-)government, with the aggravating factors of the genocide orchestrated and motivated by the COVID-19 pandemic.

These new geographic dynamics of labour expose the magnitude of the challenges for us to understand the contradictions that inhabit the territorial conflicts of the capital vs labour relationship and the complexities that unfold from the various transversalities and intercurrences present in the world of labour in contemporary times. It will be with attention focused on the social process of struggles and class confrontations, that we will continue to strengthen research experiences, in order to place ourselves in the orbit of demands that can help to make visible the contradictions that are important in the agendas of social movements in order to decide the main lines of our actions and that of most of the other companions. Likewise, it will be with the production of texts and public intervention that will allow us to demonstrate the contents and geographical expression of the plasticity of the labour or the expropriation, in a broad sense, of the fragmentations, that is, the composite and simultaneous movements within the social structure, which indicate what we present to discuss the territorial movement of labour and class.[13]

We disagree with those who defend, comfortably situated on the academic left, that there are no subjects capable of mobilising for revolutionary struggles and actions, and that the university must fulfil the role of sheltering reformist theses, which are, of course, anti-popularist, or remain distant from the search for actions and protagonists originating from popular social organisations, where we also continued to include unions in the systemic contradictions of the capitalist mode of production and in the centrality of labour, as a potentiality of emancipatory and resistance struggles. The bets that only comply with short-term agendas, at most recognising the attributes limited to the action allowed by the institutional, electoral struggle, are increasingly less important for us.

Since then, we have dedicated ourselves to express interest, to pay attention to concrete, empirically proven cases, to approach debates, in order to insert ourselves in the dimensions of health and the environment, linked to labour, motivated, therefore, by the need to break our attention directed only to the degrading situations/conditions, peculiar to capitalism, whether in the countryside or in the cities, thus prioritising the connectivity that is established.

For example, we understand that quality of life and collective/public health are linked to the production of food without pesticides, and that urban workers and the whole society can have access to these goods. We defend that it is necessary to link the banners in defence of the quality of life, raised in the name of agroecology or agroecological practices, of the production of organic and poison-free food, in popular language, of agrarian reform, in defence of worker and collective health, in addition to, in relation to the general requirements of the proletarian condition, such as decent wages and employment.

This social totality of labour motivates us to research and address the main issues in search of solutions to improve the quality of life of workers, the democratisation of access to food, land, water, etc. We are already used to this land–water fragmentation, especially in Geography, the basis of the alienated fragmentation between Human Geography and Physical Geography, respectively, in addition to even deeper interpretations that effectively challenge us to occupy the most lucid defences of Marxian theses. For example, labour situated in the social totality of its existence, the indivisible social and ontological integrity of man, broken with the commodity form or the production of goods, based on the abstract form, which uses the estrangement that separates men from the production of goods for the satisfaction of use values.

The spoils of this historic milestone for humanity are blatant, as they separate men and women from the concrete conditions of life, social insertion and class interests, which, by way of illustration, could be generically interpreted in the contradictory relationship between capital vs labour. Let us take these references as those that will no longer support the apparent depoliticisation or deviation of central themes dominated and understood in a restricted way, but as a totality that begins in the

[13] Cf. THOMAZ JUNIOR, 2017.

health-disease process, as a determination of the way in which society appropriates nature, at any given moment.

Thus, we were able to develop this idea, based on Rigotto (2011, p. 32), when he tells us that the appropriation in question is "[…] carried out through the labour process, based on a certain degree of destructive development […] of productive forces and production relations" (Thomaz Junior 2014, p. 8).

Based on Antunes (1999), it seeks to identify the losses of minimum labour rights, which bring to the scene the part-time proletariat of the 1990s, subcontracted, outsourced, informal, together with the concerns of Bihr (1998), who adds to the sub-proletarianisation the absence of union protection.

All this, without identifying their origins, displacements, territorialities, aspirations, subjectivities, cultures, similarities/differences and capacities for struggle and resistance, class consciousness, is supported by the dimensions of the spatial structure of domination of the capital society, as well as the bonds that they establish in the territorial disputes that they carry out and keep them accustomed to the maintenance of cultural identities, etc., and projections as a class for itself.

Remaining faithful to the pre-fixed references in theoretical postulations and other references that do not allow us to advance analytically, politically and geographically is the same as not proposing to leave the place, only fulfilling the academic role and unduly occupying the attention of the interlocutors. In other words, it is still not understood very well; however, it is indicative of intellectual and theoretical lucidity, as a new service proletariat, as well as the other workers who swell the ranks of the systemic degradation of labour and who, in our view, make up the expanded working class, in these times of the twenty-first century. Of course, this is in resonance with the realisation of the fraying of the borders of the typical salary of the Taylorist/Fordist era, just to reinforce intellection, giving rise to slogans such as the new middle class, which tells us nothing or almost nothing.

It is time to insist on the idea that demands from each of us do not deviate from the most legitimate Marxist imposition, that is, from the centrality of labour, the centrality of the working class as a transforming subject, in the rupture of the extraction of surplus labour and of all forms of subordination of men and women to the yoke of capital, and that we are committed to socialism. It is time for us to make the class struggle debate the centrality of our actions, so that advances can illuminate our performance/militancy, making us much more than mere transmitters of knowledge or, at most, good debaters.

These references are essential to situate ourselves in the environment of destruction carried out by capital, at the dawn of the twenty-first century. This is the viewpoint of the determinations that consecrate the destructive movement of the productive forces in Brazil, following the proposed path, by way of example, of outsourcing without limits, or being intensely valid for all categories of workers, approved by the coup-fighting National Congress as it did under the false promise of seeking full employment, which multiplied uncontrollably.

Significant contingents of workers have also been relegated to the home office, which also challenges and weakens the social and political identity of workers and the working class, increasing the legion of those marginalised from the protective, regulatory and inspection mechanisms of the State, such as the Regulatory Norms (RN), inspections by the Public Ministry of Labour (MPT) and also the actions/fines carried out by the MPT, in parallel with the incidence of trade union organisations, etc. (Thomaz Junior 2018a).

In addition, we must pay attention to the qualitative dimensions or the implications of these actions on workers' subjectivity, as they are driven to formally detach and/or distance themselves from the political representation of unions, federations, centrals, that is, they deeply impact workers and the working class.

19.4 Labour and the Working Class in the Digital Age, in Times of COVID-19

More than the denial of labour rights (protective, jurisprudential), the dissemination of platformisation imposes the total lack of protection of workers, counter-reforms, etc., based on an understanding of the imperatives of flexibilisation, workers are treated as instruments orchestrated by the State and the interests of capital, companies and the bourgeoisie.

In our research, we observed that, in Brazil, impoverishment is based and amplified, clearly, quickly and in an uncontrolled stage, to the misery of the working class, without it being possible to clearly demarcate the privileging of sectors of activity. Thus, it is in line with Antunes' (2018, p. 45) understanding that, "[…] if today's youth are lucky, they will be servants". That is, the robotisation of production processes, or the triggering of the Fourth Industrial Revolution or Revolution 4.0, and artificial intelligence technologies, the deepening of Information and Communication Technologies (ICT) are based under the guise of barbarism, such as call centres, banks, car assembly plants, charcoal plants,[14] irrigated perimeters (II),[15] in the northeast of the country.

It is as if we could translate this social framework into the levels of inequality that are characteristic of India, but that are expressed on all continents, through stricter labour control practices that impose the mark of modernity in this century. Namely from the Taylorist-Fordist-Toyotist foundations–and under the guidelines of technological development and the ICT fad–and already in the second decade of the twenty-first century, through digital platforms (whether in the form of crowdworking, home office, on demand, or delivery/Uberisation).

An example that serves us for the moment in Brazil would be the activities of food delivery (food) carried out by business conglomerates (IFood, Rappi, etc.), which

---

[14] Place designated for the processing of wood burning with the objective of producing charcoal that will be directed to the activation of ovens, as well as other uses. As a rule, charcoal plants are places identified for total non-compliance with labour and slave labour rights in Brazil.

[15] Denomination that designates a form of organisation of agricultural production (commoditiesfor export or not), as a rule financed by the State, related to certain portions of land that rely on mechanised irrigation, which prioritise business segments to the detriment of family farming.

move R$ 25 billion/year and use approximately 5 million workers, without any employment relationship. In return, the workers must have the instruments of work, starting with motorcycles, bicycles, jackets, bags, protective masks, etc. Statistics show that the amount of people linked with job insecurity who work on apps and platforms represent more than five million workers. Uber alone accounts for 700,000 drivers/workers.

This mantra is consummated under procedures and decisions programmed by capital, apart from respect for workers' rights and the environment. Even less, it grants to workers the exclusive subjects of the creation of wealth arising from this system that produces useful goods and their transmutation into goods, the personal/collective usufruct, proportional to the time spent on labour in correspondence with the final price. In fact, for capital, the object of desire is the increasing appropriation of necessary labour, in order to fatten the portions of surplus labour, the subordination of land profit, etc.

These expedients historically linked with the valorisation of capital or the need to valorise value, through the extraction of surplus labour, surplus value, the creation of more money, speculative mechanisms, profit, etc., have always been destructive production forces, greatly weakening the reproduction of workers, considering the wide range of work/occupational expressions already mentioned above, of the original, traditional communities, the environment (mineral reserves, protected areas, conservation units, springs, aquifers, forests, soils, cultivation areas, etc.). At the limit, we are faced with an uncontrollable destruction machine. As Mészáros (2015) states, capital is irreformable, in line with systemic irrationality and vulnerability.

If these changes intrinsic to the forms of production were not enough, the labour relationships linked to them, the disrespect for the rights and life of workers also intensely affect the social and political identity of these subjects, with repercussions on the organisational entities, on the social movements or in subjectivity, and, more specifically, in the working class. That is, in addition to increasing the legion of men and women marginally included in production environments (in all its extension), in the name of the (destructive) development of the productive forces, this procedure has always been associated with disrespect for protective, regulatory and inspection mechanisms of the State, in favour of workers, with meanings that are mixed in tenuous demarcations between registered precarities, and informal, detached, outsourced employment, etc., enshrined under the insignia of flexibilisation.

As an example, the total outsourcing (of end and middle activities) is a demonstration that the architecture of the 2016 coup provided a safe path for the loss of rights and the weakening of labour relations. In fact, these instruments are added to the election of Bolsonaro's (anti-)government in 2018. Amazingly, the cowardice of the Federal Supreme Court (STF) gave validity to the aberration of total outsourcing, which opened the way for the federal government to institute Decree No 9.507/2.018, which allows outsourcing in the public sector. We are facing a process of devastation/destruction of important achievements of the working class, such as the corrected minimum wage, in addition to the Broad National Consumer Price Index (INPCA/IGBE), the productivity bonuses from the Gross Domestic Product (GDP), the 13th yearly salary, among others.

All this takes place under the prescription of financial capital and the bourgeois hegemony that marks the class pact in Brazil, after the 2016 coup, through constitutional reforms (labour, social security), cemented with the election of Bolsonaro's (anti-)government of a militaristic, oppressive, misogynist, fascist, necropolitical nature, and a new composition of the State, sustained in the scope of a social, class pact, whose main line is the promise of the greatest devastation of social and labour rights ever recorded in the country.

There is a Brazil that everyday departs from the pores of marginal inclusion and that explodes in contradictions; however, the intellectuals on duty are not seeing it. It is from this Brazil that the COVID-19 pandemic has revealed the rubble of abyssal inequalities. However, before shedding light on article titled "The worst of the virus is yet to come",[16] it is important to state that what was already bad is getting worse for workers who have been flexible since before and for those who won this patent that overlaps with the manipulative contamination intrinsic to the functioning of capitalist society. Its tools, routines and nefarious foundations are many, such as the marginal inclusion of expressive contingents of workers, whether formal or informal employees, entrepreneurs, freelancers, of all kinds, etc. In other words, with the addition of the effects of COVID-19, which has already amounted to 470,000 dead and 17 million infected in Brazil, everything is pointing towards a progressively explosive and bloody reality for workers, society and the environment in general.

The life force of this scheduled destruction, already defined as pandemonium,[17] alludes to the nefarious environment, nourished by what is most vile and backward, in the national business community, in the pact of bourgeois fractions, in the reactionarism of a wide ideological spectrum, of professional assassins entrenched in the groups of militiamen and so many others who join, either in the media groups, or under the banner of "agro is pop", etc., in order to shield the reality of the facts, the manipulations, the gambling and the patronages of all kinds, practised in broad daylight.

The harmony of all these ills reflects the lack of investments and scrapping of hospital equipment in the Unified Health System (SUS)—directly affected by Constitutional Amendment No 95 (EC) of 2017, which froze public expenditure for 20 years—and not meeting the demands of hiring servers. If the overload of manipulations, excesses and denial of the pandemic, social isolation and the boycott of vaccines were not enough, we witnessed the dismantling of the Ministry of Health, inhabited by incompetent, denialist leaders, disaffected by life, led by Jair Bolsonaro, a self-confessed practitioner of genocide, the worst and most perverse president of the Federative Republic of Brazil of all time.

The undermining of historical social fissures, in the face of the severity imposed by the COVID-19 pandemic, has depleted on an uncontrolled scale the number of unemployed, underemployed, uprooted and hungry. This situation was aggravated by the stoppage of payments and the reduction of the Emergency Aid Payments and the Emergency Employment and Income Maintenance Benefit (Bem) Payments, since

---

[16] Cf. THOMAZ JUNIOR, 2020.

[17] Cf. SOUTO MAIOR, 2020.

December 2020, which only was resumed in June 2021. However, these payments were more limited, now in the range of R$ 250.00,[18] and were restricted to those who received aid payments in 2020, and no longer R$ 600.00, which was in force from April to August 2020, and which, from September to December of the same year changed to R$ 300.00. In addition, it reaches people less directly affected by the pandemic, moving from 68.2 million people in 2020 to 46 million people in 2021, which is to say that it is more restrictive in the face of greater severity.

This subject requires specific dedication, given the significance, breadth and importance for our studies in the Geography of Labour. Even more than what was already expected, with respect to the predominant colour of the social fabric most affected by deaths, unemployment and other ills associated with the COVID-19 pandemic, it has mainly been represented by blacks and browns, who were once impoverished, residents of the periphery, slums, that is, those who have always held the stigma of marginal inclusion in the territories.

A new morphology of labour is remade, on a global scale. It does not hurt to reiterate that the working class continues to occupy a central place, within the scope of social struggles, and leading actions of resistance, under new formats and territorial expressions, in the streets and in the fields of digital/informational systems, increasingly divided (gender, race, ethnicity), with consequences and challenges of considerable magnitude for their representation bodies, whether unions, union centrals, associations, etc.

It is a fact that expands in some segments and declines in others. It happens in a variety of ways and is more heterogeneous, more exposed to the mechanisms of overexploitation and control tuned to ICT, under the baton of algorithms, the merciless slave maker and digital slaves. Though still in progress, it is already occupying us in research and orientations (undergraduate and postgraduate) related to the devastation of the flexibilisation/deprotection/denial of rights of working conditions, in the environment of digital platforms, applications, in the relational world of couriers. The reality of workers linked with the sugarcane agrohydrobusiness is also contemplated. Those workers who experience even more harmful consequences of the flexibility process are related to capital's violation of denying that COVID-19 is an occupational disease.[19] The quality of life, the health of workers and society, in general, and specifically humanity are at imminent risk.

[18] It can reach R$ 150.00, for a single-person family, and R$ 375.00, for women who provide for single-parent families, according to Provisional Measure No 1,039 (03/18/2021).

[19] Workers seek to defend themselves from this atrocity by hiring private lawyers, with the purpose of having the custody of the precautionary measure of the STF that suspended article 29 of Provisional Measure no 927/2020, which disregarded contamination by COVID-19 as non-occupational.

19.5 Final Considerations

Picking up where we started, it is necessary to break with the praxis ingrained in the academy, which idolises respect for the technical-positivist division of scientific work, in order to broaden the horizons of understanding the totality of labour beyond the workplace, in the same way as health occupational activity is not restricted to the workplace.

Problems arising from the contradiction between capital and labour are increasingly collective, but commonly seen as individual. In this sense, health problems also appear as natural, individual causes that are distant from their relationship with the organisation and labour process. It is under these references that the health of the worker is exposed to risks, diseases, marginalisation, unemployment and other marks of the systemic degradation of labour.

We agree with Laurell (1982, p. 37), for whom "[the] health-disease process is determined by the way in which Man appropriates nature at any given moment, an appropriation that takes place through the labour process, based on a certain degree of development of the productive forces and social relations of production".

We would say that the continuous making/remaking of workers, in search of specific resolutions to defend themselves from occupational diseases, unemployment/underemployment, hunger, marginal inclusion, abandonment, marginalisation, has not ceased. Yet, at a more defined level of subjectivity and ideological identity, or within the working class, in the world, it ends up signalling that labour conflicts and unrest, instead of having disappeared, diminished or even for the most pessimistic, have come to an end; they are, in fact, dispersed in different places and times.

We understand that currently, this is one of our favourite subjects, considering that, from the point of view of the Geography of Labour, if in the on-demand format (delivery services), workers compete with each other within a geographically located area, in crowdworking (home office), competition takes place worldwide, making sociability ties even more difficult (Fonseca 2020). From this, challenges multiply, considering the need to discuss the territory of trade unions' activities in the digital sphere, far beyond municipal, regional, continental borders, etc., in order to articulate struggles on a continental/transnational scale.

It is therefore with an eye on the contradictions in the dynamics of labour that we insist on the territorial movement of labour connected with the living totality of labour or its different forms of expression, beyond the borders of typical/pure wage labour, as a theoretical-empirical parameter to define what fits the working class, considering that the actions of resistance, as seen in the real world, are not limited to odd/salaried forms of work.

In other words, we have to understand the emancipatory potentialities, or the scope, meaning and Geography of its political centrality, beyond the borders of typical wage-earning, of the foundation of the pure worker/proletarian, as a theoretical-empirical parameter to define what fits in the working class, which has been redefined, at the expense of intense derealisation and precariousness of labour.

It is not possible to define with surgical precision all the steps that should be carried out, or the routines in ascending order, in order for us to reach the final emancipatory power of labour, which would be defined in its constitution as a class, beyond the thresholds of seriality, in location, longevity and revolutionary firepower.

We are certain that what has to come first is life, the preservation of life and not the concern with the economic losses of capital, with the imprisonment of public resources to be directed to the primary surplus, to the detriment of investments directed to the public health, although corporate bankruptcy was even more chaotic. The creative challenge of inventing another world must be the collective response towards the defence of humanity, against the perversities of this model of society that is in quarantine for a long withering; it is the substantive solution against the environmental catastrophe and the abyssal inequalities, which will continue to potentiate the new health crises that will come.

It is the same as saying how the inequities of capitalist society intensify the marginalisation of the most vulnerable (unemployed, underemployed, flexible/outsourced, autonomous, mistakenly called entrepreneurs or self-employed). This indicates the dismantling of the intermediate layers, which, despite still standing, they are teetering, threatened and ready to fall and join the ditch of marginal inclusion.

The expansion of research horizons to the spheres of action of public institutions, which have demonstrated concerns and concrete actions that add to ours, has positively influenced us in the even closer exercise between research and outreach. This brought us closer to the Public Ministry of Labour (PML), the Worker's Health and Reference Centre (CEREST/President Prudente), the Osvaldo Cruz Foundation and, more recently, the Working Group "Sugarcane—Impacts on Health Environmental and Rural Worker", linked with the Federal Public Ministry, of the Regional Attorney's Office (PRR/3rdRegion).

It is time to put the end of capitalism back on the agenda, as well as the emancipation of the working classes, and their responsibility in its entirety. It is time to resume the debate on socialism—although it is overdue, and therefore, it has never been so urgent—in order to illuminate the salvation of the future of humanity.

It is opportune to resume a previous reflection that we rescued from Löwy's observations (1998), when he comments that, in his opinion, the slogan "Proletarians of all countries, unite!" "[…] is the most important part of the Communist Manifesto, because it touched the imagination and hearts of several generations of militants (men and women) and workers and socialists." We allow ourselves the audacity to update the main subject of the sentence to "workers", as this flag echoes as a categorical imperative, both ethical and strategic and ontological, in contemporary times. Furthermore, our participation with the social movements involved in workers' resistance struggles, based within class conflicts, constitutes an assumption that makes a difference.

Were it not for the research activities with the corresponding actions with social organisations, in addition to participation in the public debate, we would not be able to socialise the expected results and the continuous path of collective learning, even less, we would systematise the doubts and questions that remained unanswered, with a view to composing future investigations about the world of work, under the focus of the Geography of Labour that we are helping to build.

References

Alves G (2011) Trabalho flexível, vida reduzida e precarização do homem-que-trabalha: perspectivas do capitalismo global no século XXI. In: Alves G (ed) Trabalho e saúde. A precarização do trabalho e a saúde do trabalhador no século XXI. LTr, São Paulo

Alves G (2014) O conceito de trabalho—a perspectiva histórico-ontológica. In: Alves G (ed) Trabalho e cinema, v. 4. Canal, Bauru

Antunes, R. Os sentidos do trabalho. São Paulo: Boitempo, 1999.

Antunes R (2018) O privilégio da servidão. Boitempo, São Paulo

Antunes R (2019) Proletariado digital, serviços e valor. In: Antunes R (ed) Riqueza e miséria do trabalho no Brasil IV. Boitempo, São Paulo

Antunes R (2020) Trabalho intermitente e uberização do trabalho no limiar da indústria 4.0. In: Antunes R (ed) Uberização, trabalho digital e indústria 4.0. Boitempo, São Paulo

Arendt H (2005) A condição humana. Forense, São Paulo

Bell D (1977) O Advento da Sociedade Pós-Industrial. Cultrix, São Paulo

Bihr A (1998) Da grande noite à alternativa: o movimento operário europeu em crise. Translatio nof Wanda Caldeira Brant. Boitempo, São Paulo

Bourdieu P (2008) Razões práticas. Sobre a teoria da ação. Papirus, São Paulo

Braga R, Antunes R (2009) Infoproletários—Degradação real do trabalho virtual. Boitempo, São Paulo

Brenner R (1998) The economics of global turbulence: a special report on the world economy (1958–98). New Left Review, London, no 229

Carelli RL et al (2020) O futuro do trabalho - Os efeitos da revolução digital na sociedade. ESMPU, Brasília

Chesnais F (1996) A Mundialização do Capital. Xamã, São Paulo

Filgueiras V, Antunes R (2020) Plataformas digitais, uberização do trabalho e regulação no capitalismo contemporâneo. In: Antunes R (org) Uberização, trabalho digital e indústria 4.0. Boitempo, São Paulo

Fonseca VP (2020) O crowdsourcing e os desafios do sindicalismo em meio à crise civilizatória. In: Carelli RL et al (eds) O futuro do trabalho—Os efeitos da revolução digital na sociedade. ESMPU, Brasília

Gorz A (1982) Adeus ao proletariado. Forense Universitária, Rio de Janeiro

Habermas J (1990) O discurso filosófico da modernidade. Dom Quixote, Lisboa

Heck FM (2017) O espaço e o contraespaço no circuito produtivo do capital da frigorificação de carnes no Oeste Paranaense. Thesis (Ph.D.)—PPGG/FCT/UNESP, Presidente Prudente

Huws U (2017) A formação do cibertariado. Editora da UNICAMP, Campinas

Iinstituto Brasileiro de Geografia e Estatística (IBGE) (2021) Agência de Notícias, 31 Mar 2021. Availableat: https://agenciadenoticias.ibge.gov.br/agencia-sala-de-imprensa/2013-agencia-de-noticias/releases/30391-pnad-continua-taxa-de-desocupacao-e-de-14-2-e-taxa-de-subutilizacao-e-de-29-0-no-trimestre-encerrado-em-janeiro-de-2021. Acessed on 23 June 2021

Kurz R (2004) O colapso da modernização. Paz e Terra, São Paulo

La Laurell AC (1982) salud-enfermedad como proceso social. Revista Latinoamericana De Salud, México 2:7–25

Löwy M (1998) Por um Novo Internacionalismo. Paris, RecherchesInternationales, no 52–53

Lukács J (2012) Para uma ontologia do ser social I. Boitempo, São Paulo

Luxemburgo R (1999) Reforma ou revolução? Expressão Popular, São Paulo

Marini RM (1977) Dialética de la dependéncia. Era, México

Menegat M (2006) O olho da barbárie. Expressão Popular, São Paulo
Mészáros I (2007). O desafio e o fardo do tempo histórico. Boitempo, São Paulo
Mészáros (2015) A montanha que devemos conquistar. Boitempo, São Paulo
Moreira R (2016) A geografia do espaço-mundo- conflitos e superações no espaço do capital. Consequência, Rio de Janeiro
Perpétua GM, Thomaz Junior A (2016) Revisitando o conceito de acumulação do capital: a pilhagem territorial promovida pela Veracel Celulose no Extremo Sul da Bahia. Campo Território 11:225–256
Sorry we missed you (2019) Directedby: KenLoach. Actors: Kris Hitchen, DebbieHoneywood, Rhis Stone, Paul Laverty. (101min.) United Kingdom/France/Belgium
Souto Maior JL (2020) MP 927: da pandemia ao pandemônio. São Paulo, 25 Mar 2020. Availableat: https://www.jorgesoutomaior.com/blog/mp-927-da-pandemia-ao-pandemonio. Accessed on 30 Apr 2020
Thomaz Junior A (2005) Por uma Geografia do Trabalho! (Reflexões Preliminares). Revista Tamoios, [S.l.] 1(1)
Thomaz Junior A (2009) Dinâmica Geográfica do Trabalho no Século XXI (Limites Explicativos, Autocrítica e Desafios Teóricos). 2009. HabilitationThesis—FCT/UNESP. Presidente Prudente
Thomaz Junior A (2017a) Movimento territorial do trabalho e da classe trabalhadora, desterreação e os sujeitos transcendentes/da resistência no Século XXI. Thesis (Head professor)—UNESP, Presidente Prudente
Thomaz Junior A (2017b) Degradação Sistêmica do Trabalho no Agrohidronegócio. Mercator, Fortaleza 16:1–20. Available at: http://www.mercator.ufc.br/mercator/article/view/2082. Accessed on 26 Nov 2017b
Thomaz Junior A (2018a) Geografia do Trabalho por Inteiro. Revista Pegada, Presidente Prudente 19(2):6–56. Availableat: http://revista.fct.unesp.br/index.php/pegada/article/view/6000. Accessed on 10 Mar 2019
Thomaz Junior A (2018b) Movimento Territorial do Trabalho eDesterreação do Sujeito/Classe. In: Thomaz Junior A (ed) Geografia e Trabalho no século XXI. Centelha, Presidente Prudente, pp 32–74. Availableat: http://ceget.fct.unesp.br/assets/site/pdf/Ebook_Geografia_e_Trabalho_no_S%C3%A9culo_XXI_Vol9_Especial_(1).pdf. Accessed on 10 Mar 2019
Thomaz Junior A (2019) Novos territórios da degradação sistêmica do trabalho. (Em tempos de desproteção e inclusão marginal institucionalizada). São Paulo, Terra Livre, year 34 1(52):197–240
Thomaz Junior A (2020) O pior está porvírus. Em defesa da classe trabalhadora para além da Pandemia COVID-19. Blog do OTIM. Available at: http://otim.fct.unesp.br/o-pior-esta-porvirus-em-defesa-da-classe-trabalhadora-para-alem-da-pandemia-da-covid-19/. Accessed on 8 June 2020
Thompson EP (1981) Miséria da teoria e um planetário de erros: uma crítica ao pensamento de Althusser. Zahar, Rio de Janeiro
Thompson EP (1998) Costumes em comum: estudos sobre a cultura popular tradicional. Companhia das Letras, São Paulo
Harvey D (1992) Condição pós-moderna. São Paulo: Loyola
Marx K (1982) O capital. 7. ed. São Paulo: Difel
Lessa S (1997) O reflexo como não-ser na ontologia de Lukács: uma polemica de décadas. Critica Marxista, São Paul, 4:89–112
Rigotto RM et al. (2011) Agrotóxicos, trabalho e saúde: vulnerabilidade e resistência no contexto da modernização agrícola no Baixo Jaguaribe/CE. 1a. ed. Fortaleza: Edições UFC p. 612
Thomaz Junior, Antonio (2011) Os Desafios Rumo a um Projeto para o Brasil! (Intemperismo do Trabalho e as Disputas Territoriais Contemporâneas). Revista da ANPEGE, São Paulo, 7(1):307–329

Chapter 20
A Popular Environmentalism in Defence of Life, Dignity and Territory (An Autobiographical Contribution from an Activist Geographer)

Carlos Walter Porto-Gonçalves

Abstract The Brazilian geography underwent an important turning point in the 1970s, partly as a result of a general transformation of the geography around the world and mainly due to the particularities of the historical moment which Brazilian society was traversing. An important theoretical contribution resulting from this was a radical epistemic and political change in which geography, instead of being an object of study—the geographical space—became viewed as a verb—the act/action of writing the earth: geography. The article will have an autobiographical tone due to the author's involvement in the process: a geobiography.

Keywords Geobiography · Environment · Brazilian geography · Critical approach · Epistemic

20.1 Introduction

This article recounts a wide trajectory of life intensely lived as an intellectual/activist on several complementary fronts of action, some with greater scientific involvement, such as the Brazilian Critical Geography, Environmental Thinking and Latin American Political Ecology as well as others that are more involved with the movement came to be christened as Popular Environmentalism and the struggle to redefine Agrarian Reform. There are interconnections between these two driving forces, among which I would highlight, from a theoretical and political point of view, an inflection in the case of geography, in which I produce geographic knowledge from a different perspective, in the case, from the movements/struggles/social conflicts where our study interest ceases to take precedence in the traditional purpose to which geographers dedicate themselves—the organisation of space. I am interested in geography as a verb, as an act/action to mark the earth, for people who struggle to

C. W. Porto-Gonçalves (✉)
Emeritus professor at the Fluminense Federal University, Niteroi, Brazil
e-mail: cwpg@uol.com.br

R. C. Lois González and M. A. Mitidiero Junior (eds.), *Brazilian Geography*, Advances in Geographical and Environmental Sciences,
https://doi.org/10.1007/978-981-19-3704-0_20

overcome the situation of subalternation/oppression/exploitation to which they are subjected.

The very idea of *social movement* implies, like every movement, changing place and not just as a metaphor of space. As every society is established, its social and power relations conform to its geography and the place assigned to the different social groups/classes. In a misogynist society, for example, every woman knows, consciously or unconsciously, that she cannot go everywhere in the city. The same applies to racist societies, with racial discrimination. The same applies to the social groups/classes with regard to appropriating an earning from what they produce, whether they are proletarians or owners of the means of production. From this epistemic core, I encountered questions related to the debate on social groups/classes in their own struggles. This led me to engage with other academics from other disciplines, who contributed to establish the field of Environmental Thought and Latin American Political Ecology. This provided a backdrop for the debate about the struggle for land transfigured by the territorial conflicts from my intimate relationship with the movement of the peasants (rubber tappers) and Indigenous Amazon population (Alliance of Forest Peoples), and by extension, with Latin American social movements.

Here, in this article, I have gathered some reflections on the movement that would later be coined as Popular Environmentalism. Within that field, some, like me, championed the title Eco-socialists. In 1988, I became one of the leaders in writing the "Basic Principles for an Eco-socialist Society". This document is a summary, resulting from a broad effort by activists at the time involved in the construction of the Workers' Party who sought to incorporate nature into the debate on the social and political changes underway in Brazil. The final document never achieved the importance within the party we sought, which led to almost all of its signatories abandoning the party in the following years given the developmental nature prevailing there. For the most part, those who left went to seek other ways to continue what the Manifesto professed, as did I.

At this time, I do not want to delve into the details about the name of this document, because I do not believe it is the appropriate time. After all, everything that involves the word socialism today is perceived much more by what its detractors say than its followers. This is a recognition of defeat, even if it may be partial, in the construction of this alternative. Moreover, those who advocate the end of socialism radicalised their speeches in such a way that it polarised the debate to such an extent that asserting those principles ends up reinforcing the denial thereof in the current context. However, more than ever, we must resume the debate on life in the current context. Back in 2002, in Bogotá, I gathered with other companions,[3] and we drafted another document, the "Manifesto for Life; an Ethos of Sustainability". For my part, I am calling for continuity between the two documents. At this time, at the age of 72 years, my sensitivity to life has acquired a dimension of someone who can see on the horizon that life has limits, which gives us a very specific meaning in which experiencing life is predominantly through thought. This experience of life places us before the limits that pertain to all of life, in particular, whose end is part of life that, nevertheless, is reproduced with all its strength from the limits of every life, as Edgar Morin would say (Morin 2002).

This more personal feeling, however, is at a time when the life of each one of us is seen as part of the life of everyone, when "we see ourselves as members of the same biotic community, inseparably linked by vital flows to millions of other communities of species. The life experience shared directly and simultaneously that, for the first time, at the global level, we are experiencing, is an experience of our fragility and vulnerability. The virus enables us to 'discover ourselves' as parts of a threatened species" (Araoz 2020).

Thus, our personal sense of the limit of life becomes a global and group experience. This brings me to a controversial category, that of Exterminism, brought to the debate in the early 1980s by E.P. Thompson (1985) to overcome the limitations that the Cold War imposed on the political imagination. At that time, humanity was subjected to the threat of extermination with the Arms Race, particularly with nuclear weapons. According to Ricardo Gaspar Müller, a scholar of Thompson's work, "The Exterminism of the Cold War (…) although it seems like a rational movement, in which participating agents make apparently rational decisions, at the heart of the process operates a perverse logic, a self-generation system and a generalised state of momentum in the direction of total destruction". (…) As Thompson thought, the category of Exterminism was also guided by the criticism of the principle of military strategy ironically called *Mutual Assured Destruction* (MAD) (Müller 2013).

It was not the first time that such a situation was portrayed for reflection. For example, Spanish painter Francisco Goya (1746–1828) depicted two peasants fighting on a swamp. As can be seen, the more they moved to gain territory on their opponent, the more they sank into the environment. Michel Serres (1994) brought us this memory of Goya in the opening of his book with the evocative title of *The Natural Contract*. This invites us to think beyond Social Contracts, whether it is Liberal Capitalist or Socialist, in that they both ignore the swamp metaphorically represented in Goya's work, that is, nature is a mobile ground where history moves. There is so much life in the swamp!

After all, we live in a world where not only the capitalists dispute resources among themselves, but also the workers claim to enjoy goods as they are offered within capitalist society. Gradually, we move further towards the abyss, towards Exterminism, as E.P. Thompson would say, even after the end of the Cold War with the fall of the USSR. The devastating character of Liberal Capitalism threatens all humanity with its Necropolitics and calls upon the political imagination to seek societal forms that no longer ignore life. It is a Natural Contract, as Michel Serres suggested, which like every contract, is the work of men and women setting the limits for their lives (*bios* + *polis*).

Here, we will provide the reader with a reflection of the environmental field by placing life (*bios* + *polis*) at the centre of our concerns. To this end, let us allow ourselves to be guided by a concept in the field developed by Pierre Bourdieu, analysing how the environmental field was constituted through its contradictory practices, concepts, theories and narratives (Bourdieu 2001). In making this reflection of the environmental field, we provided ideas that signal the construction of another social order where life, dignity and territory are a source of inspiration, ideas that we do not draw from our head, rather we collect them together with social groups in a situation of oppression/exploitation/subalternation in their searches to overcome this situation.

20.2 The Constitution of the Environmental Field

The environmental field has been in continuous establishment since the nineteenth century in the main countries, especially in the USA. In the 1960s, the field had an important inflection point, when it no longer remained limited to a subject among experts and a few NGOs (Sierra Nevada, UICN) and became an issue openly debated by society. These dates and geographies are relevant, given that in the nineteenth century in Europe and in the USA, a powerful process of transforming the metabolism of life was provided by the widespread use of fossil energy (coal, followed by petroleum).

It could be said that, until then, all the reproduction of life, in any quadrant of the planet and whatever society was, depended on the photosynthesis provided by the sun each day and the variety of biological productivity that the different regions of the planet produced (autopoiesis), in complementarity. In short, it was the sun each day that dictated life, and it was basically with this energy that cultures creatively moved. The Industrial Revolution (in the social and power relations through technology) caused a revolution in the metabolism of reproduction of life with the steam engine. At first, the impact was local/regional, although it was intertwined with the worldwide dynamics of the capitalist economy. With the fossil revolution, this economy would make a leap that was believed to be limitless. Since then, the production and reproduction of life have been measured by HP—horsepower—a measurement that still holds a relationship with the previous phase—the animal driving force—that is, a number that corresponds to a horse, for a job the horse no longer does (or the ox or any other driving animal), that will allow an exponential increase in the transformation of matter. In his book, The Price of Progress, Elmar Altvater insists on the fossil character implied in the development of capitalism. For his part, Marx sensed the depth of what happened when he drew attention to the fact that from then on, capital was inscribed in the metabolic circuit of production (D-M{Means of production + workforce}-M'-D'), and it no longer drew its profits only from commercial gain (D-M-D').

The whole world will be involved (uncovered) in this process, with the widespread use of the steam engine in rail transport and in seaborne shipping, and thus increasingly transporting more matter and more energy over longer distances from their centres of origin to industrial processing sites. Planetary metabolism will no longer be the same, and talk of metabolic fracture has already begun (von Liebig and Marx). The colonial character underlying this order of things that naturalises the international division of labour has been imposed since then with increasingly rising levels in the demand for matter and energy that are destined for the main countries of the world system that was based on a capitalist economy. This Great Transformation (Karl Polanyi 2012) would not have occurred without money, nature and labour (the workforce) being transformed into merchandise, which had become the motive for the new social order that had been shaped. This new social order placed the economy at the centre of life, and not the economy as Aristotle understood it, but a market economy. Until the advent of capitalism, the market was an institution that had always followed the destiny of peoples at any historical time and in any geography and was always subordinate to some ethical and moral principle. The Great Transformation that occurred, which came to be called the modern world, is that the market is free of moral and ethical values, whatever they may be. With this, it was believed that there would be no limits to the use of money, nature and labour (the workforce). Inspired by Polanyi, we could say that with capitalism, there is a Great Transformation to a market without ethics or moral values.

The technical-scientific discoveries from the nineteenth century onwards fed the myth that there are no limits to metabolic transformations and the periodic table of chemistry will be the symbolic expression of advancing knowledge of nature and its broad use according to the different industries that developed, deepening the territorial division of labour. The fate of people definitively ceased to be based on the metabolic conditions of the reproduction of life in the regions where they had been able to live until then. Natural resources, as they came to be called, which before were the conditions that made the life of social groups possible, were valued in places beyond the geobiographic context in which they were metabolically implicated (*involvement/environment*), by a distant market listed on stock exchanges controlled by increasingly powerful and alien groups to the contexts of each social group, according to the broad hierarchy that stretched from the peripheries to the centres of the system. In the end, its involvement and uncovering came to be ubiquitous. The places/regions were forced open and forced to become involved/uncovered. Thus, an unequal geography of haves and have-nots was established, according to geographies, which gradually impacted the planetary metabolic reproduction of everyone, although not all could be equally protected from damage. As Karl Polanyi said, a true satanic mill began to advance throughout the planet.

In this context, at the end of the nineteenth century, several initiatives to protect some areas considered relevant emerged in the USA and Western Europe, either because of their originality, the biological or aesthetic wealth of their landscapes, all of them basically defensive in character since it was accepted critically that the ongoing uncovering process would be inevitable. In truth, it was accepted that a particular civilising process that arose in a province of the world—Western Europe—could colonise the entire world with its values. To this end, many peoples and communities that have been uncovered over thousands of years with their knowledge woven together with natural regions, the most varied have been simply removed from their territory, expropriated. As such, this is not only genocide or ethnocide, but also epistemicide and ecocide. A form of knowledge, science, which has emerged in a specific region of the world, in the same province of the world—Western Europe—has imposed itself by ignoring other forms of local or regional knowledge, which in one way or another, could offer sources of inspiration for an intercultural dialogue;

this required denying the power relationships that made them inferior and rejected them. This is not a denial of scientific knowledge whose importance we feel in our daily life. However, science is a form of knowledge among other relevant forms, and "within science, there is no single way to do science" (Silva 2020).

The coloniality of knowledge and power, which survived the end of colonialism as a historical phenomenon, simply ignored that no one lives without *knowing how* to gather, without *knowing how* to fish, without *knowing how* to farm, without *knowing* how to shepherd, without *knowing how* to take care of their health (various medicines), without *knowing how* to protect themselves from the elements and without *knowing how* to represent aesthetically. In short, no people or community lives without *knowledge.* Taking action is inscribed in life, in its practices, in its narratives, in its language and in the history of peoples. Scientific knowledge, in the name of a so-called universalisation, ignored the multiple universalisations possible from the different cultures and peoples and their knowledge. The whole history of food production, shelter and curing methods implies knowledge. Ecocide and epistemicide walked hand in hand in the expansion of a specific civilisation driven by a desire for power and domination that has been sponsored since the Renaissance by the Pope (Treaty of Tordesillas), by money through the sword of colonial states.

Until the 1960s, a vision in defence of nature prevailed, which can be taken as pessimistic, in defence of some areas, either for their natural or landscape wealth in private or public parks and environmental conservation units, which in a way, accepted that the process of human development would be irreversible in its nature as a predator. Two main currents emerged from this first phase of the constitution of the environmental field, both in the USA, with the Preservationists, whose greatest exponent was John Muir (1838–1914), and the Conservationists, whose most well-known figure was Gifford Pinchot (1865–1946). The initiatives actually sought to raise awareness among the authorities[1] and landowners, including Pinchot, of the importance of preserving some areas or of knowing how to use them without destroying them, as proposed by Conservationists. There was no conflict, nor a question, but rather the identification of a problem for which they proposed some degree of administration to the owners and authorities.

The 1960s marked a turning point in the environmental field. Since then, the environmental problem has become an issue in and of itself, and as a matter of question, it has been through conflicts within society. In addition to the classic civil rights or social conflicts, a new order of conflict has arisen: environmental conflicts.

---

[1] In most Western languages, the literal sense of uncovered (*desenvolver, desarrollar, developper, todevelop, sviluppare, entwickelné*) meaning to remove from the wrapper, envelope or package. It has the same meaning as to unwrap or to unravel (remove from the skein). In short, to open what is closed. However, this opening is not organic, that is to say, the organism itself or something that comes from within and according to its specificities (Buss and Scheibe 1992). It is not organic. The uncovering always comes from outside and is always associated with the idea of growth, as if it was a permanent attribute. This never occurs in life. The constant demand for matter and energy was born when capital entered the metabolic circuit of the production of goods, with the use of carbon molecules with the steam machine. This led to the belief that there would be no limits to growth. Finally, capital has confused its desire with reality, and today it wants to make a further leap to continue creating the carbon market.

The way society was dealing with the conditions of production-reproduction of life was under open dispute, including resuming a debate between changes within an order or against an order. The environmental field is beginning to be established, and, like every incipient field, it is also being crossed by contradictory positions, although what constitutes it is up for debate. In the case of the environmental field, it is the fate that society(ies) gives to nature.

To accept this term—nature—implies accepting the framework of knowledge that thinks of the life of human beings as separated, in the tradition of European civilisation, and thus accepts the human/nature dichotomy, as well as the way Western universities institutionalised themselves with natural sciences separated from the human sciences. It must be considered that there are many other moulds of rationality forged by multiple cultures and ways of life whose ways of thinking/feeling/acting are not dichotomous, but relational. To provide an example, the Andean peoples (Quechua and Aymara) have a word that is poorly translated as nature: *pachamama*. However, this translation is insufficient, since its root—*pacha*—means space and time, at the same time. *Pacha* implies everything that exists in the community of humans and non-humans, whether they are alive or of other worlds, including spirituality (Estermann 2006). Non-human beings have life and have their own spirit and thus must be treated and respected. It does not hold dichotomies. *Mama* means mother. *Pachamama* is the mother of space–time. However, international organisations continue to base their analyses, norms and controls on the same US/European-centric matrix of specific knowledge as if it was knowledge. Thus, it reproduces the coloniality of knowledge and power that has been imposed on the world since the Renaissance (European).

In this sense, it is the warning that indigenous peoples and peasants made during the UNCTAD—United Nations Conference on Environment and Development—Rio 92, when they associated the year 1992 with 1492 and thus put a long time up for debate—500 years—reminding us that colonisation had disqualified them in the name of the superiority of European civilisation, Christianisation, progress and, more recently, development. Now, after 500 years, we were told when we were there in Rio de Janeiro discussing issues of which they always had knowledge to contribute, in other words, knowledge developed *with* the land, *with* the water, *with* the fauna and *with* the flora. Let us emphasise that this is knowledge *with* and not *about* or a domination *against* land, water, fauna or flora. However, they warned when we were there to discuss the environmental collapse that a particular civilisation, with its geopolitics of knowledge, its dominant economic and political system, had imposed on all other civilisations, peoples and cultures of the whole planet. The peak of this long-term process can be pinpointed to 1945, when an artefact of science—the atomic bomb—was launched on Hiroshima and Nagasaki in Japan. Since then, we have been in a new era: one before and one after Hiroshima (B.H. and A.H.). Science revealed that it was not *necessarily* at the service of life. The anthropocentrism that had claimed its place at the expense of theocentrism had come to paroxysm to determine life as we know it. Since then, humanity, in its diversity, has come to live under the spectre of death. Moreover, science was too serious to remain exclusively in the hands of scientists and professional politicians. One of the main arguments that justified technoscience was that its practical effectiveness allowed us to overcome metaphysical knowledge and establish the domination of nature (Bacon), containing issues that were hardly debated until that time. After all, the domination of nature indicates that caring for nature is not what would move scientific development. The very idea of dominating nature implied taking human beings out of it. Dominating implies making the being that is dominated into what the dominator wants to make of it, regardless of it being in its own nature.

Lastly, the environmental field acquired a new feature in the 1960s. Most important is the turning point suffered in the environmental field, which meant going beyond the limits of the state and the owners, as the Preservationists and Conservationists had been doing until then, making the environmental issue into an interest of society as a whole. Consider that, in liberal societies, the limits imposed on most members of society are given by the private property institute itself, which, as its own denomination suggests, implies private citizens. Private property citizens, most of them, suffer the effects of decisions by private owners. Therefore, the argument that environmental issues must be resolved beyond the borders of states without making the same reasoning about the limits of properties that are also overtaken by the same practices is paradoxical.

Let us remember that the word "politics" in Greek derives from *polis*, which was originally the name of the wall that separated the city from the countryside. *Polis* was the wall, the limit and only then went on to designate what was within the limit. Thus, the *polis* became the non-countryside, the city, from which politics and citizenship emanate. In this sense, it is interesting to consider the origin of the word forest, from the Latin word "forestis", which is an adjective derived from foris ("outside"), which appears in the expression silva forestis (literally, "outside forest"). This was the name given to the territories in which only the king was allowed to exploit hunting and the extraction of wood. Therefore, they were outside the boundaries of the community area (the silva communalis) according to the etymological dictionary (https://www.dicionarioetimologico.com.br/floresta/). Finally, the countryside and the people from the countryside—the peasants—are alien to citizenship. They are slaves, foreigners and women. Prejudices still prevail today. The Greeks considered politics the most sublime of the arts (Aristotle) and made men be valued as "public men" and devalued women, as we still call them "public woman", which in this case, is synonymous with prostitute. All of this invades the environmental field when life in the countryside and peasants is seen as inferior, a life that implies a dense relationship with the metabolism of life, or when the relationship of domination of the city over the countryside is blindly accepted, or what would justify colonial domination as if they were *land nullius*, land of no one, demographic voids. However, vast areas of the planet were properly occupied with peoples and cultures whose practices were of creative coexistence with the other conditions of production-reproduction-inspiration of life. One example: In the region of the present Amazon, there have been people for 19,000 years B.P (before the present), in present-day Colombia, with the Chiribiquete cultural formation. At the time, the Amazon region was covered by vast savannahs (closed, flat, savannahs) and not by the immense forest that covers it today. It was from certain havens, when the last glaciation was receding between 18,000 B.P. and 12,000 B.P., that the various species begin to occupy this vast area of approximately 800 million hectares, which today is the largest forest extension in the world. Therefore, there were people before the forest. Today, over 200 peoples, ethnic groups and nationalities still live there. Let us not forget, there is no life without knowledge. This forest is not a "foris", where "forestis" came from, and it is not at the service of the king. Moreover, as the American anthropologist/archaeologist Darell Posey has well characterised, it is a humid *tropical cultural* forest—a hybrid concept that does not fit in the US/European-centric tradition that separates nature from society (culture).

What constitutes the epistemic-political core of the environmental field is precisely what questions the limits, at the heart of politics, of the relationship between society and nature. Even with this formulation, we are accepting, albeit provisionally, the idea that society and nature are seen as such in the US/European-centric mould. Thus, polis and politics, city and citizenship, are words/concepts impregnated with politics, as an art of establishing limits which, in modern terms, would imply recognising everyone as equals to establish the limits and thus recover the deep meaning of *demos*—of the people—in the establishment of rules—*cracy*. Reinventing politics becomes necessary to debate life in its dual sense of *bios* and *polis*.

That is what the environmental field brings as a public debate. It is worth remembering that the environmental field has done this together with a number of social movements that started to assert themselves starting in the 1960s. Other movements have been added that have already fought, not only for rights that have already been established, but also for the rights to have and formulate other rights. It is notable that these movements that arose at that time are movements that are made from social groups that have been/are assimilated to nature. After all, modern society authorises the domination of nature, according to the principles of the enlightenment. Thus, the subordination/oppression/exploitation is justified because, *by nature*, they are fragile, because *by nature*, they are emotional and non-rational, unlike men. The same could be said of the social groups that work with their hands, bodies that were butchered, since they were used as labour, as if working with their physical strength did not involve not only their metabolism, that same nature, but also, they did not stop thinking because they worked with their hands. They are almost always seen as proletarians, which implies seeing them as offspring that need to be reproduced, just as flesh needs salt—in this case wages that should guarantee their reproduction—as well as the flesh of tomorrow, the prole/proletariat. The other, non-urban peoples, such as who peoples who coexist with life—with the forests, savannahs, tundra, steppes, water, rivers and glaciers—with the metabolism provided by photosynthesis, by the sun—they were called wild, which means that they would be *from the jungle*, and therefore, from nature. As such, the domination/oppression/exploitation of them by the urban areas that, by nature, are always superior is once again justified. With this, they forgot about the word *urbanum*, from Latin, which meant the groove made by the plough in the earth and thus the inscription of humans in the metabolism of life. There would also be a relationship between homo and humus. Moreover, Europeans thought of themselves as not having colour. After all, "coloured people" are non-white, as it is referred to colloquially. Indeed, not Europeans in general, but Europeans "from above" spoke as if they were not from a specific place, as if they were from a non-place, they were atopic, universal, with their "hybris from Point Zero", as the Colombian philosopher Santiago Castro-Gómez phrased it.

We cannot ignore the fact that in the post-war period (1945), a revolution in the media made possible the expansion of the means of exploiting desires, through radio and television, with marketing and advertising techniques, and thus the powerful industrial machine. This is especially true in the USA, whose monopolies have been strengthened since the 1929 crisis and by their connections with state power in the war industry, the Industrial-Military Complex, will begin to spread the principles of a way of life, *the American way of life*, that was believed to be limitless. Since then, a process has been launched which will later be known as the Great Acceleration, where demand for raw materials and energy acquired levels is never seen in the whole of human history. Not without meaning, criticism of consumerism and war waste were two pillars of the new environmental field that was formed in the 1960s. Even business sectors have mobilised themselves in the name of the "limits of growth", such as the Club of Rome, which is a contradiction in terms. They accuse their critics, saying it is impossible to have a society without growth founded on the principles of private property, capital accumulation and profit. The human species would be the only one, in accordance with the principles of the liberal economy that would have no limits on growth, which contradicts the very idea of life, in any way of life, which implies saying that the capitalist liberal economy is, in essence, anti-life.

Since 1945, we have not only been under the spectre of death A.H.—after Hiroshima—and seen the greatest demand for matter and energy, but also we have also seen the greatest expropriatory process ever lived by rural populations and traditional peoples and communities in the same period of time (50–60 years) throughout the history of humanity. Consider the relationship between these three phenomena: the Great Acceleration of demand for raw materials and energy, the action of large monopolistic corporations that exponentially increased their production capacity during the war and the deterritorialisation of billions of inhabitants. The planet's urban population rose from 1.4 billion in 1960 to approximately 3.8 billion in 2015. The rural population, though declining in percentage terms, went from 1.8 billion to 3.4 billion in the same period! Let us avoid the Malthusian ideology, which finds many supporters in the environmental field, since the demand for matter and energy goes far beyond the rate of population growth. The cities that have formed since then are far from the cities that the US/European-centric urban-centric model has set as the horizon of progress, civilisation and development. This is not the case.

Today, of every ten urban inhabitants in the world, seven are in Latin America, the Caribbean, Africa or Asia. Of every ten inhabitants of these same cities on the periphery of the world's system, no less than seven are in their precarious urban suburbs. That is the great socio-geographic phenomenon of the contemporary world that is even growing in the central countries. Since the 1970s, with the recent advance of the new phase of liberalism, capital has been freeing itself from the ties of the state since the unilateral breach of the Bretton Wood contract by the USA. This is especially true after the bitter Chilean experience led by General Augusto Pinochet's dictatorship with the help of the Chicago Boys and the money released from political controls that triggered a deep expropriatory process invading indigenous and peasant territories and their living spaces. In addition, the satanic mill that is the consumer of matter and energy will show its harmful social and environmental effects, confirming what Karl Polanyi had warned us of: money, as an instrument of exchange, cannot be a commodity itself, because this opens up room for an economy, or rather, for a society where the economy has no ethical or moral principle. Labour and nature were also exposed to widespread commodification in the name of the principles of the liberal economy. Thus, they were exposed to a principle, that of scarcity, which, as one of the leading liberal economists of the twentieth century, Robert Triffin, taught us, is the opposite of the concept of wealth. According to Triffin, the liberal economy has nothing to say about the concept of wealth, because its fundamental concept is its opposite: scarcity.

In *physis*, in nature, in *pachamama*, whatever name you want to give to the space–time where life is made from itself—autopoiesis—where we make ourselves, where life is given in the complementarity between cold and heat, between wet and dry, between the earth–water–air, between the tropical regions and the temperate and cold regions, between the plains and the plateaus, between erosion and sedimentation, finally, where the concept of scarcity does not make sense, the wealth of each place is the very nature of that place/region whether it is dry or humid, cold or hot, plain or plateau. This is the richness of each place. It was this diversity, complementarity and reciprocity that inspired social groups, peoples, ethnic groups and civilisations to create some form of mutual aid, some principle of reciprocity between themselves and others, as Russian geographer Piotr Kropotkin rightly pointed out in his book *Mutual Aid.* It should be noted that Kropotkin (2009), inspired by Charles Darwin, developed a unique scientific effort to remove the impoverished reading from his master who, at the time, saw his work being reduced to the Hobbesian principle of the struggle of all against all, the struggle for survival and affirmation of the strongest. Kropotkin's *Mutual Aid* acquires great importance in the present times, since Kropotkin did not oppose "the law of the strongest", the "mutual aid", but insisted on understanding this dialectic of the metabolism of human and non-human life pointed out by Darwin. The pandemic that threatens us today calls us all to feel/think/act in solidarity, since the life of each person depends on the life of all.

Although the environmental field from the 1960s has been expanded beyond the state, technicians and especially the engineers and owners, it will gradually be reinstitutionalised through initiatives by the UN and neo-governmental organisations. The 1972 Stockholm Conference on the Environment and the Club of Rome, for example, showed how powerful groups are beginning to reorganise to address the questions coming from the streets. It may be useful here to recount the Gramscian concept of the restricted state (executive, legislative and judicial) and an enlarged state that indicates that power relations, as part of social relations, extend to so-called civil society, since they do not end at the factory door, farm gate or office entrances. Power relations cross these boundaries, as well as enter homes, companies and communities. Since then, this expanded state has grown even further through organisations that claim to be non-governmental, which grow together with the ideology of the minimum state, which rightly implements the idea of less government: they are non-government, they are non-governmental. Since then, these neo-governmental organisations that are said to be non-governmental grow exponentially and broaden the radius of action of corporations and states that finance them by mediating the interests of large sectors of society by diminishing their capacity for self-organisation, and ultimately, their political role. At first, many organisations that were said to be civil and non-profit making still acted in support of their own agendas arising in the field of social struggles, including an environmental agenda of their own.

However, in the last 50 years, since 1972, with the Stockholm Conference, and the leap in Rio de Janeiro with the UN Conference on Environment and Development in 1992, with the Earth Charter and other documents, they are once again encouraging development. This development is now being encouraged with the epithet of sustainable development (Brundland Report 1987). This development really started to be questioned in the 1960s and has already been renamed as sustainable development, ethnodevelopment, local and integrated development. Finally, as a noun, "development" was given an adjective that does not put it in check, as if we can debate the adjective but not the noun "development".

History seems to be stubbornly developing from its contradictions, and in the midst of this satanic mill of development, devastation and conflict emerged Popular Environmentalism. In the Brazilian Amazon, still in the 1980s, this Popular Environmentalism was consistent with the Forest Peoples Alliance, led by AiltonKrenak, DaviKopenawa Yanomami and Chico Mendes. It was from the bowels of this world, which put into practice the most intense and extensive expropriatory process ever lived in a time equal to that which has come since the post-war years, which gave birth to Popular Environmentalism. It is not only in resistance to this expropriatory process, but in the name of its forms of existence, decolonising itself and affirming itself with its own values, R-existence, therefore.

It was in the Amazon through the indigenous/peasant movement that, in 1990, with two great marches in Bolivia and Ecuador, a new horizon of philosophical and political meaning for life was affirmed with the slogan "struggle for life, dignity and territory", which are distinct from freedom, equality and fraternity imposed by the geoculture of the second modern coloniality that these social groups experienced and with/against which they are confronted. It is important to emphasise this character that these groups do not know modernity, even if they have experienced it as colonial violence, suffering and a counterweight. Thus, it is a thought that *feels*, which lends all meaning to the Colombian intellectual Orlando FalsBorda that, with his actions, participation and research, developed his thesis of feel thinking. It is important to highlight the geography of these two marches for life, dignity and land that set out from the Ecuadorian Amazon and the Lower Lands of the Bolivian Amazon and went to Quito and La Paz, respectively. After all, they are marches that started from the peripheral regions of peripheral countries of a peripheral region, Latin America/*AbyaYala*. They brought life from an Amazon threatened by the extractive colonial capitalist involvement/uncovering. Therefore, they defend human and non-human life. They defend the dignity of beings as they are, "neither calques nor copies, but heroic creations", and thus revive Peruvian intellectual Jose Carlos Mariategui (1894–1930), and the philosophy of Venezuelan intellectual Simon Rodrigues (1769–1854), who said "we invent or we mess up". In short, they are worthy of being

the Yuracaré, Tsimane, Mojeños-Trinitarians, Quechua, Aymara, Guarani, peasants of diverse cultural formations and Quilombolas/Cimarrons/Pallenqueros/Cumbes. However, such a rich and differentiated dignity is not a mere cultural identity that can be asserted without the material conditions for it, hence the territories for the territorialities to assert themselves. It is not only a cultural struggle, even if it is. Thus, they speak no more of a national state, but of a plurinational state; they do not talk of an alternative *of* development, but of alternatives *to* development so that other forms of well-being can assert themselves as equally legitimate in their differences, in terms of equality, in an anti-colonial, anti-racial and anti-patriarchal manner. The defence of non-human life is based on the principle of the rights of nature. Finally, more than a crisis of capitalism, we see ourselves in the face of a crisis imposed by a civilising process that wanted to dominate nature and to subject most of humanity to the designs of those who made themselves private owners and used the prerogatives imposed by their right—the Roman, updated as a liberal—to accumulate wealth, and still submitted the meaning of life to an abstract metric—money—to ultimately accumulate capital. What has now been proposed to us by the "lower class" of the peripheral regions of peripheral countries of a peripheral subcontinent are ideas that point to other horizons of meaning. At this time (2021), the struggle for life, dignity and territory is updated in the most peripheral countries, Chile, of this peripheral region, Latin America/*AbyaYala*, where the role of women is felt, where the Mapuche indigenous people live (in Mapundungun, People, *Che*, from the earth, Mapu) on the periphery of the ends of the world, *Araucania*. Thus, a geography emerges that comes from the streets and countryside (battlefields) and reveals a geography as a verb, doing trials of worlds *with*, *against* and *beyond* the state (Anibal Quijano). Finally, the fact that space is being openly questioned in its scales and territories clearly indicates that the policy it contains is being denatured. Men and women are reinventing themselves from their ethnic-racial and class conditions, especially those social groups struggling to overcome the situation of subalternation/oppression/exploitation they experience and from the experience of feeling/thinking/acting, as highlighted by E. P. Thompson in The Poverty of Theory.

The environmental field has undoubtedly been so imposed that, today, there is no social sector that no longer refers to it, even if they do so in contradictory perspectives among themselves, which requires a refined analytical criticism to distinguish the chaff from wheat. After all, though the environment has been destroyed more and more, there has never been more talk about it, as there has been in the last 60 years. This puts into question the very environmental field that would be serving the environmentalists more than the cause on whose behalf it was constituted. The major international summits where the environmental issue is, in some way, being debated are increasingly reduced to the climate issue (why?) and to its related carbon market. It has served much more as a large market where large NGOs negotiate funds with the big corporations that dominate the agenda and, there, seek to mediate their own interests, many times doing so in the name of social groups in situations of oppression/exploitation/subalternation through consensus that end up withdrawing from these social groups more exposed to deterritorialisation their creative potential to dissent and offer other horizons of meaning for life, including other knowledge that, as we have seen, was disregarded by the geopolitics of knowledge that imposed a science of US/European-centric origin that ignored other wisdom, other knowledge.

Perhaps it is time, finally, to decolonise thought/action, as the peoples of the forest, especially the indigenous peoples, did in Rio de Janeiro, when they associated 1992 with 1492, providing a long-term reference that is not usually part of big corporations' business portfolios with the timings of their annual balance sheets. However, this time is one of the many times that inhabit our space at the same time (Milton Santos and Marc Bloch), whether as a coloniality that survives colonialism, racism, patriarchalism, epistemicide and its ecocide. If there is still any doubt about the meaning of this satanic mill, the coronavirus pandemic invites us to take seriously the ongoing metabolic fracture and overcome the ways of doing/thinking/acting that a particular civilisation imposed on all civilisations, peoples and cultures, which humanity, always plural, invented. Anyone who knows the evolution of life, this larger category that involves all of us sentient beings, inspires us to build horizons of meaning for life that recognise that evolution and diversity have always gone hand in hand and that reciprocity and mutual aid can be the basis of institutions that, even recognising that the violence of "all against all" is part of evolution. They recognise that it is mutual aid that will allow the existence of the other, of the different beings, and ultimately, the existence of life. After all, even the predator no longer exists without its prey. Many indigenous peoples and peasant communities have the habit of feeding and not of fighting what the US/European-centric paradigm calls pests. This is not a logic of war, of fighting them, but of recognising that they also have a right to life. Ultimately, it is not about domination, but about caring for nature and the other. Thus, it is possible to see diversity together with equality as long as we overcome the relations of power, of domination, which pass through them, which go through us. What should this be called? However much it can guide us, it can also lead us to closures, but anyone who knows that nature not only inspires us, but also the peoples and their cultures with their own names for the Common Good, Socialism, the Degrowth and its multiple interferences. What we do not lack are inspirations that, however, require us to decolonise thinking, admitting that beyond the egocentrism of "*I think, therefore I am*" (R. Descartes) that gave birth to modern thought, it is necessary to take the place of the body and the place of feeling *as* the locus of life, as a source of a feel thought (FalsBorda).

Therefore, as every living body, we understand that there is no life without pores, without opening to the other, which indicates our incompleteness, since as humans (1) we only live by the language that, as such, is the construction of common sense and (2) we live only with our relations with the environment, such as the metabolic conditions of production-reproduction-inspiration of life. In short, where the body of each person feels/knows that his ontological existence is not prior to the relationships that constitute us/that we constitute with our actions. We are the environment that we confirm/that conforms to us (*bios* + *polis*). In their struggles to overcome this situation, the social groups in the situation of oppression/exploitation/subalternation are offering us other horizons of meaning for life. I invite everyone to walk with them.

References

Araoz HM (2020) La pandemia como síntomadel Capitaloceno: economía de guerra. In: La Tinta, Argentina
Bourdieu P (2001) O poder simbólico. Ed. Bertrand Brasil, Rio de Janeiro
Brundtland G (1987). Report of the World Commission on Environment and Development: our common future. United Nations General Assembly document A/42/427. New York
Buss MD, Scheibe LF (1992) Mineração, Meio Ambiente e O Caráter Paradoxal do Conceito de Des(-) Envolvimento. In: 37. Cong. Bras. Geologia, São Paulo. Boletim Dos Resumos Expandidos, vol I. SBG, São Paulo, Sp, Brasil, pp 607–608
Estermann J (2006) Filosofía andina Sabiduría indígena para un mundo Nuevo, 2nd edn. Ed. ISEA, La Paz
Kropotkin P (2009) A Ajuda Mútua, um fator de evolução. A Senhora Editora, São Sebastião
Morin E (2002) O Método, 2. A Vida da Vida. Porto Alegre. Ed. Sulina
Müller RG (2013) Exterminismo e Luta de Classe em E. P. Thompson: primeiras reflexões. Revista Projeto História, São Paulo, no 48
Polanyi K (2012) A subsistência do homem e ensaios correlatos. Ed. Contraponto, Rio de Janeiro
Serres M (1994) Contrato natural. Lisboa. Ed. Instituto Piaget
Silva JSS (2020) A hybris do 'ponto zero' e o 'autismo científico'. In: Racismo Ambiental. Consultar. https://acervo.racismoambiental.net.br/2013/11/20/a-hybris-do-ponto-zero-e-o-autismo-cientifico/
Thompson EP et al (1985) Exterminismo e Guerra Fria. Ed. Brasiliense, São Paulo

Chapter 21
Decolonisation Challenges of the Brazilian/Latin American Geography/ies

Rogério Haesbaert

Let us not forget this crystalline truth:
the value of man depends on where he is.
(Milton Santos 1996*, p. 8, translation)*

Abstract The theoretical trajectory of Brazilian Geography has been strongly influenced by the hegemonic thought in the European and North American context. Some authors such as Josué de Castro have, nevertheless, presented the seeds of approaches that are today labelled "decolonial". With the end of the military dictatorship, the renewal promoted by the rise of historical materialism inaugurated a new period that is much more attentive to the spatiality of power and territory. The geographer Milton Santos played a notable role and reinforced a change in geographical thinking, with an influence throughout Latin America. More recently, the dissemination of so-called decolonial approaches has created new conditions for international and interdisciplinary dialogue, particularly by focusing on the acknowledgement of "local" or "regional" knowledge and the political commitment to social change, not only in analytical-critical terms, but also in practical-participative terms. As such, territory as an instrument for struggle against the coloniality of power (Eurocentric binary thought, patriarchalism, racism) is central for many subaltern groups, allowing us to speak about an openness of Brazilian and Latin American geographies (plural) to a fertile exchange with their ancient Euro-North American roots.

Keywords Brazilian geographies · Decolonial approaches · Coloniality of power · Territory

---

Rogério Haesbaert: Geographer, full professor in the Postgraduate Programme in Geography at the Fluminense Federal University (Niterói, Rio de Janeiro), and coordinator at NUREG (Centre for Territory Studies and Globalisation Resistance).

R. Haesbaert (✉)
Universidade Federal Fluminense, Niterói, Brazil
e-mail: sul.roger@gmail.com

R. C. Lois González and M. A. Mitidiero Junior (eds.), *Brazilian Geography*, Advances in Geographical and Environmental Sciences,
https://doi.org/10.1007/978-981-19-3704-0_21

This article intends to analyse, albeit briefly, the relations between Brazilian (and, by extension, Latin American) Geography and the decolonial approaches that have marked our social sciences in recent decades—remembering that one of the central issues for decolonial thought, widely geographical, is the role of the "locus" (the "place" or, more broadly, the space) in which one lives and in which knowledge is produced. The text is organised into four parts: a general introduction situating the specificity of Brazilian Geography(ies) in the (re-)reading of the main epistemological currents of the discipline, the main decolonial elements present in this(these) Geography(ies), the expressive conceptual example of the territory and some final considerations. The central issue, therefore, revolves around the controversy over the possibilities and challenges of a "Brazilian" geographic reading in the geo-historical context of the deep coloniality of power experienced in the Latin American context.

21.1 Questioning Replicant Logic and Eurocentrism

As its designation indicates, the perspective of decolonial thought firstly implies the questioning of the mostly (neo)colonial order that encompasses us within a capitalist—also called modern-colonial—world-system that is shaped by profoundly unequal and propagating dynamics of centre-to-(semi)periphery relationships. Although this dynamic is much more manifold, contradictory and ambiguous than that which is normally propagated, the centre–periphery duo, read in an eminently relational manner, still represents heuristic value, particularly when it comes to the coloniality of economic and political-military power. However, as this coloniality must be viewed at the intersection—or rather, through the intersectionality—of its multiple dimensions and intertwined subjects, there is no doubt that a coloniality of knowledge is also clear—one that imposes the knowledge or epistemes of the "centre", particularly from parts of Western Europe, in its universalising pretence.

Readings about a Geography with Brazilian roots can oscillate between a vision that exaggerates the coloniality of this geographic knowledge, a kind of mimetic reproduction of central geographies and an excessively endogenous perspective, which recognises the existence of an "eminently" Brazilian Geography. Obviously, this is not a question of simplistic ideas in a mere transposition—mild and decontextualised—of the philosophical-geographic bases from the "centre" to the "periphery". More recent views even acknowledge the existence of a "Brazilian school of Geography" (e.g. Silva 2012, when analysing the connections between French and Brazilian geographical thought) or respond positively to the question about the existence of a "Brazilian Geography" (Moreira 2010)[1]. In any case, care must be taken not to fall into an interpretation which, explicitly or implicitly decolonial, suggests the mistake of an inversion (even a hierarchical one) between the "North" and "South" poles.

[1] The main names selected by Moreira in order to search for this answer are Josué de Castro, Aziz Ab'Saber, Carlos Augusto Monteiro, Bertha Becker and Milton Santos.

A term such as "Brazilian Critical Geography" is a recurring one, even among many Latin American neighbours, as they highlight the importance of the influence of Brazilian geographical thought in other countries. Without deepening the discussion, however, these designations almost always end up meaning simply a geographical thought located in Brazilian territory—that is, coming from Brazilian institutions or, at most, produced by geographers of Brazilian nationality.

In fact, it is very difficult to establish the "limits" of a "Brazilian Geography" as, in a sense, our country also manifests its semi-peripheral condition in geographic knowledge: receiving influence and dialogue with different matrices of thought coming from the "centre" while at the same time influencing other peripheral contexts within so-called Latin America. Perhaps associated with an "anthropophagic" tradition (Andrade 1995 [1928]), a mark of our culture is this great openness to the reception and (often) "swallowing" of different external influences. It cannot be denied, however, that sometimes, under the easy attraction of fads, this "anthropophagy" can in practice become an unreflective incorporation of theories and worldviews forged in Europe or the USA.

During part of the military dictatorship period, some centres of Brazilian Geography[2] were marked by the entry of the Anglo-North American-based "New Geography", also called Quantitative or Pragmatic Geography (due to its links with planning), with a neo-positivist philosophical foundation. This is often seen as the most explicit attempt to replicate models forged in central countries, without due criticism beginning with its application to realities as different as those of Brazilian/Latin American peripheral spaces. Works such as Lamego's (2014), however, show important nuances in this process, which is a long way from effectively shaping a quantitative Geography in Brazil—even in the official institution that most welcomed it, the Brazilian Institute of Geography and Statistics (IBGE).

According to Lamego (2014, p. 27), quantitative Geography, even within the scope of the IBGE, was never hegemonic nor was it tied to a "preconceived [political] plan", "in which Brazilian quantitative geography would play a role in a project of political or economic control" (2014, p. 27, translation), as some Marxist critics, hastily, led us to believe. Even quantitative Geography, therefore, did not only coexist with other approaches (more widespread than itself) but "hybridised" between different "matrices and their specific versions in a very peculiar dynamic" that did not constitute a simple reproduction of models coming from abroad. The different geographic currents, at IBGE as elsewhere, were "assimilated, adulterated and recontextualised in constitutive [geo]historical contingencies". (Lamego 2014, p. 2–3, translation).

With the "slow, gradual and secure" (in official language) opening of the military regime, which began in 1978, the renewal promoted by the emergence of historical materialism inaugurated a new period, much more attentive to the contradictory and conflicting space of power relations. The role of geographer Milton Santos, who returned from abroad in 1978, was notable and strengthened a shift in geographic

[2] While the military dictatorship corresponds to the period between 1964 and 1985, it can be said that the relevance of quantitative Geography in Brazil, especially via IBGE—but also with an important centre at UNESP-Rio Claro (SP)—extended from 1969 to 1978 (Lamego, 2014).

thinking, with influence throughout Latin America, broadening the analysis towards a reading of the context of domination and exploration within what we today call the modern-colonial world-system[3]. His book "Por uma Geografia Nova" [For a New Geography] (Santos 1978), published in English in 2021 (demonstrating its relevance to the present), was a milestone, playing with the journey from a neo-positivist "New Geography" to a "New Geography" with Marxist influences.

In this way, a new epistemological relationship was gradually drawn up, with greater exchange and recognition on the part of partners in the "centre", initially within the scope of Iberian heritage and then also in relation to the French, English and North Americans. The diffusion of post-colonial and/or decolonial approaches has given rise to new conditions for international and interdisciplinary dialogue, especially by favouring the recognition of so-called local or regional knowledge and political commitment, not only in analytical-critical terms, but also in practical-participatory terms with effective engagement in social transformation.

Thus, both regional analyses, renewed towards the recognition of new regionalising subjects (Haesbaert, 2010), and territory as an instrument of struggle, allow us to speak about an opening of Brazilian and Latin American Geographies (now effectively in the plural) for a fertile exchange with their former Euro-North American matrices. We will now deal with some of the decolonial challenges facing us, their confrontation by some geographers and their incorporation—explicit or implicit—into a key concept, namely the concept of territory.

21.2 The Decolonial Challenge and Brazilian Geography

Initially, despite all of its geographic/geopolitical foundation,[4] the emergence of the decolonial approach in Latin American thought did not have direct involvement from geographers. One of its most important initial milestones is found in the work of the Antillean thinkers Frantz Fanon and AiméCésaire who, in the 1950s and 60s, highlighted the strength of colonialism and racism in the Latin American and African social formation. The renewing critical thinking of these authors—especially when they re-dimensioned the weight of racism in our social formation—was followed by other thinkers initially with a Marxist base, such as the Argentinian Enrique Dussel (and his philosophy of liberation, even in the 1970s) and the Peruvian sociologist Aníbal Quijano (and his coloniality of power in the early 1990s). These approaches, however, can be considered within a path that has much older roots, such as those in the work of José Carlos Mariátegui, who inspired Quijano with his proposal for a socialism with indigenous roots (Mariátegui 2008 [1928]), or that of Rodolfo Kusch

[3] Given the ambivalences of modernity, built on the interplay between autonomy and heteronomy (as the philosopher Cornelius Castoriadis said and as analysed in Haesbaert, 2002), perhaps it would be preferable to speak of "colonial capitalist".

[4] "The locus of enunciation (...) shows the need to think of knowledge as geopolitical" (Mignolo 2003, p. 21).

in the geoculture of his "dialectic of the mestizo continent" (Kusch 2000 [1953], translation).

It is interesting how these two classical authors, although along quite different theoretical lines (materialist/Marxist in Mariátegui, symbolic/phenomenological—and Heideggerian—in Kusch), end up emphasising the spatial, geographic dimension, either through the "new regionalism" advocated by Mariátegui, which is based on land redistribution, indigenism and territorial decentralisation of the State, or a culture based on the "soil", or being seen as a "spiritual support point" for rooting, as proposed by Kusch (2000:110). This kind of immersion in geography reveals itself in multiple ways in the foundation of decolonial thinking, demonstrated today in expressions such as "geopolitics of knowledge" and "frontier thinking". Although they have been mentioned more recently by authors such as Walter Mignolo and Ramón Grosfoguel, it is worth noting the pioneering role of Gloria Anzaldúa and her frontier narrative in "Borderlands" (Anzaldúa 2016 [1986]).[5]

A kind of more explicit formalisation of the decolonial approach, however, comes mainly from the creation of the *Modernidad-Colonialidad* group, in the late 1990s, led by Latin American intellectuals based in the USA such as Walter Mignolo and Ramón Grosfoquel. For Escobar (2003), it was about the formulation of a "colonial/modernity research programme" in a significantly different perspective. Among the factors that would have made a mark on his genealogy are

> … Liberation Theology since the 1960s and 1970s; the debates in Latin American philosophy and social science on notions such as a philosophy of liberation and an autonomous social science (e.g. Enrique Dussel, Rodolfo Kusch, Orlando FalsBorda, Pablo Gonzáles Casanova, Darcy Ribeiro); dependency theory; debates in Latin America on modernity and postmodernity in the 1980s, followed by discussions on hybridity in anthropology, communication, and cultural studies in the 1990s; and, in the United States, the Latin American subaltern studies group. (Escobar 2003, p. 54, translation)

There are, therefore, a range of influences in the dialogue with the politically libertarian and epistemologically critical ideas of the central countries, mixed with influences from African–American "peripheral" thinking. The degree of connection to the thought of and/or affiliation with academic institutions in central countries and the level of distance from direct political action, or even belonging to a specific cultural group, have recently weighed on the debate over the legitimacy of the decolonial movement. The use [in Portuguese and Spanish] of the Anglicism "decolonial" rather than "anti-" or "*des*colonial" is criticised, at times in an exaggerated manner, and a more incisive presence among social movements is advocated, thereby strengthening the transformative political engagement in defence of subordinate groups. This is the position of, for example, Silvia Rivera Cusicanqui.

[5] "Through her own (inseparable) thinking-acting, Anzaldúa (2016, p. 136) synthesises the border condition in the figure of the "new mestiza", capable of developing "tolerance towards contradictions, (...) ambiguity", learning "to be Indian in Mexican culture, to be Mexican from an Anglo point of view" and playing "juggling games with cultures". This "pluralist mode of operation" (2016, p. 136)—which I would call transterritorial – also appears, as we will see later, in the proposal of Silvia Rivera Cusicanqui, another mestiza, a descendant of Bolivian Aymaras.

Brazil, which has only in recent decades turned more emphatically towards its Spanish-speaking neighbours, ended up, for reasons that are not very clear, entering the decolonial intellectual scene relatively late. This may be linked to its semi-peripheral and, therefore, also "colonising" (or "sub-imperialist") role on the Latin American scale, in addition to the weaker influence of subaltern and/or indigenous knowledge (although not in the Afro case) in its cultural formation. Even so, it is important to remember that, among the thinkers who marked the "decolonial genealogy", Arturo Escobar himself refers to at least one Brazilian: Darcy Ribeiro. Long before him, however, in the 1920s, we must once again remember Oswald de Andrade in his proposal for an anthropophagic or cannibal view based on Brazilian culture—which can, without a doubt, be extended to most of Latin America.

With more specific regard to Brazilian Geography with a decolonial bias, on the other hand, it seems to have kick-started this foray into Latin America, particularly through the work of Geographer Carlos Walter Porto-Gonçalves, who for nearly two decades has been dialoguing with the ideas of the *Modernidad-Colonialidad* group. More recently, a work that explicitly defends the decolonial renewal of critical thinking is that of Valter do Carmo Cruz (2017), with an emphasis on the collection "Geography and decolonial thought", organised in partnership with Denilson de Oliveira. In some sense, I include myself in this group, due to the exchange that we have had for several years within the same Geography department at the Universidade Federal Fluminense[6] and through the publication of the book "Território e descolonialidade" (Haesbaert 2021).

In addition to this more explicit relationship, however, it is also important to remember, within the geographical scope, how much a series of previous readings had already revealed a clear concern for decolonising, both within the scope of the so-called traditional, more empirical Geography and in critical Geography with a Marxist framework. The theoretical trajectory and institutional formalisation of Brazilian Geography starting from the influence of hegemonic currents in the European and North American contexts is well known, marked in the first half of the twentieth century by the dominance of the so-called Regional Geography affiliated to the French School of Geography (such as defined in Berdoulay 2017 [1981]). However, some classical authors such as Josué de Castro (although not formally a geographer)[7] made contributions that can today be largely considered as decolonial approaches, in line with critical thinking from a Brazilian/Latin American perspective. Ferreti (2019, 2020) makes an important reinterpretation of the anti-colonial

---

[6] This included, for example, the joint organisation (with sociologist Carlos Vainer, from UFRJ) of the IVth Meeting of the Latin America and Coloniality of Power Professorship in 2013, bringing together some of the great names in this debate such as Anibal Quijano, Alberto Acosta, Catherine Walsh and Edgardo Lander.

[7] In this text, when we identify a "geographer", we refer to an author's central contribution linked to issues that are considered geographical, and not to his/her academic-institutional training—if that were the case we would, in addition to a classic author like Josué de Castro, be excluding the most well-known intellectual of Brazilian geography, Milton Santos, whose university education was in Law.

perspective present in works such as those by Josué de Castro and Manuel Correia de Andrade.[8]

The most renowned Brazilian geographer, Milton Santos, also reveals a series of elements throughout his work that are consistent with a decolonial proposal, thinking about Geography from a Brazilian/Latin American perspective. After a brief analysis of this decolonial bias in his work, carried out in Haesbaert (2021), recently, at a seminar in his honour,[9] I ended up carrying out a broader assessment of this legacy. In this presentation, I highlighted some of the propositions that mark his intellectual trajectory as a Latin American or "Third World" (a term present even in the title of one of his books) observer, which took place not only in relation to the Brazilian and Latin American context, but also in relation to the world scale, as is evident in a very personal reading of globalisation (Santos 2000a).

The presence of decoloniality in Milton Santos' work appears, ranging from the influence he revealed to have received from the work of authors such as Josué de Castro, to the recognition of his legacy in an "epistemology of the South" (as the only geographer present in the collection of the same name, organised by Sousa Santos and Menezes 2010). His extensive experience in several countries of the so-called global South (study trips to Côte d'Ivoire, Senegal and Mali, teaching in Tanzania and Venezuela) led him to an acute perception of the non-subordinate character present in the peripheries and to the certainty (sometimes somewhat idealised) of the transforming power of the poor; these "slow men" are capable of subverting the unequal verticality of global capitalism with its counter-rationalities moulded in spontaneity, creativity and solidarity—precursors, he believed, of a "popular period of history" (Santos 2000a). In one of his first works, "Marianne em preto e branco", Milton Santos stated that he was not "one of those who think that Brazil and Brazilians only need to learn from other worlds—they have nothing to teach, including Europe: in fact, we already have a lot to teach" (1960, p. 86, translation). The very definition of space in "underdeveloped countries" would not be "comparable to the definitions adopted in developed countries, which often appear generic" (Santos 2009 [1971], p. 122).

Undoubtedly, one of the greatest Miltonian contributions to decolonising thought, articulating a theory beginning from the so-called periphery (or, in his words, from the "Third World"), occurred in the economic sphere, through the theory of the "two circuits of the urban economy of underdeveloped countries" (the [translated] subtitle

[8] On Manuel Correia de Andrade, based on works such as "A Guerra dos Catanos", Ferretti says: "Andrade's analyses of the historical revolts in the Northeast and the abolitionist and quilombola struggles throughout Brazilian history make an important contribution to Spivak's questions about the voice of the subordinate: Brazilian subordinates spoke through their acts, exercising a centuries-old agency and influence through revolt, construction of community and different levels of negotiation that marked the entire history of Brazil. (...) Crucially, Andrade provided examples of how to engage in critical reflections on subordinate spaces, despite the limitations of colonial archives(...)." (2019, p. 1646).

[9] Seminar "Milton Santos, 20 years—A period, a crisis: globalisation and territorial instability—a critical reading of the situation from Milton Santos", held from 22–24 June 2021. Presentation available at https://www.youtube.com/watch?v=Z3wCsV3LY8g

of his book "O espaçodividido"). He was fully aware of this when he identified his proposal as "a new contribution to the search for a theory of space and urbanisation that is sorely lacking in the Third World" (Santos 1979, p. 9, translation). Thus, corroborating his innovative contribution, he states: "the recognition of the existence of the two circuits requires a new discussion of established theories, such as that of the economic base (urban exports), central places [by Walter Christaller] and growth poles (…)." (1979, p. 16, translation).

On the other hand, despite being widely claimed in defence of blackness, the issue of racism, so dear to decolonial thinking as one of the main marks of the coloniality of power beginning from Latin American slavery, was only made explicit in some of his later works.[10] Another central issue of decoloniality, which is linked to the criticism of patriarchalism and the resulting gender inequality—so present in our social (or socio-spatial, as Milton Santos would say) formations—was also little addressed. This feature reminds us of the criticism made by one of the precursors to the geographical approach to gender, Doreen Massey, against two of the greatest geographers with a Marxist foundation, David Harvey and Edward Soja, especially in her article "Flexible Sexism" (Spanish version: Massey 1992).

This does not mean that historical materialism, which so influenced the work of Santos, Harvey and Soja, and which is one of the mainstays of decolonial approaches, was not concerned in its origins with what is now called intersectionality, the focus of phenomena which is not subordinate to a central factor or dimension (sometimes a priori, as is often the case with economics and/or social classes), but is in the complexity of the intersection between socio-economic class, ethnicity, gender, generation and disability. It should be not forgotten that to this combination of dimensions is added today, especially through the decolonial perspective, the spatial dimension, the geographically contextualised knowledge, because "(…) the spatial perspective, the context, is a key factor in understanding intersectionality in a situated and dynamic way" (Rodó-Zárate 2021, p. 63, translation).

Marx himself, today intensively revisited, reveals, mainly in the final phase of his work, the questioning of a historically linear/progressive and Eurocentric universalism (embedded in a teleological idea of revolution based on the development of capitalist productive forces). Anderson (2019), for example, highlights the multilinear view of history in a Marx concerned with the intersectionality between class, nation and race, in addition to some relevant reflections on gender. While maintaining the centrality of the question of capital and class structure, Marx would overcome a "univocity or exclusivity", incorporating a conception of totality that, in Anderson's

[10] In one of his main interviews, Milton Santos reveals that he has a "very complicated relationship with the black world", due to a life story that has always made a mark on him, beginning with an "education to be a man of the court, a man of a full social life". Therefore, despite the "blackness present as a result of the physical condition", he states: "I'm not a militant in anything, be it politics or racialism" (Santos, 2000b, p. 88, translation). However, with his outstanding role as one of our rare black intellectuals with international projection, the influence of his name on Brazilian black movements is growing—as can be seen, for example, in the "rounds of conversation" of the 'Milton Santos, 20 Years' Seminar, already mentioned here, whose schedule and links are available at https://www.miltonsantos20anos.com.br/c%C3%B3pia-programa%C3%A7%C3%A3o-1.

words, "offered considerable scope for particularity and difference" to the point of, sometimes, taking "these particularities—race, ethnicity or nationality—determinant of totality" (2019, p. 355, translation). There is no doubt, however, that in practice a large portion of the so-called communist political movement spread and often even radicalised ideas from a Marxist Eurocentric economics-based model with pretensions to universality. Thus, as Aimé Césaire warned us in his letter of exclusion from the Communist Party, they forgot the equally "legitimate and fruitful" alliances, imposed by time and space, far beyond mere class alliances.

It is in this line of situated and explicitly decolonial intersectionality that we can position another Brazilian geographer, already mentioned, Carlos Walter Porto-Gonçalves, whose work we also analyse in Haesbaert (2021, particularly pages 150–155). Porto-Gonçalves combines a decolonial analysis with the political engagement in resistance movements—which he prefers to call r-existence movements (resistance signifying defence of one's own existence)—of subaltern groups, especially the so-called traditional peoples, starting with rubber tappers. His thesis on rubber tappers in Acre in the struggle for the establishment of extractive reserves (Porto-Gonçalves 2003) and his proximity to leaders of the movement (such as Chico Mendes) is the main reference in this *praxis* or (political) theory–practice intersection. Although the influence of the so-called decolonial movement was not explicit at the time, the work is guided by a clearly decolonial proposal. In this work, Porto-Gonçalves identifies different matrices of rationality in the appropriation of nature, in the geographical production of space—or rather, territoriality—in particular what he calls "material/symbolic appropriation characteristic of rubber-stamping" or "indigenous-caboclo" territoriality (2003, p. 113, translation). We can also assert that the intersectionality present in Porto-Gonçalves' works assumes not only the social dimension of space, but its deep intermingling with the dynamics of nature. Porto-Gonçalves seeks to reassess the concept of territory in the light of the Latin American reality and, in particular, the strength of the r-existence of subaltern groups. In his words:

> We will not understand the emergence of social movements with strong emancipatory potential in Latin America and the Caribbean if we remain prisoners of analytical categories designed from a specific reality such as Europe, no matter how much we have become accustomed, with Eurocentrism, to the idea of a thinking that is universal and therefore valid under any circumstances. (…) among ourselves, coloniality has always been part of our modern formation. (…). Thus, saying 'coloniality' is also saying that there are other subalterned matrices of rationality resisting, r-existing (…). Here, more than resistance, which means resuming a previous action and is, therefore, always a reflex action, what we have is r-existence - that is, a way of existing, a certain matrix of rationality that acts in the circumstances, even re-acts from a topos - in short, from a place of its own, both geographical and epistemic. In fact, it acts between two logics. (Porto-Gonçalves 2013 [2006], p. 169, translation, author's italics)

Porto-Gonçalves reveals a fundamental element in order to think about geographic-conceptual decolonisation in the Latin American context. He reminds us of the strength of territory as a category of practice (as I prefer to call it), produced amidst the multiple struggles for r-existence of various subaltern groups. The category of practice thus overlaps with the simple intellectually proposed analytical category, often deprived of the indispensable link with the use that is made of it in everyday practices. In addition, it is important to remember that the territory in the Brazilian/Latin American scope is widely disseminated as a normative category, used in hegemonic policies, with emphasis on those that are linked to security in the face of indiscriminate violence that is projected throughout the continent—as was the case with the "Territories of Peace" in Brazil or the recent "Territorial Control Plan" in El Salvador.

21.3 An Example of Conceptual Decolonisation: Territory and Multiplicity of De-territorialisation Processes

The conceptual example in which the decolonial reading in Brazilian/Latin American Geography is likely most evident is that of territory, accompanied by the debate on the processes of de-re-territorialisation that constitute it. Through the geographies focused on territory, some of the most important specificities of Brazilian and Latin American geographic thought can be revealed.

Therefore, at the same time there is a challenge of Eurocentrism in relation to the predominance of a view of territory centralised in the power of the state and of hegemonic groups and an appreciation of the territorial practices and knowledge of subaltern groups. As we have seen, this is not a brand-new perspective, since it can be found beginning with the so-called classical geographers, but the decolonial approach undoubtedly reinforces it (a lot), sealing the inseparability between de-re-territorialisation as a theory and as a practice and the consequent engagement of the intellectual with the effects (or bases) of his knowledge in the practices of r-existence.

Some problems, however, can also be mentioned. The most widespread of these is the one involving an exaggerated amplitude with which the concept of territory ends up being used. Analytically, mainly due to the strong influence of Milton Santos' ideas, territory can be equated with the very concept of space. Santos (2000b) states that "actually, I abandoned the search for this distinction between space and territory. (…) I use one or the other, alternatively, defining in advance what I mean by each of them" (2000b, p. 26, translation).

In fact, for greater conceptual rigour, even as social groups increasingly broaden their meaning, it is important to maintain the conceptual tradition of the territory focused on space-power relations or, in other words, on the issues that refer to the spatialisation of relations of power. Our conception of power and politics, however, requires an expansion in the light of the action of these groups, demonstrating the complexity with which "power" and "politics" are redefined, in the inseparable interweaving with economy and culture (and increasingly, also, with the so-called forces of nature).

Thus, speaking of power through the territory starting with the Brazilian/Latin American geographic reality requires highlighting, at the same time, the force of the (para)military and state violence of the dominant groups and the deep sense that symbolic and even affective relationships play in the imposition and/or transformation of these relations. This is, as we proposed a long time ago, a game in which power is realized between the dynamics of domination and (re)appropriation of space (in the terms of Lefebvre 1986).[11]

It is true that this symbolic dimension of power—and, by extension, of territory—was already indicated, albeit with less emphasis, in the work of a European author, Gottman (1973), and his "iconography" of territory. To clearly translate the idea that European (thinking that starts from Europe) does not necessarily imply being Eurocentric (imposing this thought as hegemonic and/or that which should be universalised), we also have the example of the French geographer Joël Bonnemaison (1981). With a dense work carried out among the native peoples of the island of Tanna, in Melanesia, and beginning from the lived space of these groups, he conceived the territory in a broad manner as a space of existence, defining the very "being" of these peoples, in a way that recalls the "territorial ontologisation" (Escobar 2015) attributed to many traditional Latin American peoples.

Another dilemma, therefore, in the decolonising and/or representative proposals of a Latin American specificity in the debates on territory is the extent to which the category of intellectual analysis should be immersed in the practical use and meaning of its effective users.[12] Although territory is seen and exercised by so many

---

[11] Domination tends to give rise to purely utilitarian and functional territories, without allowing the attainment of a true socially-shared meaning and/or a relationship of identity with the space. Thus, territory must be seen from the perspective not only of a politically-structured *dominion* or control, but also of an *appropriation* that incorporates a symbolic 'identity' dimension—and, one might say, depending on the group or social class we are referring to—an emotional dimension" (Haesbaert 1997, p. 41, author's italics). "(...) The territory, relationally speaking, (...) results from the differentiated interaction between the multiple dimensions of power, from its more strictly political nature [linked to State power] to its more properly symbolic character, passing through relations within so-called economic power, inseparable from the legal-political sphere" (Haesbaert 2004, p. 93). "We could say that territory, as a relationship of domination and appropriation of society-space, unfolds along a continuum that ranges from a more 'concrete' and 'functiona' political-economic domination to a more subjective and/or 'cultural-symbolic' appropriation" (Haesbaert 2004, p. 95–96).

[12] Latin American territorial struggles have a lot to teach us, but it is obviously not sufficient to take this concept as a simple transposition of its practical use, its 'native' construction—because the purpose of its use can also be to confuse, or it manifests itself in an excessively-manifold and mutable manner, making a process of understanding difficult. As Gramsci (2011) stated, fighting both blind passion and sectarianism, "the popular element 'feels', but does not always understand or know; the intellectual element 'knows', but does not always understand and, even less, 'feels'" (p. 202). As intellectuals, even though recognising ours as one among several forms of knowledge (and not as a great "law of history", as Gramsci sometimes still believed) in what Boaventura de Sousa Santos calls "cognitive justice" or "ecology of knowledge", it is also necessary to note the specificity of this knowledge, the limits of the (non-) rationality in which it is immersed, and the extent to which it can contribute to the re-reading and/or strengthening of these struggles and the social processes involved in them. Just as the field of ideas about territory is not a simple reflection of the multiple-sphere of concrete actions, these can also be remade in the light of new analytical conceptions about territory. The truth is that the boundaries between categories of analysis, practice and standardisations are always relative and, although endowed with relevant specificity, they are always open to their mutual reconfiguration" (Haesbaert 2021:348–349).

groups as an existential space, inhabited in an integrated/integrating manner (via the inseparability of nature and culture), for the purpose of a minimum analytical rigour, indispensable to our investigations, we propose keeping the main focus of territory within space-power relations, even though it is part of the rather broad notion of power already expressed here.

This leads us to the observation that territorialising oneself while exercising power over a space, "governing" a space (in the broad sense of Foucault's "governamentality" [Haesbaert 2014; Cruz 2020]) also implies "maintaining control" for the care of and the emotion and/or identification constructed with and through this space. Consequently, to de-territorialise oneself is to lose this power, this control, as well as losing the ability to care for and affect and be affected by the space made 'our' territory. Today, increasingly, it is through care and (symbolic and emotional) identification with a territory that one conceives and/or reveals the empowerment of a group.

Another issue that helps to understand the reasons behind the emergence of territory, and the processes of de-re-territorialisation as central concepts for decolonising in the Latin American context concerns the intensity with which our people were extinguished or subordinated through violent processes of territorial expropriation. The pattern of (neo-)extractivist capitalist accumulation that is still reaffirmed today is marked by what some native peoples call "terricide", the most violent form of deterritorialisation, as it implies not only the expulsion and/or degradation of the land but, with it, the destruction of an entire cosmovision and way of life.

Thus, unlike many intellectuals from central countries who saw a de-territorialisation that was generic or much more related to hegemonic groups, attributed to the increase in geographic mobility, here de-territorialisation was read, above all, from the space of subaltern groups, as a process of territorial insecurity (Haesbaert 2004). We therefore assert that

> Surprisingly, (…) the more specifically-social perspective, which the debate on de-territorialisation should prioritise, is practically unaddressed. It is likely that this neglect, linked to the critical reading generally implied by the issue, in turn linked to the growing exclusion (or precarious inclusion) promoted by contemporary capitalism, should be associated with the fact that these discourses are fundamentally shaped from the central countries. For it is precisely from another point of view, the 'peripheral', that we would like to highlight here the approach that links deterritorialisation and exclusion [or precarious inclusion](…). (Haesbaert 2004, p. 312, translation)

Faced with a Latin American reality where de-territorialisation as territorial precariousness and socio-spatial instability are so visible, it was also possible to identify what I called "human agglomerates of exclusion". In these spaces of profound exclusion – or, in José de Souza Martins' sociological reading, of precarious inclusion – it is clear that the struggle for territory is both a struggle for access to land, as a basis for material reproduction, and a struggle for recognition and/or maintenance of a cultural – and in this case, also territorial – identity.

Much of what was claimed for the dominant classes and groups as deterritorialisation actually referred to processes that I proposed to call multi- or transterritorialisation – although this is obviously not a prerogative of the hegemonic groups. Transiting through multiple territories can be a strategy that is both spontaneous, by free choice (as seen with the richest), and compulsory, as a dynamic of survival of the most insecure, forced to leave their territory/territoriality due to material insecurity and/or violence. A modern-colonial world of supposedly-exclusive state territories and well-defined borders would be like this …

… giving way to the world of multiple territorialities activated according to the interests, the moment, and the place in which we are. Here we see a positive angle (the concomitant experience of multiple 'territories' and identities) and a negative angle (the fragility and instability of our relationships with others and with the environment) (p. 35). We could say that today, a fundamental trait in 'post-' or 'neo-' modernity is human multi-territoriality, where, according to the space/time in which we find ourselves and the interests at stake, certain identities [territorialities] are activated to the detriment of others. (Haesbaert 1997:42, translation)

However, this concept of multi- or transterritoriality, later unfolded in other works (Haesbaert 2004, 2014), must be considered with due care, so as not to replace the classic exclusivist vision of a single territorial logic to be universalised—that of the nation state—with an endless array of territorialities at our disposal, so that we can activate them whenever we want or need them. Just as deterritorialisation is not reduced to the processes of territorial insecurity and to subaltern groups, being extended to the "broad" sense of destruction or abandonment of a territory (by/for any subjects), the re-territorialisation of a multi- or transterritoriality of course does not only happen in a positive and manifold sense.

What we call multi- or transterritoriality—and what others, like Vanier (2008) designate as interterritoriality—can be closely associated with the processes of transculturation (Ortiz 1999) or interculturality (Walsh 2009), so dear to decolonial thinking.[13] Closely related, one of the most widespread decolonial debates is the one that involves miscegenation as a fundamental feature of Latin American culture. Some of Milton Santos' latest works evoke the idea of miscegenation within our context. For him, "Latin American cities are mestizos" (Santos 2006, p. 25, translation), and, in a broader sense, there would be a "false globality" that, based on an "illusory ideal of whiteness" forgets "that miscegenation and appreciation of the local are on the same plane as the realisation of globality" (p. 20, translation). By "Europeanising", Latin America would achieve an "unprecedented combination in world history"—a specific type of modernisation, one might say, marked by a "hybrid model—a hybrid, bifrontal, ambiguous discourse" (Santos 2006:19 & 20, translation).

The recognition of this transcultural and mestizo Latin America (a characteristic that obviously cannot be generically extended to the entire continent) involves one of the great controversies of the decolonial approach, which ranges from vehement criticism of miscegenation as a whitening instrument (as in the conception of miscegenation as an apparatus of coloniality of power [Carelli, 2020]) to miscegenation as

---

[13] On the relationship between transterritoriality and transculturalism see Haesbaert & Mondardo (2010), and on multi-inter-transterritoriality and interculturality see Haesbaert, 2021 (especially pp. 337–338).

an instrument of social transformation, as in the Aymara conception of the "*Ch'ixi* world" defended by Bolivian sociologist Silvia Rivera Cusicanqui. Inspired in the literal sense of *ch'ixi* as "scratched ash formed from an infinity of black and white dots that are unified for perception but that remain pure, separate", this would be:

> … a way of thinking, speaking and perceiving that is based on the multiple and the contradictory, not as a transitory state that must be overcome (as in dialectics), but as an explosive and contentious force, which enhances our capacity to think and act. Thus, it opposes the ideas of syncretism, hybridity and the dialectic of synthesis, which are always in search of the one, the overcoming of contradictions through a third element - harmonious and complete in itself.[14]

Faced with effectively constructed spaces, we must be careful, however, not to exaggerate in this decolonising reading of geographic phenomena through the bias of multiplicity and, consequently, of multi-/transterritoriality. Thinking initially from a view on the Brazilian and Latin American urban peripheries, the complexity of the issue is clearly perceived. The territories/territorialities—and, consequently, the forms of power which constitute them—neither can nor should always be interpreted through a multiple bias. In view of the enormous socio-economic and cultural inequality in Latin America, there are countless space moments in which the authoritarianism of strict territorial fixations imposes itself, such as those of criminal and/or paramilitary groups (or even the institutional apparatus of the state and/or the Church) that establish rigid boundaries of circulation for the subaltern.

Although we may not have a territorial confinement there in the normalizing form of disciplinary societies (more typical of central countries), which still believed in the shaping of a (new) order, nor the extreme closure of the confinement of colonial slavery, this is, at the very least, a question of processes of territorial bio-/necropolitical containment of poverty (seen as unwanted or even criminalised). Directly or indirectly, an attempt is made, if not to confine (in theory, to discipline) the most precarious, then to at least bar and/or contain their mobility and/or proliferation in certain spaces (Haesbaert 2014).

Finishing this item, we can summarise the specificity of the approach (decolonial, in a broad sense) of territory in the Brazilian/Latin American context through the following points:

> the breadth of use of the concept, both analytically as well as practically and normatively, to the point that it is sometimes equated with the concept of space, although generally anchored in the strength of power relations in its various dimensions (from economic-state and/or military hegemonic power to the symbolic-identity empowerment of subordinated groups);
>
> the diffusion of the 'territory' designation in everyday practices, as a category of practice and a political tool among different social groups, led by native peoples in their struggles for autonomy (to the point that "whoever says 'autonomy' also says 'territory'" (Svampa 2016, p. 105,[15] translation) – this diffusion, as we have seen, helps to explain the strength of the analysis of (re-) territorialisations of subaltern groups in our context.

---

[14] Source: https://www.brasildefato.com.br/2018/04/26/a-experiencia-de-uma-sociologia-que-se-tece-por-meio-da-paixao-e-do-coletivo (accessed: 20 April 2019).

[15] Without forgetting that territory also has a strong binding role in the urban peripheries of Latin America, to the point of sociologist-activist Tiaraju d'Andrea affirming that "perhaps territory

the emphasis on territory defined by the use (Santos) and symbolic appropriation (Haesbaert, Porto-Gonçalves) of the subjects who build it, well beyond the more functional and/or "technical" and hegemonic-state predominant territory view in Anglophone Geographies. (see, for example, Elden 2013, when conceiving territory as "political technology")

As a result of these attributes, there are contributions in the debate on territory that can be identified even more evidently as decolonial contributions, as is the case of the aforementioned "ontologisation of territory"—formulated by Arturo Escobar (2015) especially from his coexistence with the Afro-descendants of the palenques (former refuges against slavery) on the Colombian Pacific coast, and the concept of body territory—based mainly on the action of the indigenous feminist movement (Zaragocin 2018; Haesbaert 2021, Chap. 4: "Territory as r-existence: from the body-territory to the territory-body [of the Earth]", translation).

21.4 Some Final Considerations

It can be said that Brazilian Geography or, more broadly, Brazilian-Latin American Geographies manifest themselves in different ways in the face of the decolonising challenge that fights Eurocentric universalism (at the same time as it re-evaluates readings of critical-libertarian knowledge that comes from the "centre"), the presuppositions of economism (through a situated intersectionality) and the separation between theoretical reflection, practical sensitivity and political commitment. These challenges are complex, but as we have seen, to a greater or lesser degree they are not a prerogative of current decolonial thought. This means a long path accumulated in the history of geographical thought, from the so-called traditional Geography with more empiricist bases to Marxism and phenomenology, and which today culminates in this amalgamation of proposals that make decolonisation, at the same time, an intellectual and political approach and a perspective of seeing and inhabiting the world.

If the roots of decolonisation are located in the distant past, historically marked by decisive events in the struggle for the expansion of autonomy, particularly of subaltern classes and social groups, our horizon of expectation is projected towards a more distant future, in the hope that setbacks will be avoided and that the struggles that involve society-space in all its diversity are affirmed. But if the approach of the multiple is one of the foundations of decoloniality, one must be careful not to

consciousness is (...) a new form of class consciousness". With the fragmentation of the "working class", he notes, poor populations began to search for their reinforcement and a certain unity (in diversity) by identifying with the territory, life and living space, with the positive redefinition of terms such as "periphery", thus seeking to overcome stigmas and fight for autonomy, as occurs with many cultural groups anchored in the territory. At the same time, however, other heteronomous territorialising subjects arise, such as neo-Pentecostal churches and organised crime (in the case of São Paulo, the PCC), "which also fulfil the role of organising a territory that is socially disaggregated". Source: https://outraspalavras.net/resgate/2021/06/28/tiaraju-dandrea-periferia-brasileira-alem-dos-cliches/ (accessed: 2 July 2021).

underestimate the accumulated reflection that accounted for macro-readings such as that of the modern-colonial world-system (from Braudel and Wallerstein, at least) that so many lessons led us to the understanding of the entire space forged by capitalist globalisation.

One of the great lessons of decolonial reading, as we have seen, is that which refers to the debate on intersectionality. Centred on the key issue of (class) inequality, it deals with expanding the range of combinations of the dimensions in which it takes place—crossing class, gender, ethnicity, generation and functional multiplicity. With the appreciation of concrete contexts as our guide, these are the ones which will define the game in the intertwining of these different dimensions—a game that is therefore articulated in a specific history and geography,[16] as we have seen here in the case of Brazil and Latin America. In this context, together with a prodigal nature, we perceive an extremely predatory, extractive capitalism, in which very diverse classes and cultural groups r-exist, in a unique blend that united the indigenous, European, African and, oftentimes, also Asian contributions (as in the case of the Japanese in Brazil, the Chinese in Peru and the Indians in the Antilles and the Guianas). A large part of this "mestizo" heritage was due to colonising violence. On the other hand, reflected practices of mixing or, as we have seen, of living with differences—"anthropophagic" or, better still, *ch'ixi*—are also a mark of our continental space.[17]

Geographically, in concrete terms, the challenges of decolonising or "de-capitalising" territories marked by the authoritarianism of a patriarchal-slavery past, whose legacy is still intensely present, are also immense. In any case, we have seen that the struggle for autonomy, both among indigenous peoples and in the metropolitan peripheries, is also a struggle for the defence of territories, in search of a politically empowering unity in diversity. Thus, we have multi-scale r-existence initiatives that are articulated around the defence of specific and delimited territories (often conquered via legal loopholes within the hegemonic logic of nation states) as well as around the demand for free movement between different territories (as in the struggle of the South American Guarani indians for the recognition of a "transterritorial condition" between Paraguay, Brazil, Bolivia and Argentina).

Therefore, it has never been more important to know the unity-multiplicity relationship within the Brazilian/Latin American space, at the intersection of all its dimensions. At the same time, scrutinising our geo-historical specificity (a Geography full of historicity) and our insertion in the world-system is an indispensable condition in order for decolonisation, as an intellectual and as a practical-political

[16] As stated by Rodó-Zárate (2021): "There are issues that cannot be explained by capitalism or the relations of production, just as there are issues that cannot be explained by gender or race. (…) Starting from a situated perspective of intersectionality, in which place has a constitutive role, the debate is not whether there is an axis that explains more at a global level, but rather which axes are necessary in order to understand a situation of concrete inequality". (p. 97).

[17] Rivera Cusicanqui goes as far as proposing the "idea of a *ch'ixi* world as a possible horizon of emancipatory transformation. By living in the midst of opposing mandates, creating links with the cosmos through allegories, the *ch'isi* balance, contradictory and at the same time interwoven, of the irreducible differences between men and women (or between Indians and mestizos, etc., etc.) would make another world possible". [other worlds, I would say] (2018, p. 56).

perspective, to impose itself in the geographical universe, without forgetting that the coloniality of power is projected on different scales, from internal colonialism (the exercising of which begins with action on our own body-territories) to heteronomous links with the hegemony of central states.

Perhaps the greatest decolonising challenge of Brazilian/Latin American Geography(ies) comes close to that which has long been sought after by classical "regional" geographers, albeit very often in a rather controversial way: the integrated reading of space, without universalising presuppositions, which includes the dynamics of nature as an inseparable dimension. The first great geographical mistake of modern Eurocentric coloniality was the dissociation between society and nature, the first and most disastrous hierarchising dualism. The geographical decolonising work (re)starts with this great undertaking—defending the natural-cultural diversity of territories as a fundamental principle of struggle, against the terricide that puts the very existence of our lives on the planet at risk.

This is why it is important to recognise the decolonising force of other knowledges ('saberes'), such as that disseminated by native people in the defence of territories where "human" power does not intend to supersede or subjugate the "forces of nature"—which we are also made up of. They propose to us a new, decolonised/decolonising Geography, one that prioritises knowledge and the defence of life territories in all the multiplicity (and at the same time strategic unity) of the pluriverse of cosmovisions that (still) enrich our world.

References

Anderson KB (2019) (2000) Marx nas margens: nacionalismo, etnia e sociedades não ocidentais. Boitempo, São Paulo

Andrade O (1995) (1928) A Utopia Antropofágica, 2ª. edn. Globo, Rio de Janeiro

Anzaldúa G (2016) (1987) Borderlands/La Frontera: la nuevamestiza. Capitán Swing Libros, Madri

Berdoulay V (2017) (1981) A Escola Francesa de Geografia: uma abordagem contextual. Perspectiva, São Paulo

Bonnemaison J (1981) Voyage autour du territoire. *L'Espace Géographique X*(4)

Carelli L (2020) Arqueología del mestizaje*:* colonialism y racialización. Ed. UFRO, Temuco; CLACSO, Buenos Aires

Cruz VC (2017) Geografia e pensamento descolonial: notas sobre umdiálogo necessário para a renovação do pensamento crítico. In: Cruz VC, Oliveira DA (orgs) Geografia e giro descolonial: experiências, ideias e horizontes de renovação do pensamento crítico. Letra Capital, Rio de Janeiro

Cruz VC (2020) Da produção do espaço ao governo do espaço: deslocamentos metodológicos para uma abordagem territorial. In: Limonad E, Barbosa JL (orgs) Geografias, Reflexões Conceituais, Leituras da Ciência Geográfica, Estudos Geográficos. Max Limonad, São Paulo

Elden S (2013) The birth of territory. Chicago University Press, Chicago & London

Escobar A (2003) Mundos y conocimiento de otro mundo. El programa de investigación de modernidad/colonialidad latinoamericano. Tabula Rasa 1

Escobar A (2015) Territorios de diferencia: la ontologia política de los "derechos al territorio". Cuadernos de Antropología Social 41

Ferretti F (2019) Decolonising the Northeast: Brazilian subalterns, non-European heritages and radical geography in Pernambuco. Ann Am Assoc Geogr 109:5

Ferretti F (2020) Subaltern connections: Brazilian critical geographers, development and African decolonisation. Third World Q 41:5
Gottman J (1973) The significance of territory. University Press of Virginia, Charlottesville
Haesbaert R (2002) Questões sobre a (pós)modernidade. In: Haesbaert R (ed) Territórios Alternativos. Contexto, São Paulo
Haesbaert R (2004) O mito da desterritorialização: do "fim dos territórios" à multierritorialidade. Bertrand Brasil, Rio de Janeiro
Haesbaert R (2010) Regional-Global: dilemas da região e da regionalização na Geografia contemporânea. Bertrand Brasil, Rio de Janeiro
Haesbaert R (1997) Des-territorialização e identidade: a rede "gaúcha" no Nordeste. EdUFF, Niterói
Haesbaert R (2014) Viver no limite. Bertrand Brasil, Rio de Janeiro
Haesbaert R (2021) Território e Descolonialidade: sobre o giro (multi)territorial/de(s) colonial na América Latina. CLACSO, Buenos Aires
Kusch R (2000) Obras Completas. Fundación Ross, Rosario
Lamego M (2014) O IBGE e a Geografia quantitativa brasileira: construindo um objeto imaginário. Terra Brasilis (Nova Série) 3
Lefebvre H (1986) (1974) La production de l'espace. Anthropos, Paris
Mariátegui J (2008) (1928) Sete ensaios de interpretação da realidade peruana. Expressão Popular, São Paulo & CLACSO, Buenos Aires
Massey D (1992) Sexismo flexível. Sociológica 18(7)
Moreira R (2010) O pensamento geográfico brasileiro 3: as matrizes brasileiras. Contexto, São Paulo
Ortiz F (1999) (1940) Contrapunteo cubano del tabaco y del azúcar. Ariel, Barcelona
Porto-Gonçalves C (2003) Geografando nos varadouros do mundo: da territorialidade seringalista (o Seringal) à territorialidade seringueira (a Reserva Extrativista). Edições Ibama, Brasília
Porto-Gonçalves CW (2013) (2008) Lareinvención de los territorios: la experiência latino-americana y caribeña. In: Porto-Gonçalves CW *Territorialidades y lucha por el territorio em América Latina.* Unión Geográfica Internacional, Lima
Rodó-Zárate M (2021) Interseccionalidad, desigualdades, lugares y emociones. Bellaterra, Manresa
Santos M (2006) (1996) Por uma epistemologia existencial. In: Lemos A, Silveira ML, Arroyo M (orgs) Questões territoriais na América Latina. USP, Buenos Aires: Clacso& São Paulo
Santos M (1978) Por uma Geografia Nova. Hucitec, São Paulo
Santos M (1979) O espaço dividido: os dois circuitos da economia urbana dos países subdesenvolvidos. Francisco Alves, Rio de Janeiro
Santos M (1996) Por uma Geografia cidadã: por uma epistemologia da existência. Boletim Gaúcho de Geografia 21
Santos M (2000a) Por uma outra globalização: do pensamento único à consciência universal. Record, Rio de Janeiro
Santos M (2000b) Território e sociedade: entrevista com Milton Santos. Fundação Perseu Abramo, São Paulo
Santos M (2009) (1971) O trabalho do geógrafo no Terceiro Mundo. EdUSP, São Paulo
Santos M (1960) Marianneem preto e branco. Livraria Progresso, Salvador
Silva JB (2012) França e a Escola Brasileira de Geografia: verso e reverso. Edições UFC, Fortaleza
Sousa Santos B, Menezes M (orgs) (2010) Epistemologias do Sul. Cortez, São Paulo
Svampa M (2016) Debates Latinoamericanos: indianismo, desarrollo, dependencia y populismo. Edhasa, Buenos Aires
Vanier M (2008) Le Pouvoir des Territoires: Essai sur l'Interterritorialité. Economica & Anthropos, Paris
Walsh C (2009) Interculturalidad, Estado, Sociedad: luchas (de)coloniales de nuestraépoca. UniversidadAndinaSimón Bolivar & Abya-Yala, Quito
Zaragocin S (2018) La Geopolítica del útero: hacia una geopolítica feminista decolonial en espacios de muerte lenta. In: Cruz, D. and Bayon, M. (eds) *Cuerpos, territorios y feminismos.* Abya-Yala & Estudios Ecologistas del Tercer Mundo, Quito

Chapter 22
Brazilian Feminist Geographies: Occupying Space, Resisting Negation and Producing Challenges to Geography

Joseli Maria Silva and Marcio Jose Ornat

Abstract This chapter aims to understand the emergence of feminist geographies and the geographies of sexualities after the 2000s in Brazil. It evidences the epistemological traditions that hampered their expansion as well as the strategies of resistance adopted by a group of researchers who call themselves feminist geographers. The analysis was carried out based on the idea of the geopolitics of knowledge, on the understanding that scientific output occurs against a background of unequal distribution of economic, political and symbolic power among the agents operating in the scientific field. We constructed our arguments based on the data stored in the Brazilian Geography Observatory at the State University of Ponta Grossa, which currently has around 24,000 scientific papers published in online journals that are included in the Qualis System of Coordenação de Aperfeiçoamento de Pessoal de Nível Superior (Coordination for the Improvement of Higher Education Personnel; CAPES). In addition to the quantitative data gathered regarding authorship profile, themes and volume of scientific output, we interviewed researchers with greater impact in the area of feminist geographies and the geographies of sexualities to understand the difficulties they face in their day-to-day research.

Keywords Feminist geographies · Sexualities · Epistemology · Geopolitics of knowledge

22.1 Introduction

We are part of an epistemological current that is not widely endorsed in Brazil—feminist geographies. This peripheral position in the field of power relations in scientific output challenges us to face the limits and boundaries of geographical knowledge hegemonically established at various scales.

When we debate with the Brazilian geography community, we need to prove that what we create as feminist scientific knowledge is a body of knowledge specific to the

J. M. Silva (✉) · M. J. Ornat
Universidade Estadual de Ponta Grossa, Ponta Grossa, Brazil
e-mail: jmsilva@uepg.br

R. C. Lois González and M. A. Mitidiero Junior (eds.), *Brazilian Geography*, Advances in Geographical and Environmental Sciences,
https://doi.org/10.1007/978-981-19-3704-0_22

field of geographic science, and therefore, we are always stretching the boundaries of what is established as a problem to be solved by geography. When we hold discussions with international feminist geography in which our approaches are accepted, we have to struggle for the knowledge produced by feminists from the global scientific periphery to be valued.

Therefore, engaging in feminist geography from Brazil has involved a perpetual and complex struggle waged permanently at different, intricate scales. Our position in the global geopolitics of the production of geographic knowledge has allowed us to establish the common thread of this chapter, which intends to raise the profile of women's work in Brazilian geography and to show the potential of feminist epistemological criticism of the hegemonic forms of understanding the contemporary situation. For us, scientific output occurs against a background of unequal distribution of economic, political and symbolic power among agents and institutions that work in the scientific field. Thus, Brazilian geography has its own geography, which positions knowledge in a hierarchical manner, according to the elements of power that are seized by certain operators of science.

We built our argument based on data stored in the 'Brazilian Geography Observatory' (OGB) of the State University of Ponta Grossa, which currently boasts a collection of 17,636 articles, published between 1974 and 2015, in 90 scientific journals maintained by Brazilian geographical entities at all levels of the ranking established by the Brazilian government for the evaluation of scientific journals under the Qualis System[1]. In addition to the quantitative data collected on authorship profile, volume and classification, twenty interviews were also carried out with female and male researchers with a high impact of citations in the area, considering the number of articles produced registered in the Lattes curriculum and the h index.

In the first section of this chapter, we show that Brazilian geography is characterised by a strong and recent presence of women. In the second section, we prove that this female presence in geography has not enabled a significant expansion of studies on gender, sexualities and racialities. And, finally, we bring feminist criticism to the incorporation of elements such as love and the ethics of care to understand the complexity of the contemporary reality.

22.2 The Recent Feminisation of Brazilian Geography

Women taking their place at universities is a recent achievement in Brazil. In the 1980s, women represented 30% of Brazilian university students, and in the 2000s, women represented 50% (Marques 2012). de Oliveira Cesar and Silva (2021) point out that the process of feminisation of geography is also recent, when analysing the

[1] The Brazilian government, through the 'Coordination for the Improvement of Higher Education Personnel' (Qualis-CAPES System) uses metrics to qualify scientific journals in Brazil, organizing a ranking of quality attribution according to the following order of strata: A1, A2, B1, B2, B3, B4, B5. The evaluation base was the triennium 2013-2015.

gender profile of the sixty postgraduate courses in geography existing in Brazil in 2017. The student body is made up of 49.8% women and 50.2% men at the master's level, and 45% women and 55% men at the doctoral level. When considering lecturers on postgraduate courses in geography in Brazil, the proportion of women is 38.6% and men, 61.4%. A total of 31.2% of coordination positions on postgraduate courses in geography in Brazil are held by women, and 68.8% by men.

Although administrative and teaching structures are important for the progression of women in universities, academic prestige in the scientific field is attained through scientific publications and citation rates, especially when they are made possible by the scientific journals best positioned in the ranking established by the Qualis System, created and maintained by the Brazilian government's Coordination for the Improvement of Higher Education Personnel (CAPES).

An analysis of the sexual division of scientific output in terms of the publication of scientific articles was also carried out by de Oliveira Cesar and Silva (2021). Considering that women participate in scientific output as lecturers and students, they are subject to the same requirements as men in terms of producing research results. In a total of 17,636 articles published between 1974 and 2015 in 90 online scientific journals in the area of geography in Brazil, women represent 39.8% of authorship and men represent 60.2%. When examining the distribution of publications in the quality ranking established by the aforementioned Qualis System in the period 2013–2015, it is possible to conclude that women are at a great disadvantage.

Women publish less than men in all established qualification strata, but female representation is even lower in journals in higher positions. In the highest stratum classified as 'A1', women account for 30% of authorship, and men 70%. In the next stratum, 'A2', female authorship stands at 37% and male authorship, 63%. In the 'B1' stratum, women represent 39% and men 61% of all authorship. In the 'B2' stratum, female authorship represents 43% and male authorship, 57%. The 'B3' stratum is represented by 42% of women and 58% of men, 'B4' with 41% female authorship and 59% male authorship and, finally, 'B5' has 38% female authorship and 62%, male. An analysis of the composition of the editorial board of the 90 journals included in the research by de Oliveira Cesar and Silva (2021) shows a significant male advantage, with men accounting for 62% of the composition of the editorial boards, and women only 38%.

This difference in the classification of scientific output between men and women cannot be explained by the time spent by them on research and teaching activities. Pinto and Silva (2016) analysed the use of time of twenty scientists (ten men and ten women) considered to be high performance due to their levels of productivity and the impact of their theories on the field of geography[2]. Based on the adaptation of the time use methodology used by the Brazilian Institute of Geography and Statistics (IBGE), a survey was carried out of what each person did every hour of the day from Monday to Friday and at weekends. The activities reported were separated by sex, considering people with children and grouped into five different groups as follows:

[2] Productivity measured by the number of articles published per year, according to Lattes Curriculum and the impact measured by the h index.

1—teaching and research activities (hours spent on classes, laboratory experiments, guidance and administrative meetings); 2—personal care (hours spent on hygiene, food and physical activity and beauty); 3—leisure (hours spent on entertainment, parties, cinema, trips out, watching TV); 4—rest (hours spent resting and daytime sleep); and 5—family care activities (hours dedicated to cleaning, preparing meals and caring for other family members).

The results presented by Pinto and Silva (2016) show that, from Mondays to Fridays, women participate in 64.9 h of activities, and men, 60.0 h. When the activities are observed from the aforementioned groups, we found that women dedicate 55 h to teaching activities, while men dedicate 50.5 h. In the 'personal care' group, women dedicate 1.2 h and men, 1.5. In the 'leisure' group, women dedicate 1.2 h and men 5 h. In the 'rest' group, women do not engage in this activity, while men allocate 1 h. Women dedicate 7.5 h to the 'Family care activities' group, and men, 2 h.

At weekends, the pattern of women spending a higher total number of hours on activities is maintained, with 18 h of activities, with men spending 16.6 h. Women dedicate 8 h to teaching activities, while men dedicate 5.2 h on weekends. This demonstrates that women have greater difficulty in managing their workloads on weekdays. Both women and men spend 2 h on 'personal care'. Women spend 3 h on the 'leisure' group, and men, 5.6 h. Both allocate 1 h to the 'leisure' group. Women spend 4 h on the 'family care activities' group, and men, 2.8 h. When looking at the total number of hours spent on weekend activities among men and women, the difference between the two is just over 2 h. However, women spend more time at weekends than men on 'teaching activities' and 'family care activities', while men have more hours dedicated to leisure.

The relaxation of working conditions at Brazilian universities has led to various academic activities being completed in the staff members' free time. This exacerbates the disadvantages women face in allocating hours to teaching activities, since demands on women in the domestic sphere are more intense, which makes increasing the productivity of their work more challenging for women than for men. Intellectual activities depend on concentration, and this is often impossible for women in a domestic setting, especially when they have small children. This immense workload in the domestic sphere is hidden and appears in scientific statistics as low productivity, which leads to frustration and low self-esteem.

de Oliveira Cesar and Silva (2021) present an analysis of the curricula of the sixty postgraduate courses in geography and take as a point of discussion the programmes of the discipline of epistemology of geography. This discipline is relevant because it aims to discuss the understanding of geographic science, its development and fundamental concepts. Additionally, it is in this discipline that the limits of that which falls under geography are established. In addition to the contents of the discipline, the main works and authors indicated as bibliographic references were surveyed. There are a total of 1804 bibliographic references from the sixty programmes of the discipline of epistemology. 90.3% of them are of male authorship, and 9.70%, female. The 1804 references were grouped according to the frequency with which they appeared in the programmes, and of the 10 works with the highest frequency of indication, only one female reference appears, in eighth place.

The contents of the epistemology disciplines of the programmes have common characteristics. The discipline is presented as an ordered historical narrative in which one paradigm is superseded by the next, in a sequential and evolutionary manner, in addition to being full of heroic male characters. The way in which the narratives of the epistemology of Brazilian geography are constructed establishes the limits of thought developed by the scientific community, making it even more difficult for women to play a leading role in the production of geographic knowledge. The narrative of what 'true geography' is becomes truth through repetition, associated with the teaching authority of whoever perpetuates that narrative.

In addition to having worked hard to state the obvious, the same limits, problems and obstacles to be faced by women in the scientific field of geography have been pointed out by the international literature for decades (Berman 1977; Monk and Hanson 1982; Bowlby and Peake 1989; McDowell and Peake 1990; Rose 1993; McDowell 1999; Ramon and Monk 2007; da Silva et al. 2007; Silva 2009; Ramon et al. 2011). To celebrate the twenty-fifth anniversary of the journal 'Gender, Place and Culture—a Journal of Feminist Geography', a special volume was released that discussed the development of feminist geographies in the world (Blidon and Zaragocin 2019). This volume reflects important advances in gender studies and feminist geographies in various parts of the planet. But it also shows that, in the twenty-first century, these knowledge perspectives face obstacles that must be overcome in order to reduce the various forms of oppression in the scientific field.

Reaching such repetitive results across the planet is often tiring and discouraging. Even so, these indisputable data offer the opportunity to open dialogue with the academic field to dismantle the established powers. In the next section of the chapter, we show the emergence of contesting epistemological movements that problematise relations of gender and sexualities ingrained in the alleged naturalness of a male hegemony in the production of scientific knowledge.

22.3 The Fissures of Male Power in Brazilian Geography and the Tension Caused by Approaches to Gender and Sexualities

Although the number of women in Brazilian geography is substantial in quantitative terms, as evidenced above, geographical scientific output has opened up very little to studies involving diversity in terms of gender, sexuality and race. Among the 17,636 articles published between 1974 and 2015 in 90 scientific journals maintained by Brazilian geographical entities, at all CAPES qualification levels, only 238 articles were found involving gender, and 76 on sexualities.

da Silva et al. (2007) argue that the long military dictatorship that dominated Brazil between 1964 and 1985 made it difficult to expand gender studies in geography and only after the process of re-democratisation and through the effervescence of social movements in the following decades was it possible to expand analyses of gender and sexualities. In the 1980s, two articles were published focussing on the role of women and labour relations. In 1984, the article 'A Zona Canavieira da Cotinguiba e o trabalho da mulher' ('The Cotinguiba Sugarcane Area and women's work') (Silva 1984) and 'A mão e o torno: a divisão sexual do trabalho entre osprodutores de cerâmica' ('The hand and the wheel: the sexual division of labour among producers of ceramics') (Dantas 1987) were published, both in the journal Geonordeste. In the 1990s, five articles were published whose approach to women and work remains relevant, expanding the analysis to urban areas. One of the most important articles in this pioneering phase is that of Rossini (1998). That same year, da Silva (1998) introduces a discussion of gender and feminist geographies.

It is at the beginning of the twenty-first century that gender studies begin to be produced with greater intensity. Between 2000 and 2009, 59 articles were published, and between 2010 and 2015, 172 articles. Of the total of 238 articles on gender in Brazil, 97% of them were published after the 2000s. The expansion of publication of articles involving sexualities follows a similar pattern. The first 4 articles appeared in the 1990s, all of them by Miguel Ângelo Ribeiro, focussing on the areas of prostitution in Rio de Janeiro and their relationship with tourism. Between 2000 and 2009, 6 articles were published, highlighting the intensity of the approach to transvestites, prejudice and appropriation of urban territories. Between 2010 and 2015, 66 articles were published. A total of 94.7% of the geographic scientific output on sexualities was published after the 2000s, following the trend seen in gender studies (Silva and Vieira 2014; Silva and Ornat 2016).

The increase in the number of articles was made possible by the establishment of political initiatives to confront the power structures of the scientific field, such as the conquest of forums for formulating ideas and political organisation. This expansion did not take place spontaneously or due to the recognition of feminist contributions by the hegemonic sectors. Although there was already production in the area of geography involving women and gender, the group of Brazilian geographers who called their production 'feminist geographies' only emerged in the 2000s. The definitive assumption of this position occurred with a debate that took place on the sidelines of the Latin American Geographers Meeting in 2009, in Montevideo, Uruguay. Some of the people who researched gender and sexualities argued that geography was a unique science and that gender and sexualities were just 'perspectives' to be developed. Another group contested this uniqueness and claimed its own epistemological and methodological path, taking responsibility for raising the profile of feminist geographies in Brazilian geography (Silva et al. 2015).

The awareness of this scientific identity as feminist geographers was fundamental to developing deliberate actions to conquer political and scientific forums for promoting ideas. Worn out by the rejection of their articles with the justification that gender and sexualities were not themes specific to geographic science (Silva and Vieira 2014), researchers created the Revista Latino-americana de Geografia e Gênero in 2010. Such a strategy was only made possible by the popularisation of the Internet in Brazil after the 2000s, which enabled low-cost publication. Since then, this journal has published 44.9% of gender articles and 56.6% of sexualities articles in Brazil.

Another important strategy to strengthen the identity of feminist geographies was the creation of their own scientific event and the conquest of forums for the expression of ideas in other diverse events in Brazilian geography. In 2011, the I Latin American Seminar on Geography, Gender and Sexualities was organised in Rio de Janeiro, with further events held in 2014 in Porto Velho (Brazil), in 2017 in Mexico City (Mexico) and in 2019 in Tandil (Argentina). Gradually, gender and sexualities became part of longstanding national events. Two of these now offer specific sessions for the presentation of papers, the National Meeting of Graduate Studies in Geography and the International Symposium on Agrarian Geography.

These initiatives to strengthen feminist geographies in Brazil are related to the administrations of the Workers' Party (Partido dos Trabalhadores—PT) between 2002 and 2015. The social policies implemented in the country allowed the emergence of new social actors who had access to public universities in Brazil through the implementation of quota policies for places for black, indigenous and low-income students. The presence of these designated individuals strained the elitist hegemonic profile of academic settings, resulting in discussions concerning differences and the academic setting. Another factor that had repercussions for the scientific output of geography was the emergence of voices that conflicted with the canons of geography as a result of the implementation of a policy of expanding postgraduate programmes beyond the already consolidated centres of scientific output.

The formation of an academic periphery in Brazil in new spaces of scientific output in the interior of a huge country with a rich territorial diversity created opportunities for a plural and diversified geography. Furthermore, the formation of national plans for the promotion of rights relating to gender and ethnic/racial relations and sexual diversity, the actions of which permeated educational policies at all levels of Brazilian education, led to the expansion of discussions on social differences that went beyond mere class.

The reconfiguration of the 'geography' of the Postgraduate Programmes in geography in Brazil was striking (Sant'anna Neto 2012). Of the 60 courses existing in 2017 in the national territory, only 19 were created before the 2000s. A total of 61% of postgraduate programmes in geography, settings par excellence for the production of geographical scientific research, emerged at the beginning of the twenty-first century (Suertegaray 2007) and spread to inland regions of the country. de Oliveira César and Pinto (2015) point out that of the 224 groups of researchers existing in Brazilian geography, 17 carry out research on gender and/or sexualities. The vast majority of these are based in peripheral universities that have only recently instituted postgraduate programmes.

Articles on gender and sexualities published in scientific journals in the area of geography in Brazil represent only 1.8% of Brazilian scientific output. Additionally, they are mostly published in journals that do not occupy the highest positions in the ranking that establishes the quality standards of Brazilian scientific output. Only 6.3% of the production of gender articles were published in the upper strata (A1 and A2). Articles on sexualities are not represented in stratum A1, and only 7.9% of publications on sexualities are published in A2 journals. The rest of the production, both in terms of gender and sexualities, are mostly published in journals classified in the middle strata, B1 and B2.

Brazilian feminist geographies are produced by researchers based at universities that occupy a peripheral position in the organisation of the geopolitics of scientific prestige networks in Brazil and, in addition, they are published in journals without central classification in the power structures. Despite such weaknesses from the standpoint of institutional power relations, it can be said that the social context of the rise of minority rights in Brazil in the 2000s has allowed the emergence of a complex feminist geography, characterised by strong alliances between approaches to gender, sexualities and racialities, without manifesting the rifts observed in the historiography of Anglophone feminist geography. The articles from this period are mostly based on the intersectional and performative perspective of gender (Silva 2009; Souza and Ratts 2009).

The increase in scientific output contesting the privileges of gender and sexualities in the field of Brazilian geography has impacted hegemonic conceptual traditions, with structuralist bases, in particular class relations, as well as the idea of a materially visible spatial product. In the following section of the chapter, we address the alternative approaches that feminist currents have developed, proposing the idea of a space that is produced by actors embodied in interdependent relationships on multiple scales.

22.4 Feminist Epistemologies: Love as an Element of Spatial Policy and the Ethics of Caring for Life

Brazilian geography values large-area approaches, large-scale capitalist production, social movements and large material expressions of resistance. This epistemological characteristic prioritises processes, makes agents disembodied and constructs politics as something specific to the public space. Although other scales are simultaneously forming this space illuminated by Brazilian geography, they have been left in the shadows, rendered invisible by the phallocentric power of knowledge production.

Feminist epistemologies have challenged the idea of pure and abstract rationality in the understanding of spatial reality and have adopted the assumption of embodied human experience, simultaneously personal and social, involving space and time of embodied everyday life and full of passions. People live concretely guided by beliefs and feelings and make choices based on what they believe to be right. By acting daily, even without reflecting, people confront obligations, responsibilities, feelings of guilt and shame, and these embodied experiences are supported by ethics. Ethics is a creative and negotiated process of ways of being in the world that guarantee the principles of 'living better', this process being relational and political.

Contemporary western ethics has been challenged by the feminist philosopher Butler (2016), questioning what we call human life. What a society considers to be a human life depends on its social contracts and power frameworks. Its concern is, above all, to institute ethics under which humanity can create social forms of viable and liveable lives, taking into consideration the precariousness of all human life. The idea of 'precarious life' is a perspective developed by Judith Butler (2006) to highlight the social forms in which we frame and value what we consider to be a life. It starts from the inherent vulnerability of our social existence, in the sense that all lives can be eliminated at will, or by accident and, therefore, human life is never guaranteed in itself, independently, hence the sense of precariousness of each and every life. Thus, Butler's (2006) idea of 'precariousness' implies understanding the fact that we are inexorably connected to each other and that human life is, in a way, in the hands of others. Precariousness implies exposure to the other, both to those we know and to strangers. We are definitely related, and interdependent beings and precariousness are inherent to human life, which derives from the fact that all human life is vulnerable to potential destruction.

However, precariousness is not an ontological condition, but a social condition from which the principles of politics emerge. In this sense, as a society, we build different levels of appreciation of what is considered life and how we conceive some lives that we are unable to perceive as wounded, lost or lamented, simply because they are not considered lives. Even if all human life is equally defined by precariousness, the contexts of power through institutions of different natures maximise the precariousness of life for some, while minimising it for others, and this is profoundly spatial in nature.

It is socially that we conceive what life is and if some lives are not conceived as lives according to certain epistemological frameworks, they may be wasted, and their loss may not be mourned by a society. Precariousness for Butler (2006) is instituted and produced politically in the establishment of populations and social groups that suffer from a lack of social support, becoming differentially exposed to violence and death. So, the heightened state of precariousness that some lives experience as opposed to others imposes on us obligations to critically question the conditions under which it becomes possible to perceive a life as precarious and to explore ways to reduce precariousness in order to promote the conditions for the fruition of life.

The prosperity of human life in a world of differential precariousness demands that we consider our lives as relational and interdependent. This means thinking collectively about the protection of life and the conditions for maintaining life that increase or decrease precariousness in given spaces and times. Normative schemes produce a framework of beings whose lives are recognised and of lives that do not receive this recognition. Thus, the social recognition of what constitutes a life is variable and dependent on the conditions we create for this recognition to occur, as 'the condition of being recognised precedes recognition' (Butler 2016, 19). If precariousness is unevenly distributed, the conditions that sustain a 'liveable' life must be questioned as a political element of an interdependent social and spatial existence. 'Liveability' is the capacity of society to sustain social life in the inherent conditions of precariousness and in the social, political and spatial management of precariousness.

If, on the one hand, liveable life is sustained by basic material conditions such as food and shelter, on the other hand, it is affected by the conditions of social intelligibility, i.e. it is dependent on the normative conditions that socially shape who the subject is whose life is recognised as life. Judith Butler's philosophy seeks to rethink life, to criticise the normative discourses of intelligibility through which life is currently construed, which promote the unequal distribution of precariousness.

The concept of 'human life' is not a given but is normatively constructed and sustained in ethics and politics. To make the world liveable, it is essential that we think critically about the mechanisms of production and reproduction of what is 'human' in contexts of precariousness and in the ever-present possibility of constituting lives that do not matter as human lives. So, by 'dehumanising' certain human lives, we are making their spatiality precarious. Precarious spaces are formed from dehumanised lives, those that do not deserve to be preserved from the standpoint of socially instituted power arrangements.

The maintenance of life has been considered by feminist philosophy, which has developed theories about the 'ethics of care'. Robinson (1990) argues that, unlike masculine philosophy, which focuses on the autonomous, abstract and rational subject, feminists think of subjects in interdependent relationships and accountable to each other for the maintenance of life. According to her, it is easy to think about the obligations to care for family members or friends, as close people influence each other. However, people beyond our families are also vulnerable and affected by our actions and choices.

The ethics of care, therefore, has implications beyond the sphere of personal and private relationships, including in the public and political sphere. Robinson (1990) proposes that we should think about the ethics of care in broad-scale relationships, since we live in a globalised world and are connected in every way. According to her, the ethics of care asks not only the 'reason' why we should care about others with whom we share the world, but also 'how' we should care and 'how' we promote relationships of personal and social care.

The ethics of care is not a given but involves learning to listen and be attentive and receptive to the needs and suffering of others, involving the understanding of how relationships are built and how difference is perceived and maintained through institutions and structures in societies. Bringing the ethics of care into the public space does not mean taking away from the state its responsibilities so that care is exercised by poorly paid caregivers, but rather making this institution assume primary responsibility for maintaining structures (political, social and economic) that support the principles of caring for life, as argued by McDowell (2004).

A society that in its structures of values has care as a civilising foundation cannot maintain the idea that care is an exclusively female task, an undervalued type of work or a secondary moral value. On the contrary, the fact that care is a central element of human life needs to be adopted by social and political institutions. Toronto (1993) argues that care, in addition to being fundamental, is also necessary for maintaining individual safety and collective well-being in everyday life. Care forms a large part of our interpersonal relationships and involves justice. According to her, care is a human activity that must consider everything we do to maintain and correct our 'world' so that we can live in it in the best possible way, and this world includes our bodies and the environment—in short, everything that is intertwined in the complex life-sustaining network. This proposition for understanding care has been expanded in several ways, far beyond traditional areas such as health and social welfare assistance. Currently, the ethics of care guides theories concerning politics, the environment and justice at different spatial scales, as Victoria Lawson (2007) argues.

Care is so fundamental to our condition of mutual existence on the planet that we are often unaware of it in daily life as an essential element of our relationships and our collective responsibility. Barnes (2012) proposes that we should think about care in three different, interrelated ways. First, care is a way of conceptualising personal and social relationships. Second, care comprises a set of values, or moral principles, which provide a direction for our conduct. Third, care needs to be thought of as a practice involving the people who provide care and those who receive care. Therefore, the time invested in care practices between people is not always equal, generating power structures that also unequally distribute advantages and disadvantages.

Held (2006) argues that, although there are various ways of thinking about the ethics of care, there are recurring characteristics among the versions she analysed. The first, usual characteristic in approaches to the ethics of care is the irrefutable moral relevance of meeting the needs of others, in particular those for whom we take responsibility, not only for family dependents, but also for other people beyond the private space. The second common feature of the ethics of care is that emotion plays a fundamental role in understanding what morality recommends. Sympathy, empathy, sensitivity and the ability to react are considered moral emotions that need to be cultivated to determine what morality recommends. The third characteristic of care concerns consideration of the concrete challenges of real people, rejecting the concept of abstractions and the universalisation of moral problems. People in relationships of care are acting together and with each other, and the well-being of a relationship of care involves cooperative well-being between them, as well as the relationship itself.

The fourth characteristic of the ethics of care that persists in Held's (2006) analyses is that, like many feminist thoughts, there is a re-conceptualisation of traditional notions of the public and the private. The traditional view, embedded in the dominant moral theories, considers the family a private sphere, outside politics, in which the government should not intervene. Feminists have shown how men's greater social, political, economic and cultural power has structured the 'private' sphere to the disadvantage of women and children, making them vulnerable to domestic violence as well as leaving women economically dependent on men and subject to a highly unequal division of labour in the family.

Hegemonic moral theories consider the life of the public space to be relevant to morality, but private space and intimate relationships are not important. The ethics of care addresses moral issues among unequal and dependent people, contrary to the dominant idea of the meeting of two free, rational and equal people. Inequalities and interdependence are also at the base of the conception of public space, resulting in the problem of relationships that are often established in an involuntary and emotional way. No one at birth chooses their sex, gender, colour, class, sexual orientation and nationality, but even so, people experience these characteristics in relationships of interdependence and inequalities that mark human experience.

The fifth characteristic put forward by Held (2006) is fundamental for the conception of the ethics of care. It has to do with a person's way of understanding. In the conceptions of dominant moral theories, people are understood as rational and autonomous individuals who establish social cooperation to increase the chances of each of them achieving their ends, and at the same time, the objectives of society. Under this hegemonic conception, people are first conceived as distinct individuals and only then establish relationships. Under the ethics of care, people are conceived as relational and interdependent, morally and epistemologically, from birth and throughout life, as it is relationships that form their identities.

Butler (2017) acknowledges the impossibility of human existence outside of relationships, but also warns of the need to criticise those same relationships that form individuals. Relationships are sources of moral identity. People share the world and commit to each other, and with some of them, develop bonds, commitments that are central to their lives, substantiated in asymmetrical care relationships.

The willingness to care has been questioned by feminist thought, arguing about inequalities in care practices for human life. Rose (1994) argues that care work is gendered and racialised and this is perfectly demonstrated in the statistics regarding 'time-invested' in care, with black women being those who most benefit humanity by caring for life.

The politicisation of the ethics of care has been an arduous path for feminist thinking, as philosophers need to deal precisely with the elements that underpin masculine arguments attributing female inferiority. After all, from the male perspective, the body 'naturally' prepared for reproduction and the care of life and emotion as a 'natural' moral weakness were the premises that validated the idea that care was a 'naturally' female task and the energy that results in the expenditure of time for the care of life would be love.

While love is unquestionably important to human life, it needs to be explained. Love has been considered an effect, something mysterious, impenetrable and little addressed by science, with the exception of the feminist strand. The establishment of the politicisation of experiences of love has been a concern of this strand to build elements of resistance to male power. Toye (2010) argues that it is difficult to communicate the meaning of love, and there is no definition of love per se. Feminist philosophers have been seeking scientific conceptions of love since the 1970s. When analysing this trajectory of understanding love, Jónasdóttir and Ferguson (2014) combine the analyses into three interpretative paths: love as disillusionment or ideology, love as an element of ethics and epistemology and love as a social and biomaterial human power.

Love conceived as an ideology was centred on the idea of romantic love, one of the elements of gender relations between heterosexual couples, imbued with male power and the exercise of patriarchy (Toye 2010). According to Langford (1999), the experience of modern western love is understood by erotic, spiritualised passion, exclusively focussed on romantic codes and this has formed the basis of our social identity, playing an important role in maintaining the pattern of social arrangements, notably the mononuclear heterosexual family.

Love as an ethical and epistemological element can be seen in the production of Toye (2010) who has proposed the politicisation of love and criticised the substitutive forms of understanding by the feminist approach, such as the idea of care. For her, although love implies care, it is more than that. She proposes that love is an ethical concept that names not only a particular qualitative relationship between a particular self and the other, but a particular distance or spacing between them.

What is at stake for her is the status of the subjects of the relationship, as it is this spacing that allows the dynamic existence of two subjects constantly involved in processes of transformation and becoming, and both parties have this right. Based on the philosophy of Luce Irigaray, she says that it is necessary to replace the expression 'I love you' with 'I love with you', arguing that the meaning of love between subjects of equal status is the sharing of love experiences. Thus, the beings of the action are not objects of desire, but subjects that trigger by choice and responsibility the emotion that identifies the relationship.

The idea of love as an ethical element was also developed by Hooks (2000) who understands love as a form of existence, a choice and responsibility for the world. Thus, love is intention and action rather than a feeling, and she even suggests that love should be a verb and not a noun. According to her, love can be understood as a social and biomaterial human power. She understands that love is a distinct human capacity, a creative and recreative energy that, alone or associated with other capacities, such as work for example, allows people to act intentionally to improve and change their lives and the conditions of existence.

If love is not something given and definitive, but rather takes place in a force-field in which different conceptions of love are being disputed, it is necessary for feminist thought to create alternative definitions from those produced by male knowledge that mark hierarchical gender relations. An interesting alternative narrative was constructed by Jónasdóttir and Ferguson (2014), Jónasdóttir et al. (2011) and Jónasdóttir (2011a, b). This series of texts explore love as an element that structures sexualities and genders, both as identities and in embodied and socially structured material actions, through values and powers that interact and create temporally and spatially distinct organisations. Therefore, for this group of authors, love is political, understood as a creative and reproductive human capacity that can be exploited.

Gender relations, love practices and the uses of the power of love in the process of production and reproduction of people constitute important elements of social understanding, as Jónasdóttir and Ferguson (2014) argue. Thus, Jónasdóttir (2011a, b) creates an interrelated perspective between culture and materiality to understand love in the patriarchal organisation of heterosexuality, deeming it crucial to understand this hegemonic form that oppresses all other forms of sexual love.

In these works, Jónasdóttir (2011a, b) makes a strong approach to historical materialism, stating that the economy (the production of means of subsistence) and sexuality (the production of life) are internally related, although distinct, social processes. They are not separate structures, economic production on the one hand and social reproduction on the other, related only externally. She claims that not only human work, but also human love must be understood as a practical sensory activity and a world-creating capacity, allowing the analysis of reality on multiple spatial scales complementary to each other.

Jónasdóttir's analysis (2011a, b) argues that, in capitalist societies, exploitation of the capacity for work is connected with the exploitation of the human capacity for love, which translates into time spent on the tasks of caring for life. For her, human capacities for love and work are living sources of energy that can be released to construct the world.

To summarise her long career of studies on love, Jónasdóttir (2014) summarises her central statements about 'love' or the 'power of love' in eight related points. 1—human love is comparable to work in certain fundamental aspects, but cannot be reduced to it; 2—love is a human activity resulting from relational flows of forces and processes of interactions and transactions, occurring in sociosexual relationships; 3—the 'power of love', like the 'power of work', is an alienable and exploitable human social force; 4—love sustains the contemporary capitalist structure of the 'formally egalitarian patriarchy' because the system is dependent on the exploitation of the power of women's love by men; 5—human love can be qualified as a 'practical sensory human' activity, as well as work in Marxist theory; 6—love offers a 'type of sociability' that enhances 'world-creating capabilities' when it contains genuine reciprocity between subjects; 7—love is important as a unique transformative power in social life, a fundamental causal human power and an agent of change, which can be used both for freedom and for the domination of some over others; and 8—love is not a monocausal power, but intersects with the social creation of cultural norms and values, beliefs and knowledge, or the establishment of political (including judicial) forms and institutionalised treaties.

Love as energy that potentiates action and creates a material world has its geographies. Morrison et al. (2012) make an important contribution by encouraging us to think about the relationship between love and its spatial dimension. For them, love is experienced bodily, both when considered as energy, as just a physical/chemical sensation, and even when it exists in relationships of care and erotic desire.

Based on Jónasdóttir's (2014) idea about love as energy, social force and practical sensory human activity, we may make the claim that geographic space can be understood as the result of bundles of embodied relationships of affection, mediated by the processes of power interactions that trigger specific actions and materiality in a permanent process of reconstruction. Love as a spatial process is an underexplored field and has great research potential. After all, if love implies bodies, energies and relationships, it certainly cannot do without geographic space, and in times of pandemic, this challenge is urgent and necessary.

22.5 Closing Remarks

In this chapter, we highlight the growing feminisation of Brazilian geography and emphasise that the mere bodily presence of women is insufficient to produce geographies that make gender relations and sexualities visible in this scientific field. We point out that the hegemonic versions of geographic science are only destabilised when we consider the geographic nature of science itself, marked by privileges of gender, sexualities and racialities that allow the concentration of economic, political and symbolic power, and we act deliberately to dismantle them. This often involves tiring work for us feminists to state the obvious. Claiming that women work longer than men to sustain life on the planet and that their presence is recognised and valued, may bring prospects of transformation in the ways in which we conceive concepts and ways of interpreting the geographic space. The idea of love as energy and the ethics of care, which implies responsibility for the lives of humans and non-humans, is a proposition that has been seen as being of lesser importance in a world increasingly driven by technologies and the market. However, there is no future without taking care of the present, and only a geographical imagination capable of conceiving a space in which reason/emotion, mind/body and nature/society are considered interdependent will enable us to create a world in which no human life is wasted.

References

Barnes M (2012) Care in everyday life: an ethic of care in practice. University of Bristol Press, Bristol

Berman M (1977) Facts and attitudes on discrimination as perceived by AAG members: survey results. Prof Geogr 29(1):70–76. https://doi.org/10.1111/j.0033-0124.1977.00070.x

Blidon M, Zaragocin S (2019) Mapping gender and feminist geographies in the global context. Gend Place Cult 26(7–9):915–925. https://doi.org/10.1080/0966369X.2019.1636000

Bowlby S, Peake L (1989) The challenge of feminist geography. J Geogr High Educ 13(1):104–106. https://doi.org/10.1080/03098268908709063

Butler J (2016) Quadros de guerra: quando uma vida é passível de luto? Civilização Brasileira, Rio de Janeiro

Butler J (2006) Vida precária: el pode del duelo y la violencia. Paidós, Buenos Aires

Butler J (2017) Relatar a si mesmo: crítica da violência ética. Autentica Editora, Belo Horizonte

da Silva SM (1998) Geografia e Gênero/Geografia Feminista—O que é isto? Boletim Gaúcho de Geografia 23:105–110

da Silva V, Maria S, Lan D (2007) Geography and gender studies: the situation in Brazil and Argentina. BELGEO 3(3):371–382. https://doi.org/10.4000/belgeo.11227

Dantas BG (1987) A mão e o torno: a divisão sexual do trabalho entre os produtores de cerâmica. Geonordeste 4(1):17–29

de Oliveira César, TR, Pinto VA (2015) A produção intelectual da Geografia Brasileira em torno das temáticas de Gênero e Sexualidades. Revista Latino-americana de Geografia e Gênero 6(2):119–132. https://doi.org/10.5212/Rlagg.v.6.i2.0008

de Oliveira Cesar TR, Silva JM (2021) Geografia brasileira, poder, gênero e prestígio científico. Revista da Anpege 17(32):244–258. https://doi.org/10.5418/ra2021.v17i32.12473

Held V (2006) The ethics of care: personal, political and global. Oxford University Press, Oxford and New York
Hooks B (2000) All about love: new visions. Harper Perennial, New York
Jónasdóttir AG (2011a) What kind of power is 'Love Power'? In: Jónasdóttir A, Bryson V, Jones KB (eds) Sexuality, gender and power: intersectional and transnational perspectives. Routledge, New York, London, pp 45–59
Jónasdóttir AG (2011b) ¿Qué clase de poder es 'el poder del amor'? Sociológica 26(74):247–273
Jónasdóttir AG, Ferguson A (2014) Love: a question for feminism in the twenty-first century. Routledge, New York, London
Jónasdóttir AG, Bryson V, Jones KB (2011) Sexuality, gender and power: intersectional and transnational perspectives. Routledge, New York, London
Jónasdóttir A (2014) Love studies: a (re)new(ed) field of feminist knowledge interests. In: Jónasdóttir A, Ferguson AG (eds) Love: a question for feminism in the twenty-first century. Routledge, New York, London, pp 11–30
Langford W (1999) Revolutions of the heart: gender, power and delusions of love. Routledge, London and New York
Lawson V (2007) Geographies of care and responsibility. Ann Assoc Am Geogr 97(1):1–11. https://doi.org/10.1111/j.1467-8306.2007.00520.x
Marques F (2012) Limites na diferença. Revista Pesquisa FAPESP 196, June. https://revistapesquisa.fapesp.br/limites-da-diferenca/
McDowell L (1999) Gender, identity and place: understanding feminist geographies. University of Minnesota Press, Minneapolis
McDowell L (2004) Work, workfare, work/life balance and an ethic of care. Prog Hum Geogr 28(2):145–163. https://doi.org/10.1191/0309132504ph478oa
McDowell L, Peake L (1990) Women in British geography revisited: or the same old story. J Geogr High Educ 14(1):19–30. https://doi.org/10.1080/03098269008709094
Monk J, Hanson S (1982) On not excluding half of the human in human geography. Prof Geogr 34(1):11–23. https://doi.org/10.1111/j.0033-0124.1982.00011.x
Morrison C-A, Johnston L, Longhurst R (2012) Critical geographies of love as spatial, relational and political. Progr Human Geogr 37(4):505–521. https://doi.org/10.1177/0309132512462513
Sant'anna Neto JL (2012) Os desafios da Pesquisa e Pós-Graduação em Geografia no Brasil. Revista Discente Expressões Geográficas 8(8):14–28
Pinto VAM, Silva JM (2016) Produção científica e geografia: devassando o poder da invisibilidade de gênero do fazer científico. Terra Livre 47(2):52–78
Ramon MDG, Monk J (eds) (2007) Feminist geographies around the world. Specialissue, Belgeo 3. https://doi.org/10.4000/belgeo.11103
Ramon G, Dolors M, Ortiz A, Pujol P (2011) Universidade e gênero na Espanha: trajetórias acadêmicas de docentes na geografia. In: Silva JM, da Silva ACP (eds) Espaço, gênero e feminilidades ibero-americanas. Todapalavra, Ponta Grossa, pp 43–63
Robinson F (1990) Globalizing care: ethics, feminist theory, and international relations. Westview Press, Boulder-Colorado
Rose G (1993) Feminism and geography: the limits of geographical knowledge. Polity Press, Cambridge
Rose H (1994) Love, power and knowledge: towards a feminist transformation of the sciences. PolityPress, Cambridge
Rossini RE (1998) As geografias da modernidade. Geografia e gênero: mulher, trabalho e família. O exemplo da área de Ribeirão Preto—SP. Revista Do Departamento De Geografia 12:7–26. https://doi.org/10.7154/RDG.1998.0012.0001
Silva AF (1984) A Zona Canavieira da Cotinguiba e o trabalho da mulher. Geonordeste 2(1):52–57
Silva JM, Ornat MJ (2016) Wake up, Alice, this is not wonderland! Power, diversity and knowledge in geographies of sexualities. In: Brown G, Browne K (eds) Companion to geographies of sex and sexualities. Routledge, London, pp 185–195

Silva JM, de Oliveira Cesar TR, Pinto VA (2015) Gênero e geografia brasileira: uma análise sobre o tensionamento de um campo de saber. Revista da ANPEGE 11(15):185–200. https://doi.org/10.5418/RA2015.1115.0007

Silva JM, Vieira PJ (2014) Geographies of sexualities in Brazil: between national invisibility and subordinate inclusion in postcolonial networks of knowledge production. Geogr Compass 8(10):767–777. https://doi.org/10.1111/gec3.12165

Silva JM (2009) Ausências e silêncios do discurso geográfico brasileiro: uma crítica feminista à geografia eurocêntrica In: Silva JM (ed) Geografias Subversivas: discursos sobre espaço, gênero e sexualidades. Todapalavra, Ponta Grossa, pp 25–54

Souza LF, Ratts AJ (2009) Espaço, cultura e poder: gênero e raça em análise na geografia. Ateliê Geográfico 3(1):107–120. https://doi.org/10.5216/ag.v3i1.6257

Suertegaray DMA (2007) Rumos e Rumores da Pós graduação e da Pesquisa em Geografia no Brasil. Revista da ANPEGE 3(1):17–31

Toronto JC (1993) Moral boundaries. A political argument for an ethic of care. Routledge, London and New York

Toye ME (2010) Towards a poethics of love: poststructuralist feminist ethics and literary creation. Feminist Theor 11(1):39–45. https://doi.org/10.1177/1464700109355212

Chapter 23
Association of Brazilian Geographers (AGB): The Construction of a Geography of Struggle

Charlles da França Antunes and Paulo Alentejano

Abstract The history of science is necessary for the process of doing science. The history of science is not just the result of strictly scientific processes. In this regard, our intention with this text is to look at the Association of Brazilian Geographers (AGB) and its role in professional training, its importance in the elaboration, appropriation, and dissemination of ideas of/in geographical science in Brazil, and in the definition of the agenda of struggles of the geographical community itself in Brazil. These processes are the result of what the history of geography itself has handed down to us: the great proximity between the AGB and the transformations perceived in the science of geography and Brazilian society in the years subsequent to 1934. In this regard, we discuss the *ways*, *means,* and *processes* which have given the AGB a prominent role within this scientific community in Brazil, and thereby attempt to understand to what extent and in what way the AGB and the movement surrounding it have been responsible for consolidating Brazilian geography, for a geography of struggle. It can also be asserted that without the existence, ideas, practices, interventions, and transformations of the AGB, since 1934, upon its creation; since 1946, with the holding of its assemblies in different parts of the country; since 1972, with the holding of its first mass event; since 1979, with its more radical and democratic transformation, the history of geography in Brazil would have been very different. The AGB, born in the previous century, is an expression of the phenomenon of the social history of geography in Brazil. Its emergence and subsequent development are the result of a specific conception of geography, of a long institutional history within Brazilian geography. The AGB is not only the bearer of a certain professional institutionality, it is a moving entity, but also a space where science is developed as knowledge in the service of the transformation of society and the geographical space, in line with the philosophy of praxis. This is why in recent years the AGB has been a growing space with an active and vibrant presence of social movements

C. da França Antunes (✉) · P. Alentejano
Universidade do Estado do Rio de Janeiro (DGEO/FFP/UERJ), Rio de Janeiro, Brazil
e-mail: charllesdafranca@gmail.com

P. Alentejano
e-mail: paulinhochinelo@gmail.com

R. C. Lois González and M. A. Mitidiero Junior (eds.), *Brazilian Geography*,
Advances in Geographical and Environmental Sciences,
https://doi.org/10.1007/978-981-19-3704-0_23

which, in dialogue with the geographies, day by day are constructing more committed geography, a geography of struggle.

Keywords Association of Brazilian geographers · Brazilian geography · History of geographical thought · Democratic construction · Political struggle

23.1 Introduction

Scientific associations play an important role in the formation and consolidation of scientific fields. In Brazilian geography, this history could not be different. Understanding Brazilian geography, its directions and especially its changes in directions, is an intellectual movement that is imperatively subject to analysis in the Association of Brazilian Geographers (AGB).

The present text aims to bring a little of this history and discuss the *ways*, *means*, and *processes* which have given the AGB a key role within this scientific community in Brazil, and thus, seek to understand to what extent and in what ways the AGB and the movement around it were responsible for the consolidation in Brazilian geography of a geography of struggle.

Every history has its beginning. The beginning of the associative history that brings together scholars and those interested in geographical thought in Brazil does not occur with the Association of Brazilian Geographers (AGB), since this associative path in geography had its beginning, even if very embryonic, in the so-called geographical societies.[1] However, it is in the AGB that this associative history and representation of a community has a past, and mainly, a present and future, and is confused with the institutionalisation of university geography in Brazil post-1930, and then, therefore, with Brazilian geography itself.

Considering geographical associations as institutions/movements means conceptualising them as material and organisational structures, where social relations and, therefore, power relations are formed (Zusman 1996). In this way, institutions can and should be studied in their explicit and implicit social functions, that is, what is established in their formal objectives, manifested, for example, by their statutes, but also in those never clearly expressed, such as, for example, the impact on the geographical community of their actions as a movement/initiative.

---

[1] Geography societies are institutions of the nineteenth century and are inextricably linked to the colonialist and imperialist action that took place during this period, even more markedly between the last quarter of the 1800s and the first two decades of the twentieth century, with a strong European predominance and among European countries with notable hegemony such as France, Germany and England. The first of these societies was founded in 1821 in Paris, followed by Berlin in 1828 and London in 1830. The first among the Latin American societies were that of Mexico, in 1833, and that of Brazil, in 1838 (Rio de Janeiro Geography Society—SGRJ). In Brazil, the Brazilian Historic and Geographic Institute (IHGB) and the Rio de Janeiro Geography Institute (SGRJ) can be understood as the beginning of the associative history of geography. They are understood as institutions that brought together, at first, scholars and those interested in geography, whether bearing the title and denomination of geographers or not (Zusman 1996).

The AGB is one of those entities that, due to its long existence, could already have an indisputable presence in the history of scientific institutions in Brazil, but due to some peculiarities of its trajectory, occupies a prominent place among them, starting with its contribution to the transformations in geographical science. The AGB was innovative in its own foundation, since the national name for associations and entities created outside the federal capital at that time (Rio de Janeiro) was not common and usual. The AGB may be inscribed in those institutions discussed by Moreira (2008) when he presents the existence of three forms of geography—where, in short, the first geography is the real one of our empirical surroundings; the second geography is the theoretical and conceptual one of our discourses; and the third geography is that of our habits, customs, mentalities, and institutional experiences. The first changes quickly, the second seeks to accompany it in its change and the third tends to act as a brake on the changes of the second. When considering that the AGB is inscribed in the analysis presented, and recognising it as part of the third geography, it is also important to recognise its differentiated insertion over the decades of its existence, that is, how the process of mentalisation of habits and customs materialised in its institutional practices, and thus understand its own and specific definition as a kind of power of force that can intervene as a brake or a start-up in the reciprocal encounter and in the destiny of the second in relation to the first geography (Moreira 2008).

Performing a reading of these institutions/associations, and in this case, in particular the AGB, from the historical context in which they were formed allows not only to give a new meaning to their role in relation to the problem presented above, but also in relation to the political projects in progress, in the historical contexts in which they were formed. It is the AGB that, in our understanding, made a difference in this associative history in Brazilian geography. First, the AGB initially went through its transition phase between the end of the nineteenth century and the beginning of the twentieth century, assuming national and foreign heritages, and at the same time going beyond the limits of these existing institutions in Brazil. Second, the AGB is the association that effectively articulated with the other institutions that formed the bases of Brazilian geography in the post-1930s, and which some authors would call "modern". Furthermore, it is the AGB that, throughout its history, has articulated the basis of the so-called community of geographers in Brazil (students, professors, and technicians), from its foundation to the present day, and lastly, it is the AGB transformed into movement that also transforms thinking/doing/producing into geography in Brazil, where the AGB is not only the bearer of a certain professional institutionality, of a being in movement, but a space where science develops as knowledge at the service of the transformation of society and geographical space, in line with the philosophy of praxis.

23.2 From Foundation to Democratic Refoundation

The AGB arose in a context of Brazilian geography called by many the period of modern institutionalisation of geographical science, alongside a set of other institutions that already existed or that emerged soon after its creation, and which would shape not only said period, but the possibility of producing and thinking about geography in Brazil from the perspective of a given scientific modernity. The establishment of an institutional apparatus dedicated to this discipline dates from the 1930s with the organisation of university courses in geography in São Paulo (1934) and Rio de Janeiro (1935 and 1938),[2] the regulation of the discipline in basic education in some states, the founding of the Association of Brazilian Geographers (1934), the creation, by the state, of the Brazilian Council of Geography (1937) and the Brazilian Institute of Geography and Statistics (1938). This is the initial formation scenario that allows understanding the formation/consolidation process of the AGB as an association of representation and struggle in Brazilian geography. Such acts, interconnected, quickly form a community of geographers in the country, precisely in the period of the history of this science in our country that combines with the existence of these aforementioned institutions. Thus, the role of the AGB, in our understanding, was fundamental for the occurrence of this formation process of the geographical community, which is understood not only through the existence of those trained in the discipline, but, above all, through the contact, articulation, and relationship between professionals and students of geography and their peers. Therefore, the AGB, for its associative character and for its practices of meetings, either through the historic weekly meetings and assemblies, in the National Meetings of Geographers, Brazilian Congresses of Geographers and National Meetings of Geography Teaching, or through its publications, played a fundamental role in geographers being able to meet and value each other, thus forming a community, which is an essential fact for the formation and consolidation of the scientific field of geography in Brazil.

The AGB was, during its initial decades of existence, a point of articulation between society and universities—forming the scientific association–universities–society triad. Among the institutions that endorsed a given modernity in the production of knowledge, of a certain worldview, the AGB was the one that, due to its organisational characteristics, played a more active role in the universities' dialogue with society, but also with the state with society, by often assuming official policies and forms of manifestation, of governments, which in the end were also the projects of certain fractions of society represented by an elitist discourse.

> (...) until the 1960s, Brazilian universities were limited to training a small number of liberal professionals, initially exclusively from the traditional elites, and after 1930 also from the middle classes, but while maintaining the elitist character of social privilege (Fernandes 2020: 87).

---

[2] The University of São Paulo, in 1934, with its Faculty of Philosophy, Sciences and Literature, as well as the University of the Federal District, in 1935, transformed shortly afterwards into the National Faculty of Philosophy at the University of Brazil, in 1938, and today the Federal University of Rio de Janeiro.

It is important to note that the projects of society, manifested mostly at universities and that were established from this relationship, were primarily those where the discourse of the elites and their expressions dictated the pace of the movement. The AGB was founded on 17 September 1934, in the city of São Paulo. Scholars from different political backgrounds and training participated in this creation, but only those who had in common the fact that they were considered "local notable people and all francophones".[3] They were geologist and professor at the Polytechnic School of the USP, Luís Flores de Moraes Rego; historian and librarian and, from 1935 onwards, Director of the Municipal Library, Rubens Borba de Morais, and lawyer and student of the History and Geography course, Caio Prado Junior; who together and led by Pierre Deffontaines—a French geographer who had the task of leading the creation of the USP Geography course—formally founded the Association of Brazilian Geographers (AGB).

The relationship between the AGB and the São Paulo elite, then dominant in the 1930s, and the association's engagement with the state's political and economic project were hallmarks of the association's early days. Geography, therefore, would play an important role in the implementation of the country's modernisation and progress project.[4] The role of geography and the newly created AGB are directly related to the meaning of the foundation of the University of São Paulo and its functioning in the 1930s, 1940s, and 1950s. This was undoubtedly the profile assumed by the AGB in its first decades of existence—if on the one hand it formed a community of people interested in and trained in geography, on the other it articulated very closely with the discourses of the local elites (especially in São Paulo), or with what appeared to be a need for knowledge about the territory of São Paulo and a State project for Brazil.

In 1936, the same professor Pierre Deffontaines, now in Rio de Janeiro, founded the Rio de Janeiro version of the AGB, which built its first years of existence in a strong and intimate relationship with members of the National Council of Geography, but also with the then created University of the Federal District (UDF) and with its replacement, the University of Brazil (UB). In this case, the proximity is not so much with a regional elite, but with the very structure of the national state that was strengthening at the time.

The foundation of the AGB also has a strong connection with the then created Chair of Geography at the University of São Paulo. Thus, from the beginning, the Association and the Chair have nurtured each other, but the creation and long-term

[3] The elitist character is seen in focus. Lucchesi (2015:7) reinforces our position regarding the elitist character of this training when he states that "(…) It should be noted that, at the beginning of the twentieth century, the elitist character prevailed even in Europe, which had already had higher education since the beginning of the thirteenth century. From Antiquity to the mid-twentieth century, higher education was intended only for the social, economic, and intellectual elite".

[4] For Zusman (2001:19–20), geography "sheltered in its epistemological particularity (as a globalising knowledge) and supported by the scientificity guaranteed by the ascription to the French School (as a conceptual framework with legitimacy supported by the disciplinary tradition) presents itself as a knowledge with potentiality to give legitimacy to the process of territorial enhancement and regional development".

existence of an association of Brazilian geographers could not be limited to the scope of a chair at a university institution. However, the chair—its conception at the university and its social composition, present in recent universities in Brazil—reproduced its hierarchical and selective structure within the AGB. Over the first two years, the AGB's membership was built, through the appointment of new members by the then full members, in such a way that they are individuals in direct relationship with other individuals, whether for academic, political, or even merely personal reasons, which would guarantee the possibility of being nominated and accepted to belong to this new and select social group.

Thus, the social structure inspired by the system of chairs in universities was present in the structures of the AGB for many years, from the categorisation of members, to the prominent place of some associates, to the determination of who could or could not form the association's membership.

> The chair was not presented only as a structural core for the preservation and strengthening of school models or archaic educational habits. It was, in essence, the bastion of conservative thought and behaviour... (Fernandes 2020: 247).

During the period between the founding of the AGB in 1934 and its first major reform in 1945,[5] the one called by Monteiro (1980) as that of the "*implementation of scientific geography*" is the period in which the AGB had an incipient and irregular activity. It remained within a regional scope, practically restricted to the states of São Paulo and Rio de Janeiro, although there were some associates from other states in its membership. During that time, its activities took place mainly within the USP, involving its participants in an essentially academic, intense, but internal daily life; and inside the National Council of Geography (CNG), where the geographers from Rio de Janeiro would meet. Thus, the AGB came into being through scientific meetings, articulations for the creation of Local Centres in municipalities in these two states (but also in other nearby states), and the publication of studies in journals created for this purpose[6] or in other existing ones.[7]

The statutory reform of 1945 completely changed the association's form of organisation and created the conditions for action to take place beyond the borders of the states of São Paulo and Rio de Janeiro. In 1945, Brazilian geography was not only restricted to the two training institutions—USP and UDF (the current UFRJ), as twelve courses, in public and private universities, and in several parts of Brazil had already started their activities.[8] It is the beginning of the period called by Monteiro (1980) as "*the AGB's crusade of national diffusion*" and by Scarim (2001) as the "*second generation*", where the AGB would begin its nationalisation process with

[5] In 1936, there was a small reform of the Statute, but it did not introduce significant changes in the entity's institutionality. Throughout its history, the AGB has carried out seven statutory reforms.

[6] *Revista de Geografia* (Geography Journal) (1935–1936), *Revista da AGB* (AGB Journal) (1941–1944).

[7] *Revista Brasileira de Geografia* (Brazilian Journal of Geography), from the IBGE, for example.

[8] "(…) from 1800 to 1889, 14 higher education establishments were created, which is 0.16 per year; 64 more from 1890 to 1929, which is 1.6 per year; and 338 more from 1930 to 1960, which is 11.3 per year" (Fernandes 2020: 89).

the creation of regional sections,[9] holding annual assemblies in different parts of the country and important fieldwork, marked by the perspective of integration of the Brazilian territory and the construction of knowledge about this territory, carrying out a true "crusade" of scientific dissemination and professional diffusion of geography throughout the national territory. Twenty-five assemblies were held in the most diverse states of the country. The publication of Annals and Journals and participation in international events are also aspects of greatest relevance for understanding the construction of the AGB's national project (Fig. 23.1).

The institutional model of the AGB built in 1945 and reinforced in 1963 lasted until 1970, when, among other changes, the annual assemblies were replaced by biennial meetings—National Meeting of Geographers (ENG) and congresses every 10 years—Brazilian Congress of Geographers (CBG). In this statutory reform of 1970, the AGB underwent changes in its form of organisation, more specifically in its form of general meeting, where events started to count on the participation of several hundred interested parties, while reaching thousands of participants in the future. The Regional Sections of the AGB already existed in several states of Brazil, and the country, immersed in the corporate and military dictatorship,[10] witnessed an expansion of university courses, mainly in private institutions,[11] and a reform of higher education[12] which was characterised by external supervision and by a conservative modernisation project, which, in fact, was the university's entrepreneurship.[13] The corporate and military coup of 1964 eliminated trade union and popular movements, silenced dissenting voices in the academic community, compulsorily retired scientists, and dismissed, exiled, persecuted, arrested, tortured, and killed students and professors.[14] However, the first decade of the authoritarian regime instituted from 1964 onwards in Brazil did not seem at first glance to create any form of obstacles to

[9] In the 1945–1970 period, 6 (six) regional sections and 8 (eight) local centres were created in 11 (eleven) different states of Brazil.

[10] The corporate and military dictatorship, resulting from a movement of conservative forces, among them the National Armed Forces, the Church and the bourgeoisie, was established through a coup d'état, on 31 March 1964, instituting a regime that lasted 21 years.

[11] As a result of this policy, we would witness a surprising expansion of higher education institutions. In 1968, there were 41 universities and 397 isolated establishments and federations; in 1985, when the military government ended, 68 universities (20 of which were private) and 791 isolated establishments and federations (606 of which were private) were already in operation. In 1968, there were 1712 courses offered in Brazil, and in 1985, it jumped to 3923 courses offered (Lopes 1997).

[12] From a legal point of view, the university reform process was sealed with the enactment of Law 5540, of 28 November 1968, which dealt with the organisation and functioning of universities in the country, and its content contains the rancidity of the conservative guidelines that surrounded the entire reform process (Fernandes 1978).

[13] (…) Lastly, a third negative element was introduced in the universities: the conception that education is a commodity. The student would not know the value of education if they did not pay for the course. This idea germinated with the MEC-USAID agreements, with which it was intended to strangle public schools and allow for the expansion of commercialised education" (Fernandes 1978:106).

[14] See Motta, Rodrigo Patto Sá. *As universidades e o regime militar*. Jorge Zahar Editor (2014).

Fig. 23.1 Ordinary General Assemblies (AGO) of the Association of Brazilian Geographers (AGB) 1945–1970

the organisation of the AGB and its activities, which reinforces the very conservative and elitist character that the association had at that time. Even after the 1964 coup, the AGB, represented by the professors and other researchers who were associated with it, did not seem to have been ostensibly reticent in the face of the new regime, although some voices had manifested themselves against it, quite promptly. This definition of association apparently and relatively exempt from the turbulence of contextual debates, as if it could follow its course independently of the individual human and collective hardships of society, gave rise to a vision of the association as an alienated and conservative structure.

Despite the maintenance of hierarchical structures in the entity, in a way that recalled the elitist, despotic and gerontocratic structure inspired by the chairs (Fernandes 2020), represented in the figures of full members, now tenured members, honorary members, cooperating members, and corresponding members, the inflection of the model was already noticeable, as an expressive set of the increasingly larger contingent of participants had their capacity stunted or reduced in the decision and action processes. Such tensions, roughly speaking, implied the distinction between the tenured members who were at the "top" of the knowledge pyramid—producing research—and the other participants, related to the teaching and/or learning of geography (Lourenço 2004). This hierarchical structure, combined with the absence of internal democracy (reflections of the conditions established in the universities and in Brazilian society), and the inability of the association to establish the necessary dialogues to attend to the political, economic, and social issues that were effervescent in the country were the combined elements that spurred the most significant changes that would take place at the AGB, and consequently, in Brazilian geography.

The movement that was structured in Brazilian geography in the late 1970s and advanced through the 1980s, entitled and known as the *Movimento de Renovação* (Renewal Movement) or more precisely the *Movimento de RenovaçãoCrítica da Geografia Brasileira* (Critical Renewal of Brazilian Geography Movement), was characterised by a strong political and ideological content and allowed for understanding of the dynamics and complexity of the relationship between science and society. And that was perfectly right, given that we were in the midst of a struggle for the political opening of the country, for the reconquest of the right of assembly/articulation, for the resumption of left-wing social movements and political parties, and in short, for the re-installation of the democratic state and rule of law. In the case of geography, four spaces of struggle were prioritised. The first of these spaces was the universities; the second was the AGB; the third space was the "official geography", and in this space, the focus was the Brazilian Institute of Geography and Statistics (IBGE), and the fourth space was that of the schools, in the dispute over curricula, contents, and the very presence of the geography that is taught.

However, it was at the AGB that these transformations had the greatest impact in terms of its structuring and political project. The transformation/renewal of geography movement that was underway from the end of the 1970s first impacted the AGB and, a little later, the geography departments. Here again, it is time to resume the society–AGB–universities triad and dialogue with Fernandes (2020) to understand this moment in which organised segments of Brazilian society are acting in denial of the current social order.

> Institutions change before the social order as a whole, and it is often the changes that occur in them that create the fermentation necessary for change in the orientation of spirits, for calibration of the collective will according to strongly divergent inspirations and for dynamisation of tensions and conflicts of global society. In this respect, the universities anticipate, as a social microcosm that lives with greater freedom and with greater relative intensity, the historical destiny of global society: it first absorbs the action of emerging historical and social forces, first experiences its political significance, and first tests its power of denial of the social order. (Fernandes 2020: 355/356).

The Renewal Movement which geography has experienced is certainly unique in the set of sciences in Brazil. The uniqueness comes from the historical view of the process, a view that involves the various subjects/actors who are part of this history. Two characteristics are fundamental for understanding the uniqueness of the recent renewal of geography. The first refers to the form/process, which has its emblematic reference in 1978, despite the fact that we agree with Bosi (1992) when he states that "dates are the tips of the iceberg". This renewal is both epistemological and political. The criticisms that were made and the dissatisfactions that ended up generating the rift were not only about what epistemological statute geography should be produced under in universities and schools, referring to a scientific perspective, but also about which and for whom geography would be produced, contemplating a clear project of society, in which geography should participate. The second characteristic refers to the agents of this renewal. Unlike other sciences in Brazil, geography did not have academics as the main actors in the scientific transformation—students and teachers of basic education were the new and main actors who entered the scene in this process. Geography is perhaps the only science that, in Brazil, in its recent history, has gone through such a radical process of transformation of thinking/producing without the exclusive, or even main, direction of Academia. This production of the new, based on critical theoretical assumptions, especially in Brazil, did not, in principle, find fertile ground for its construction in the university environment. Thus, we agree with Moreira (2008) when he states that:

> It took time to realise that ingrained habits, which are what actually react to the imperative of new needs for change, were things of the second geography. Therefore, it is specific to the scope of the second, not the third, although the latter does have an enormous apparatus of institutionality. This is due to the lack of awareness of the nature of these habits in the 1970s. (Moreira 2008:106)[15]

The Geography Departments of Brazilian universities, for the most part with conservative and authoritarian postures, only much later would incorporate into their agenda the debate on the new directions that were being followed in Brazilian geography and, from that, materialise, through their curricula and publications, the production of this geography that received the mark of criticism. Moreira (2008) corroborates this analysis when he states that:

[15] For Moreira (2008), the three geographies are practice, theory and institutionality, respectively.

> Therefore, we understand that we must look at the context of the third geography. To change the habits and customs historically ingrained in the short-term vision of the child. To play in the contextual game of interests. To put mentalities in line with a reality that now transpires beyond the old appearance. To know how to deal with the inertia of institutions. Hence, the debate shifts, and in a tone of aggressive combativeness, from the focus of theory to that of the curricula and departments of universities. (Moreira 2008:107)

The AGB lived through what would become one of the most expressive moments of strength of those who demanded the democratic emergence of the association. Gathered initially during the 3rd National Meeting of Geographers, in the city of Fortaleza/CE, in 1978, and later in the Extraordinary General Assembly, in São Paulo/SP, in 1979, coming from the most diverse parts of the country, university professors and secondary school teachers, undergraduate and postgraduate students, full and cooperating members, and even those who were not or could not be members in their fullness of rights, initially broke down barriers, broke with customs and routines, and later intensely debated each point of what would be the statute that would remain in force, in essence and for the most part, to this day.

> It was necessary for the holders of power to anticipate events, maintain control of the context, while reorganising the State machine. The "opening" was born. The course taken by the movement, until then latent, was already taken, however. Even in the most placid of places comes the breath of the resurgence of the struggles of those who have been repressed for so long. Even the AGB. (AGB 1980:1)

The rift of 1978 and 1979 confronts the geographical institutionality linked to the dictates of the leaden regime and an applied geography, focussed on planning, based on systemic and locational theories and with strong connections with the interests of those who made the *Brazilian economic miracle*[16] happen, which would be fought vehemently by those who proposed a true *epistemological guerrilla war*. There was, in the years before the meeting in Fortaleza, a kind of silent war within Brazilian geography, a war that expressed the very hard struggle carried out in that state of exception and Institutional Acts.[17]

The radicalisation of those students and professors transformed the AGB into what is now considered, from the point of view of the possibility of participation and representation, one of the most democratic scientific associations in the country, where horizontality and democracy—from the conditions of being a member, of representation and of deliberation of its organisational guidelines and struggles—are the hallmarks that define it.

---

[16] If there was any "miracle" in a relatively prolonged period of growth with little inflation, it was basically explained by: (a) the way in which the distribution of income was "disciplined" by replacing collective bargaining in the labour market with a "wage policy" that was rigid, centralised and—from the point of view of capital accumulation—perfectly effective, and (b) the increasing international integration of capitalist economies, causing changes in the global division of labour, which ended up providing a large influx of foreign capital into Brazil (Singer 1977:10).

[17] The Institutional Acts (AI) were measures enacted by the corporate and military dictatorship instituted after the 1 April 1964 coup. In assuming constitutional power, the military junta or the indirectly elected presidents enacted 17 institutional acts, which determined the revocation of mandates, exiles, imprisonments, censorship and other arbitrary acts. The 1968 AI-5 was considered the most hard-hitting of all AIs.

The rifts and radical changes caused at the AGB in the sum of the actions and events of 1978 and 1979, and which transformed the association, left profound marks not only on the AGB itself, but also on the Brazilian geographical community as a whole. Many key professionals of Brazilian geography, until then active and participants in the events, dissatisfied with the statutory changes and with the new directions of the AGB and geography, left the entity, requesting their disaffiliation.

> The difficulties are not few: the reaction of the old social base of the AGB; the weight of the old image of "Official geography"; the old structural and material dependence of the AGB on the fiefdoms that are the Geography Departments and the official planning bodies that employ geographers; the old practice of personalism, medalism, favouritism, mandarism, the old idea that the AGB exists to promote and legitimise certain "powers that are constituted" (...) The 5th ENG, more than the 4th (1980) and the 3rd (1978), reveals such a spectrum with extraordinary clarity. However, it reveals something even more important: real changes are taking place. A new social base is being forged, coming from the new generations of geographers, from the midst of the 1st and 2nd year professors who have long been separated from the AGB, from the students always placed at the forefront of the political struggles of Brazilian society, from the return of the AGB members who feel a connection to the new movement. The guiding axes of the "praxis" of geography and the AGB are being reoriented. (AGB 1982:7–8)

The AGB emerging from the struggles, disputes, and statutory transformation is now organised in the decentralised form of Local Sections (with a municipal base), with a National Executive Board and decisions taken collectively from the so-called Collective Management Meetings (RGC).

> The changes embody a democratic and representative AGB, directed towards the bigger matters of work and education in Brazil, for the collage of those who make geographical knowledge their work with the others who live from their work. Implicitly, they call for the practice of line pluralism, both in terms of scientific thought and political and ideological stance. In this short period that passed from the III ENG to the IV ENG, the responses varied from those who made efforts towards the construction and strengthening of the democratic process to those who went to create spaces elsewhere for the continuation of the elitist and anti-democratic practice of before. (AGB 1980:7)

An important hallmark of these transformations carried out in the association is the inexistence of differentiation among its members. Where before there was a categorisation of associates—honorary, full, and collaborating, today there are only members, indistinctly between students and professionals, from the youngest to the most experienced and renowned, and this is the rift with the structure of chairs imported from universities. Once again, we use the parallel with universities, mediated by Fernandes (2020), to understand the importance of eliminating the categorisation of members in the AGB.

> ...an integrated and multifunctional university will never be achieved without eliminating the Brazilian standard of higher education and, with it, the irresponsible and unlimited power of the full professor. (Fernandes 2020: 122/123)

Thus, renewed, organised, and combative forms of scientific entities emerged in the wake of other social, popular, and union movements in the late 1970s.

> As in so many other similar entities, in the AGB the big question of the role of sciences and scientists in this process emerges. It takes shape, breaks chains, and materialises in the 1979 Statutes, triggering an intense and heated debate. It converges, after all, on the demand for a ninth "praxis": that of the popular and democratic movement. In short, everything converges on a new practice: collective management. (AGB 1982:7)

Thus, the rebirth of the union movement arises with the new salaried categories, including the teachers of basic education and higher education, who started to organise their class entities throughout the country, and with this directly impacting the AGB. What was at stake at that moment was the reorganisation of workers and the struggle against the consequences in the lives of these workers, including those in education, throughout the long years of the dictatorship's economic policies. In the years that followed the process of political opening in Brazil, it was possible to perceive this change in the position of educators, and this is what Arroyo (1980) highlights.

> (...) The organisation of educational work on a business basis led the orderly public teachers to feel that they were not civil servants, but a labour force sold to an employer called the State. This new reality adds a very important element in the structure of education and in the defining of its directions. (Arroyo 1980:17)

The construction and reformulation of their own resistance forums were decisive for the education workers in this struggle, and public manifestations at scientific or class events started to become more frequent—and this would happen with the AGB and its main events. The change in the entity's form of organisation and its democratisation would make the National Meetings of Geographers a more expressive and more significant event in its political content, where the academic character not only ceases to be exclusive but also transforms itself towards a deeper critique of its thinking/doing. The combination of the historical events of the 1980s, which nurture each other—transformations in Brazilian society and transformations in the AGB's way of being—would make the association an interesting movement of organisation of associates—professors and students (now without regimental differentiation)—that would begin to practically and concretely visualise the relationships between politics and science, struggle and geography, research, and social transformation materialised in the association and in its events.

The academic environment in Brazil and that of geography could not be different, and it involves the recent historical meaning of being an intellectual/teacher/wage earner in a peripheral society, and all this at the same time, all combined in the same individual.

Understanding the movement of resistance and transformation in Brazilian geography and in the AGB necessarily involves the way in which the subjects of this everyday thinking/doing of science, work, and association would somehow build their social and institutional self-recognition. Brazilian society changes, geography changes, the AGB changes, the universities change, all together and combined, in different temporalities and with different impacts, but in a movement of transformation.

> These facts and trends in the awareness and practice of producers and teachers, whether we like it or not, constitute central determinants of a new Brazilian educational context. (...) a new awareness and new practice are taking place in the lower classes, as their struggle and organisation is trying to change their condition as a worker, as a labour force. (...) The problem of the division of labour, classifications, expropriation from the worker of their knowledge and their intellectual work, will be part of this awareness. (...) (Arroyo 1980:19)

A significant part of the geographical community was not oblivious, for example, to the process of increasing social inequality in the country, to the increase in problems on the outskirts of large cities or in the Brazilian countryside. The conflicts that took place in the ENGs at the end of the 1970s and the incessant search for a more committed geography, capable of critically analysing the social reality, point to this new stance of geographers in our country that will have impacts on the AGB and its events.

23.3 From Democratic Refoundation to the Dilemmas of the Pandemic

The National Meetings of Geographers would, from 1980 onwards, incorporate what in a certain way happens in Brazilian society, and consequently in Brazilian geography and in the AGB itself. The topics of the ENGs change, incorporating into the agenda the important issues of Brazilian society; the structure of the Meetings changes, with an increase in the participating public; the subjects in dialogue with the AGB change, with the prominent participation of those who for a long time were on the periphery of the ENGs and with the new actors that enter the AGB scene, both internal and external to Brazilian geography. The creation of the *Terra Livre* journal, in 1986; the presence of texts/theses as a basis for discussion and positioning of the association in the ENGs; the election of professors without academic degrees and students to the management of the AGB; the approximation with the social movements of the city and the countryside; the increase in concerns about the teaching of geography at school—which culminated in the holding, in 1987, of the I National Meeting of Geography Teaching, known as "Fala Professor"; and the creation of its commissions—urban and agrarian matters, teaching, professional matters—are examples of the important transformations that took place at the AGB in the early 1980s, which combine with the first years of political reopening in Brazil, with the first years of the association's new statute and the redefinition of its objectives.

That movement that began at the end of the 1970s, and took place through most of the 1980s, transforming the AGB and Brazilian geography, had in 1988, one of its points of reference. During the VII National Meeting of Geographers (ENG), an important discussion occupied the political attention of the association—the affiliation of the AGB to the Unified Workers' Central (CUT), which would occupy the agenda for the next two years (Fig. 23.2).

Fig. 23.2 National Events (Congresses and Meetings)organized by the AGB/1954–2019

The VII ENG was held in 1988, in Maceió, the capital of the state of Alagoas, then devastated by a flood caused by the combination of heavy rains with increasing environmental degradation, associated with social segregation that made many people among the poorest population in the city homeless. This local context was associated with a national scenario of increasing political mobilisation due to the debates about the drafting of the constitution that would succeed the one enacted by the corporate and military dictatorship, as well as the first presidential elections that would take place the following year. The AGB reacted to the local context by promoting a demonstration in the streets of Maceió, occupying the entrance to the State Government Palace and publishing a study on conditions in the city. It was the first time that Brazilian geography was present on the streets, in an organised manner.

But the main debate at the VII ENG was about the affiliation of the AGB to the Unified Workers' Central—CUT, which at the time was the most left-wing Central Trade Union and the most representative within the country's political scenario. Led by sectors linked to the Student Movement (ME) and basic education teachers and opposed mainly by university professors, the proposal was approved by a small margin of votes in a plenary session with around 800 participants.

The mere existence of the discussion on the agenda of a national event organised by the AGB is already proof of the change in attitude towards the national reality assumed by it, but the transformation intensifies when it is seen from the deepening of the relations between the growing labour movement and associations of a markedly scientific character, where a given scientific field can be identified as a partisan political act, in the expanded sense of taking sides, taking a position on the most general, emerging, and important issues of society, as indicated by Rodrigues (2004).

> However, the most important thing to emphasise is that the intention to join a central trade union, at a time of growing social mobilisation, demonstrated the geographers' interest in participating in the general movement of society. (Rodrigues 2004:9)

However, in the election for the National Executive Board of the Association of Brazilian Geographers (DEN-AGB) that took place the following day, which basically opposed the same political forces, the result was the opposite, with the victory of the candidates that had manifested themselves against affiliation to the CUT. Some maliciously attributed this turn of events to the excessive celebration of the students on the night in Maceió.

Thus, the AGB left Maceió strongly divided and between the ENG in Maceió and the subsequent one, held in Salvador, in 1990, a large part of the debates revolved around the reasons for not affiliating to the CUT: little commitment by the DEN-AGB to making it effective, given its opposing position in the ENG, or bureaucratic obstacles posed by the CUT, related to the hybrid composition of the AGB—professors, students, technicians, researchers, etc. The dissemination by the AGB-Rio of the manifesto "O pau da barraca" (The tent pole) and the fierce response from the DEN-AGB were the apex of this polarisation. In a text written a few years later, the then President of the AGB presented a more contemporaneous reading of the struggle that took place at the time.

> One of the emblematic issues of the AGB's participation in the movement of society was the approval, in 1988, in the plenary session in Maceió, of the AGB's affiliation to the Unified Workers' Central – CUT. The Unified Workers' Central could not accept our affiliation, as the CUT is a union branch and the AGB brings together students who are not considered able to be represented in a union or a central union. The participation of a CUT representative in Salvador, in 1990, ended an internal controversy that took place for two years, that is, the

> idea that there was no commitment on the part of the National Executive Board to comply with one of the deliberations of the Assembly in Maceió. (Rodrigues 2004:9)

The change in the global and Brazilian context between 1988 and 1990 hit the AGB hard. The defeat of the left-wing camp in the 1989 Brazilian presidential election and the collapse of the Soviet bloc paved the way for neoliberal hegemony and bewildered the left. The governments of Collor, Itamar, and FHC[18] implemented the neoliberal agenda in Brazil throughout the 1990s, with privatisation, commercial and financial liberalisation, and the removal of workers' and retirees' rights. The field of education saw an advance in the business-based logic in management and evaluation, enrolment in private education, and the scrapping of public schools and universities, therefore…

> If the territorial siege is an allegory, the economic siege is a palpable reality. In part, the economic siege is associated with the territorial siege exercised by private higher education, but above all it results from changes in the social security system for civil servants, as well as the adoption of public policies of financial contingency, which mean cuts in the number of postgraduate scholarships, reduced support for research and funding for public universities. Each of these factors would be, in isolation, a reason for concern; combined, they have a devastating effect. (Menezes 2000:43)

In public universities, the neoliberal logic was present, above all, with the stimulus towards productivism (the infamous Lattes Curriculum[19] was created in this period and soon became an instrument for the control and measurement of academic production) and the search for external funding through public–private partnerships and tenders. Konder (1983) strongly criticises this model centred on curriculum individualism when he states that…

> Modern society needs organisation and efficiency. To get a job, to get a promotion, to form a career, the subject needs to show off their qualities, and flaunt their successes. There are already even manuals that teach citizens how to prepare their curriculum vitae". (...) "The curriculum vitae is the tip of the iceberg: it is the most ostensible element of an ideology that envelops us and educates us in the principles of the capitalist market; it is the expression of an ideology that instils in our heads that "racehorse mentality" referred to by the writer Dóris Lessing. We must not confess the high failure rate of our existences, because we must be "competitive". Camões, the brilliant Camões, the author of so many wonderful lyrical poems, could not put in his curriculum vitae the famous line: "I always got my discourse wrong". (Konder 1983:125)

Early criticism by Tragtenberg in a premonitory text from the late 1970s:

---

[18] Fernando Collor was the first president elected directly after the corporate and military dictatorship, in 1989. Itamar Franco was his vice president and took over after Collor's resignation, on the eve of his impeachment, in 1992. And Fernando Henrique Cardoso (FHC) was the second directly elected president, in 1994.

[19] The Lattes Platform, created in 1999, is a virtual curriculum system created and maintained by the National Council for Scientific and Technological Development (CNPq/MCT), through which it integrates curriculum databases, research groups and institutions into a single information system, in the areas of Science and Technology, operating in Brazil. The Lattes Curriculum has become a national standard for recording the academic and professional life of students and researchers across the country, and is now adopted by most development institutions, universities, and research institutes throughout the country.

> The (current) university appropriation of knowledge is the capitalist conception of knowledge, which constitutes capital and takes shape in university habits. Universities reproduce the dominant capitalist mode of production not only through the ideology they transmit, but through the servants they form. (Trangtenberg 1990:4)

Although it cannot be said that the neoliberal logic came to dominate the AGB, expressions of this were manifested, with: (1) the increasing depoliticisation and academicisation of the ENGs, in which the most heatedly political discussions were gradually being put aside, while debates of a more academic nature gained centrality; (2) the advance of academic productivism in the ENGs, with the exponential growth in the number of works presented, which went from 182 in the ENG of 1990 to 1110 in the ENG of 2000, in the wake of the implementation of the Lattes Curriculum and the change in the funding policy for the students' attendance at the meetings, now conditional to the submission of works; (3) the removal of the AGB from the basic education teachers' unions; (4) the removal of basic education teachers from the AGB; (5) the topical inflection of the *Terra Livre* journal towards academicism—from 1986 to 1988, of the 6 issues published, only 1 had a more academic profile; of the 17 issues published between 1989 and 2004, on the other hand, 9 had a more academic profile, 5 had a more explicitly political character and 3 focussed on the history of the AGB.

But the disputes within the entity did not cease and among the expressions of anti-neoliberal resistance in the AGB we can mention: (1) the creation of Spaces for Dialogue (EDs)—later renamed Spaces for Dialogue and Practices (EDPs)—at the National Meeting of Geography Teaching "Fala Professor", in Curitiba (1999)/ENG in Florianópolis (2000), with the aim of fighting productivism and allowing a re-engagement with basic education teachers; (2) the resumption of the protagonist role of the Student Movement (ME) in the AGB, exemplified by the *Movimento Sem Crachá* (Movement Without a Badge) (ENG 2002) and the *AGB para Quem*? (AGB for Whom?) (CBG in 2004); and (3) the growing articulation with social movements carried out in particular by Working Groups (GT), created in the Local Sections of the AGB.

The EDs were proposed in view of the growing dissatisfaction of the most left-wing sectors of the AGB with the advancement of the academicist/productivist logic in the ENGs in which quick presentations of work without any qualitative debate were multiplying. The idea with the EDs was to create spaces where, instead of the sequential presentation of numerous works without any expansion of the debate about them, topical discussion groups would be formed that would take place over 9–12 h over 3–4 days, in order to allow for the in-depth debate of the topics of the works grouped in each space. It was also a way of inhibiting the increasingly generalised practice of multiplication of works by university professors who became co-authors of works of their students and with that they submitted dozens of works in the meetings, in a practice then known as "fattening the Lattes". As the rules of the EDs stipulated the presence during all sessions, the old practice of migrating from room to room "presenting" works was curbed. And, lastly, it also represented a way of attracting basic education teachers back to the AGB meetings, based on a less academic format of presentation of works that also contemplated didactic experiences and not only research like that carried out at universities.

In addition to the EDs, the Working Groups (GTs) represented another important space of opposition to the neoliberal logic in the AGB. Although they were not created in this period, dating from the 1980s, they played an important role in the period, as a channel of dialogue between the AGB and the social movements, contributing to the struggles for agrarian and urban reforms, against socio-environmental impacts, and to free, quality public education. According to the definition on the AGB's webpage, the GTs are:

> forums where participants articulate around topics that require Reflection, Discussion, Accumulation, Intervention – with a clear political sense. This is the movement of the GTs, oriented towards supporting the taking of positions by the entity. This materialises the historical role of the AGB as an instrument of intervention in society, based on the contributions of geography and its associates. This is how, in recent years, the GTs (Working Groups) have been gaining a central role in strengthening the permanent work and political action of the Local Sections and consequently the National AGB. It should also be noted that regardless of the topic and spatiality of the action, the collective production of positions is the essence of this form of participation, and this has been the focus of the proposal of the recent AGB managements.

Among the activities carried out by the GTs, we can mention the organisation of round tables, lectures or courses on the topics of action of the GTs, the articulation with social movements, unions and other social organisations, the proposition of institutional questions (to public bodies or private entities, according to the situation) or lawsuits (with the public prosecutor, e.g.), the formulation of documents (such as official letters, booklets, manifestos, and collections of texts).

The VI Brazilian Congress of Geographers (CBG),[20] in 2004, was marked by the reaction to the "professionalisation" of the ENGs and the defeat of a certain, albeit embryonic, proposal of "NGO-isation"[21] of the AGB, representing a milestone in the resumption of a class perspective in the AGB, associated with the growing opening to new issues such as race, gender, and sexuality.

In 2002, the ENG in João Pessoa had already announced the exacerbation of tensions caused by meritocratic influences and, in a way, contaminated by the advance of neoliberal practices within universities and the AGB, expressed in the emergence of the *Movimento Sem Crachá* (MSC), a rebellion by postgraduate students against what they considered high costs of enrolment/accommodation/participation in the

---

[20] The Brazilian Congresses of Geographers are events organised by the AGB, every ten years. The first was held in 1954, in the city of Ribeirão Preto—SP.

[21] It is important, however, to consider the phenomenon in a broader political context, since NGO-isation has proved to be a way of reducing the autonomy of social movements and a possibility of expanding State control over these movements. In the words of ROY (2004:3) "in the long term, NGOs are accountable to their funders, and not to the people among whom they work." Here, we are not talking about NGOs that were involved with social movements, such as, for example, in the process of opening the dictatorship and the redemocratisation of the country; we are referring to organisations that, in the most recent scenario, are linked to a more conservative, or even liberal, action, to the so-called corporate social responsibility, incorporating the strictest sense of the meaning of the third sector.

ENG, which made it difficult for large segments of students to participate in the Meeting, in a context in which the democratisation of access to public universities was being rehearsed, with the creation in 2001 at the Rio de Janeiro State University (UERJ) of the first racial quota system at a Brazilian university.

But the MSC was limited to the specific questioning of access to the ENG infrastructure. In the following years, the MSC evolved into the *AGB praquem?* movement. In addition to continuing to protest against the elitism of the ENGs, the ME realised that the very meaning of the AGB was at stake, in the face of proposals for statutory reform that sought to recategorise the AGB (a throwback to the past of distinction between associates disguised as a response to a requirement of the Confea-Creas System[22]), as well as a so-called "NGO-isation" of the entity under the argument of the need to raise funds to carry out its activities.

Thus, the *AGB praquem?* movement emerged in the CBG in Goiânia as a fundamental political force to stop certain presented proposals for statutory reform, in addition to opposing the process of "professionalisation" of the organisation of the ENGs, expressed in the outsourcing of the organisational activities of the Meeting to private companies. Unfortunately, the movement was diluted in the following years, in spite of some of its members having joined local boards and the AGB Board itself. If the recategorisation, professionalisation and "NGO-isation" were rejected, other contradictions continued to hover over the entity, since neoliberal hegemony was expanding throughout the whole of Brazilian society and in the universities, with evident impacts on the AGB and its events.

Among the main contradictions of the period, we can highlight.

1. gigantism of the ENGs x privatisation of universities—on the one hand, the ENGs grew in size in the wake of the expansion of higher education and the strengthening of postgraduate studies which multiplied the undergraduate and postgraduate courses in geography, increasing the number of professors and students interested in participating in the meetings; on the other hand, public universities, where the ENGs have historically been held, with rare exceptions, pressured by the cut in resources and by the privatist ideology itself, started to charge for spaces and services that were previously not charged for, from auditoriums to accommodation and food; as a result, the ENGs tended to become increasingly expensive and this deepened the contradictions within the entity;
2. as the AGB grew closer to social movements, the entity became increasingly present in social struggles and ensured the participation of representatives of different movements in round tables and other activities of the ENGs, and also incorporated the realisation of street activities during the meetings; on the other hand, the expansion of productivism and academicism in universities alienated

[22] The Confea-Creas System is formed by Confea—Federal Council of Engineering and Agronomy and by Creas—Regional Councils of Engineering and Agronomy. It was created in 1933, and it is the official body responsible for the recognition and oversight of the professional practice of a series of professional categories, including geographers, who were legally recognised as a professional category in 1979 and included in the Confea-Creas System.

segments of the entity, under the accusation of its growing politicisation to the detriment of the academic quality of its meetings;

3. as the AGB grew closer to the most diverse social movements, which brought new vigour to the entity, on the other hand, the multiplication of movements had a negative effect, by attracting militants from the entity, resulting in increasing difficulty in replenishing the member base; with the multiplication of social movements, many students and professionals who until then had the AGB as their privileged space of political action, started directing their activities towards them and stopped working in the AGB;
4. in the face of all these contradictions, the 2010s were marked by extreme difficulty in forming National Executive Boards (DENs)—on no less than two occasions the ENG ended up without the election of a Board (in the 2014 CBG and in the 2018 ENG), forcing extraordinary assemblies to be held on dates subsequent to the meetings; and also because of the difficulty in determining the venue of the subsequent ENG—on four occasions throughout the decade (for the 2010, 2012, and 2018 ENGs and for the 2014 CBG); in such a way that only in the 2016 ENG was there an election of the DEN and determination of the next ENG, as provided for by statute; in other words, the 2010s were marked by enormous difficulty in maintaining the entity's institutional normality.

Another relevant issue is the relationship that has been established between the AGB and the National Association of Postgraduate Studies and Research in Geography (Anpege), as well as between the AGB and the Organising Committees of area meetings in geography (Symposium on Urban Geography—Simpurb, National Meeting of Agrarian Geography—ENGA and National Symposium on Agrarian Geography—SINGA, etc.). On the one hand, we can identify a competition between the AGB and the Anpege and between the ENGs and the area meetings, with criticisms of an alleged excessive politicisation of the AGB and the loss of academic quality of the ENGs, as opposed to the higher academic quality of the area meetings and the Anpege. The creation of the Anpege itself is largely motivated by the vision of many professors who headed the geography departments, as well as the increasingly numerous postgraduate courses, such that the AGB would have become a more political entity than academic and, therefore, it would be necessary to create another entity of a more eminently scientific nature, linked to postgraduate courses, which would be the space par excellence of scientific production in Brazilian geography. The creation of area meetings entails a similar direction, added to the assessment that the ENGs would not entail more specific debates in each area.

On the other hand, we can also observe a spread of the AGB culture to the area meetings and to the Anpege, either with the adoption of the EDPs' modality, the closeness to social movements and the carrying out of street activities,[23] as in the

[23] Activities included in the schedule of events that take place outside the boundaries of the universities where the meetings take place, taking to streets and squares of the cities where the events are being held, allow for interaction of the meeting and congress participants with the local population. Social movements that are partners of the AGB are often invited to join these activities, which can be marches, rallies, etc.

case of *SINGA naRua* (SINGA on the Street), held both in the Singa in Curitiba, in 2017, and in Recife, in 2019.

The outbreak of the COVID-19 pandemic hit the AGB at a critical time. After a decade of intense difficulty in electing the DENs and deciding on locations for the ENGs, the entity saw half of its DEN resign in 2019 without further explanation and the DEN being re-formed in an environment of many internal conflicts less than a year from the holding of the 2020 ENG. With a strong tradition of face-to-face activities, including the quarterly holding of RGCs, the entity had enormous difficulties in defining a new methodology for internal debates given the impossibility of holding face-to-face RGCs and the need to suspend the 2020 ENG, in which a new election for the DEN would be held. Also, from a financial point of view, the pandemic represented a blow to the AGB, since the entity's cash flow revolves around the ENGs, either by obtaining funding from government agencies, or by the payment of annual fees by associates, which is significantly larger in the ENG years than in the other years. The AGB, which became renowned for putting Brazilian geography on the streets, found itself immobilised.

23.4 Conclusion

The AGB was born under the banner of conservatism emanating from the Universities' dominant academic logic in the 1930s and linked to the elites and the state, and remained so until the turn of the 1970s to the 1980s. The political effervescence of the 1950s and 1960s, with the debates on basic reforms, had little penetration in the entity, in such a way that the 1964 coup and the establishment of the corporate and military dictatorship hardly affected its dynamics. It was only in the context of the struggles against the dictatorship and for the country's redemocratisation in the late 1970s that the traditional hierarchical and conservative structures of the AGB were shaken and radically transformed. In the wake of the emergence of the Critical Renewal of Geography Movement, the AGB became widely democratised in the 1980s, decisively contributing to the very transformation of geography carried out in Brazil, in line with the social struggles that were amplified in the country in that decade and that led to the Constitution of 1988 and the direct elections of 1989. However, the defeat of leftist forces in the country and the collapse of the Soviet bloc converged on the affirmation of neoliberal hegemony and the AGB was impacted by the new dynamics imposed on the universities and Brazilian society in this context of privatisation, dismantling of the public universities, and the expansion of meritocracy and productivism. It resisted the neoliberal wave and reinvented itself in the 2000s, getting closer and closer to social movements and struggles, despite the contradictions arising from the continuity of the deleterious influences of neoliberalism and new contradictions generated by the increasing gigantism of the ENGs. It had great difficulty in maintaining its institutional structure in the 2010s, whether in electing DENs or determining locations for the ENGs to be held. And it suffered drastically from the impacts of the COVID-19 pandemic in 2020.

Thus, the AGB, which was born enclosed, and then opened up to the streets, found itself reclusive again. And like everyone impacted by the pandemic, it was forced to reinvent itself in order to keep moving. We hope that this new reclusion is only temporary and that the AGB will soon put Brazilian geography back on the streets.

References

AGB (1980) Anais do 4º Encontro Nacional de Geógrafos. AGB, Rio de Janeiro

AGB (1982) Anais do 5º Encontro Nacional de Geógrafos. AGB, Porto Alegre

Arroyo MG (1980) Operários e educadores se identificam: que rumos tomará a educação brasileira? Educação & Sociedade (Impresso), N 5 v 2, pp 5–23

Bosi A (1992) O Tempo e os Tempos. In: Novaes A (org) Tempo e História. Companhia das Letras, São Paulo, pp 19–32

Fernandes F (2020) Universidade brasileira: reforma ou revolução? Expressão Popular, São Paulo

Fernandes F (1978) A condição de sociólogo. HUCITEC, São Paulo

Fernandes F (1989) O desafio educacional. Cortez/Autores Associados, São Paulo

Konder L (1983) O Curriculum Mortis e a Reabilitação da Autocrítica. Revista Presença, nº 1, Rio de Janeiro

Lopes MIS (1997) A resistência no regime militar: o docente/cientista frente à política de ciência e tecnologia. DoctoralThesis, Universidade Estadual de Campinas

Lourenço C (2004) A AGB e a Geografia: notas históricas da Seção Minas Gerais. Anais do VI Congresso Brasileiro de Geógrafos. AGB, Goiânia

Lucchesi MAS (2015) Universidades Estaduais Paulistas: breve história em perspectiva comparada. In: V Congreso Nacional e Internacional de Estudios Comparados em Educación, 2015, Buenos Aires. Educación y Futuro. Debates y desafios en perspectiva internacional. Buenos Aires: Sociedad Argentina de Estudios Comparados em Educación (SAECE)

Menezes LC (2000) Universidade sitiada: a ameaça de liquidação da universidade brasileira. Editora Fundação Perseu Abramo, São Paulo

Monteiro CAF (1980) A Geografia no Brasil (1934–1977): avaliação e tendências. Série Teses e Monografias, no 37. Instituto de Geografia, USP, São Paulo

Moreira R (2008) As três geografias: refletindo pelo retrovisor sobre os problemas de toda mudança. Boletim Paulista De Geografia, São Paulo 88:97–114

Motta RPS (2014) As universidades e o regime militar. Jorge Zahar Editor

Rodrigues AM (2004) Contribuição da AGB na construção da geografia brasileira—uma outra Geografia sempre é possível, vol 1, no 22. Terra Livre, São Paulo

Roy A (2004) A resistência sob o risco da "ONGuização. Le Monde Diplomatique Brasil, ano 2, n.21, out. Available at: https://diplomatique.org.br/a-resistencia-sob-o-risco-da-onguizacao/. Accessedon: 10 Mar 2021

Scarim PC (2001) Coetâneos da Crítica: uma contribuição ao estudo da renovação da geografia brasileira. 2v. Dissertation (Master in Geography), Universityof São Paulo, São Paulo

Singer P (1977) A crise do "Milagre": interpretação crítica da economia brasileira, 4 edn. Paz e Terra, Rio de Janeiro

Tragtenberg M (1990) A Delinquência Acadêmica. In: Sobre Educação, Política e Sindicalismo, vol 1, 2ª ed. Editores Associados; São Paulo, Cortez, (Coleção teoria e práticas sociais)

Zusman PB (1996) Sociedades geográficas na promoção do saber a respeito do território: estratégias políticas e acadêmicas das instituições geográficas na Argentina (1879–1942) e no Brasil (1838–1945). Dissertation (Master in Geography), Universidade de São Paulo, São Paulo

Zusman PB (2001) Na procura das origens da AGB ... Boletim Paulista de Geografia, São Paulo, 78:7–32

Epilogue

Vladimir Kolosov
Ex-Presidento of IGU 2012–2016
Chair of the Centennial of the Geography.

The author of these lines had a chance to serve for 14 years on the Executive Committee of the International Geographical Union (IGU), one of the oldest international scientific associations, which celebrated its centenary in 2022. One of the main concerns of the IGU has always been to ensure its status as the only truly global organization that unites the national geographical communities of all countries—not only of the northern hemisphere, but also the states of Asia, Africa and Latin America. Like other members of the Executive Committee, I was always worried about the question why Brazil, one of the largest countries in the world, whose universities and other scientific institutions employ thousands of professional geographers, the birthplace of such outstanding theorists and thinkers as Milton Santos, does not fully participate in the activities of the IGU. And in general, until recent years, South American geographers were poorly represented in the IGU. Some colleagues said that the reason was that in many South American countries, the IGU is still considered an elitist organization, as it was at the beginning of the last century, and it is always dominated by Western scientists. Others pointed to a deep ideological split in the Brazilian geographic community and others to the lack of public funding for most universities, which does not allow Brazilian geographers to participate widely in the activities of the IGU. Be that as it may, Brazil either formally returned to the IGU, having paid a minimum fee, or its membership was again suspended, and I rarely had the opportunity to meet with Brazilian geographers.

It was only after reading this book that I realized what Brazilian geography is, how different it is from European and American geography, more or less familiar to me, or from Russian I know well. Brazilian geography is more than geography, it is a broad view of the world, humanity and its future. Brazilian geography is to a certain extent a guide to action, a rationale for an active position in life and at the same time a denial of many of the foundations of Western human geography. It was only

R. C. Lois González and M. A. Mitidiero Junior (eds.), *Brazilian Geography*, Advances in Geographical and Environmental Sciences,
https://doi.org/10.1007/978-981-19-3704-0

after reading this book that it became clear to me why a significant part of Brazilian geographers regard at least many areas of foreign or Western human geography as a kind of intellectual game. This work is not so much about Brazil and not even about Brazilian geography, but about a new (or partly forgotten) critical geography.

The genre of this book is difficult to determine. Some chapters look more like philosophical treatises, others like an introduction to critical geography in the Brazilian or South American version for those who are not familiar with it. Part of the sections reveals the approach of Brazilian geographers to the burning topics of our time—their understanding of the causes of poverty, urban inequality, lack of housing in cities, landlessness of rural residents and the environmental crisis. One of the chapters is on the teaching of geography in secondary schools, a raison-d'être of the university geography. Some chapters introduce the reader into dramatic ideological discussions about the essence of geographical science. Finally, there are in monograph reviews of the state of the art in individual branches of human geography and complete bibliographic descriptions. I would especially like to note the chapter on the legacy of Milton Santos. Of course, his name is known in different parts of the world and certainly to everyone who is interested in modern South America. However, it seems to me that the scale of the figure of M. Santos, the country is still not sufficiently appreciated outside South America.

The very title of the book—"Brazilian Geography"—contains the question of whether it is original or whether it uses, replicates and adapts the concepts of Western—European and American geography. The authors emphasize that the answer cannot be absolutely unambiguous, since science cannot develop in isolation anywhere. The book gives good reason to assert that an original geographical "school" is shaped in Brazil. For many reasons analysed by the authors, this discipline began to develop in the country later than in Western Europe or North America. Therefore, modern Brazilian geography at first borrowed a lot from the French and German "schools". Many major Brazilian geographers studied and began their scientific careers at European universities. However, already after two or three decades, it became self-sustained. It could hardly have been otherwise. A vast and extremely diverse territory in terms of geographical conditions, complex spatial and temporal patterns of its settlement and economic development, an ethno-cultural mosaic, flashy socio-geographical contrasts and contradictions have made geography in Brazil an extremely significant and relevant science.

Studying the problems of their country, the authors express sharp civic feelings. In their opinion, despite the achievements of recent years, Brazil is the "land of the past": a country with rich natural resources, and a large demographic potential has not been able to change its role in the international division of labour as a supplier mainly of raw materials and agricultural products and most importantly, to get rid of mass poverty. According to the book, more than half of the country's inhabitants suffer from chronic malnutrition. Only 47% of the working-age population is employed, and the sources of income for millions of workers are becoming less and less regular, and they are deprived of elementary guarantees and are on the verge of survival. Informal and part-time employment is on the rise. Like in many countries of the Global South, economic hardship is inextricably linked with socio-political problems—the systematic use of violence, the fragility of state institutions, the expropriation of the territory of indigenous peoples, political upheavals.

The strengthening of contrasts between the Global North and South, between the countries of the south, and especially the aggravation of internal contradictions, has become a factor in the divergence or even split both between national geographical "schools" and within them. Within the walls of Brazilian universities, for many years, another dramatic conflict, probably not known to everyone and visible from the outside, has been unfolding—between revolutionary practice called for by supporters of critical geography and "traditional" geographical science. This conflict directly resulted in the contradictions and activities of the country's geographical organizations, their attitude towards foreign colleagues and associations, including the IGU.

The authors believe that critical geography should be based on a creatively developed Marxist theory. It cannot be dogmatic and equally applicable to the countries of the "world core", semi-periphery and periphery and must consider the forms of exploitation that have developed in different regions of the world, in particular in South America. The authors are convinced that the Marxist theory understood in this way not only has not lost its significance in the post-industrial world, but has proved its high prognostic and practical value. The reality of many countries of the Global South, including Brazil, clearly shows that the proletariat is not shrinking but, on the contrary, its numbers and proportion in the population are increasing, and only its composition and forms of exploitation are changing.

The collapse of the USSR and the fall of the Berlin Wall became another factor in the ideological split in geography, as it led to a decrease in the influence of Marxism in the countries of the "world core". In Brazil, most authors reject neopositivist and postmodern geography, the "cultural turn" and other paradigms of Western (world?) human geography, as they lead researchers away from fundamental problems and their genesis. According to M. Santos and his followers, the functional approach to territorial planning developed by such scholars as F. Perroux or J. Friedman is aimed at smoothing out territorial contradictions and conflicts, and not resolving such problems as the unfair international division of labour, poverty and unemployment, the persistence of deep territorial contrasts at different interconnected spatial levels—global (between the north and south), regional (intra-state, between the centre and periphery of each country) and local (between different parts of cities and urban agglomerations).

In opposing in fact the mainstream of modern social geography, Brazilian geographers criticize, as they say, the standardization of the discipline. Indeed, theoretical approaches, norms of publication and evaluation of scientific results, formed in the United States and other Anglo-Saxon countries, now make part of science management in many countries of the world, including Russia. State institutions encourage scholars to publish in prestigious high-ranking English-language international journals, monopolized by several transnational publishing concerns. Such journals welcome the papers based on the concepts worked out by geographers from

English-speaking countries, built along strictly defined principles and structure, but most importantly, be related to research topics and interests of Western scientists. At the same time, taxpayer-funded papers in these journals are available to researchers only for a high fee, journal subscriptions are expensive even for wealthy Western universities, while the profits of publishing monopolies are growing rapidly. The "geopolitics of scientific knowledge" that has developed in this way is criticized by Western scholars themselves.

Defending the right to a different, self-sustained path of development of geography, the authors of the book emphasize that inequality is inherent in the capitalist method of production. It is impossible to deal with the acute global problems without solving the problem of the exploitation of labour—a source of added value. The modern period of the history is defined in the book as the "Capitalocene", which can lead, if nothing changes, to the self-destruction of mankind. The expansion of capital undermines the foundations of its reproduction because of the aggravation of environmental problems, the withdrawal from material production to speculative operations, the rejection of the welfare state, the shrinking of the middle class and universal deregulation.

Brazilian critical geographers proclaim as their goal the creation of a "metageography", the pivotal axis of which should be an approach to space as a condition, mean and result of the reproduction of social relations, an integral element and instrument of accumulation due to the functioning of the markets of land, labour and capital. Its spatial implications are enhanced by competition among the participants in these markets, leading to the centralization and concentration of capital. The development of such a geography, according to critical geographers, would make it possible to overcome the fragmentation of the discipline as a result of the emergence of ever new narrow directions and break out of the trap of finding solutions by means of regional policy, territorial organization and the use of other state institutions. The general goal is to build a world free of capital.

A critical look at the current extremely unjust world order and its dramatic contradictions based on the Brazilian material sounds very convincing. What is next? What are the alternatives, how to achieve a breakthrough, how to draw conclusions from the failure of European and other utopian projects and the collapse of "real socialism" in the Soviet Union and other countries? Many of the best minds of mankind are struggling with this task, and its solution, of course, goes far beyond the scope of one book or one scientific community. The authors of the book (or at least many of them) believe that geographers cannot remain outside observers, doing purely academic research. From their perspective, Marxist critical geography makes it possible to reveal the causes of the current situation, and geographers are obliged to join the struggle for its radical change, to convey their views to social movements and to participate in them, helping to define their goals, strategy and tactics. The second part of the title of the book speaks for itself—"Geography on the streets". The chapter on the history of the Association of Brazilian Geographers and the struggle of a significant part of its members for their vision of geography is so dramatic that it could probably form the basis of a novel.

All in all, the book offers a panorama of geographical research in one of the largest countries in the world—its connection with the social life, an unusual way of development, original approaches and contradictions among geographers. And most importantly, the monograph makes the reader thinking deeply about what is happening to our increasingly turbulent world, what the acceleration of social change will lead to.

GPSR Compliance

The European Union's (EU) General Product Safety Regulation (GPSR) is a set of rules that requires consumer products to be safe and our obligations to ensure this.

If you have any concerns about our products, you can contact us on ProductSafety@springernature.com

In case Publisher is established outside the EU, the EU authorized representative is:

Springer Nature Customer Service Center GmbH
Europaplatz 3
69115 Heidelberg, Germany

Batch number: 10732076

Printed by Printforce, the Netherlands